Automated Systems in the Aviation and Aerospace Industries

Tetiana Shmelova
National Aviation University, Ukraine

Yuliya Sikirda
Kirovograd Flight Academy of the National Aviation University, Ukraine

Nina Rizun
Gdansk University of Technology, Poland

Dmytro Kucherov
National Aviation University, Ukraine

Konstantin Dergachov
National Aerospace University - Kharkiv Aviation Institute, Ukraine

A volume in the Advances in Mechatronics and Mechanical Engineering (AMME) Book Series

IGI Global
DISSEMINATOR OF KNOWLEDGE

Published in the United States of America by
IGI Global
Engineering Science Reference (an imprint of IGI Global)
701 E. Chocolate Avenue
Hershey PA, USA 17033
Tel: 717-533-8845
Fax: 717-533-8661
E-mail: cust@igi-global.com
Web site: http://www.igi-global.com

Library of Congress Cataloging-in-Publication Data

Names: Shmelova, Tetiana, 1961- editor.
Title: Automated systems in the aviation and aerospace industries / Tetiana Shmelova, Yuliya Sikirda, Nina Rizun, Dmytro Kucherov, and Konstantin Dergachov, editors.
Description: Hershey, PA : Engineering Science Reference, [2019] | Includes bibliographical references.
Identifiers: LCCN 2018036333| ISBN 9781522577096 (h/c) | ISBN 9781522577102 (eISBN)
Subjects: LCSH: Aeronautics--Technological innovations. | Aerospace industries--Automation. | Automatic pilot (Airplanes) | Aeronautics--Study and teaching--Data processing.
Classification: LCC TL553 .A847 2019 | DDC 629.1--dc23 LC record available at https://lccn.loc.gov/2018036333

This book is published in the IGI Global book series Advances in Mechatronics and Mechanical Engineering (AMME) (ISSN: 2328-8205; eISSN: 2328-823X)

British Cataloguing in Publication Data
A Cataloguing in Publication record for this book is available from the British Library.

All work contributed to this book is new, previously-unpublished material. The views expressed in this book are those of the authors, but not necessarily of the publisher.

For electronic access to this publication, please contact: eresources@igi-global.com.

Advances in Mechatronics and Mechanical Engineering (AMME) Book Series

J. Paulo Davim
University of Aveiro, Portugal

ISSN:2328-8205
EISSN:2328-823X

Mission

With its aid in the creation of smartphones, cars, medical imaging devices, and manufacturing tools, the mechatronics engineering field is in high demand. Mechatronics aims to combine the principles of mechanical, computer, and electrical engineering together to bridge the gap of communication between the different disciplines.

The **Advances in Mechatronics and Mechanical Engineering (AMME) Book Series** provides innovative research and practical developments in the field of mechatronics and mechanical engineering. This series covers a wide variety of application areas in electrical engineering, mechanical engineering, computer and software engineering; essential for academics, practitioners, researchers, and industry leaders.

Coverage

- Intelligent Navigation
- Computational Mechanics
- Design and Manufacture
- Computer-Based Manufacturing
- Tribology and surface engineering
- Bioengineering Materials
- Mechanisms and machines
- Autonomous Systems
- Control Methodologies
- Nanomaterials and nanomanufacturing

IGI Global is currently accepting manuscripts for publication within this series. To submit a proposal for a volume in this series, please contact our Acquisition Editors at Acquisitions@igi-global.com or visit: http://www.igi-global.com/publish/

The Advances in Mechatronics and Mechanical Engineering (AMME) Book Series (ISSN 2328-8205) is published by IGI Global, 701 E. Chocolate Avenue, Hershey, PA 17033-1240, USA, www.igi-global.com. This series is composed of titles available for purchase individually; each title is edited to be contextually exclusive from any other title within the series. For pricing and ordering information please visit http://www.igi-global.com/book-series/advances-mechatronics-mechanical-engineering/73808. Postmaster: Send all address changes to above address.

Titles in this Series

For a list of additional titles in this series, please visit: www.igi-global.com/book-series

Handbook of Research on Green Engineering Techniques for Modern Manufacturing
M. Uthayakumar (Kalasalingam University, India) S. Aravind Raj (Vellore Institute of Technology, India) Tae Jo Ko (Yeungnam University, South Korea) S. Thirumalai Kumaran (Kalasalingam University, India) and J. Paulo Davim (University of Aveiro, Portugal)
Engineering Science Reference • copyright 2019 • 403pp • H/C (ISBN: 9781522554455) • US $275.00 (our price)

Non-Conventional Machining in Modern Manufacturing Systems
Kaushik Kumar (Birla Institute of Technology, India) Nisha Kumari (Birla Institute of Technology, India) and J. Paulo Davim (University of Aveiro, Portugal)
Engineering Science Reference • copyright 2019 • 303pp • H/C (ISBN: 9781522561613) • US $215.00 (our price)

Advanced Fuzzy Logic Approaches in Engineering Science
Mangey Ram (Graphic Era University (Deemed), India)
Engineering Science Reference • copyright 2019 • 488pp • H/C (ISBN: 9781522557098) • US $245.00 (our price)

Recent Advancements in Airborne Radar Signal Processing Emerging Research and Opportunities
Amir Almslmany (Independent Researcher, Egypt)
Engineering Science Reference • copyright 2018 • 190pp • H/C (ISBN: 9781522554363) • US $145.00 (our price)

Stochastic Methods for Estimation and Problem Solving in Engineering
Seifedine Kadry (Beirut Arab University, Lebanon)
Engineering Science Reference • copyright 2018 • 275pp • H/C (ISBN: 9781522550457) • US $205.00 (our price)

Design and Optimization of Mechanical Engineering Products
K. Kumar (Birla Institute of Technology, India) and J. Paulo Davim (University of Aveiro, Portugal)
Engineering Science Reference • copyright 2018 • 347pp • H/C (ISBN: 9781522534013) • US $235.00 (our price)

Managerial Approaches Toward Queuing Systems and Simulations
Salvador Hernandez-Gonzalez (Tecnológico Nacional de México en Celaya, Mexico) and Manuel Dario Hernandez Ripalda (Tecnológico Nacional de México en Celaya, Mexico)
Engineering Science Reference • copyright 2018 • 311pp • H/C (ISBN: 9781522552642) • US $195.00 (our price)

IGI Global
DISSEMINATOR OF KNOWLEDGE

701 East Chocolate Avenue, Hershey, PA 17033, USA
Tel: 717-533-8845 x100 • Fax: 717-533-8661
E-Mail: cust@igi-global.com • www.igi-global.com

Table of Contents

Detailed Table of Contents

system was developed at Flight Academy of the National Aviation University, Ukraine. Its key differential peculiarity is focused on major stages of the decision-making process of the tower controller instead of visualization from the tower window.

The implementation of the aviation tasks, including the preparatory ones, is a difficult, complex task requiring from the aviation personnel a high level of general, technical, and specialist knowledge and a wide range of skills, appropriate to the type of technical tools and systems being at the disposal of the aviation personnel as well as the complexity and difficulty of their tasks. Particular importance is currently attached to the development of training devices used in the basic training and in-service training of the aviation personnel. The authors have referred to the role of simulators in achieving the desired level of specialist competence by flight personnel, including, among others, such issues as aircraft simulators development, and their application in the aviation training; aircraft simulators classification in accordance with current aviation regulations; essential functions performed by aviation simulators; flight simulator is an essential tool for basic and in-service training of the aviation personnel.

In this chapter, the authors present an approach to the extrinsic environs diagnostics based on using visual information collected by autonomous robots. The possibility of utilizing a computer vision for the purpose of rational control implementation in the condition of the full or partial uncertainty is investigated. In the study, the combined hardware and software computer vision tools were verified. The models, algorithms, and codes for solving the local tasks of obstacle identification and mutual location kinematic parameters estimation have been developed. A series of computational and in-kind experiments that illustrate a practical possibility of implementing the navigational environment diagnosis is carried out with the aim to select a rational flight path.

The aim of this chapter is to utilize the upkeep hypothesis of on-condition, condition checking. It proposes support ideas: for example, advanced upkeep, proactive upkeep which is bolstered by solid checking and finding strategies to enhance the effectiveness of upkeep extraordinarily; so there must be decision support system framework to develop a present-day flying upkeep framework. Maintenance DSS can give powerful choice help to aeronautics support. It additionally makes strides in upkeep effectiveness and controls support costs. The flying MDSS still needs further research on the accompanying subjects.

Since current airplanes have numerous frameworks and complex structures, numerous flight parameters need to be screened. Step-by-step instructions to assemble an effective database and research criteria for judging the adequacy of ongoing information should direct further research. Since continuous flight status information has cost issues, how to set up the data download criteria should direct further research.

In this chapter, the author presents a human factors problem for automation: why, when, and how automation has been introduced in the aviation domain; what problems arise from different ways of operating; and the possible countermeasures to limit faulty interaction between humans and machines. This chapter is divided into parts: definition of automation, its advantages in ensuring safety in complex systems such as aviation; reasons for the introduction of on-board automation, with a quick glance at the history of accidents in aviation and the related safety paradigms; ergonomics: displays, tools, human-machine interaction emphasizing the cognitive demands in high tempo and complex flight situations; illustration of the AF 447 case, a crash happened in 2009, which causes are linked to faulty human-machine interaction.

A solution is proposed for the task of controlling a group of UAVs moving along a given route. The group is considered as a limited-size formation consisting of n-agents moving relative to the leader, which allows us to treat the group as some aggregate with the center of motion. The quantitative composition of a group can change while maintaining the integrity of the group. The chapter proposes the use of smooth laws governing the motion of a group. The safety of motion is ensured by introducing into the law of control the components equivalent to the creation of attractive and repulsive fields.

In this chapter, the authors propose an approach to using a heterogeneous team of unmanned aerial vehicles and remote sensing techniques to perform tactical forest firefighting operations. The authors present the three-level architecture of the multi-UAV-based forest firefighting monitoring system; features of patrolling, confirming, and monitoring missions; as well as functions of UAV in such missions. The authors consider an infrastructure for the UAV ground support and equipment used for the UAVs control. The method of the data integration into a fire-spreading model in a real-time DSS for the forest fire response is proposed. The proposed approach has been tested with the multi-UAV team that included three drones for the patrol missions, one helicopter for the confirmation mission, and one octocopter for the monitoring mission. The performance of such multi-UAV team has been studied in the laboratory conditions. The result of the experiment has shown that the proposed approach provides required credibility and efficiency of fire prediction and response.

In this chapter, the authors present a methodology for developing a model-tracing cognitive tutor. The methodology is based on Bayesian probabilistic networks for generating pedagogical interventions. The presented probabilistic model increases fidelity assessment due to its ability of independently diagnosing the degree of mastery for every knowledge component involved in students' actions; fidelity assessment in education is the ability to represent students' cognitive states as close as possible for analysis and evaluation. The cognitive tutor was developed to promote a self-regulated learning approach with an open learner model. The open learner model let students change the learning flow by changing the assigned tasks. The authors explain in detail the structural construction and employed algorithms for developing a model-tracing cognitive tutor in the domain of fault-tolerant systems. Preliminary results and future work are also discussed to assess effectiveness of the proposed approach and its implication in actual educational programs.

In this chapter, the authors present an imitation model for calculating the weight and centering of the airship, taking into account the errors that arise in the process of determining these parameters. Also in this chapter there is the analysis of existing means, methods, and methods for determining the weight and alignment of the airship in domestic airlines and air companies. The authors made the research of the sources of errors and identified their types. This chapter provides a new model of the takeoff weight and balance of the aircraft determining process, as well as the weight distribution of passengers law. These materials allow simulating one of the main operations in the preparation for departure process.

Aviation and aerospace systems are complex and concurrent and require special tools for their specification, verification, and performance evaluation. The tool in demand should be easily integrated into the standard loop of model-driven development. Colored Petri nets represent a combination of a Petri net graph and a functional programming language ML that makes it powerful and convenient tool for specification of real-life system and solving both tasks: correctness proof i.e. verification and performance evaluation.

Preface

The book *Automated Systems in Aviation and Aerospace Industries* is devoted to the development and operation of automated control systems, which in aviation are used in-flight and navigation systems, in Flight Management Systems, in Air Traffic Control systems movement, in training complexes. These systems belong to real-time systems, where the available time for making and implementing decisions is set by the speed of the flow of controlled processes. The degree of human participation in these systems is different, but almost all of them perform the functions of control and management of objects and require an assessment of their functional safety. The book is devoted to various applications of automated systems in aviation technology, which currently determine the effectiveness and safety of manned and unmanned aerial vehicles (UAV).

The main goal of the automated control systems is to assist the crew, the dispatcher to overcome the difficulties encountered in solving professional problems. An automated system must effectively cope with making erroneous decisions or actions caused by fatigue, reduced performance, an underestimation of the complexity of the situation and its capabilities, failures of components and systems, and other problems encountered in flight.

Nowadays using information technology is a very actual topic in the aviation and aerospace industries. Support for the safe functioning of the aviation system and aerospace industries is one of the most important scientific and technical problems. Statistical data show that human errors account for up to 80% of all aviation accidents (International Civil Aviation Organization [ICAO], 2004). According to International Civil Aviation Organization (ICAO) data, comparing with 2011, in 2012 the number of incidents in the world decreased on 21%, in 2013 – on 13%; the number of fatalities in 2012 decreased on 10%, in 2013 – on 55% (ICAO, 2013a). Consequently, the least number of deaths has been fixed in 2012-2013, starting from 2004. As a result of the decreased number of incidents, and increased number of departures, the frequency of incidents in the world in 2012 has been reduced to 3,2 events per million departures and in 2013 – to 2,8 events per million departures. This is the lowest value since the ICAO started to monitor the frequency of incidents in the world (ICAO, 2013b). Aviation systems cannot be wholly free from dangerous factors and connected with the risks, while, the elimination of aviation events and serious incidents continues to be the final goal of human activity in the sphere of aviation safety. Neither human activity nor systems created by it guarantee a total absence of operating errors and their consequence. In such a way, safety is a dynamic characteristic of aviation with the help of which risk factors for flight safety should steadily decrease. It is important to note that adoption of effective indices of ensuring flight safety is frequently influenced by internal and international standards and also by cultural features. So, ICAO constantly develops and improves proactive, based on the risks evaluation, methods, directed on the farther decrease in a number of aviation events in the world and to use information technology for support operators. Also, ICAO encourages aviation communities to recognize the importance of adherence of the single global approach for safety improvement and monitoring.

Recommendations of the ICAO are the correct using of information technologies in aviation systems. Modern computer systems in aviation: Automatic System, Decision Support System (DSS), Expert System, Artificial Intelligence Systems, etc. How computer systems are used in aviation and influence on operators described in this book. Evolution of human factors (HF) models show that important to continue developing the innovations in the aviation and aerospace industries for improving effectiveness. There are stages of the evolution of the HF's models (ICAO, 2002, 2004, 2013a, 2013b, 2014) associated with the appearance of new system components in activities of human-operators (H-O) such as to improve professional skills of H-Os; cooperation in the team; the influence of culture on H-Os; balance model between safety and effectiveness and balance model between automatization and control of operators for automatic system (Table 1). It is important to implement automation in aviation systems and to match the capabilities of the human operator.

In this book, we may read about applications of information technologies in aviation systems and presented a problem of using of automatic systems in aviation and the application of the balanced approach, the implementation of a balanced model between automatization and control of operators for an automatic system.

Table 1. Evolution of human factors models

Years	Models	Content of Models	Stage
1972	SHEL	Software (procedures) - Hardware (machines) - Environment - Liveware	I stage Professional skills Interaction Errors
1990	Reason's "Swiss Cheese Model"	Active errors - Latent errors - Windows of opportunity - Causation chain	
1993	SHELL	Software (procedures) - Hardware (machines) - Environment - Liveware - Liveware (humans)	
1999	CRM	Crew - Resource - Management	II stage Cooperation in team Error detection
2000	TEM	Threat and Error - Management	
2000	MRM	Maintenance - Resource - Management	
2004	SHELL-T (SHELL-Team)	Software (procedures) - Hardware (machines) - Environment – Liveware - Liveware (humans) - Team	
2004	SCHELL model and CRM	Software (procedures) – Culture - Hardware (machines) - Environment - Liveware - Liveware (humans)	III stage Culture Safety Error prevention
2004	LOSA	Line - Operation - Safety - Audit	
2009	HEAD	Human - Environment - Analysis - Design	
	PBA	Performance-Based Approach	
2010	HFACS	Human Factors - Accident - Classification - System	
2013-2017	SMS Balance Models	Safety Management System The balance models between: Safety and Efficiency Automated system and control Facilitation and security requirements etc.	IV stage Safety / Efficiency / Minimization of errors
2016-2018	CDM	Collaborative Decision Making (CDM) System-Wide Information Sharing and Management (SWIM) Flight & Flow Information for a Collaborative Environment (FF-ICE)	V stage Collaborative DM

International Civil Aviation Organization business plan (2017-2019) includes methods of effective management in future such as the efficient organization, through modernized, automated tools and management systems, including corporate performance management; implementation of automated electronic tools and completion of management systems and dashboard in aviation; automated database-to-database exchange etc. But importance to support provisioning the most balanced - solution within a given set of priorities, when different options are available, but each option has different advantages and disadvantages in terms of performance impact, that is "balanced total system approach" (ICAO, 2016).

Automation is seen as one of many resources available to the human operators — controllers and pilots alike — who retain the responsibility for management and direction of the overall Air Traffic Management system. Additionally, unexpected or unplanned events must be a required part of planning and design when considering the systems that would replace the cognitive and adaptive capabilities of controllers or pilots. The development of training for automated systems is more difficult than for non-automated systems. One of the primary challenges in developing training for automated systems is to determine how much a trainee will need to know about the underlying technologies in order to use automation safely and efficiently. It is recommended that task analysis techniques be used as the basis of the design for training in automated systems. Course development based on a task analysis can be somewhat more time-consuming than traditional training development techniques.

As technologies continue to advance, Artificial Intelligence (AI) will play bigger roles in our futures. AI has already been incorporated in many industries; the aviation industry has already successfully implemented the early stage of automation for example in daily flight operations with flight management systems or autopilot. New kinds of flight operations, such as UAVs, commercial space travel and the personal air vehicles of the future, are making the problem even more difficult. The modern aviation system dynamically adapts and evolves on a regular basis to meet demands with real-time information sharing. But as the system rapidly increases its complexity, we are reaching our limit. Traditional automation cannot handle such 'information explosions'. Human cognition and performance may need to be further augmented with the support of AI (International Air Transport Association [IATA], 2018).

The book will consider the use of information technology in aviation such as the design of new systems (Decision Support Systems, Expert Systems, Artificial Intelligence Systems, Computing System, etc.), the operation of information system in aviation and aerospace; training of personnel in aviation too. The development of training for automated systems in aviation and aerospace is more difficult than for non-automated systems. One of the primary challenges in developing training for automated systems is to determine how much a trainee will need to know about the underlying technologies in order to use automation safely and efficiently.

The target audiences of the book *Automated Systems in the Aviation and Aerospace Industries* are scientists, teachers, pupils, students and postgraduate students, air crash investigators, scientists, engineers and managers, aviation operators (pilots, dispatchers, pilots of UAVs, etc.).

Scientists from 11 countries (Ukraine, China, India, Turkey, Poland, Egypt, Mexico, Azerbaijan, Kuwait, Italy, Pakistan) presented their scientific achievements in the field of the use of automated systems in aviation and aerospace industries. This book is a collection of seventeen chapters that could cover problems about the advantages of automated systems in aviation and aerospace industries encompasses. The chapters present tasks from different science fields like using artificial intelligence systems and decision support systems in the aeronautics and aerospace industries; using automated systems in aviation personnel training; autonomous flying robots and many applications UAVs in civil and military tasks solution; role of automation on a modern airplane; models of calculating the weight and centering of the airship; using of Petri net, GERT-net, Big Data and theory of catastrophes in aviation.

Chapter 1, "Machine Learning and Text Analysis in an Artificial Intelligent System for the Training of Air Traffic Controllers," presents the application of new information technology in education for the training of Air Traffic Controllers (ATCs). Machine Learning, Multi-Criteria Decision Analysis, and Text Analysis as the methods of Artificial Intelligence for ATCs training have been described. The authors have made an analysis of the International Civil Aviation Organization documents for modern principles of ATCs education. The prototype of the neural network for evaluating the timeliness and correctness of the decision making by ATCs has been developed. The new theoretical and practical tasks for Simulation and Pre-Simulation Training have been obtained using Expert Judgment Method. The methodology for sentiment analyzing the airlines' customers' opinions has been proposed. In addition, the examples of Artificial Intelligence Systems and Expert Systems by the authors, students and colleagues from National Aviation University, Ukraine and Gdansk University of Technology, Poland have been proposed.

In Chapter 2, "Development Specifics of the Tower Controller Intelligent Training System," the activity of the tower controller is connected with the implementation of a rather complex set of measures to ensure take-off and landing operations. In connection with this, the aviation specialist should be possessed by the number of professional competencies, which are formed by mastering the theoretical material and consolidating it in practice. Frequently, the initial consolidation in practice is performed by using automated training systems, especially intelligent training systems are in a great demand. Therefore, this chapter is devoted to the specifics of the development of such type aviation-focused system, which are given on the example of the intelligent training system "Tower Controller". The prototype of this system was developed at Flight Academy of the National Aviation University, Ukraine. Its key differential peculiarity is focused on major stages of the decision-making process of the tower controller instead of visualization the view from the tower window.

Chapter 3, "Simulators as an Essential Tool for Shaping the Competence of the Aviation Personnel," is about the implementation of the aviation tasks, including the preparatory ones, is a difficult, complex task requiring from the aviation personnel a large level of general, technical and specialist knowledge and a wide range of skills, appropriate to the type of technical tools and systems being at the disposal of the aviation personnel as well as the complexity and difficulty of their tasks. This causes that particular importance is currently attached to the development of training devices, used among others in the basic training and in-service training of the aviation personnel. The authors have referred in a special way to the role of simulators in achieving the desired level of specialist competence by flight personnel, including, among others, such issues as: aircraft simulators development and their application in the aviation training; aircraft simulators classification in accordance with current aviation regulations; essential functions performed by aviation simulators; flight simulator is an essential tool for basic and in-service training of the aviation personnel.

In Chapter 4, "Environments Diagnosis by Means of Computer Vision System of Autonomous Flying Robots," the authors present an approach to the extrinsic environs diagnostics based on using visual information collected by autonomous robots. The possibility of utilizing a computer vision for the purpose of rational control implementation in the condition of the full or part uncertainty is investigated. In the study, the combined both hardware and software computer vision tools were verified. There were worded local tasks on dropping obstacle things uncertainty appearing during the autonomous robot flight. The models, algorithms, and codes for solving the local tasks of obstacles identification and mutual location kinematic parameters estimation have been developed. There is carried out a series of computational and in-kind experiments that illustrate a practical possibility of implementing the navigational environment diagnosis with the aim to select a rational flight path.

The aim of Chapter 5, "Decision Support Systems in Aeronautics and Aerospace Industries," is to utilize the upkeep hypothesis of on-condition, condition checking, it proposes support ideas, for example, advanced upkeep, proactive upkeep which bolstered by solid checking and finding strategies to enhance the effectiveness of upkeep extraordinarily, so there must be decision support system framework to develop present day flying upkeep framework. Maintenance DSS (MDSS) can give powerful choice help to aeronautics support. It additionally makes strides upkeep effectiveness and controls support costs. Notwithstanding, the flying MDSS still needs further research on the accompanying subjects. Since the current airplane has numerous frameworks and complex structure, numerous flight parameters need to screen. Step by step instructions to assemble an effective database and research criteria for judging the adequacy of ongoing information should direct further research. Since continuous flight status information downloaded identify with cost issues, how to set up the faulty data download criteria should direct further research.

Chapter 6 is "AF 447 as a Paradigmatic Accident: The Role of Automation on a Modern Airplane." The AF447 crash has raised a number of questions about the human-machine interface. This event involved a crew flying at cruising altitude that lost completely the control of the aircraft diving for three minutes and a half until crashing on the Atlantic Ocean surface. Why three experienced pilots lost the situational awareness? Why weren't they able to recover an upset aircraft? Was this the only crew involved in such an event? From the Nineties, the loss of control inflight has become one of the main cause of the accident. Maybe the technological revolution that took place in the Eighties at the aircraft design stage played a role in causing this typology of the accident. The chapter analyzes the safety issues, the human contribution and the possible solutions to this problem.

Chapter 7 is "Synthesis of the Laws of Motion Control of a UAV Group With Natural Obstacles." A solution is proposed for the task of controlling a group of UAVs moving along a given route. The group is considered as a limited-size formation consisting of n-agents moving relative to the leader, which allows us to treat the group as some aggregate with the center of motion. The quantitative composition of a group can change while maintaining the integrity of the group. The paper proposes the use of smooth laws governing the motion of a group. The safety of motion is ensured by introducing into the law of control the components equivalent to the creation of attractive and repulsive fields.

In Chapter 8, "Forest Firefighting Monitoring System Based on UAV Team and Remote Sensing," the authors propose an approach to using a heterogeneous team of unmanned aerial vehicles and remote sensing techniques to perform tactical forest fire-fighting operations. The authors present the three-level architecture of the multi-UAV-based forest fire-fighting monitoring system, features of patrolling, confirming, and monitoring missions, as well as functions of UAV in such missions. The authors consider an infrastructure for the UAV ground support and equipment used for the UAVs control. The method of the data integration into a fire-spreading model in a real-time DSS for the forest fire response is proposed. The proposed approach has been tested with the multi-UAV team that included 3 drones for the patrol missions, 1 helicopter for the confirmation mission, and 1 octocopter for the monitoring mission. The performance of such multi-UAV team has been studied in the laboratory conditions. The result of the experiment has shown that the proposed approach provides the required credibility and efficiency of fire prediction and response.

In Chapter 9, "Preliminary Sizing and Performance Calculations of Unmanned Air Vehicles," preliminary sizing and performance calculations of UAV in relation to its propulsion system are explained.

Starting with the mission profile of the air vehicle, which is one of the important design drivers, the UAV is sized for the requirements. The requirements in design consist of a set of target values e.g. payload amount to be carried, range, endurance time, cost, etc. Additionally, the parameters within a mission profile such as cruising altitude and speed of aircraft affect engine type, power level required, fuel quantity and therefore general dimensions and the gross weight of the aircraft. It is often an iterative process to size the air vehicle, and engine together. UAV is designed in a loop of calculations in which sizing, flight performance, and engine performance are done for each phase of the mission or flight profile to satisfy the overall design mission and requirements.

In Chapter 10, "Unmanned Aerial Vehicles Application for Military Geoinformation System (GIS) Tasks Solution," authors present the results of UAV application for military GIS tasks solution. The results of the visual modeling of the revealing process of invisible military objects from one of a point of observation on the mountainous terrain of the Azerbaijan Republic by using UAV are presented. The observation conditions between two points of the selected mountain terrain during battle operation have been investigated using GIS technology. The quantitative method of the invisible areas assessment and military objects in mountainous terrain are developed and offered by using UAVs. The numerical estimation method of a task support success of UAV reconnaissance flight in mountainous battle condition has been offered and considered. Using UAVs for the purpose of orthophotomap making of the terrain and combat control the detailed 3D-model has been constructed.

In Chapter 11, "Automated System of Stabilization and Position Control of Aviation Equipment," the author presents the problems of design of the robust automated system for stabilization and control of platforms with aircraft observation equipment. The mathematical model of the triaxial stabilized platform is developed. The procedure of synthesis of robust stabilization system based on robust structural synthesis is represented. The above-mentioned procedure uses loop-shaping approach and method of the mixed sensitivity. The matrix weighting transfer functions are obtained. The optimization programs in MatLab are developed. The developed procedures are approved based on the results of simulation by means of the appropriate Simulink model. The obtained results can be useful for unmanned aerial vehicles and aircraft of special aviation, which are used for monitoring of technical objects and aerial photography. The technical contributions are procedures of the robust controller design represented as the flowchart. The proposed approach is validated by application of the theoretical suppositions to the concrete example and appropriate simulation results.

In Chapter 12, "Information Technologies for Learning Principles of Fault-Tolerant Systems," authors present a methodology for developing a Model-tracing cognitive tutor. The methodology is based on Bayesian probabilistic networks for generating pedagogical interventions. The presented probabilistic model increases fidelity assessment due to its ability of independently diagnosing the degree of mastery for every knowledge component involved in students' actions; fidelity assessment in education is the ability to represent student's cognitive state as closest as possible for analysis and evaluation. The cognitive tutor was developed to promote a self-regulated learning approach with an open learner model. The open learner model let student's o change the learning flow, by changing the assigned tasks. Authors explain in detail the structural construction and employed algorithms for developing a Model-tracing cognitive tutor in the domain of fault-tolerant systems. Preliminary results and future work are also discussed to assess the effectiveness of the proposed approach and its implication in actual educational programs.

In Chapter 13, "Model of Calculating the Weight and Centering of the Airship With Measurement Error," the authors present an imitation model for calculating the weight and centering of the airship, taking into account the errors that arise in the process of determining these parameters. Also in this chapter, there is the analysis of existing means, methods and methods for determining the weight and alignment of the airship in domestic airlines and air companies. The authors made the research of the sources of errors and identified their types. This chapter provides a new model of the take-off weight and balance of the aircraft determining process, as well as the weight distribution of passengers' law. These materials allow simulating of one of the main operations in the preparation for departure process.

Chapter 14 is "Modeling With Colored Petri Nets: Specification, Verification, and Performance Evaluation of Systems." Aviation and aerospace systems are complex and concurrent that requires special tools for their specification, verification, and performance evaluation. The tool in demand should be easily integrated into the standard loop of model-driven development. Colored Petri nets represent a combination of a Petri net graph and a functional programming language ML that makes it powerful and convenient tool for specification of a real-life system and solving both tasks correctness proof – verification and performance evaluation. This chapter studies the basic and advanced features of CPN Tools – a powerful modeling system which uses a graphical language of colored Petri nets. It goes from basic elements through declarations of color sets and functions to peculiarities of hierarchical design of complex models and specification of timed characteristics. The authors accomplish the chapter with a real-life case study for avionics applications: the first example represents a simplified altitude switch model that works via protocol ARINC-429 for studying fault-tolerance issues; the second example considers facilities for performance and QoS evaluation of switched Ethernet network that can be easily converted to AFDX standard.

In Chapter 15, "Analysis of Aviation Incidents Using the GERT-Network Taking Into Account the Psychological Characteristics of the Operator," the authors considered the construction of network models of aviation incidents taking into account the psychological characteristics of the operator. In particular, the GERT- network of aviation disaster was built. Also was presented applying of software to calculate the elements of the GERT-network. The using of network models in the analysis of aviation incidents allows visualizing the cause-and-effect relationship in aviation incident and gives the opportunity to obtain quantitative characteristics of the incident for subsequent analysis.

In Chapter 16, "Applications of the Elementary Theory of Catastrophes in Aviation," the author considers the connection between the mathematical theory of catastrophes and certain characteristics of the aviation system. The basis for this connection is the similarity of the concept of "catastrophe" in aviation and mathematics. With small perturbations of some systems, their state changes little. At small perturbations of other systems, a sharp transition occurs to another state. This transition is called a jump or a catastrophe. A catastrophe can also be called a loss of stability. Comparison of the notion of "stability" in aviation and mathematics is part of this study. The mathematical representation of the simplest aviation systems and their disasters is shown in elementary examples. It does not use a deep mathematical apparatus, as well as special studies of aviation systems. The aim of the work is to direct the aviation specialist's eye to the mathematical theory of catastrophes and to look the mathematician's attention to the problems of aviation safety.

Chapter 17 is "Use of Big Data in Aviation: New Opportunities, Use Cases, and Solutions." The expansion of data and its efficient handling is becoming a more popular tendency in recent times bringing new difficulties to learn new avenues. Data analytics can be done more proficiently with the availability of distributed architecture of "Not only SQL" NoSQL databases. Technological advancements around us are changing very rapidly and a major shift is being carried out, a switch from relational to non-relational world. More precisely we are talking about the shift from traditional relational database models to non- relational database models. When moving from relational to the non-relational models, database administrators face common problems due to the fact that NoSQL is a No-Schema database. Logical mapping of the procedure from relational to the non- relational models is complex and it is not just a standard process. The purpose of conducting this research is to propose a mechanism by which the schema of a relational database management system and its data can be transformed into big data by following some standardize guidelines. This model can be quite useful for relational database administrators by enabling them to give attention to logical modeling rather than procedural writing for each and every SQL to NoSQL transition. In this chapter, we studied both models and focus our research to provide a set of rules and framework that can.

Much useful and interesting information from scientists, engineers, real pilots, and air traffic controllers, etc. are presented in all chapters.

REFERENCES

International Air Transport Association (IATA). (2018). *Artificial intelligence (AI) in aviation exploring: The fundamentals, threats and opportunities of AI in the aviation industry*. Montreal: Author.

International Civil Aviation Organization (ICAO). (2002). Human Factors Guidelines for Safety Audits Manual (1st ed.). Doc. ICAO 9806-AN/763. Montreal: Author.

International Civil Aviation Organization (ICAO). (2004). *Cross-Cultural Factors in Aviation Safety: Human Factors Digest Nº 16. Circ. ICAO 302-AN/175*. Montreal: Author.

International Civil Aviation Organization (ICAO). (2012). *Manual on Flight and Flow Information for a Collaborative Environment. Doc. ICAO 9965-AN/483*. Montreal: Author.

International Civil Aviation Organization (ICAO). (2013a). Safety Management Manual (SMM) (3rd ed.). Doc. ICAO 9859-AN/474. Montreal: Author.

International Civil Aviation Organization (ICAO). (2013b). *State of Global Aviation Safety*. Montreal: Author.

International Civil Aviation Organization (ICAO). (2016). *Business Plan (2017-2019)*. Montreal: Author.

Introduction

The book *Automated Systems in Aviation and Aerospace Industries* is devoted to the development and operation of automated control systems, which in aviation and Aerospace Industries are used. These systems belong to real-time systems, where the available time for decision making is set by the speed of the flow of controlled processes. The degree of human participation in these systems is different, but almost all of them perform the functions of control and management of objects. The main goal of the Automated Systems is to assist the crew, the dispatcher to overcome the difficulties encountered in solving professional problems. An automated system need effectively cope with making erroneous decisions or actions caused by fatigue, reduced performance, an underestimation of the complexity of the situation and its capabilities, failures of components and systems, and other problems encountered in flight.

International Civil Aviation Organization documents recommend the use of new information technologies to solve problem tasks and support aerospace system operators. Developing of Artificial Intelligent Systems such as Expert Systems, Decision Support System considering new concepts in aviation need with using modern information technologies and modern courses such as Data Science, Big Data, Data Mining, Multi-Criteria Decision Analysis, etc. Development of new aviation systems using modern information technologies is presented in the book *Automated Systems in Aviation and Aerospace Industries*.

The book will consider the use of information technology in Aviation and Aerospace such as the design of new systems (Decision Support Systems, Expert Systems, Artificial Intelligence Systems, Computing System, etc.), the operation of Information system in Aviation and Aerospace; Training of personnel in Aviation and Aerospace. The development of training for automated systems in Aviation and Aerospace is more difficult than for non-automated systems. One of the primary challenges in developing training for automated systems is to determine how much a trainee will need to know about the underlying technologies in order to use automation safely and efficiently. The target audiences of the book *Automated Systems in the Aviation and Aerospace Industries* are scientists, teachers, pupils, students and postgraduate students, air crash investigators, scientists, engineers and managers, aviation operators (pilots, dispatchers, pilots of UAVs, etc.).

Scientists from 11 countries (Ukraine, China, India, Turkey, Poland, Egypt, Mexico, Azerbaijan, Kuwait, Italy, Pakistan) presented their scientific achievements in the field of the use of automated systems in aviation and aerospace Industries. This book is a collection of seventeen chapters that could cover problems about the advantages of automated systems in aviation and aerospace Industries encompasses. The chapters present tasks from different science fields like using artificial intelligence systems and decision support systems in the aeronautics and aerospace industries; using automated systems in aviation personnel training; autonomous flying robots and many applications UAVs in civil and military tasks solution; role of automation on a modern airplane; models of calculating the weight and centering of the airship; using of Petri net, GERT-net, Big Data and theory of catastrophes in aviation. Much useful and interesting information is presented in all chapters by scientists, engineers, pilots, and air traffic controllers, etc.

Chapter 1
Machine Learning and Text Analysis in an Artificial Intelligent System for the Training of Air Traffic Controllers

Tetiana Shmelova
https://orcid.org/0000-0002-9737-6906
National Aviation University, Ukraine

Nina Rizun
https://orcid.org/0000-0002-4343-9713
Gdansk University of Technology, Poland

Yuliya Sikirda
Kirovograd Flight Academy of the National Aviation University, Ukraine

Vitaliy Lazorenko
National Aviation University, Ukraine

Volodymyr Kharchenko
National Aviation University, Ukraine

ABSTRACT

This chapter presents the application of new information technology in education for the training of air traffic controllers (ATCs). Machine learning, multi-criteria decision analysis, and text analysis as the methods of artificial intelligence for ATCs training have been described. The authors have made an analysis of the International Civil Aviation Organization documents for modern principles of ATCs education. The prototype of the neural network for evaluating the timeliness and correctness of the decision making by ATCs has been developed. The new theoretical and practical tasks for simulation and pre-simulation training have been obtained using expert judgment method. The methodology for sentiment analyzing the airline customers' opinions has been proposed. In addition, the examples of artificial intelligence systems and expert systems by the authors, students and colleagues from National Aviation University, Ukraine and Gdansk University of Technology, Poland have been proposed.

DOI: 10.4018/978-1-5225-7709-6.ch001

INTRODUCTION

In recent years we have observed large changes in the education system for training of air traffic controllers and other aviation professionals due using new information technology in aviation. One of the scientific and methodical fields which gained goals as a result of this was Data Analysis and Decision Making (DM) under various names: Statistics, Big Data, Data Mining, Machine Learning, Text Mining, Sentiment Analysis, Intelligence Data Analysis, Artificial Intelligence System, Knowledge Discovery, etc.

The aviation system is a complex system that requires investigation of the human contribution to safety and an understanding of how human performance may be affected by its multiple and interrelated components such as technical, political, physical, social, economic, informational, training culture etc. (International Civil Aviation Organization [ICAO], 2004, 2009, 2016b). The Human Factor (HF) remains a major cause of aviation accidents. That is why the current system of education needs to improve. The HF as a term requires a precise definition because when it is used in everyday life, it often encompasses all aspects of human activity (ICAO, 1998, 2002). People are the most flexible, adaptable and important element in the aviation system and the most vulnerable in terms of opportunities effect on its activity. At the initial stage of development of aviation, many problems have been associated with exposure to human noise, vibration, heat, cold and acceleration forces. But optimization of the human role in complex systems is related to all aspects of human activity, such as DM processes and knowledge; design configuration displays, controls and equipment cockpit and cabin; maintaining communications and software; preparation of plans and maps.

Air traffic controllers (ATCs) – are the experts in Air Traffic Management (ATM) for dynamic, integrated management of air traffic and airspace. The ATM system is a system that provides air traffic management through the collaborative integration of humans, information, technology, facilities and services, supported by air, ground and/or space-based Communications, Navigation and Surveillance (CNS). Statistical data show that human errors account for up to 80% of all causes of aviation accidents (Leychenko, Malishevskiy, & Mikhalic, 2006; Aviation Accident Statistics, 2018). Safety in aviation will continue to be highly dependent on the reliability of Air Traffic Service (ATS). This reliability will be directly linked to the ability of ATS planners and managers to incorporate the many lessons of history regarding the impact of HF on controller performance. According to the list of factors, performance includes quality of selection criteria, initial and recurrent training, and retention programs for ATCs (ICAO, 2008; European Organization for the Safety of Air Navigation [Eurocontrol], 2004a, 2004b). Quality of aviation training is important to decrease the influence of the HF.

BACKGROUND

Pilots and ATCs are working together for ensuring the safely, economically and efficiently flights of aircrafts (ACs). The pilots in crews are in communication with the ATC. ATCs are responsible for the order of movement of different types of ACs (manned and unmanned ACs, helicopters, etc. Control by the dispatcher includes absolutely all stages of the movement of the AC: from taxing it from the parking lot before take-off to taxiing to the parking lot after landing.

That is, in the Air Navigation System (ANS) operates the principle of "dual operator": the pilot – the ATCs (Figure 1). Actual is the study of the regularities of the activities of both operators and their teams both in the process of fulfilling their professional duties and in social life. The authors present concep-

Figure 1. Air Navigation System operates the principle of "dual operator"

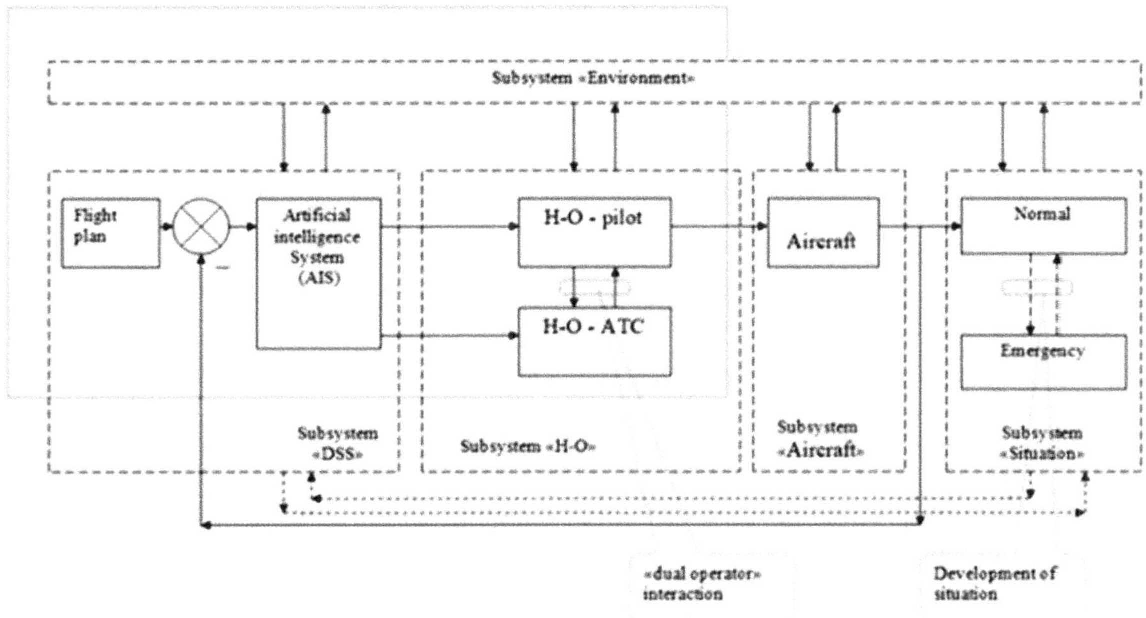

tual models of Decision Support Systems (DSS) and Expert Systems (ES) for human-operator (H-O) of ANS, such as ATCs, for training, in-flight emergencies, etc. The authors present Artificial Intelligence Technology and models of ES for H-O education in ANS.

The ANS is a complex purpose-oriented highly organized Human-Machine System (HMS) with a hierarchical structure of control of a stochastic system, with distinctive features which can be considered:

- Availability of the purpose of operation: ensuring flight safety by maintaining at the required level or improving the performance of the H-O – AC, the regularity and efficiency of air transportation;
- Presence of operators acting as a management link assessing the compliance of the results of the system's intended purpose and the DM with regard to the need for management actions;
- Availability of subsystems for the collection, transmission, and processing of information of H-Os, ACs, environment, etc.;
- Availability of control bodies;
- Availability subsystem to predict and prevent DM.

Because controllers have an incredibly large responsibility while on job and make countless real-time decisions on a daily basis, the ATCs profession is consistently regarded around the world as one of the most mentally challenging careers, and can be notoriously stressful depending on many variables (equipment, configurations, weather, traffic volume, traffic type, human factors, etc.). Documents present vision of an integrated, harmonized, and globally interoperable ATM system (ICAO, 2005). The Flight & Flow Information for a Collaborative Environment (FF-ICE) describes the information environment in support of that vision. The key aspects include support for a Performance-Based Approach (PBA), Collaborative Decision Making (CDM), System-Wide Information Sharing and Management (SWIM) by trajectory. Aviation specialists should have good professional training and knowledge of new concepts and technologies. ICAO recognizes the preparation of ATCs in the way of Knowledge Obtaining and Skill Acquisition.

The role of H-O is very large, so it is necessary to "strengthen" him. The concept of accident causation (J. Reason's Model) shows that training includes as defenses from the accident and stands among the technologies, regulations. If ATC will be well preparation is unlikely that it will lead to unforeseen situations, namely the lack of "holes" how the Swiss-Cheese Model (ICAO, 2008).

PROFESSIONAL TRAINING OF AIR TRAFFIC CONTROLLERS

Simulation Training is an important part of ATCs professional training (Eurocontrol, 2004a). It is a complex of existing forms and methods of training, in which students through the implementation of appropriately formulated complex tasks and exercises under the guidance of the instructor develop skills and practical application of the theoretical provisions of several disciplines (Eurocontrol, 2004b, 2015). The aim of training is to improve ATCs work and refinement practical skills of ATS in standard situations, potentially conflict situations (PCS), in the special conditions and in the flight emergencies. The quality and number of exercises, objective evaluation of exercises influence on the effectiveness of Simulation Training.

Simulation Training is a model of communication where the student, either individually or in the group receives information through a media at a rate (Eurocontrol, 2000). The combination of these elements defines the training event.

Media is the physical means by which an instructor or a training designer communicates a message. One media can use several supports we are going to define the media related to simulation but shall not attempt to make an exhaustive list of the many types of support and educational materials. It has defined the following five media (Eurocontrol, 2000):

1. Real Equipment.
2. High-fidelity Simulator (HI FI SIM).
3. Simulator (SIM).
4. Part-Task Trainer (PTT).
5. Other Training Device (OTD).

It is used any of the three rates of learning although most of the exercises are in real time (Eurocontrol, 2000):

1. Self-paced Learning.
2. Time-restricted learning.
3. Real Time.

According to the recommendations of European Organization for the Safety of Air Navigation (Eurocontrol) and to optimize effectiveness of Simulation Training, theoretical and practical training combined from the beginning of the training using the Pre-Simulation (Pre-SIMUL). The process of learning starts with getting by the student Skill Acquisitions, then performance Part-Task Practice and continues Simulation (Eurocontrol, 2004b, 2015).

Eurocontrol has defined several types of Simulation and has differentiated between pre-simulation and simulation exercises, and it has added the notion of Guided Simulation (Eurocontrol, 2000, 2004b, 2015). Simulation (SIMUL) is a provision of knowledge, skills, and attitudes by means of a representation of air traffic responding to the student action as real air traffic. Types of Simulation include Individual Simulation (IND SIMUL), Team Simulation (TEAM SIMUL) and Group Simulation (GROUP SIMUL). Simulation always includes Briefing (Brief), Debriefing (Debrief), and Tutoring. Guided Simulation (GSIMUL – Guided SIMUL) is the extensive interaction between the student and the computer in the form of questions, feedback, comments, instructions, and assessment. This guidance assumes the existence of a theoretical model against which the student can be compared. Guided Skills Acquisitions (GSA – Guided SA) – Skill Acquisitions (SA) with interactive assessment, comments, and guidance – are actual at this time. Guided PTP (GPTP) is a Part-Task Practice (PTP) accompanied by comments, display results, assessment of the student and the ability of feedback (Eurocontrol, 2004b, 2015).

The system of ATCs Simulation Training is characterized by low level of objective evaluation of exercises (Eurocontrol, 2004, 2015). It's connected with the development of sufficient exercises at a given difficulty level. Increasing the number of exercises leading to significant growth an amount of instructional and methodical staff and requirements for their professional skills as well as time for developing the appropriate exercises of a given complexity. The most time spent on modeling of air situation in accordance with the set of objectives in the exercise, verification of exercise to meet at a given level of difficulty, verification graph to meet a planned workload and the possibility of conflict-free exercises.

The correctness is the main criterion for evaluating the quality of the exercises (Eurocontrol, 2000, 2004b, 2015). In modeling of the air situation using inverse task of generating dynamic air situation at a given exercise intensity, takes into account the complexity of air traffic, presence, and quantity of PCS, flight emergencies (Ministry of Transport and Communications of Ukraine, 2007). The timeliness is another important criterion for evaluating the quality of the exercises (Eurocontrol, 2000, 2004b, 2015). Taking into account as correctness as timeliness can be done with the help of Artificial Neural Network (ANN). Automation of the process of exercise verification can reduce the time for its preparation (saving up to 80% of the time).

To unify an approach to the process of preparation of aviation specialists it is recommended to utilize a principle of Competency-Based Training (CBT) promulgated by ICAO (ICAO, 2012). Competency-Based training has been applied in many different contexts and professions and therefore, it is understandable that there are many different definitions of what competence and competency-based training are. Consequently, the Contracting States, along with their approved training organizations and operational units, are able to structure and conduct their Air Traffic Control (ATC) training courses in a manner that is appropriate to their regulatory context and their operational, technical and organizational environment. CBT is a consistent dimension of human performance that is used to reliably predict successful performance on the job. Performance of an ATC is the combination of knowledge and skills application during the DM process, Standard Operating Procedure (SOP) execution and during other cognitive processes for solution strategy development when dealing with emergencies.

Knowledge ICAO is recognizing as outcome of the learning process. There are different types of knowledge: declarative (facts, raw data), procedural (categorized/contextualized, application of conditional if-then rules), strategic (synthesis, inference to guide resource allocation for decision-making, problem-solving, behavioral action), adaptive (generalization, innovation, invention). Skill is an ability to perform an activity or action. It is often divided into three types: motor, cognitive and metacognitive skills.

CBT for ATC shall be developed in accordance with Instructional Systems Design (ISD). The training program shall be composed of integrated theoretical and practical instruction and program shall include a continuous evaluation acceptable to the authority. ISD is a type of formal approach to training where the goals of the training are carefully determined often from various types of assessments of the learners, goals are established to address the results of the assessments, various methods of training and learned are developed and designed to achieve those goals, and evaluation plans are established the measure the quality of the training and extent of achievement of the goals. ADDIE (Analysis, Design, Development, Implementation, and Evaluation) is one of the most popular ISD models and includes next goals:

1. *Analyze* the organization and individual needs and then identify training goals which, when reached, will equip learners with the knowledge and skills to meet the organization and individual needs. Usually, this phase also includes identifying when training should occur and who should attend as learners.
2. *Design* a training system that learners and trainers can implement to meet the learning goals. This phase typically includes identifying learning objectives (which culminate in reaching the learning goals), what strategies and activities are needed to achieve the objectives, what resources (money, supplies, facilities, etc.) might be needed, any lessons and the sequence of lessons, etc.
3. *Develop* a training "package" of resources and materials, including, e.g., designing webinars, developing audio-visuals, graphics, manuals, etc.
4. *Implement* the training package, including to deliver the training by implementing the strategies and conducting the activities, sharing feedback about the program and training methods, administering tests, modifying the design of the training and its materials based on feedback from participants, etc. This phase can include administrative activities, such as copying, scheduling facilities, taking attendance data, billing learners, etc.
5. *Evaluate* the training, including during and after implementation of training. Evaluation is of the design of the training program, usage of the resources and the results gained by participants in the program.

The progression of ATCs training has been structured into three phases: Initial Training, Unit Training and Continuation Training (Figure 2).

Regulatory requirements of ATCs preparation are included in documents for education, for example, in personnel licensing categorizes there are six ATCs ratings that may be endorsed on ATCs license, or record (ICAO, 2016b):

1. Aerodrome control rating (ADC);
2. Approach control procedural rating (APP);
3. Approach control surveillance rating (APS);
4. Area control procedural rating (ACP);
5. Area control surveillance rating (ACS);
6. Approach precision radar control rating (APRC).

Figure 2. Progression of ATCs training (Eurocontrol, 2004b)

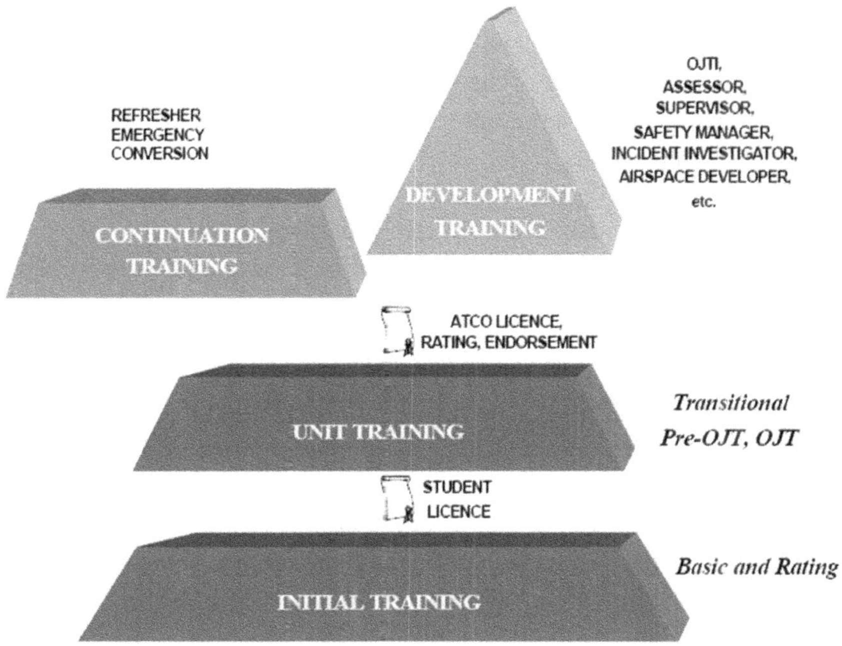

There is a general requirement that all ATCs must demonstrate knowledge in the following subjects (Table 1).

Educational methods to be applied are lecture, lesson/demonstration, practical classes, laboratory work, taxonomy.

To assess theoretical knowledge of trainee, Eurocontrol is recommending for application predefined levels of taxonomy (Eurocontrol, 2004b):

Table 1. Subjects and descriptions for knowledge by ATCs (ICAO, 2016a)

Nº	Subject	Description
1	Air Law	Rules and regulations relevant to ATC.
2	ATC Equipment	Principles, use and limitations of equipment used in ATC.
3	General Knowledge	Principles of flight; principles of operation and functioning of aircraft, engines, and systems; aircraft performance relevant to ATC operations.
4	Human Performance	Human performance including principles of threat and error management.
5	Meteorology	Aeronautical meteorology; use and appreciation of meteorological documentation and information; origin and characteristics of weather phenomena affecting flight operations and safety; altimetry.
6	Navigation	Principles of air navigation; principle, limitation, and accuracy of navigation systems and visual aids.
7	Operational Procedures	ATC, communication, radiotelephony, and phraseology procedures (routine, non-routine and emergency); use of the relevant aeronautical documentation; safety practices associated with flight.

- **Level 0:** 'To be aware of'.
- **Level 1**: Requires a basic knowledge of the subject. It is the ability to remember essential points; trainee is expected to memorize data and to restore it.
- **Level 2**: Requires an understanding of the subject sufficient to enable the trainee to discuss intelligently. The individual is able to represent for himself or herself certain objects and events and to act upon these objects and events.
- **Level 3**: Requires a thorough knowledge of the subject and the ability to apply it with accuracy. The trainee should be able to make use of his/her repertoire of knowledge.
- **Level 4**: The ability to establish a line within a unit of known applications following the correct chronology and the adequate method to resolve a problem situation. This involves the integration of known applications in a familiar situation.
- **Level 5**: The ability to analyze new situations in order to elaborate and apply one or other relevant strategy to solve a complex problem. The defining feature is that the situation is qualitatively different to those previously met, requiring judgment and evaluation of options.

In education need evaluation criteria of skills, both interim and final progress checks (ICAO, 2016a, 2016b). This necessity is absolutely undisputed for nearly all-institutional learning processes (Table 2).

Mentioning errors (E) throughout the table should be understood in the manner of error itself (slips and lapses), mistakes (knowledge-based errors), mismatches, inadequacy, etc. In general, the responsibility of instructor is to organize and to conduct SA by a trainee with the help of the simulator. More in details, the instructor starts with determination activity of trainee's theoretical and practical levels. In order to make judgments, instructor must possess excellent knowledge of many of theoretical and practical subjects, such as Air Code; Flight rules; Regulation on airspace utilization; ATC rules, guidance, regulations, directives, and instructions; Procedures of corresponding ATC units, etc. Instructor evaluates trainee on all his activities, beginning from trainee's readiness of task execution ending by feedback analysis and provision of further recommendations. Completeness, adequacy, and quality of

Table 2. Evaluation of the technological operations

Nº	Technological Operation	Evaluation Mark		
		Excellent	Good	Satisfied
1	Acceptance of duty on the workplace	0 errors	1 error	2 errors
2	Phraseology adequacy	≤1 error	2-3 errors	4 errors
3	Coordination with adjacent units	≤1 error	2-3 errors	4 errors
4	The regularity of flights	0 errors	1 error	2 errors
5	The accuracy of aircraft's positions determination	0 errors	1 error	2 errors
	By distance	±5 kilometers		
	By azimuth	±4 degrees		
	By direction finder	±2 degrees		
6	Correspondence of decision making to given situation	0 errors	1 error	2 errors
7	Provision of proper separation and general safety issues	0 errors	1 error	2 errors and more
8	Adequacy of console operations	≤1 error	2-3 errors	4 errors and more

mentioned above results depend in direct on quality of instructor. Furthermore, personal aspects of an instructor, his ability to manage preparation process, the capability to perform SA process forecast, talent to apply the individual strategy of SA for the trainee, etc. To manage the position of outcomes it is forced to structure specific filtering tasks which are mandatory to be done by the trainee, in order to guaranty not exceeding the predefined level of preparation. In other words, it's ensuring correspondence of trainee SA level with requirements of certain technical operation predefined criteria. Media tools that may be applied to satisfy the goal are next (ICAO, 2016a):

- Part Task Trainer (PTT);
- Other Training Devises (OTD);
- Radar Trainer (RAD);
- Automated Work Place "Situation" (AWP SIT).

The process of SA by trainee from the point of view of technological operations could be divided into five categories (Table 3).

With an eye to establishing flexibility of preparedness level control, it is better to split the management into three levels based on time: Strategic level (new tasks planning phase), Pre-tactical level (planning phase based on previous results of E) and Tactical phase (management of recommendations in a day of training). As far as ICAO is aiming ATCs activities on a safety provision, it is vitally important to ensure an adequacy of media tools we are applying for SA process. To define adequacy we should use outcomes of experts' assessment on training tools' elements as their probability to be reflected adequately in certain Media Tool (1):

$$E_{adequacy} = \sum_{i=1}^{n} T_i \cdot R_i \cdot C_i \cdot N , \quad (1)$$

where i – is a required procedure or skill planned to be gained; T_i – is a period of use of i-th function in Media Tool ($t_{sec}/60$); R_i – is a relative rank of importance of i-th function in Media Tool; C_i – is a relative rank of correspondence of i-th function in Media Tool; N – is a general quantity of procedures for certain task.

To improve the evaluation of the quality of ATCs education the use of smart machines employing Artificial Intelligence and Machine Learning (neural-expert system, fuzzy-logic system, etc.) was pro-

Table 3. Category of SA by technological operations and correspondence to types of task

N°	Category of SA by Technological Operations	Brief Description	Type of Task
1	Short-term memory practice	Training of our short-term memory for the required amount of data.	AWP SIT
2	"Eye" practice	Positioning of aircraft training.	OTD/RAD
3	Procedures and phraseology	Typical procedural algorithm training. React to pilots' requests.	OTD/PTT
4	Conflict resolution	Situation analysis. Conflict detection. Separation provision.	PTT/AWP SIT
5	Decision-making	Situation analysis. Strategy application. Issues on regulation and efficiency of traffic.	PTT

posed. In this context, a special class of learning methods for ANN called Deep Learning has proven to be particularly successful. Deep Learning is a specialization of Machine Learning (ML), which in turn is a subdiscipline of Artificial Intelligence, a major branch of research in computer science. ML is a class of methods of Artificial Intelligence, whose characteristic feature is not a direct solution to the problem, but learning in the process of applying solutions to many similar tasks. The ML, on the one hand, was formed as a result of the separation of the science of neural networks into methods of learning networks and the types of topologies of their architecture; on the other hand, it incorporated methods of mathematical statistics. To construct such methods are used the next courses:

- Mathematical statistics;
- Numerical mathematical methods;
- Optimization methods;
- Probability theory;
- Graph theory, etc.

ML is a field in the computer science, but it differs from the traditional computing approaches. In traditional calculations, examples of which are Petri nets and cellular automata, algorithms are sets of explicitly programmed instructions used by computers to perform calculations or solve a problem. But ML algorithms allow computers to learn on input data and use statistical analysis to output values that fall within a certain range. Due to this, computers become more productive in building models from sample data, which makes it possible to automate the decision making processes based on the given information.

A hybrid neural-expert system is a perspective direction of development of a neuron-information technology (Komashynskiy & Smirnov, 2002). ANN has many advantages compared to traditions and knowledge-based of diagnostic systems (Arkhangelskiy, Bohaenko, Grabowskiy, & Ryumshyn, 1999; Suzuki, 2013). It can be trained on examples, work in real-time, secure and tolerant of errors. With the help of ANN diagnoses state of the patient, performs the prediction on the stock market and the weather forecast, makes decision on granting the loan, diagnoses condition of equipment, guided operation of the engine, etc. (Komashynskiy & Smirnov, 2002; Gardner & Dorling, 1998; Goedeking, 2010). ANN is being created by the serial and parallel association of individual neurons. A neural network is grouped in two classes according to the type of connections: straight directional network (which links don't have loops) and recurrent network (with feedback connections). (Komashynskiy & Smirnov, 2002; Arkhangelskiy, Bohaenko, Grabowskiy, & Ryumshyn, 1999; Suzuki, 2013; Gardner & Dorling, 1998; Goedeking, 2010; Borovikov, 2008). The most common among straight directional networks are single-layer and multi-layer perceptrons, cognitron and Radial Basis Function (RBF) networks; among the recurrent network can be distinguished Hopfield, Boltzmann and Kohonen networks (Arkhangelskiy, Bohaenko, Grabowskiy, & Ryumshyn, 1999; Suzuki, 2013).

Using of neural networks and neuro-fuzzy systems is appropriate for solutions many problems in aviation, where it is necessary to process a large amount of fuzzy information and solving difficult formalized multiparameter nonlinear tasks and problems, namely in the case of decision making about departure and selection of an optimum alternate aerodrome landing (Kharchenko, Shmelova, Artemenko, & Otryazhiy, 2011); choosing the optimum place of forced landing (Kharchenko, Shmelova, Sikirda, & Gerasimenko, 2011); estimation of efficiency of alternative variants of flight completion in the event of an extraordinary situation (Shmelova, Yakunina, & Yakunin, 2013); diagnostic erroneous actions of the operator of ergatic aviation system in flight emergencies (Shmelova, Yakunina, Moyseenko, & Grinchuk, 2014), etc. It is proposed to use the neural network model of evaluation correctness and timeliness of decision making by ATCs in case of PCS during the performance of exercises in the Simulation Training through pre-training studying.

ES use for DM Multi-Criteria Decision Analysis (MCDA). MCDA is a discipline aimed at supporting decision makers who are faced with making numerous and conflicting evaluations (ICAO, 2009). One of the methods for solving multi-criteria decision problems is Expert Judgment Method (EJM).

Many problems with the application of these methods are considered in the "Decision Making" discipline, which is compulsory for the training of aviation specialists (pilots, controllers), since it contains mathematical methods and applied information technologies. So, DM includes Machine learning too. Training topics in the course "Decision Making":

1. Analysis and synthesis of aviation HMS "pilot-aircraft".
2. EJM. Quantitative estimation of the complexity of the aircraft flight stages.
3. EJM. Quantitative estimation of the significance of the Landing System (GNSS, ILS, VOR, etc.) (ATSEP) / Quantitative estimation of the complexity of procedures operators during the working process (ATC).
4. The Simplex method (diet problem, finance distribution on advertising).
5. The Transportation problem. Airplane distribution along the routes.
6. The Network planning. DM by H-O under Certainty: ATC (emergency situation – ASSIST); ATSEP.
7. DM under Risk in the emergency situation using the Decision tree.
8. DM under Risk: construction of airport with using MS Excel.
9. DM under Uncertainty: tasks of cost/profit (supply of Lufthansa).
10. DM under Uncertainty: tasks of choosing optimal landing aerodrome in the emergency situation.
11. Dynamic programming. Tasks: the problem of minimum cost way; the traveling salesman problem; the task of climb/take-off of the aircraft.
12. Neural networks. Neural network for admission of the student to Simulation Training by the number of hours and level of training.
13. Fuzzy logic. Quantitative estimation of the risk.
14. Characteristics of the H-O in ANS. Sociometric method (compatibility of the group). Socionic method (Test).
15. Correlation and regression. Compare the results of correlation and regression analysis of Eurocontrol data.
16. Systems (Man-Machine, DSS, ES, Neuron system) (classification, analyzes, compare, examples of systems).

For example, contents of Training N°1 "EJM. Quantitative estimation of the complexity of the aircraft flight stages and Training N°2 "Deterministic models. Network planning. DM in the emergency situation".

1. Theory 2 hours (or 1 hour).
2. Practice 2 hours (or 1 hour).
3. Individual work 2 hours (or 1 hour).
4. Presentation 2 hours (or 1 hour).

Training N°1 "EJM. Quantitative estimation of the complexity of the aircraft flight stages:

1. *Theory.* Basic of EJM for ANS (Classification of methods of DM. The algorithm of EJM. The matrix of individual and group preferences. Coordination of experts' opinion. Multi-criteria decision problems.
2. *Practice.* Tasks: Quantitative estimation of the complexity of the aircraft flight stages – determination of significance (complexity) of the phases of the flight of the aircraft (Figure 3).
3. *Individual work.* DM in the Team. Lottery-choice of tasks for the team:
    - Quantitative estimation of the complexity of the aircraft flight stages – determination of significance (complexity) of the phases of the flight of the aircraft;
    - Quantitative estimation of the complexity of the navigation parameters during aircraft flight;
    - Quantitative estimation of the significance of the Landing System (GNSS, ILS, VOR, etc.);
    - Quantitative estimation of the complexity of procedures operators during the working process
        - Determination of controller's workload for aircraft service, three different types of simulator training exercises: Control Zone (CTR), Terminal Control Area (TMA) and Control Area (CTA) with various type of complexity analyzed.

Figure 3. The process of applying solutions to many similar tasks in the quantitative estimation of the complexity of the aircraft's flight stages

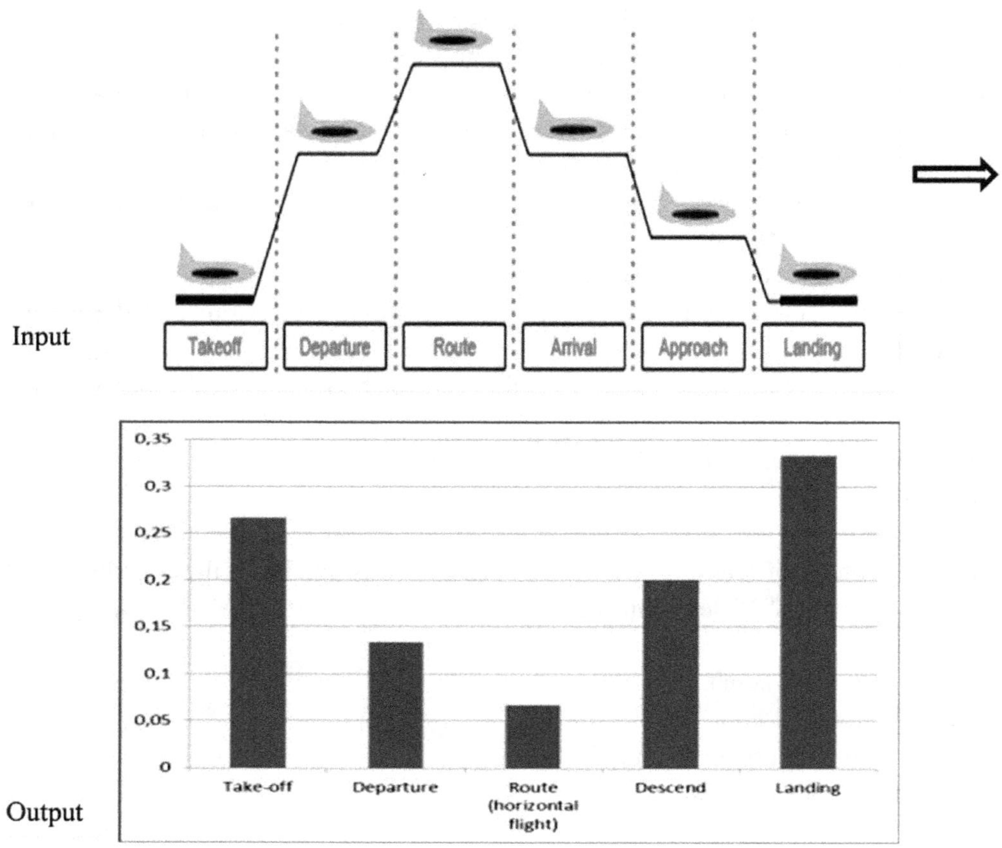

- ◦ Quantitative estimation of the HF problem;
- ◦ Select of the sources of projects financing, etc.

4. *Presentation.* Training N°2. “Deterministic models. Network planning. Decision making in the emergency situation”.
   a. *Theory.* DM under Certainty. Method of network planning. Algorithm for constructing network graph.
   b. *Practice.* The deterministic models for H-O (air traffic controller, pilot, engineer, etc.). “ASSIST”, Flight Manual, etc. (Figure 4).
   c. *Individual work.* DM in the Team. Lottery-choice of tasks for the team:
      - DM under Certainty by H-O in-flight emergencies (for ATCs). Example;
      - DM under Certainty by H-O for repair of systems (for ATSEP). Example;
      - DM under Certainty by H-O for control/repair of systems (for UAV). Examples.
   d. Presentation.

Figure 4. The process of applying solutions to many similar tasks in DM under Certainty

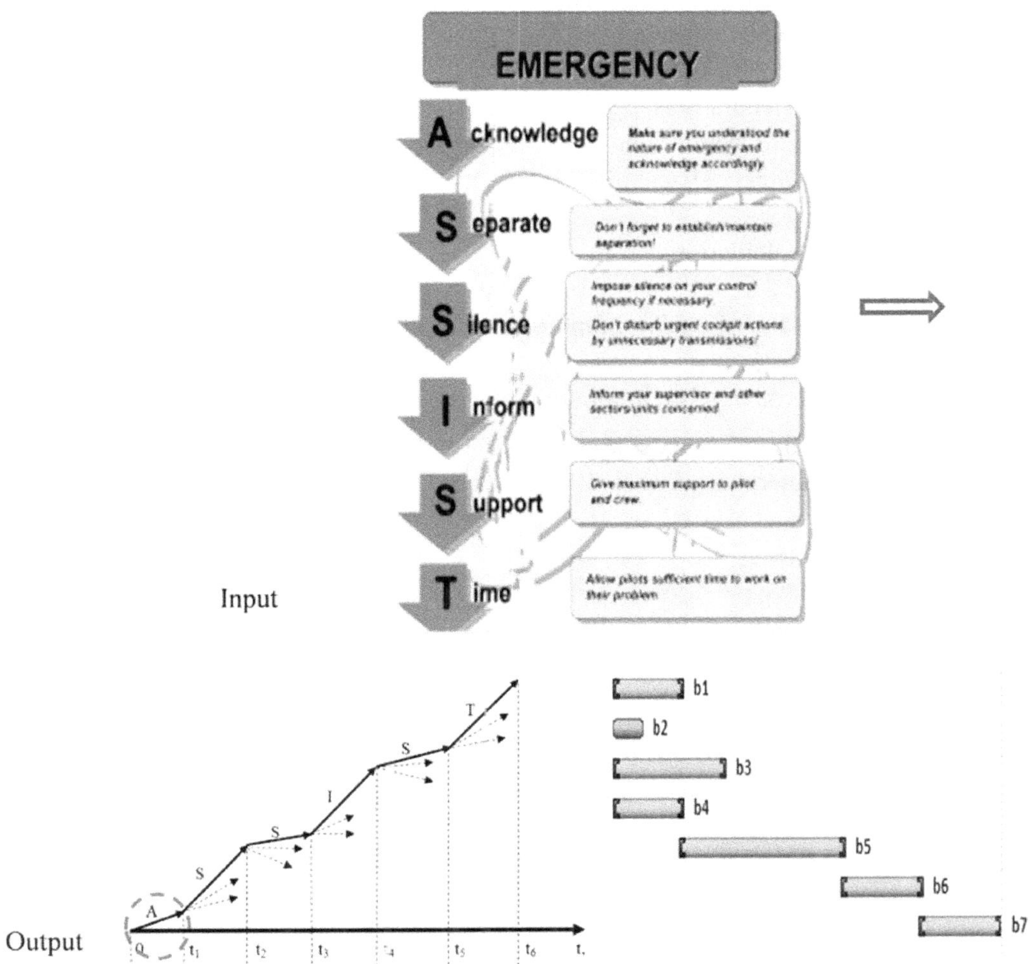

Estimation of Multi-Criteria Decision

MCDA, ML methods with the help of EJM use for estimation of the tasks complexity during the Pre-Simulation Training. It has been determined that a flight stage, every zone, where air traffic service is provided, has the different technology of work and individual instructions for procedures.

Block-scheme of the algorithm of EJM is shown in Figure 5. Main steps of EJM – is a simple approach for calculation:

1. Questionnaires for experts.
2. The matrix of individual preferences, R_i.
3. The matrix of group preferences, R_{ij}.
4. Experts' group opinion (sample average, arithmetical mean), R_{grj}.
5. Coordination of experts' opinion for each factor:
    a. Dispersion for each factor, D_j.
    b. Square average deviation, σ_j.
    c. Coefficient of the variation for each factor, ν_j.
6. Coordination of experts' opinion for all factors:
    a. Kendal's coordination coefficient, *W*.
    b. Spearman's correlation coefficient, R_s.
7. The significance of the calculations:
    a. Kendal's coordination coefficient *W* – criterion χ^2.
    b. Speaman's correlation coefficient R_s – Student's *t*-criterion.
8. Weight coefficients, w_i.
9. Graph of results of the calculation.

The algorithm of EJM for the ATCs Pre-Simulation Training:

1. Questionnaires for experts: *m* – is a number of experts, $m \geq 30$.
2. The matrix of individual preferences: R_i – is a system of preferences of *i*-expert, $i = \overline{1, m}$.
3. The matrix of group preferences $R = R_j$ (2):

$$R = R_j = R_{gr} = \frac{\sum_{i=1}^{m} R_i}{m}. \tag{2}$$

where $R = R_j = R_{gr}$ – is an experts' group opinion, $j = \overline{1, n}$; *n* – is a number of factors; *m* – is a number of experts; R_i – is a system of preferences of *i*-expert.

4. Coordination of expert's opinion.
    4.1. Calculation of dispersion *D* (3):

Figure 5. Block-scheme of the algorithm of EJM

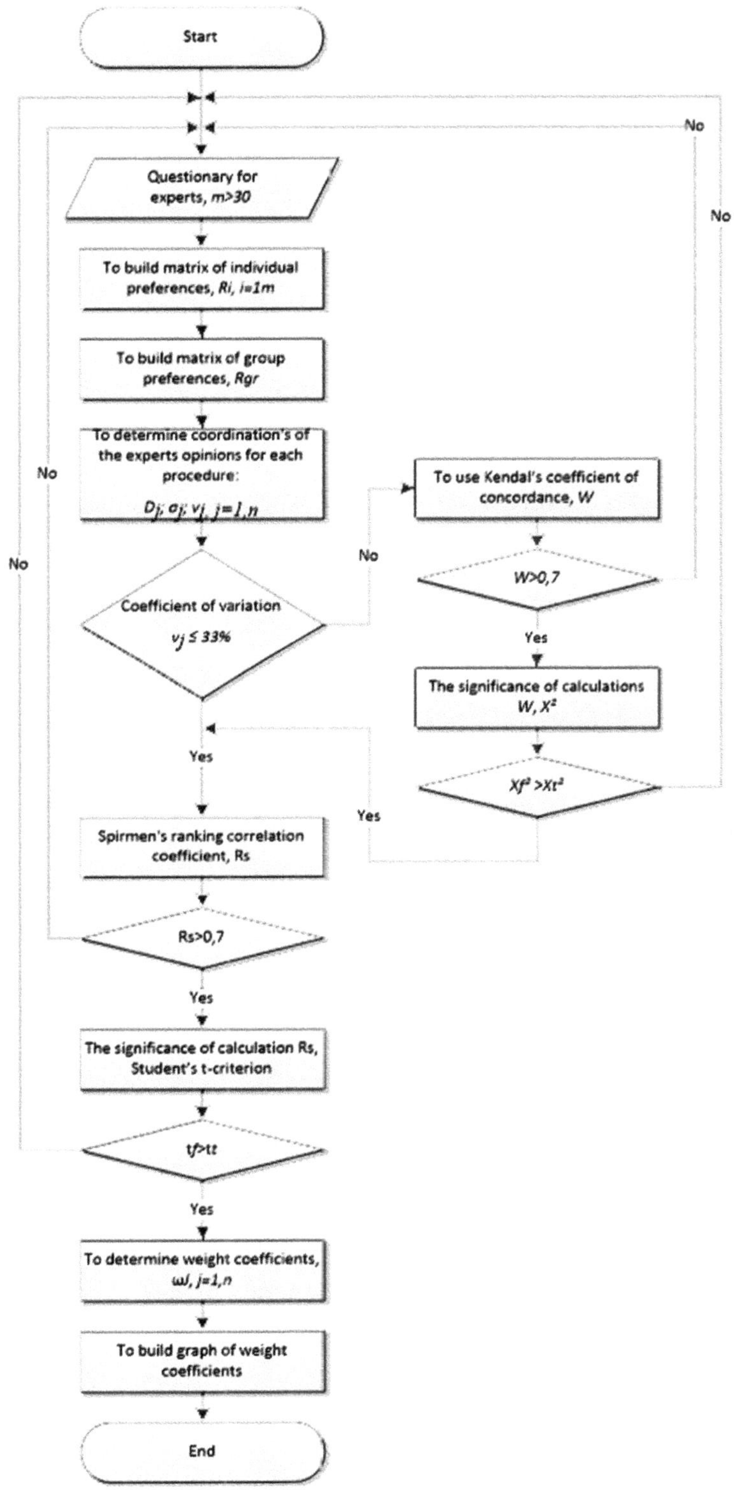

$$D = \frac{\sum_{i=1}^{m} (R_{gr} - R_i)^2}{m-1}. \quad (3)$$

4.2. Calculation of square average deviation σ (4):

$$\sigma = \sqrt{D}. \quad (4)$$

4.3. The coefficient of the variation ν (5):

$$\nu = \frac{\sigma}{R_{gr}} \cdot 100\%. \quad (5)$$

If $\nu \leq 33\%$ then the opinion is concerted and system of expert group has been obtained. If $\nu > 33\%$ then it is necessary to calculate Kendal's concordation coefficient W (6):

$$W = \frac{12S}{m^2(n^3 - n) - m\sum_{i=1}^{m} T_i}; \quad (6)$$

$$S = \sum_{j=1}^{n} (\sum_{i=1}^{m} R_{ij} - \overline{R})^2;$$

$$T_i = \sum_{i=1}^{m} (t_i^3 - t_i);$$

$$\overline{R} = \frac{1}{n} \sum_{j=1}^{n} (\sum_{i=1}^{m} R_{ij}),$$

where S – is a generalized dispersion; *ti* – is a number of the same ranks in the *i*-row which fixed the *i*-expert.

Kendal's concordation coefficient must be within the limits $0,7 < W \leq 1$. If $W < 0,7$ it is necessary to repeat the interrogation.

5. Compare the system of preferences R_{gr} and R_i, $i = \overline{1, m}$ with the help of rating correlation coefficient R_s (Spearman's coefficient) (7):

$$R_s = 1 - \frac{6\sum_{j=1}^{n}(R_{gr} - R_i)^2}{n(n^2 - 1)}. \quad (7)$$

6. The significance of the calculations.
   6.1. The significance of the calculation of Kendal's coordination coefficient *W*, criterion χ^2 (8):

$$\chi_f^2 = \frac{S}{\frac{1}{2}m(n+1) - \frac{1}{12(n-1)}\sum_{j=1}^{n} R_j} > \chi_t^2, \quad (8)$$

where χ_f^2 – is a factual value of the variable; χ_t^2 – is a table value of the variable.

   6.2. The significance of the calculation of Spearman's coefficient R_s with using Student's *t* – criterion (9):

$$t_{critical} = R_s\sqrt{\frac{n-2}{1-R_s^2}} > t_{st},. \quad (9)$$

where *n* – is a number of factors; t_{st} – is a tabulated value while the number of degrees of freedom $f = n-2$ and error $\alpha = 5\%$.

7. Weight coefficients w_j of *j*-factors (10):

$$w_j = \frac{Cj}{\sum_{j=1}^{n} C_j}; \quad (10)$$

$$\sum_{j=1}^{n} w_j = 1,$$

where *n* – is a number of factors; $C_j = 1 - \frac{R-1}{n}$; $C_j = 1 - \frac{R-1}{n}$ – are the estimates, $j = \overline{1,n}$.

8. Graphical presentation of the weight coefficients.

EJM is a structured communication technique, originally developed as a systematic, interactive forecasting method which relies on a panel of experts. The experts answer questionnaires in two or more rounds. After each round, a facilitator provides an anonymous summary of the experts' forecasts from the previous round as well as the reasons they provided for their judgment. Thus, experts are encouraged to revise their earlier answer in light of the replies of other members of their panel. It is believed that during this process the range of the answers will decrease and the groups will coverage towards the "correct" answer. Finally, the process is stopped after a predefined stop criterion (e.g. the number of rounds, achievement of consensus, and stability of results) and the mean or average scores of the final rounds determine the results.

Sentiment Analyzing the Airlines' Customers' Opinions

Since the early 2000s, the proliferation of computer technology has contributed to the fact that the procedure for creating information content has become available to almost everyone. The content of such information resources as social networks, feedback collection services, web forums, and blogs is actively formed by the users themselves.

Consolidated subjective experience of individual users is a source of valuable information (Aman & Szpakowicz, 2008). For example, a collection of user feedback about a specific airlines institution contains information about the level of professionalism of the staff, the quality of service. However, in order to extract such information, it is necessary to analyze the impressive volumes of text data. A special section of computer linguistics is devoted to solving this problem – automatic analysis of the tonality of the text (sentiment analysis or opinion mining) (Bijuraj, 2013; Rizun & Taranenko, 2018; Ruzin & Kucharska, 2018).

The initial goal of the methods of tone analysis was the classification of documents, and later of the sentences, according to a given scale of tonality, usually a two-point (positive-negative) or three-point (neutral or contradictory tone is added). However, for many applied problems such a general tonality analysis at the level of the whole text or even individual sentences is clearly not enough. Often, instead of a general assessment of tonality, a more detailed study of the expressed views on specific aspects is required. For example, for a medical facility, such aspects can be: service, equipment, location, etc.; for the airlines, such aspects can be: prices, conditions, timetable, etc.; for the object educational institution possible aspects: the educational program, the quality of training, the cost, etc. Therefore, over time, the initial formulation of the task of the analysis of tonality has acquired a more detailed formulation and has emerged as a separate problem of aspect-based sentiment analysis, which is to automatically determine the views of the user, expressed in the text, with respect to specific aspects of the entity being examined.

Methods of aspect-emotional analysis of the text are developed within the framework of two machine learning approaches: with the teacher and without the teacher (Bijuraj, 2013). In the approach based on training with the teacher, a marked collection of documents is needed, which lists examples of emotional expressions and aspect terms. The process of preparing such a training collection is costly both in terms of time and effort, which is a significant limitation of the practical use of methods based on this approach.

The methods of teaching without a teacher allow you to avoid dependence on training data. For their work, you also need a body of documents, but the preliminary markup is not required. Within the framework of this approach, the probabilistic-statistical regularities of the text are found and on their basis, the key subtasks of the aspect-emotional analysis are solved: the identification of aspect terms and determination of their tonality. However, such methods require complex tuning to a given domain. For example, the Topic Modelling method based on Latent Dirichlet Allocation (LDA) in its original form is not able to effectively detect aspect terms, therefore, its additional adaptation and adjustment of the correspondence of the identified topics to the target set of aspects is required.

Texts are complex objects containing objects of the material and spiritual world of a person expressed in language, connections between them, components of emotionality. Therefore, the above-mentioned automatic text processing procedures are performed within the framework of some formal models, representing a more simplified and accessible for computer processing description of the source text. In terms of the field of machine learning, such a representation is called an attribute description of the object. Two main models of text representation became widespread: 1) Boolean model; 2) Vector model. The Boolean model is the simplest, in which the text is represented as a set of words (bag of words), called terms. Initially model has a dictionary V of n terms of a certain subject area. To represent a piece of text or a document p, the occurrence of each term $v \in V$ in this document p is checked. The result is a binary term incidence vector (Liu, 2012) (11):

$$\vec{x} = (x_v), x_v = \begin{cases} 0, v \notin p \\ 1, v \in p \end{cases}, \forall v \in V. \tag{11}$$

For any sufficiently representative collection of documents, the size of V is, as a rule, at least several tens of thousands of terms. Specific documents appear to be only a small subset of such terms (of the order of a few hundred), so the resulting document vectors turn out to be very sparse. The representation of a single term, for example, a search query, is the limiting of such cases when only one component of the incidence vector is nonzero.

Despite its simplicity, such a model proves to be quite successful in solving some problems, for example, information retrieval or spam filtering. It is entirely logical to assume that the documents relevant to the query contain query terms. Then the degree of similarity of the document and the query can be calculated using Boolean operations. Similarly, for example, when filtering spam, only the fact that there are certain words in the text that is well reflected by this model is important.

However, the excessive simplification that the Boolean model produces is not always useful. For example, in a more complex task – the thematic classification of text documents - not only the presence of the term is important, but also the frequency of its occurrence, both in a specific document and in the entire collection. A more complex model that allows taking into account such features is the vector space model (Luhn, 1957).

In the case of a Vector model, texts are also represented as vectors in the n-dimensional space of terms: $\vec{x} = (x_1, x_{2,} ..., x_n)$, where $n = |V|$ is the size of the dictionary. However, each component xi, instead of the binary sign of the presence of the term sets its weight, or significance, in this text. In the simplest case, such weight can be the frequency of occurrence of a term in the text. However, it is equally important to take into account the prevalence of the term, which is expressed by its global frequency in the collection of texts.

The most common group of methods of weighting terms – TF.IDF (Term Frequency-Inverted Document Frequency). There are three assumptions underlying the TF.IDF (Manning, 2009; Rizun, 2018):

- Multiple occurrences of a term in a document is no less important than a single (TF assumption);
- Rare terms are no less important than frequent terms (IDF assumption);
- In the case of the same number of terms, long documents are no more important than a short (normalization assumption).

Thus, in the weighting method TF.IDF, the general expression for calculating the weight of the *i*-th term *t* in the *j*-th document is (12):

$$t_{i,j} = L_{i,j} \cdot G_i \cdot D_j, \tag{12}$$

where $L_{i,j}$ – is the local weight of the *i*-th term in the *j*-th document; G_i – is the global weight of the *i*-th term in the collection of documents; D_j – is the normalization for the *j*-th document.

Each component (12) can be calculated from different formulas. For example, for a normalized scheme TF.IDF, such formulas take the forms (13), (14), (15):

$$L_{i,j} = f_{i,j}, \tag{13}$$

where $f_{i,j}$ – is the frequency of the *i*-th term in the *j*-th document;

$$G_i = \log \frac{N}{N_i}, \tag{14}$$

where N – is the number of documents in the collection; N_i – is the number of the *j*-th documents in which the *i*-th term is encountered.

The Cosine normalization for the document vector is:

$$D_j = \frac{1}{\sqrt{\sum_{i=0}^{m} (L_{i,j} \cdot G_i)^2}}, \tag{15}$$

where m – is the number of the *i*-th terms in the *j*-th document.

Then, using (13), (14) and (15), the general formula for calculating the weight of the i-th term in the j-th document is (16):

$$t_{i,j} = \frac{f_{i,j} \cdot \log \frac{N}{N_i}}{\sqrt{\sum_{i=0}^{m} (f_{i,j} \cdot \log \frac{N}{N_i})^2}} \tag{16}$$

For automatic classification of texts by tonality, two main approaches are used: knowledge-based approach and machine learning. At the first approach, classification is carried out by means of the rules made by experts in the subject area. Because of the high labor-based approach, the knowledge-based approach is relatively rare in practice.

Peter Turney in his work (Turney, 2002) extracts of some tech-hundred sets of words (for example, applying a-tion in conjunction with nouns, bunk-being, combined with verbs), and to assess their then – tonality used Altavista search system. Two entries were submitted to the input of this system for each set of words. In response to the first, the number of documents in which the given set meets in the same context as the word "good" is returned; in response to the second – the number of documents in which the given set meets in the same context as the word "bad". The dial tone is determined by the highest number of documents in the answers.

To determine the orientation of the word and its weight estimation, lexical databases (WordNet, Roget's Thesaurus, etc.) are also used (Turney, 2002). The WordNet database consists of nodes (words) connected by edges (a synonym for relations). As a distance between two words in WordNet, you can count the number of edges of the shortest path between two nodes that represent the given words. Thus, to determine the tonality of a word, it is necessary to calculate the distance from a given word to an oppositely oriented one (for example, "good" and "bad").

According to Rizun (Rizun et al., 2017; Rizun & Taranenko, 2018; Rizun & Waloszek, 2018), the dictionary of Sentiment Lexicon, compiled by experts manually, was used to determine the tone of individual words. In the dictionary, each word and phrase is associated with the orientation of the key (positive/negative) and strength (in points). The tone of the document was determined by counting the points included in the document words of emotional vocabulary.

The author's method, proposed by Salton (1989), is based on a lexical approach: to determine the tonality of texts, a dictionary of estimated words is used, each of which has a numerical weight that determines the degree of significance of the word. The method of working with the dictionary closest to the Sebastiani (2002), however, there are significant differences: firstly, the dictionary proposed by the authors is created on the basis of a statistical analysis of the collection of training texts (training collection); secondly, the weight of evaluation words is determined with the help of a genetic algorithm and work with the training collection.

According to the method, proposed by Sebastiani (2002), a vocabulary of the valuation vocabulary is used to determine the tonality of the text. This dictionary consists of evaluation words – words expressing an emotional assessment, positive or negative. Each weighting word in the dictionaries is weighted, which determines the strength of the expressed evaluation. For positive words, positive weight values are used, negative values for negative ones. In the dictionary, there are also word-modifiers and words expressing negation. They are not evaluative, but only change the weight of the estimated words that follow in the text.

The tone of the text is determined on the basis of the calculation of the weights of the appraisal words included in it. For each text T from the training collection, two weights are counted, the first of which is equal to the sum of the positive appraisal words from the dictionary included in the text, the second of which is the sum of the negative evaluation words (17):

$$W_T' = \sum_{i=1}^{N_C} \left| w_i \right|, \tag{17}$$

where $W_T^{!}$ – is the weight of text *T* for tonality *C*; w_i – is the weight of the *i*-th word evaluation; N_C – is the number of estimated bigrams of tonality C in the text T.

All T_i texts are placed in a two-dimensional estimated space (positive tonality – negative tone) in accordance with their scales $W_T^{!}$. To classify texts according to the linear function (18):

$$f\left(W_T^{pos}, W_T^{neg}\right) = W_T^{pos} + W_T^{neg} \quad (18)$$

where W_T^{pos} – is the positive weight of the text T; W_T^{neg} – is the negative weight of the text T.

If the value of the function *f* is greater than zero, the text is positive, otherwise – negative.

In order to use these algorithms in the ATCs training the following methodology of sentiment analysis the airlines' customers' opinions is proposed (Figure 6).

As a tool for technical implementation of the methodology the R-language and Python are proposed.

Illustrative Examples

Estimation of Pre-Simulation Training Tasks Complexity

ATCs Simulation Training is one of the major components in the preparation of ATC students. Before beginning the exercises on the simulator (different zones and difficulties) intended ATC they must have appropriate knowledge, according to theoretical practices – Pre-Simulation Training. Theoretical knowledge is the basis for the admission of the ATCs to Simulation Training. Since the student must master before this list of disciplines, which is regulated of Educational Plan (Curriculum). Education plan – is the educational program with additional information. It details (for each exercise) the type of events that occur during the period of training necessary teaching materials, methods of preparation. It also contains the time scale, which will be performed the exercise.

Figure 6. The methodology of sentiment analysis the airlines' customers' opinions

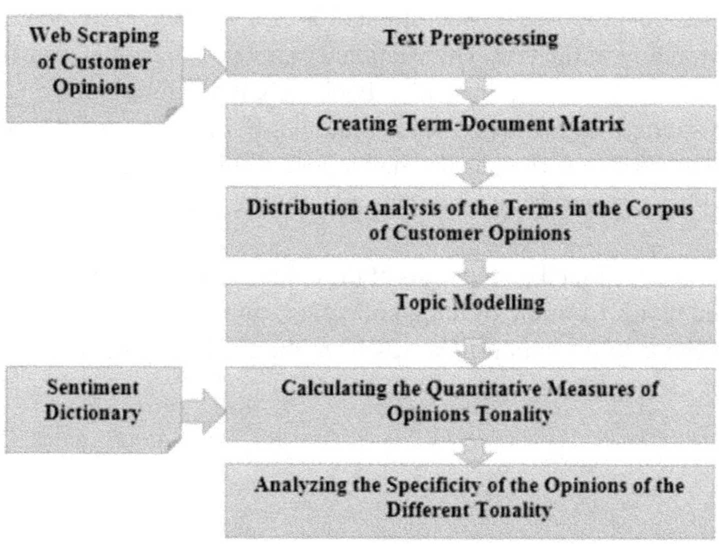

According to Eurocontrol's requirements, the Basic Training of student consists of two phase:

1. Theoretical education.
2. Practical education (practical training on Simulator).

The purpose of basic education is to provide students the basic knowledge and practical skills of ATM needed for further rating training to obtain a certificate ATC student.

Program of ATC Basic Training has the list of discipline such as ATM, Navigation, Human Factors, etc. Basic Training includes to Pre-Simulation Training and determines whether a student ready to Simulation Training or not (Figure 7).

Firstly, the student should fix to learn a number of these disciplines. After that it's seen: if student has acceptable level in this disciplines he/she ready to Simulation Training, if no – not ready. Simulation Training included three zones according to normative documents: CTR, TMA, and CTA (ICAO, 2007). Each of zones has the list of exercise with different difficulties, it depends only from the number of the lesson. The organization of the airspace over given area should be arranged so that it corresponds to operational and technical considerations only. In addition, aerodromes, where ATC is provided, should be designated as controlled aerodromes.

According to ICAO requirements, the Ukrainian ATS airspace has been classified and designated for following classes: class C and class D (Ministry of Transport of Ukraine, 2003) (Table 4).

For estimation complexity of the exercises in the system of Pre-Simulation Training, it is necessary to find weight coefficients, which characterize the complexity of airspace zones. Assessments of the

Figure 7. The interconnection between Pre-Simulation Training of student and Simulation Training

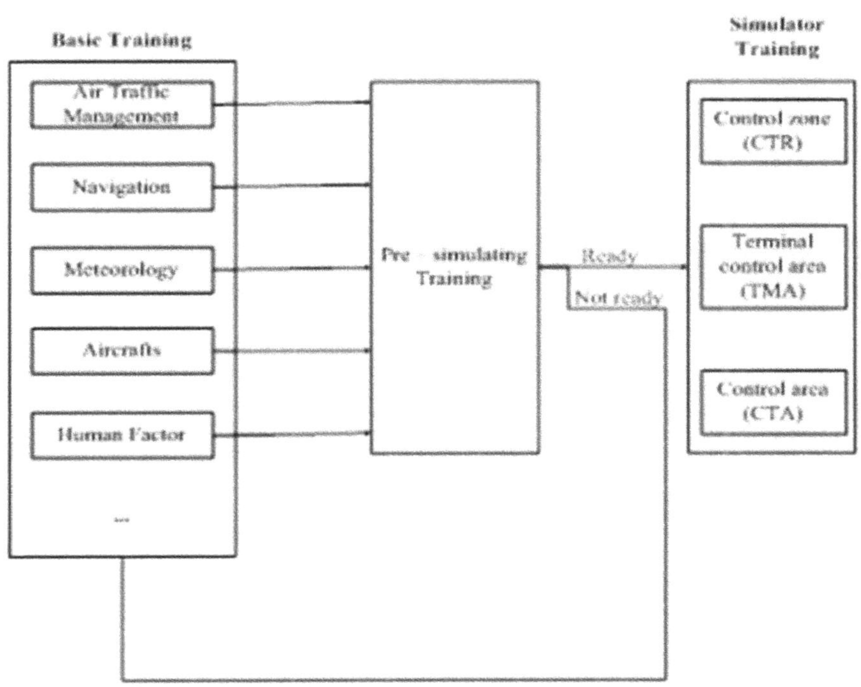

Table 4. Implementation of airspace classes in Ukrainian airspace

N°	Class of Airspace	Zones of Airspace	Restrictions
1	D	CTR	1500-2900 m
2	D (except Boryspil TMA, where class C is applied due to high traffic volumes)	TMA	D (1500-2900 m) and C (2900 m – Fl 660)
3	D and C	CTA	D (1500-2900 m) and C (2900 m – Fl 660)

complexity of airspace zones with using the algorithm of estimation of the complexity of the tasks in the Pre-Simulation Training have been obtained (Shmelova, Lazorenko, & Bilko, 2014):

1. Questionnaires for experts – ATCs with working experience.
2. The matrix of individual preferences. Evaluation of complexity of airspace zones (CTR, TMA, and CTA): R.

R_i has been obtained, $i = \overline{1,m}$, where m – is a number of experts; R_i – is a system of preferences of i-expert.

For example, $R_i = R_{iTMA} \succ R_{iCTA}, R_{iCTR}.\ 1, \mathrm{m}$

3. The matrix of group preferences for CTR has been obtained:

$$R_{grCTR} = \frac{\sum_{i=1}^{m} R_{iCTR}}{m} = 2{,}64\,.$$

Average value R_{gr} for TMA and CTA have the similar calculations (Table 5).

4. Concordance of expert's opinion.
   4.1. Calculation of dispersion D:

$$D_{!\,TR} = \frac{\sum_{i=1}^{m} (R_{grCTR} - R_{iCTR})^2}{m-1} = 0,401099\,.$$

Calculations for TMA and CTA would be the same variant (Table 5).

   4.2. Calculation of square average deviation σ:

$$\sigma_{CTR} = \sqrt{D_{CTR}} = 0{,}633324\,.$$

Table 5. The matrix of group preferences

Coordination of Expert's Opinion	CTR	TMA	CTA
	x_1	x_2	x_3
R_{gr}	2.642857	1.142857	2.214286
D_i	0.401099288976	0.131868480769	0.335164576356
σ_i	0.633324	0.363137	0.578934
ν_i, %	23.9636	31.77445	26.14542

Calculations for TMA and CTA would be the same variant (Table 5).

4.3. The coefficient of the variation ν:

$$\nu_{CTR} = \frac{\sigma_{CTR}}{R_{grCTR}} \cdot 100\% = 23,9636\%.$$

Variation for TMA and CTA have the similar calculations (Table 5).

If $\nu_{\mathrm{CTR,TMA,CTA}} \leq 33\%$ then the opinion is concerted and the system of expert group has been obtained. For example, $R_{gr} = R_{TMA} \succ R_{CTR} \succ R_{CTA}$.

Calculations show that opinion is concerted and it is necessary to obtain weight coefficients of complexity for airspace zones.

5. Weight coefficients of complexity (Table 6):

$w_1 = w_{\mathrm{CTR}}$ – weight coefficient for CTR zone;

$w_2 = w_{\mathrm{TMA}}$ – weight coefficient for TMA zone;

$w_3 = w_{\mathrm{CTA}}$ – weight coefficient for CTA zone.

6. Graphical presentation of weight coefficients of CTR, TMA, and CTA (Figure 8).

Calculations have been performed with using computer program MS Excel.

Integrated estimation of tasks Q_{jl} with j-level of complexity in air traffic control for n-zones (19):

Table 6. The results of the calculation of weight coefficients

N°	Zone	R_{gr}	C_j	w_j
1	CTR	2.642857	0.453333	0.226161
2	TMA	1.142857143	0.953333	0.475786
3	CTA	2.214285714	0.596667	0.298053

Figure 8. Weight coefficients of zones' importance

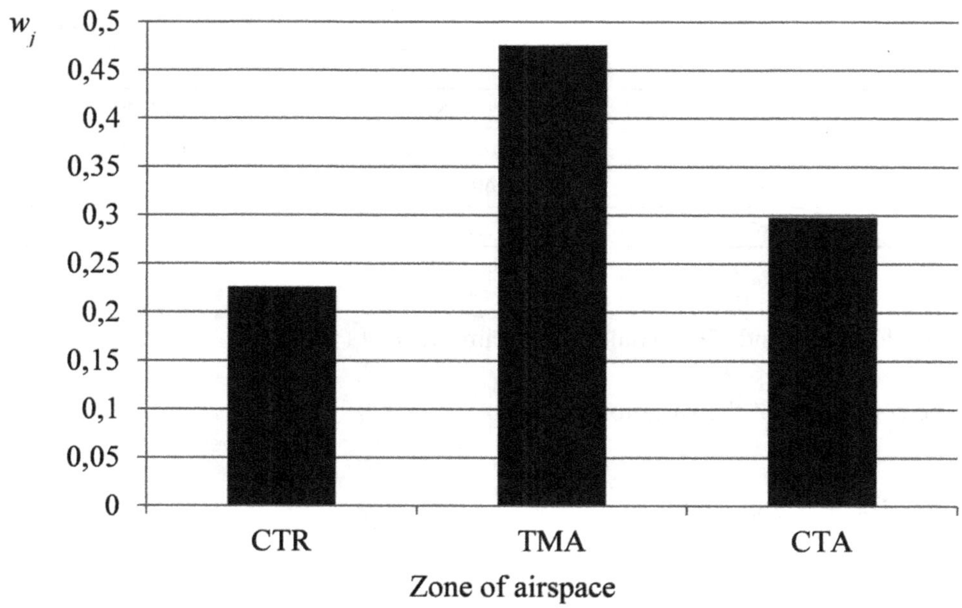

$$Q_{jl} = \sum_{j=1}^{n} \sum_{l=1}^{L} w_j q_l, \tag{19}$$

where q_l – is an estimation of the task with given complexity and type of airspace zone (CTR, TMA, and CTA); w_j – is a weight coefficient (complexity of airspace zone); l – is a level of the task.

So, we can see that the most difficult zone of airspace according to the opinion of the expert is the Terminal Control Area (w_{TMA} = 0.475786).

In the future, ATC instructor would take into account this expert's opinion for determination of a task according to difficulty. Integrated estimation for implementation of the task with complexity characteristic of airspace zone has been calculated with the help of the additive aggregation method. For automating estimation of Pre-Simulation Training on initial stage the neural network of multilayer type has been built (Shmelova, Lazorenko, & Bilko, 2014).

Neural Network Model for Admission of ATC Students to Simulation Training

Contracting States shall take the appropriate measures to ensure that ATC student does not constitute a hazard to air navigation. A Contracting State shall not permit ATC student to receive instruction in an operational environment unless that ATC student holds a current Class 3 Medical Assessment. Before issuing an ATC license, a Contracting State shall require the applicant to meet the requirements of and the requirements of at least one of the ratings. Unlicensed State employees may operate as ATCs on condition that they meet the same requirements.

Four-layers neural network model for evaluating correctness and timeliness of decision making by air traffic controller during the Pre-Simulation Training has been built and its parameters have been obtained (Shmelova, Sikirda, Zemlyanskiy, Danilenko, & Lazorenko, 2016). The first layer (input) includes exercises that perform student to solve potentially conflict situation, the second layer (hidden) depends physiological characteristics of student, the third layer (hidden) takes into account the complexity of the exercise depending on the number of PCS, the fourth layer (output) – is assessment of student during performance of exercise. Neural network model also has additional inputs (Bias) that including restrictions on calculating parameters.

Figure 8 represents the neural network for admission of ATC student to Simulation Training by the number of hours and level of training. It is the multilayer perceptron with three hidden layers:

- The input: are the disciplines of Pre-Simulation Training;
- The first layer (input): is a calculation of hours on theoretical training in accordance with the student's knowledge evaluation;
- The second layer: are the restrictions on the given number of hours (hidden layer);
- The third layer: are the restrictions on passing hours (hidden layer);
- The fourth layer: – are the restrictions on student's assessments (hidden layer);
- The output – is an admission / not admission of ATC student to Simulation Training.

Threshold activation functions have been obtained according to requirements of hours and level of training marks (discipline) in compliance with the assessment criterion of tasks.

ATC student's Pre-Simulation Training consists of 5 disciplines (number depends from the teacher):

1. Navigation (NAV).
2. Air Traffic Management (ATM).
3. Meteorology (MET).
4. Human Factors (HF).
5. Aircrafts (ACTFTs).

For each discipline, the student has a task (test) that he must perform in a given time. Therefore to be admitted he/she needs deal with three barriers or three hidden layers (Figure 9):

1. The specified number of hours in the study of disciplines, which regulates as educational plan (T).
2. Time of task/test completed (t_{compl}).
3. Test mark (w_i).

If a student is unable to cope with these three hidden layers he\she returns to the initial stage. And try to pass again for admission to training.

All positions of assessment criteria have been prescribed and must be used in a proper way as given below:

1. To take a duty and workplace preparation.
2. An ability to follow the prescribed standard phraseology (excepting tasks with emergency and urgency situations).
3. Coordination with adjacent ATC units and other kinds of aerodrome service provision units.
4. Handling of procedural control.

Figure 9. Neural network for admission of ATC student to Simulation Training by the number of hours and level of training

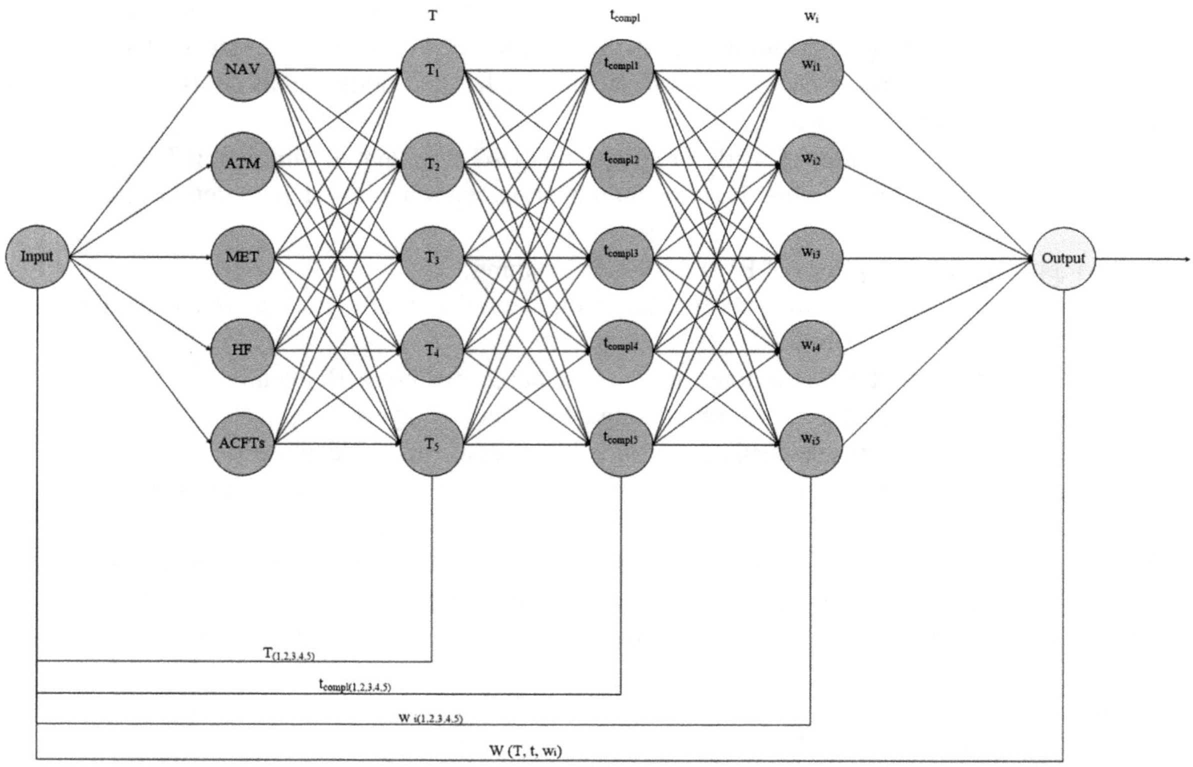

5. Handling of visual control.
6. Daily flight plan conduction.
7. Execution of traffic massages timesheets.
8. Timeliness and accuracy of decision making in ATC.
9. Compliance with safety in ATC.
10. Performing of console operations.

Some of the criteria might not be used in order to which are chosen by a supervisor before training. During Pre-Simulation Training it is necessary to take into account that abilities and skills of ATC are evaluating feedback.

The applicant shall have completed an approved training course and not less than three months of satisfactory service engaged in the actual control of air traffic under the supervision of an appropriately rated air traffic controller. The applicant shall have:

1. Satisfactorily completed an approved training course;
2. Provided, satisfactorily, under the supervision of an appropriately rated ATC.

The applicant shall have demonstrated, at a level appropriate to the privileges being granted, the skill, judgment, and performance required to provide a safe, orderly and expeditious ATS (ICAO, 2012, 2016a, 2016b).

With the help of Microsoft Excel, the neural network model that correctly maps the input to the output using historical data has been created. Historical data for the building of neural network table for admission of student:

1. Dendrites (input): Navigation, Air Traffic Management, Meteorology, Human Factors, and Aircrafts. It is a set of main disciplines, which ATC student must know before Simulation Training.
2. Ranks (R_i): previously obtained with the help of EJM.
3. Estimates (C_i): necessary value for obtaining weight coefficients.
4. Weight coefficients (w_j): weight coefficient for each discipline.
5. Specified number of hours in study of disciplines which regulates by Educational Plan (T).
6. Estimated number of hours for preparation ($w_j x_i$).
7. Time of task/test completed (t_{compl}).
8. Time for tests in minutes.
9. Maximum time for tests.
10. Admission of time for testing.
11. Test mark (w_i)
12. Minimum score.
13. Admission according to the mark.
14. True/False: a logic function of Microsoft Excel.

The model can be used to produce the output – admission to the ATC Simulation Training (when output is unknown). Firstly, there is a measurement – student readiness for all criteria of Pre-Simulation Training tests. The student has attended the sufficient number of lessons; runs out of time passing tests and got a positive assessment (Figure 10).

Figure 10. The student is ready to Simulation Training (screenshot of Microsoft Excel)

№ п/п	Input (dendrites)	Ranks Ri	Estimates Ci	Weight coefficients (*synapse*) wj	Specified number of hours in study of disciplines, which regulates as educational Plan (T)	Estimated number of hours for preparation wj Xi (*axons*)	Time of task\test completed (tcompl)	Maximum time for tests (min)	Admission of time for testing	Test mark (wi)	Minimum score	Admission according to mark
1	Navigation	2	0,8	0,266666667	90	24	55	55	Ready	90	60	Ready
2	Air Traffic Management	1	1	0,333333333	89	29,66666667	55	55	Ready	80	60	Ready
3	Meteorology	3	0,6	0,2	233	46,6	55	55	Ready	70	60	Ready
4	Aircrafts	4	0,4	0,133333333	162	21,6	55	55	Ready	80	60	Ready
5	Human Factor	5	0,2	0,066666667	611	40,73333333	40	55	Ready	60	60	Ready
	Minimum Estimated coeff	161,2	3	1		162,6			ИСТИНА			ИСТИНА
					Function of activation	Ready						

Figure 11. The student isn't ready to Simulation Training according to discipline "ATM" (screenshot of Microsoft Excel)

№ п/п	Input (dendrites)	Ranks Ri	Estimates Ci	Weight coefficients (*synapse*) wj	Specified number of hours in study of disciplines, which regulates as educational Plan (T)	Estimated number of hours for preparation wj Xi (*axons*)	Time of task\test completed (t_{compl})	Maximum time for tests (min)	Admission of time for testing	Test mark (wi)	Minimum score	Admission according to mark
1	Navigation	2	0,8	0,266666667	90	24	55	55	Ready	90	60	Ready
2	Air Traffic Management	1	1	0,333333333	70	23,33333333	55	55	Ready	80	60	Ready
3	Meteorology	3	0,6	0,2	233	46,6	55	55	Ready	70	60	Ready
4	Aircrafts	4	0,4	0,133333333	162	21,6	55	55	Ready	80	60	Ready
5	Human Factor	5	0,2	0,066666667	611	40,73333333	40	55	Ready	60	60	Ready
	Minimum Estimated coeff	161,2	3	1		156,2666667			ИСТИНА			ИСТИНА
					Function of activation	Not ready						

If the student did not attend all lessons on ATM, he/she is not ready to Simulation Training. Despite that the other criteria – specified number of hours in studying disciplines and got the good marks of the test performance (Figure 11).

When the student runs out of time – tests Navigation and Aircrafts – 60 minutes, but he/she knew all the answers and got positive marks, also is not ready (Figure 12).

If the student got a bad mark on Human Factors he isn't ready to Simulation Training (Figure 13).

This automatically process shows ready student or not ready to simulator training, using three types of barrier – hidden layer in the neural network model.

Figure 12. The student isn't ready to Simulation Training according to time for tests (screenshot of Microsoft Excel)

№ п/п	Input (dendrites)	Ranks Ri	Estimates Ci	Weight coefficients (*synapse*) wj	Specified number of hours in study of disciplines, which regulates as educational Plan (T)	Estimated number of hours for preparation wj Xi (*axons*)	Time of task\test completed (t_{compl})	Maximum time for tests (min)	Admission of time for testing	Test mark (wi)	Minimum score	Admission according to mark
1	Navigation	2	0,8	0,266666667	90	24	60	55	Not ready	90	60	Ready
2	Air Traffic Management	1	1	0,333333333	89	29,66666667	55	55	Ready	80	60	Ready
3	Meteorology	3	0,6	0,2	233	46,6	55	55	Ready	70	60	Ready
4	Aircrafts	4	0,4	0,133333333	162	21,6	60	55	Not ready	80	60	Ready
5	Human Factor	5	0,2	0,066666667	611	40,73333333	40	55	Ready	60	60	Ready
	Minimum Estimated coeff	161,2	3	1		162,6			ЛОЖЬ			ИСТИНА
					Function of activation	Ready						

Figure 13. The student isn't ready to Simulation Training according to mark (screenshot of Microsoft Excel)

№ п/п	Input (dendrites)	Ranks Ri	Estimates Ci	Weight coefficients (*synapse*) wj	Specified number of hours in study of disciplines, which regulates as educational Plan (T)	Estimated number of hours for preparation wj Xi (*axons*)	Time of task\test completed (tcompl)	Maximum time for tests (min)	Admission of time for testing	Test mark (wi)	Minimum score	Admission according to mark
1	Navigation	2	0,8	0,266666667	90	24	55	55	Ready	90	60	Ready
2	Air Traffic Management	1	1	0,333333333	89	29,66666667	55	55	Ready	80	60	Ready
3	Meteorology	3	0,6	0,2	233	46,6	55	55	Ready	70	60	Ready
4	Aircrafts	4	0,4	0,133333333	162	21,6	55	55	Ready	80	60	Ready
5	Human Factor	5	0,2	0,066666667	611	40,73333333	40	55	Ready	59	60	Not ready
	Minimum Estimated coeff	161,2	3	1		162,6			ИСТИНА			ЛОЖЬ
					Function of activation	Ready						

Estimation of Multi-Criteria Decision With the Help of Expert Judgment Method and Recommendations for Description of ATC Procedures

In the areas of high intensity of air traffic, the main aspect of efficiency for determining the number of ATC sectors may be the controller's workload in the control zone. As workload, is used the measure in terms of complex criteria, which takes into account the number of AC, which are simultaneously present in the zone and consumption which controllers spend on the AC service during the defined time interval (for one hour, during their shift, etc.).

The complexity of ATC workload has been estimated with the help of EJM's algorithm and MS Excel program in Graduate Student's Degree Thesis "Evaluation of workload of control zone using the method of expert assessments" Stasiuk Oksana (Shmelova, Sikirda, Assaul, Stasiuk, 2015). During the preparation of homework she has compared approach procedures that include:

- Coordination of information with adjacent sectors;
- Control of climbing aircraft;
- Control of descending aircraft;
- Control of transit aircraft.

The Center provides radar coverage over vast distances at set altitudes. Depending on the location, aircraft should contact center between 8000 and 41000 feet (FL80-FL410). The center handles mainly IFR flights. This is because VFR traffic doesn't regularly tend to fly at great altitudes. A Center typically accepts traffic from and passes traffic to, a Terminal Control Center, like an Approach, or of another Center. Most Centers are operated by the national governments of the countries in which they are located.

The Center will normally provide these following instructions:

- Give "fly direct waypoints";
- Altitude changes and descents;

- Give STAR's if needed;
- Instruct pilots to fly into a holding pattern;
- Give speed instructions;
- Give IFR clearances for pilots departing at uncontrolled fields and flying IFR.

A Control Center can be subdivided into Areas and then into sectors. Each sector will be working on a radio frequency. Each sector also has secure landline communications with adjacent sectors, approach controls, areas, ARTCCs, flight service centers, and military aviation control facilities. These landline communications are shared among all sectors that need them and are available on a first-come, first served basis. Aircraft passing from one sector to another shall be handed off and requested to change frequencies to contact the next sector controller. Sector boundaries are specified by aeronautical charts. The communications between pilots and ATCs are done via radio, usually VHF AM bands (between 118 MHz and 137 MHz). The frequencies are published in aeronautical charts and manuals. As a normal procedure, when a pilot is told to change to another controller, the new frequency will be announced. In addition, Center controllers also are connected electronically to other Centers. This connection allows controllers to be linked to electronic traffic management system. Radar is used to monitor and give instructions to the traffic. Usually, Center controllers give "direct to waypoint" instructions. This waypoint can be GPS fixes or Radio Navigation beacons (NAVAIDS) like VOR's or NDB's. Center controllers also assign Standard Terminal Arrival Routes (STARs) and/or the runway for landing at the destination airport. In the Airspace over the oceans there is no Radar Coverage, so the procedures to manage the air traffic are different from the normal airspace. In Oceanic Airspace controllers estimate the position of the traffic from pilot reports and computer models, as is impossible to have a picture of what is happening. For Oceanic flights pilots need a special clearance, also the communications are done using HF radio because of its greater propagation over long distances.

The first step of EJM are the matrixes of individual and group preferences (Tables 7, 8).

Figure 14 shows that the highest percentage of ATC at Area Control Centers corresponds to descend procedures.

Example of Sentiment Analyzing the Airlines' Customers' Opinions

The main definitions used in this research:

NORMALIZATION: Transform various forms of the same term into a common, "normalized" form.

TOKENIZATION: Segmenting the character stream into linguistic units.

STEMMING: Removing affixes and suffixes to allow the retrieval of syntactic and morphological variations of query terms.

Table 7. The matrix of individual preferences of the expert number 1

Procedures	Transit	Climb	Descend	Coordination	Σr	*R*	*R*
Transit		0	0	0	0	4	4
Climb	1		0.5	1	2.5	1; 2	1.5
Descend	1	0.5		1	2.5	1; 2	1.5
Coordination	1	0	0		1	3	3

Table 8. The matrix of group preferences that has been obtained using MS Excel

Experts	Procedures			
	Transit	Climb	Descend	Coordination
1	4	1.5	1.5	3
…	…	…	…	…
30	4	2	1	3
R_{grj}	3.8	1.6	1.4	3.2
D_j	0.075	0.05	0.05	0.075
σ_j	0.273861	0.223607	0.223607	0.273861
ν_j	7.206876	13.97542	15.97191	8.558165
C_j	0.3	0.85	0.9	0.45
w_j	0.12	0.34	0.36	0.18

Figure 14. Estimation of multi-criteria decision with the help of EJM and recommendations for the description of ATC procedures

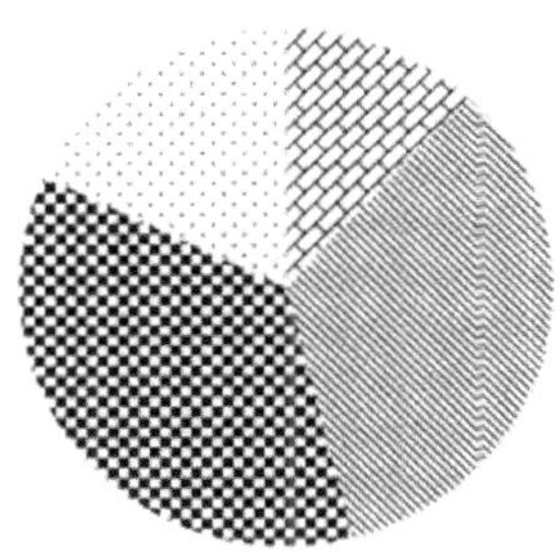

LEMMATIZATION: Use a vocabulary and morphological analysis of words to remove inflectional endings only and return the base or dictionary form of a word.

DOCUMENT TERM MATRIX: Vector model of text representation. Kind of matrix that describes the frequency of terms that occur in a collection of documents that belong to the corpus.

STOP WORDS: An alternative or a complementary way to eliminate words that are (most probably) irrelevant for corpus analysis. Stop-words are those words that (on their own) do not bear any information/meaning.

CORPUS: The collection of text or speech material that has been brought together according to a certain set of predetermined criteria.

ZIPF'S LAW: Distribution of frequency of occurrence of words tithing the Corpus.

Research Questions, raised during the research:

1. Can the Text Analysis methods help to justify and support making decisions of an expert in Air Traffic Management?
2. What are the general topics in the customer's opinions about U.S. airline?
3. What is the most common expression in these opinions?

4. What is the structure of the tonality in the customer's opinions?
5. What are the main questions that clients discuss in positive, neutral, negative opinions?

The research process followed some of the general steps of the Knowledge Discovery in Databases (KDD) plan. This KDD plan consists of some of the following steps:

- Learning;
- Creating a target data set;
- Data cleaning and preprocessing;
- Data reduction and transformation;
- Choosing the functions of data mining;
- Choosing algorithms;
- Data mining;
- Pattern evaluation;
- Use of discovered knowledge.

As a DataSet for realizing the methodology of sentiment analyzing the airlines' customers' opinions, the Twitter database about the problems of each major U.S. airline was used (Social Media Data, 2016). The examples of the opinions are presented in Table 9.

Corpus Preprocessing

Before data mining algorithms can be used, a target data set must be assembled. As data mining can only uncover patterns actually present in the data, the target data set must be large enough to contain these patterns while remaining concise enough to be mined within an acceptable time limit. That is why the first part of feature extraction is preprocessing the corpus. This typically involves several parts like removal of the punctuation marks, removal of numbers, converting the text to the lowercase, removal of

Table 9. Examples of customers' opinions in the Twitter database about the problems of each major U.S. airline (Social Media Data, 2016)

Examples of Customer's Opinions
I didn't today... Must mean I need to take another trip!
Seriously would pay $30 a flight for seats that didn't have this playing.
It was amazing and arrived an hour early. You're too good to me.
Thanks!
SFO-PDX schedule is still MIA.
So excited for my first cross-country flight LAX to MCO. I've heard nothing but great things about Virgin America.
I flew from NYC to SFO last week and couldn't fully sit in my seat due to two large gentlemen on either side of me. HELP!
You know what would be amazingly awesome? BOS-FLL PLEASE!!!!!!! I want to fly with only you.
Why are your first fares in May over three times more than other carriers when all seats are available to select???
You guys messed up my seating. I reserved seating with my friends and you guys gave my seat away ... I want the free internet.

some unnecessary characters, term weighting, stemming and removal of stop words. Stop words are those common words that do not add meaningful content to the document. Some examples are the, and, but, or. Stop words can be a pre-specified list of words or they can be dependent on the context or corpus. In case of this research was decided to include to the list of stop words also terms like: will, back, come until, want, time, way, thank, well, one, can, you, who, that, to, within, without, like and *flight* (in our case).

Data preprocessing very often presents the most time-consuming phase in the whole process of knowledge discovery. The complexity of data pre-processing depends on the data sources used. However, it is an essential part of the text mining process as a whole because the preprocessing when done properly allows for the efficient knowledge extraction in the later parts of the process.

Distribution Analysis of the Terms in the Corpus of Customer Opinions

The next step in text mining which we performed in our research was the creation of Document-Term Matrix (DTM), which is a list with the occurrence of all the terms in the corpus that are remaining after the text pre-processing. It is necessary to perform the term frequency analysis in order to determine later on the importance of the occurring terms. In case of this research, the DTM of 8734 terms and 14849 opinions were created in which 100% of the rows have zeros, meaning that some words occur in one document but doesn't occur at all in some other.

In the next step, the analysis of the frequency distribution of the terms in the corpus was conducted. When we took into consideration all the words which occurred in the corpus (which we analysed) we have got, as a result, the total distribution of terms (Figure 16).

As we can see, the results of the terms frequency distribution histogram which we got for our analysis are very similar in their shape to the theoretical model of the term frequency distribution histogram with its long tail and the top part that represents around 20% of the most frequently appearing terms.

But because of the fact that the results of the frequency distribution presented in such form do not give us a clear picture of the analyzed topic, we applied the Zipf's Law which says that the frequency of a word in a given corpus is inversely proportional to its rank in the frequency table for that corpus. That is why we decided to focus on the terms that are occurring in our corpus at least 400 times or more. After the reduction of the other remaining terms, the histogram took a new form (Figure 17).

From the results presented in this graph, we could deduce that the most frequently appearing terms are: *hour, thank, cancel, help, now, delay, service, time, custom, flight, call, wait, bag, plane.*

As we all know, human being, especially in critical situations, in general, are strongly oriented towards visual messages directed to them and they react most effectively on something that is presented to them in a graphic form. One of them with the focus on the most frequently appearing words (Figure 18).

Analysis of the frequency of occurrence can be carried out not only for individual words but also for word combinations (so-called *n-grams*).

Figure 15.

```
<<DocumentTermMatrix (documents: 14849, terms: 8734)>>
Non-/sparse entries: 104599/129586567
Sparsity           : 100%
Maximal term length: 10
Weighting          : term frequency (tf)
```

Figure 16. Zipf's Law of the customers' opinions

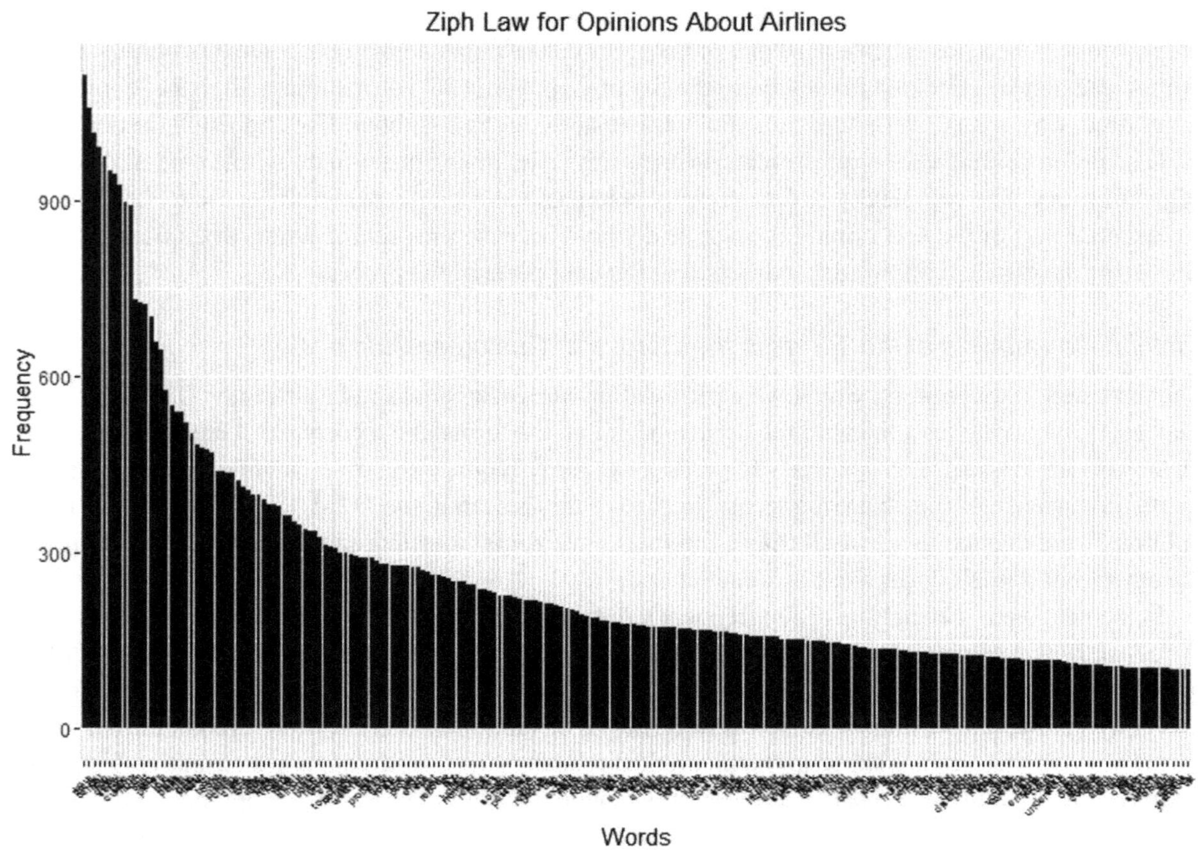

The results of this analysis are shown in Figure 19 and Table 10 (the bi-grams with the frequency greater than 70).

We can see in these bigrams mostly negative and sometimes neutral emotions. But now it is only our perception. For doing the research for the whole Corpus, the special Sentiment Lexicon should be used. But before this step, it is important to determine the main topics raised in the opinions of customers.

Topic Modeling

In natural language processing, Latent Dirichlet Allocation (LDA) is a generative statistical model that allows sets of observations to be explained by unobserved groups that explain why some parts of the data are similar. For example, if observations are words collected into documents, it posits that each document is a mixture of a small number of topics and that each word's creation is attributable to one of the document's topics. LDA is an example of a topic model and was first presented as a graphical model for topic discovery.

LDA is another method we would like to use in our research and then we move to Latent Semantic Analysis (LSA). On this first method, the various types of web browsers are based, which means that this method is intended for large collections of documents. The LDA, based on the display of the documents, randomly selects the words. Then it counts the appropriate probabilities because it confirms its

Figure 17. The distribution of the most significant customers' opinions words

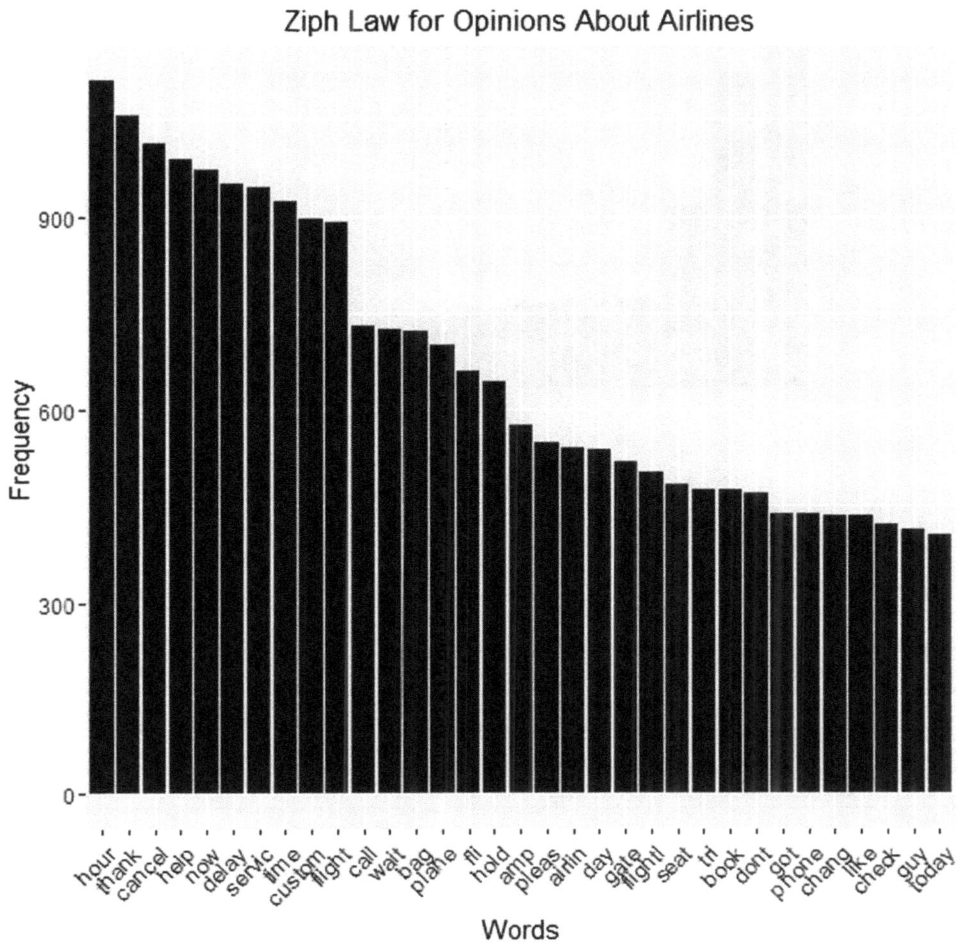

Figure 18. Most frequently appearing words in customers' opinions about US airlines

Figure 19. Bi-grams with the frequency greater or equal to 70

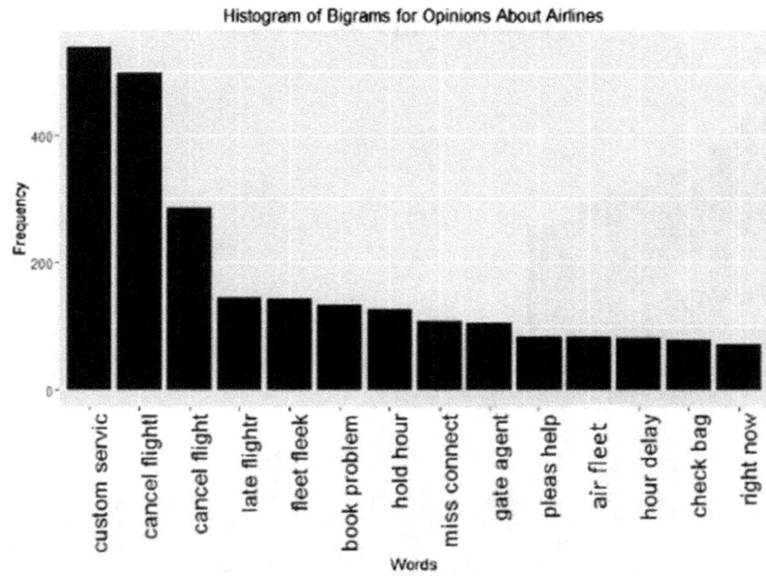

Table 10. Bi-grams with the frequency greater or equal to 70

Bi-Grams	Frequency
custom service	539
cancel flight	499
late flight	145
fleet fleek	143
book problem	133
hold hour	126
miss connect	109
gate agent	105
pleas help	84
air fleet	84
hour delay	82
check bag	78
right now	72
hour late	70

choice. Graded documents are categorized based on themes and topics, so it is possible to find similarities between them. In our analysis we check how this method works when testing our corpus (Figure 20).

Topic modeling deals with the problem of automatically classifying sets of documents (opinions) into topics. Here we have used the LDA algorithm. We have set our k as 4, which means the number of topics for which we would like to divide our corpus. Basically, the program has divided the whole our corpus into four topic categories according to the terms contained in each of the opinions.

Figure 20. The main topics in customers' opinions about US airlines

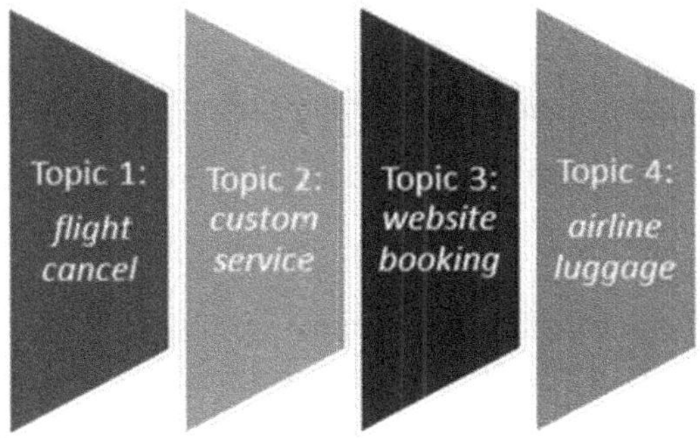

Calculating the Quantitative Measures of Opinions Tonality

To conduct the Sentiment analysis procedure were used the Sentimental Dictionaries (Breen, 2011). Calculating the quantitative measures of opinions tonality is performed according to formulas (17), (18). The main indicators after the calculation of the quantitative measures of opinions tonality are following (Figure 22):

Figure 21.

```
   Min. 1st Qu.  Median    Mean 3rd Qu.    Max.
-5.0000 -1.0000  0.0000 -0.2355  0.0000  5.0000
```

Figure 22. The histogram of quantitative measures of opinions tonality

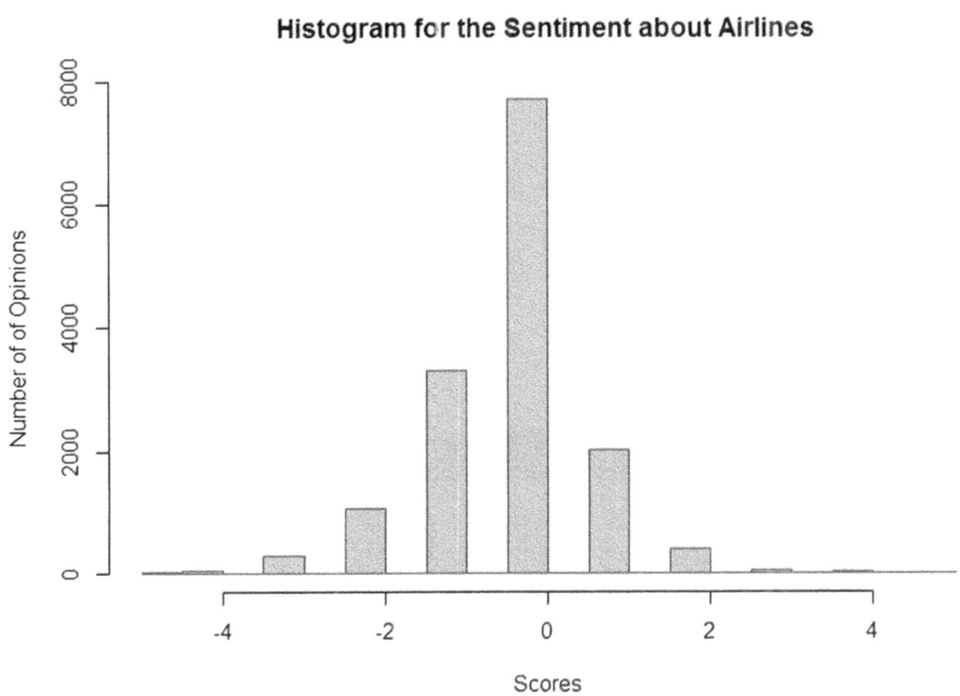

It is proposed to divide into the following three groups of Opinions with different tonalities according to the received Scores (Table 11).

Analyzing the Specificity of the Opinions of the Different Tonality

The resulting groups of Opinions with different Tonality were analyzed using the text mining techniques (Figures 23-28):

- Identified keywords and bi-grams for each group;
- Identifying the problems / issues covered in the customer's opinion.

FUTURE RESEARCH DIRECTIONS

It is relevant to adapt the developed neural network for ATCs Pre-Simulation Training of others ANS's H-Os: pilots, Unmanned Aerial Vehicles operators, flight dispatchers, rescuers, specialists of aviation security, etc. that will allow increasing the quality of their professional study.

Table 11. Groups of tonalities

Tonality	Scores	Number of Opinions
Positive	≥ 1.5	2454
Neutral	> -1.5 and ≤ 1.5	7715
Negative	≤ -1.5	4680

Figure 23. Bi-grams distribution for positive opinions

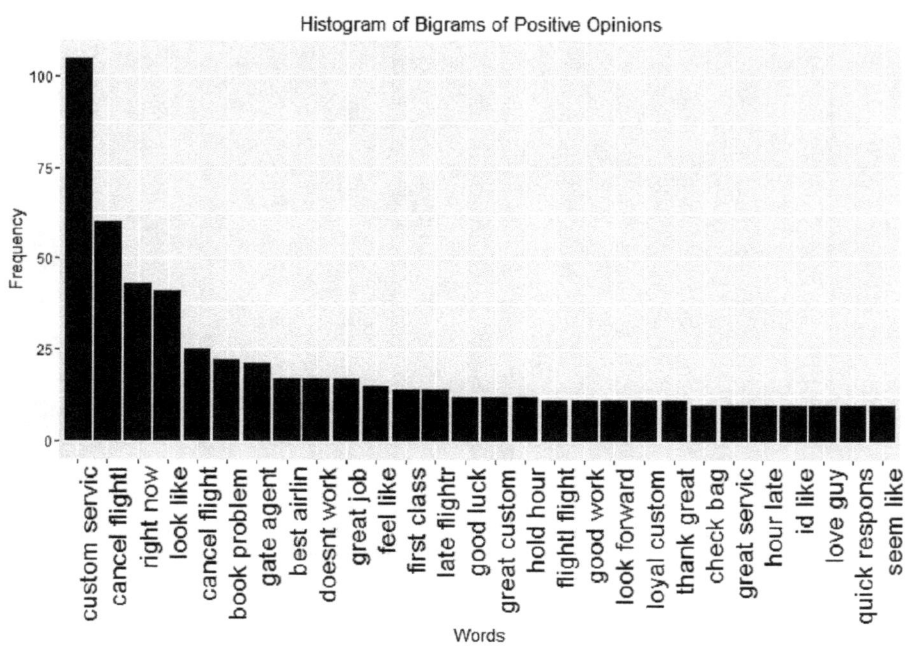

Figure 24. WordCloud for positive opinions

Figure 25. Bi-grams distribution for neutral opinions

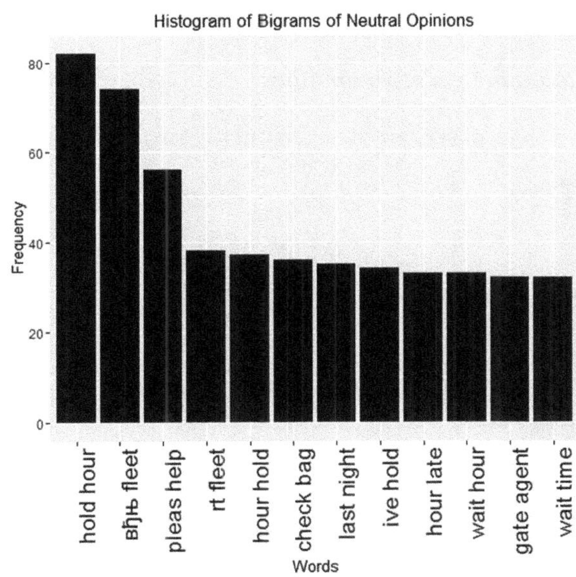

It is our belief, that the neural network models can be used for improving the professional training of specialists in any technogenous production (hydraulic engineering, chemical and military industries, gas and oil pipelines, nuclear power plants and transport, etc.).

What's Next?

On the next steps we are planning to use the neural network in the System for control and forecasting the development of emergency situations on the base of AIS / DSS / ES that taking into account the

Figure 26. WordCloud for neutral opinions

Figure 27. Bi-grams distribution for negative opinions

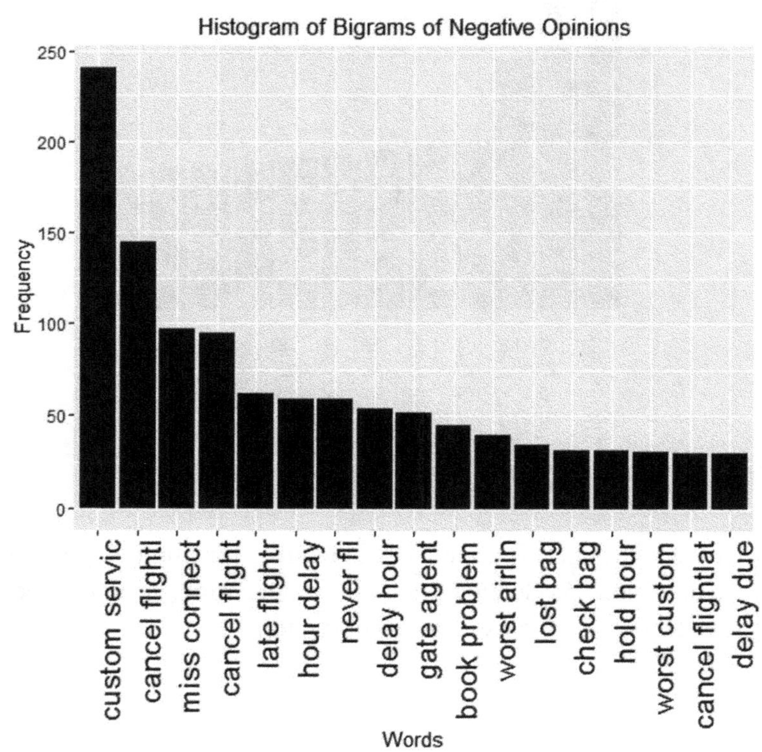

Figure 28. WordCloud for negative opinions

influence of the professional factors (knowledge, habits, skills, experience) as well as the factors of non-professional nature (individual-psychological, psycho-physiological and socio-psychological) on the decision making process by H-O of ANS.

CONCLUSION

ATM is considered to be a safe industry, with a very small number of serious incidents, accidents, and catastrophe. But not looking that air transport is one of the safest in the world, experts are trying every day to reduce the occurrence of risks to a minimum. ATC is the organization of aircraft movements in airspace in accordance with procedures defined in appropriate documents. Flight rules and organizations should be standardized in some way, so that all the pilots in the world, the same procedures, irrespective of the country or region in which they fly. ATCs held responsibility for each aircraft that serves and has information about the plane as it should be.

The main mission of Ukrainian State Air Traffic Services Enterprise (UkSATSE) is to provide an efficient and safe air navigation services in the Ukrainian airspace and in the airspace over the high seas where the responsibility for providing air traffic services is delegated to Ukraine under the international agreements, subject to the current and future airspace users' needs and operational conditions in the air traffic market both in Ukraine and European Region. UkSATSE is a base of the national Air Navigation System and the Joint Civil-Military ATM System of Ukraine. Each year they recruit new candidates for ATC. They have to pass a certain selection and approaches to certain criteria. Among UkSATSE activities safety is of the highest priority. The UkSATSE policy in the field of safety, quality of services, occupational health and safety, environmental and security management is implemented within the framework of the Integrated Management System of UkSATSE pursuant to the requirements of international standards.

Before becoming ATCs candidates must possess certain skills as a practical and theoretical education. Practical education consists of appropriate task on Simulator: change management area (CTR, TMA, and CTA); the intensity of traffic; regulation of PCS in real time. Theoretical education consists of the list of basic discipline, which ATC student must know because it is a key aspect of the preparation of the ATCs to Simulation Training. Without some knowledge, they cannot manage air traffic safely.

The system of ATCs Simulation Training is characterized by low level of objective evaluation of exercises. For improving the evaluation of the quality of ATCs training the use of smart machines employing Artificial Intelligence and Machine Learning (neural-expert system, fuzzy-logic system, etc.) has been proposed.

With the help of EJM the difficulty of ATC operations (tasks) has been determined. Using expert's opinion and criterion of weight coefficients, the complexity of airspace zones for operation on Initial Training, such as Simulator Practice, has been defined. Neural network for admission of ATC students to Simulation Training by the number of hours and level of training, which consists of five basic disciplines (Navigation, ATM, Meteorology, Human Factors, Aircrafts) and has three hidden layers (specified number of hours in study of disciplines, which regulates as educational plan; time of task\test completed; test mark), has been built. Estimation of multi-criteria decision with the help of EJM has been done and recommendations for the description of ATC procedures have been given. Sentiment analyzing the airlines' customers' opinions has been provided.

Automation of estimation of Pre-Simulation Training on the phase of Initial Training increases the efficiency of Simulation Training through interactive evaluation of timeliness and correctness of instructor's tasks performance. Performance of exercises accompanies by comments, displaying results, assessments of the students and feedback. With using information about the difficulty of ATC zones and ATC procedures instructor may increase the number of hours of training and reorganize educational plan. Application of the semantic analysis in the ATCs professional training allows increasing the efficiency of the selection of the relevant educational tasks, the formation of a sequence of exercises and its adaptation to the trainees.

REFERENCES

Aman, S., & Szpakowicz, S. (2008). Using Roget's Thesaurus for Fine-Grained Emotion Recognition. *Proceedings of the 3rd International Joint Conference on Natural Language Processing*, 296–302.

Arkhangelskiy, V., Bohaenko, I., Grabowskiy, G., & Ryumshyn, N. (1999). *Neural Networks in Systems of Automation*. Kiev: Techniques.

Aviation Accident Statistics. (2018). *National Transportation Safety Board*. Retrieved from www.ntsb.gov/aviation/ aviation.htm

Bijuraj, L. V. (2013). Clustering and its Applications. *Proceedings of National Conference on New Horizons in IT – NCNHIT 2013*, 169-172.

Borovikov, V. P. (2008). *Neural Networks. STATISTICA Neural Networks: Methodology and Technology of Modern Data Processing* (2nd ed.). Moscow: Hotline-Telecom.

Breen, J. (2011). *Twitter Sentiment Analysis Tutorial*. Retrieved from https://github.com/jeffreybreen/twitter-sentiment-analysis-tutorial-201107/tree/master/data/opinion-lexicon-English

European Organization for the Safety of Air Navigation (Eurocontrol). (2000). *Simulations Facilities for Air Traffic Control Training. HUM.ET1.ST07.3000-REP-02*. Brussels, Belgium: Author.

European Organization for the Safety of Air Navigation (Eurocontrol). (2004a). ATM Services' Personnel (2nd ed.). ESARR 5. Belgium, Brussels: Author.

European Organization for the Safety of Air Navigation (Eurocontrol). (2004b). *EATM Training Progression and Concepts*. Brussels, Belgium: Author.

European Organization for the Safety of Air Navigation (Eurocontrol). (2015). *Specifications for the ATCO Common Core Content Initial Training* (2nd ed.). Brussels, Belgium: Author.

Gardner, M. W., & Dorling, S. R. (1998). Survey and Critique of Techniques for Extracting Rules from Trained Artificial Neural Networks. *Atmospheric Environment*, *32*(14–15), 2627–2636. doi:10.1016/S1352-2310(97)00447-0

Goedeking, P. (2010). *Networks in Aviation: Strategies and Structures*. Springer. doi:10.1007/978-3-642-13764-8

International Civil Aviation Organization. (1998). Human Factors Training Manual (1st ed.). Doc. ICAO 9683-AN/950. Canada, Montreal: Author.

International Civil Aviation Organization. (2002). Human Factors Guidelines for Safety Audits Manual (1st ed.). Doc. ICAO 9806-AN/763. Canada, Montreal: Author.

International Civil Aviation Organization. (2005). *Global Air Traffic Management Operational Concept. Doc. ICAO 9854-AN/458*. Montreal, Canada: Author.

International Civil Aviation Organization. (2007). Air Traffic Management (15th ed.). Doc. ICAO 4444-ATM/501. Canada, Montreal: Author.

International Civil Aviation Organization. (2008). *Threat and Error Management (TEM) in Air Traffic Control. Cir. ICAO 314-AN/178*. Montreal, Canada: Author.

International Civil Aviation Organization. (2009). *Manual on Global Performance of the Air Navigation System. Doc. ICAO 9883*. Montreal, Canada: Author.

International Civil Aviation Organization. (2012). *Manual on the Approval of Training Organizations. Doc. ICAO 9841-AN/456*. Montreal, Canada: Author.

International Civil Aviation Organization. (2016a). Manual on Air Traffic Controller Competency-based Training and Assessment (1st ed.). Doc. ICAO 10056-AN/519. Canada, Montreal: Author.

International Civil Aviation Organization. (2016b). *Procedures for Air Navigation Services – Training (PANS-TRG). Doc. ICAO 9868*. Montreal, Canada: Author.

Kharchenko, V., Chynchenko, Yu., & Raychev, S. (2007). Threat and Error Management in Air Traffic Control. *Proceedings of the National Aviation University, 3–4,* 24-29. 10.18372/2306-1472.33.1491

Kharchenko, V., Shmelova, T., Artemenko, O., & Otryazhiy, V. (2011). *A. r. Computer Program "Choosing the Pre-Flight Information and Decision Making about Departure for Automated System of Pre-Flight Information Preparation (AS PIP)"*. Kyiv: State Department of Intellectual Property.

Kharchenko, V., Shmelova, T., Sikirda, Yu., & Gerasimenko, O. (2011). *A. r. Computer Program "Optimizing the Choice of Alternative Variant of Aircraft Flight Completion in Flight Emergencies "Prompt"*. Kyiv: State Department of Intellectual Property.

Komashynskiy, V., & Smirnov, D. (2002). *Neural Networks and Its Use in Systems of Management and Communication*. Moscow: Hotline-Telecom.

Leychenko, S., Malishevskiy, A., & Mikhalic, N. (2006). *Human Factors in Aviation: monograph in two books. Book 1st*. Kirovograd: YMEKS.

Liu, B. (2012). Sentiment Analysis and Opinion Mining. *Synthesis Lectures on Human Language Technologies*, *5*(1), 64–72. doi:10.2200/S00416ED1V01Y201204HLT016

Luhn, H. P. (1957). A Statistical Approach to Mechanized Encoding and Searching of Literary Information. *IBM Journal of Research and Development*, *1*(4), 309–317. doi:10.1147/rd.14.0309

Manning, Ch. (2009). *Introduction to Information Retrieval*. Cambridge University Press.

Ministry of Transport and Communications of Ukraine. (2007). *The Rules for Issuing Certificates of Aviation Personnel in Ukraine*. Kyiv: Author.

Ministry of Transport of Ukraine. (2003). *About Approving the Flight Rules of Aircrafts and Air Traffic Services in the Classified Airspace of Ukraine*. Kyiv: Author.

Rizun, N. (2018). Semantic Analysis and Text Summarization in Socio-Technical Systems. In T. Shmelova, Yu. Sikirda, N. Rizun, A.-B. M. Salem, & Yu. Kovalyov (Eds.), *Socio-Technical Decision Support in Air Navigation Systems: Emerging Research and Opportunities* (pp. 243–281). Hershey, PA: IGI Global. doi:10.4018/978-1-5225-3108-1.ch008

Rizun, N., & Kucharska, W. (2018). Text Mining Algorithms for Extracting Brand Knowledge. The Fashion Industry Case. In K.S. Soliman (Ed.), Innovation Management and Education Excellence through Vision 2020 (pp. 1972-1983). Milan, Włochy: International Business Information Management Association (IBIMA).

Rizun, N., & Taranenko, Yu. (2018). The Methodology of Constructing and Analyzing the Hierarchical Contextually-Oriented Corpora. *Proceeding of Federated Conference on Computer Science and Information Systems*. 10.15439/2018F69

Rizun, N., Taranenko, Y., & Waloszek, W. (2017). The Algorithm of Building the Hierarchical Contextual Framework of Textual Corpora. *Eighth IEEE International Conference on Intelligence Computing and Information System*, 366-372. 10.1109/INTELCIS.2017.8260064

Rizun, N., & Waloszek, W. (2018). Methodology for Text Classification using Manually Created Corpora-based Sentiment Dictionary. *Proceeding of the 10th International Conference on Knowledge, Discovery and Information Retrieval*. 10.5220/0006932602120220

Salton, G. (1989). *Automatic Text Processing: Transformation, Analysis, and Retrieval of Information by Computer*. Addison-Wesley Longman Publishing.

Sebastiani, F. (2002). Machine Learning in Automated Text Categorization. *ACM Computing Surveys, 34*(1), 1–47. doi:10.1145/505282.505283

Shmelova, T., Lazorenko, V., & Bilko, A. (2014). Estimation of Pre-Simulating Training Tasks Complexity. *Proceedings of the National Aviation University, No., 1*(62), 17–22.

Shmelova, T., Sikirda, Y., Assaul, A., & Stasiuk, O. (2015). Structural Analysis of Management Environment of Aviation Enterprise from the Point of Systematic Approach. *Proceedings of the National Aviation University, 2*, 3–12.

Shmelova, T., Sikirda, Yu., Zemlyanskiy, A., & Danilenko, O. (2015). Fuzzy Assessment of the Situation's Complexity during Simulation Training of Specialists in Air Traffic Services. In *Materials of International Scientific-Practical Conference "Problems of Energy Efficiency and automation in industry and agriculture"*. Kirovohrad: Kirovohrad National Technical University.

Shmelova, T., Sikirda, Yu., Zemlyanskiy, A., Danilenko, O., & Lazorenko, V. (2016). Artificial Neural Network for Air Traffic Controller's Pre-Simulator Training. *Proceedings of the National Aviation University, No., 3*(68), 13–23.

Shmelova, T., Yakunina, I., Moyseenko, V., & Grinchuk, M. (2014). *A. r. Computer Program "Network Analysis of Flight Emergency"*. Kyiv: State Intellectual Property Service of Ukraine.

Shmelova, T., Yakunina, I., & Yakunin, R. (2013). *A. r. Computer Program "Test Extraordinary Incident"*. Kyiv: State Intellectual Property Service of Ukraine.

Social Media Data. (2016). *Twitter US Airline Sentiment. U.S. Airlines Sentiment Analysis*. Retrieved from https://data.world/socialmediadata/twitter-us-airline-sentiment.

Suzuki, K. (2013). *Artificial Neural Networks – Architectures and Applications*. InTech. doi:10.5772/3409

Turney, P. (2002). Thumbs Up or Thumbs Down? Semantic Orientation Applied to Unsupervised Classification of Reviews. In *Proc. ACL-02, 40th Annual Meeting of the Association for Computational Linguistics*. Association for Computational Linguistics.

KEY TERMS AND DEFINITIONS

Air Navigation System (ANS): Is a complex of organizations, personnel, infrastructure, technical equipment, procedures, rules, and information that is used to provide of airspace users of safe, regular and efficient air navigation service.

Air Traffic Control (ATC): A service provided for the purpose of preventing collisions between aircraft and on the maneuvering area between aircraft and obstructions and expediting and maintaining an orderly flow of air traffic.

Air Traffic Controllers (ATCs): Are the coordinators of the movement of aircraft to maintain safe distances between them. Air traffic controllers typically do the following: monitor and direct the movement of aircraft on the ground and in the air; control all ground traffic at airport runways and taxiways; issue landing and takeoff instructions to pilots; transfer control of departing flights to other traffic control centers and accept control of arriving flights; inform pilots about weather, runway closures, and other critical information; alert airport response staff in the event of an aircraft emergency.

Air Traffic Management (ATM): The dynamic, integrated management of air traffic and airspace including air traffic services, airspace management, and air traffic flow management—safely, economically, and efficiently—through the provision of facilities and seamless services in collaboration with all parties and involving airborne and ground-based functions.

Air Traffic Service (ATS): A generic term meaning variously, flight information service, alerting service, air traffic advisory service, air traffic control service (area control service, approach control service, or aerodrome control service).

Analysis, Design, Development, Implementation, and Evaluation (ADDIE): Is the generic process traditionally used by instructional designers and training developers. The five phases—analysis, design, development, implementation, and evaluation—represent a dynamic, flexible guideline for building effective training and performance support tools.

Artificial Intelligence System (AIS): Was a distributed computing project undertaken by Intelligence Realm, Inc. with the long-term goal of simulating the human brain in real time, complete with artificial consciousness and artificial general intelligence.

Artificial Neural Network (ANN): Is biologically inspired computer program designed to simulate the way in which the human brain processes information. ANN gathers its knowledge by detecting the patterns and relationships in data and learns (or is trained) through experience, not from programming.

Collaborative Decision Making (CDM): Is a joint government/industry initiative aimed at improving air traffic flow management through increased information exchange among aviation community stakeholders.

Communication, Navigation, and Surveillance (CNS): Are the main functions that form the infrastructure for air traffic management, and ensure that air traffic is safe and efficient.

Competency-Based Training (CBT): Training and assessment that are characterized by a performance orientation, emphasis on standards of performance and their measurement and the development of training to the specified performance standards.

Decision Making (DM): Is the cognitive process resulting in the selection of a belief or a course of action among several alternative possibilities.

Decision Support System (DSS): Is the interactive computer system intended to support different types of activity during the decision making including poorly-structured and unstructured problems.

Document-Term Matrix (DTM): Is a mathematical matrix that describes the frequency of terms that occur in a collection of documents. In a document-term matrix, rows correspond to documents in the collection and columns correspond to terms.

Expert Judgment Method (EJM): Is the procedure for obtaining an assessment of the problem based on the opinion of specialists (experts) for the purpose of subsequent decision making (choice).

Expert System (ES): Is a computer system that emulates the decision-making ability of a human expert.

European Organization for the Safety of Air Navigation (Eurocontrol): Is an international organization working to achieve safe and seamless air traffic management across Europe.

Flight and Flow Information for a Collaborative Environment (FF-ICE): A product of the ICAO Global ATM Concept that defines information requirements for flight planning, flow management, and trajectory management and aims to be a cornerstone of the performance-based air navigation system.

Guided Simulation (GSIMUL): Is extensive interaction between the student and the computer in the form of questions, feedback, comments, instructions, and assessment. This guidance assumes the existence of a theoretical model against which the student can be compared.

Human Factor (HF): The term that describes all the many aspects of human performance which interact with their (aviation) environment to influence the outcome of events. These may be related to either the physiological or psychological aspects of human capability, both of which are able to directly affect the way in which the human operator performs in different circumstances.

Human-Machine System (HMS): Is a system in which the functions of a human operator (or a group of operators) and a machine are integrated. This term can also be used to emphasize the view of such a system as a single entity that interacts with external environment.

Human-Operator (H-O): Is the human component in the aviation system (the human subsystem).

International Civil Aviation Organization (ICAO): It codifies the principles and techniques of international air navigation and fosters the planning and development of international air transport to ensure safe and orderly growth.

Instructional Systems Design (ISD): Involves a systematic process for the assessment and development of training solutions, designed specifically for the purpose of formal training delivery. There are two widely recognized instructional design models in use today by both educational institutions and corporate training functions. The most traditional is the ADDIE model, of which there are several variations. The second is the agile model, whereby there are several variations, including rapid application development, rapid content development, and the successive approximation model.

Knowledge Discovery in Databases (KDD): Is the process of discovering useful knowledge from a collection of data. This widely used data mining technique is a process that includes data preparation and selection, data cleansing, incorporating prior knowledge on data sets and interpreting accurate solutions from the observed results.

Latent Dirichlet Allocation: (LDA): Is a generative statistical model that allows sets of observations to be explained by unobserved groups that explain why some parts of the data are similar.

Latent Semantic Analysis (LSA): Is a technique in natural language processing, in particular, distributional semantics, of analyzing relationships between a set of documents and the terms they contain by producing a set of concepts related to the documents and terms.

Machine Learning (ML): Is a class of methods of Artificial Intelligence, whose characteristic feature is not a direct solution to the problem, but learning in the process of applying solutions to many similar tasks.

Multi-Criteria Decision Analysis (MCDA): Is a discipline aimed at supporting decision makers who are faced with making numerous and conflicting evaluations.

Multilayer Perceptron Network (MPN): Is a class of feedforward artificial neural network.

Performance-Based Approach (PBA): Is a decision-making method based on three principles: the strong focus on desired/required results; informed decision making driven by those desired/required results and reliance on facts and data for decision making. The PBA is a way of organizing the performance management process.

Potentially Conflict Situation (PCS): A situation in which, subject to the absence of intervention by an air traffic controller, a violation of the separation norms (vertical, lateral, or longitudinal separation between two aircraft do not provide a safe distance between them) will occur after some time.

Pre-Simulation (Pre-SIMUL): The practice in the restricted or real time of a part of the skills necessary for the operational task in a possibly unrealistic environment (e.g., 2D aerodrome).

Radial Basis Function (RBF): Is a term that describes any real valued function whose output depends exclusively on the distance of its input from some origin.

Simulation (SIMUL): Provision of knowledge, skills, and attitudes by means of a representation of air traffic responding to any student action as real air traffic. Simulation always includes briefing, tutoring, and debriefing.

Standard Operating Procedure (SOP): Is a set of step-by-step instructions compiled by an organization to help workers carry out complex routine operations. SOPs aim to achieve efficiency, quality output and uniformity of performance while reducing miscommunication and failure to comply with industry regulations.

System-Wide Information Sharing and Management (SWIM): Is a Federal Aviation Administration advanced technology program designed to facilitate greater sharing of air traffic management system information, such as airport operational status, weather information, flight data, the status of special use airspace, and National Airspace System restrictions.

Term Frequency-Inverted Document Frequency (TF.IDF): Is a weight measure often used in information retrieval and text mining. This weight is a statistical measure used to evaluate how important a word is to a document in a collection or corpus.

Ukrainian State Air Traffic Services Enterprise (UkSATSE): Is the organization responsible for providing air traffic service (ATS) across Ukrainian airspace and in the airspace over the Black Sea where the responsibility for ATS is delegated to Ukraine by international agreements. As part of this service, UkSATSE also provides airspace management (ASM), air traffic flow management (ATFM), on-route meteorological services for air navigation (MET), and pre-flight information services (ARO/AIS).

Chapter 2
Development Specifics of the Tower Controller Intelligent Training System

Oksana Piliponok
Kirovograd Flight Academy of the National Aviation University, Ukraine

Liudmyla Dzhuma
Kirovograd Flight Academy of the National Aviation University, Ukraine

ABSTRACT

The activity of the tower controller is connected with the implementation of a rather complex set of measures to ensure takeoff and landing operations. In connection with this, the aviation specialist should possess the number of professional competencies, which are formed by mastering the theoretical material and consolidating it in practice. Frequently, the initial consolidation in practice is performed by using automated training systems, especially intelligent training systems are in a great demand. Therefore, this chapter is devoted to the specifics of the development of such type aviation-focused system, which are given on the example of the intelligent training system "Tower Controller." The prototype of this system was developed at Flight Academy of the National Aviation University, Ukraine. Its key differential peculiarity is focused on major stages of the decision-making process of the tower controller instead of visualization from the tower window.

BACKGROUND

Analysis of materials of the National Bureau of Air Accidents Investigation of Ukraine, data from the State Aviation Administration of Ukraine and the Flight Safety Foundation gives evidence of the fact that a significant number of aviation accidents and incidents occurs during take off and landing of an aircraft (Approach, n.d.; National Bureau of Air Accidents Investigation of Ukraine, 2014, 2015, 2016; State Aviation Administration of Ukraine, 2013; Statistical Summary, n.d.).

During take off engines are operated with take-off mode. There is a risk of their failure at the moment when an airplane performs take-off run, which may cause an inability to get off the ground or

DOI: 10.4018/978-1-5225-7709-6.ch002

incapability to stop on a runway. That is why, there is a point during take off, at which a pilot makes the decision whether it is better to take off or slow down. If the failure occurs after passing this point, then take off should be done under any conditions (even with one working engine).

The next danger, which is connected with take off, is an incorrect plane balance. For take off a pilot uses take-off configuration of an aircraft, which is based on a plane balance, calculated before a flight. Whether this calculation is correct can be determined only after getting off the ground.

The other hazard is the crosswind. It may hinder take-off operations and take a ship down from a runway.

A landing is also a rather complicated stage. It is caused by several factors. One of these factors is the wind, especially the crosswind and the heading wind. It makes difficult to aviate a plane. If the wind changes its direction or abruptly stops, then an aircraft may sharply lose an altitude and bump into the ground.

The second factor is a visibility. Of course, it also complicates takes off, but visual cues play a much greater role for landing. Poor visibility conditions may lead to a collision with obstacles in an area of an aerodrome or to landing outside a runway.

The third factor is a plane balance. Centering on landing differs from centering on take off. This is explained by fuel consumption during flight. The next factor is a coefficient of adhesion. After touching down, there may be a risk of skidding an aircraft or other problems, connected with a surface state of a runway.

The factors, described above, are not a complete list of those, which affect the complexity of taking off and landing. So, all of them should be taken into account for ensuring safe flight operations.

One of the participants of take-off and landing stages is a Tower controller. He ensures (Dnipropetrovsk regional branch, Zaporizhzhya air traffic service system, 2013):

- Aerodrome control service;
- Flight information service;
- Emergency service.

In the process of aerodrome control service, a Tower controller provides permissions for arriving and departing crews, as well as a various pieces of information, aimed at the maintenance the discipline in an aerodrome zone in order to prevent collisions between aircraft. Also a Tower controller interacts with the airfield maintenance and supply service on the issue of permits for works on the airfield. The airfield maintenance and supply service reports a Tower controller about destroyed surface of the runway, the absence of appropriate marking, the repair work in the aerodrome working area, etc.

The purpose of flight information service is the provision of consultations and information to ensure the safe and efficient performance of flights (Dnipropetrovsk regional branch, Zaporizhzhya air traffic service system, 2013). For example, aircraft crew may initiate the request to a Tower controller for getting meteorological information such as weather forecast for the aerodrome, the weather forecast for landing, aerodrome warning, warning about wind displacement at the aerodrome and other additional information.

Emergency service is aimed at providing a comprehensive assistance to aircraft in distress (Dnipropetrovsk regional branch, Zaporizhzhya air traffic service system, 2013). During this service, a Tower controller not only keeps the connection with aircraft in distress but also notifies the rescue service about the emergency.

In addition to the presented service, a Tower controller operates the lighting system, which is used in poor weather conditions, when the visibility on the runway is low. Tower controller also controls the operational capability of landing radio equipment and provides continuous visual observation of all aircraft manoeuvres, the movement of special vehicles and position of people on the flight field and in his control area (Dnipropetrovsk regional branch, Zaporizhzhya air traffic service system, 2013).

Suchwise, a Tower controller makes different managerial decisions, the accuracy and timeliness of which depends on his level of competence, experience and personal qualities (Pilipyonok, 2014).

An important stage of preparation a Tower controller to the work is training on the simulator. Existing simulators, which are used in Ukraine and abroad, are aimed at presenting a Tower controller's workplace as close as possible to the real one, but they do not focus on the decision-making process of an air traffic controller (Pilipyonok, 2013). Such approach provokes mistakes in his work and, as a result, may lead to aviation accidents and incidents.

Besides, a Tower controller very often trains on the simulator in front of an instructor. So, an instructor shows his personalism during a learning process. Therefore, there is a need for improved training methods, which are used for a professional preparation of a Tower controller.

Nowadays, automated training systems, especially intelligent training systems, are in a great demand. They allow adapting the learning process to peculiarities of a separate user. Consequently, the creation of such an intelligent training system for Tower controller and its implantation in the educational process can help him to improve the level of professional preparation and, as a result, to raise the effectiveness of air traffic control.

Thus, the chapter is devoted to solving an important scientific and practical task, related to the development specifics of the intelligent training system for aviation specialist, in particular, a Tower controller.

PROPOSED APPROACH

At present, humanity lives in the era of an information revolution. Every day, the amount of information, which is necessary to support and develop the survivability of human society, increases. In this regard, there is a need to learn this information efficiently and as quickly as possible. Especially, such need concerns technical specialties, because, in the context of a rapid development of information technology, they undergo the greatest changes.

Analysis of the results of scientific research of domestic and foreign authors on an aforesaid question bears evidence to solving an existing problem by development and implementation specialized intelligent training systems. Intelligent training system is a new tool, allowing form professional skills, which are necessary for a particular specialist, by organizing the educational process in an environment, adapted to peculiarities of a separate user.

The development of an intelligent training system for each profession has its own specifics. Canvass a subject of development specifics of the Tower controller intelligent training system.

Analysis of the Object of Research

Any elaboration begins with an analysis of the object of research and identification of the problem, connected with this object.

The analysis of the professional activity of a Tower controller indicates that this aviation specialist performs a rather complex set of measures to ensure take-off and landing operations, which requires:

- Fast attention switching;
- A good visual and aural memory;
- The ability to mentally operate objects in three-dimensional space;
- The ability to plan actions and quickly execute mathematical mental calculations;
- Stress and fatigue tolerance;
- The ability to take adequate and well-considered decisions in a limited time, both in normal and emergency situations.

It is connected with a constant analysis and exchange of information flows, with which a controller deals with during his shift.

Information flows, related to take-off and landing operations, circulate between different structural units, including the aerodrome control tower (Pilipyonok, 2017).

The path of an information flow is mostly outlined by the task that is being solved at the moment (Pilipyonok, 2017). Therefore, an information flow, passing through structural subdivisions, can cover them entirely or partly (Pilipyonok, 2017).

There is a certain transmission of an information flow (Pilipyonok, 2017). So, for example, a crew of an aircraft during landing at discrete moments of time communicates with Approach control unit, which subsequently transfers him to a Tower controller (Pilipyonok, 2017). After landing of an aircraft, a Tower controller hands it over to a Ground controller. The same, but in reverse order, occurs during take off (Pilipyonok, 2017). This approach ensures an uninterrupted ordered air flow (Dzhuma & Pilipyonok, 2014).

The nature of an information flow essentially depends on environmental conditions and working shifts of airport employees (Pilipyonok, 2017). The change of meteorological conditions, as well as the change of employees, shift, in each individual case, can lead to various events at the workplace (Pilipyonok, 2017). In connection with this, the functioning of the air traffic service system is determined by external and internal information, which is common to open stochastic systems (Pilipyonok, 2017).

Thus, the activity of a Tower controller is associated with a formation and exchange of information flows in air traffic service system, which may be described as an open, sophisticated, continuous-discrete, stochastic, dynamic management system (Dzhuma, Pilipyonok, & Timoshenko, 2012; Dzhuma & Pilipyonok, 2014; Pilipyonok, 2012a, 2012b). Thereby, the aviation specialist should be possessed by the number of professional competencies to carry out his functional duties.

Mentioned competences are formed by mastering the theoretical material and consolidating it in practice. Frequently, the initial consolidation in practice is performed by using aviation simulators.

Well-known training and educational systems, which are used in the world for the training of a Tower controller, are (ATCsimulator, n.d.; ATTower®, n.d.; .BAO Tower, n.d.; Dzhuma, Pascal, & Lishavskaya, 2011; Gorbunov, Eliseev, & Nechaev, 2010; Multipurpose, n.d.; NexSim, n.d.; Software, n.d.; The FIRSTplusTM, n.d.; Tower Simulator, n.d.):

- FIRSTplusTM Tower Simulator (the developer is Raytheon Company, Canada);
- TATIC TWR (the developer is Saipher ATC, Brazil);
- ATCsimulator (the developer is Russell B. Davis, USA);

- Tower Simulator (the developer is FeelThere Development Team and Wilco Publishing, Belgium);
- ATC Tower Simulator (the developer is Micro Nav Ltd, Great Britain);
- BAO Tower Simulator (the developer is Bruce Artwick Organization, USA);
- NexSim Tower Simulator (the developer is Computer Sciences Corporation, USA);
- ATTower Simulator (the developer is UFA Incorporation, USA);
- TWR Simulation (the developer is Indra, Spain);
- Multipurpose controller simulator "Expert" (the developer is Nita Company Ltd , Russian Federation);
- Air traffic controller simulator of a new generation on the basis of mixed reality (the developer is A. Gorbunov in association with MSTU CA, Russian Federation);
- MaxSim (the developer is Adacel Incorporation, Canada) and others.

Represented complexes are quite powerful, but they have a number of shortcomings. Limitations are a high cost of training, the lack of mobility in placement, a limited functionality, the need to adapt to local conditions and the main among them is the lack of emphasis on the air traffic controller's attention in the decision-making process. Such a mistake leads to errors in a Tower controller work.

Interviews, questionnaires and evaluation of statements from the tape recorder of the controller radio exchange allowed to determine the laws of occurrence mistakes during issuance of permits for an aircraft take-off and landing, and, accordingly, gave the opportunity to highlight the most vulnerable stages of the decision-making process of a Tower controller (Pilipyonok, 2017). According to the research, the most frequent errors are:

- Non-direct commands;
- Violation of the sequence of information issuance;
- Giving a partial information;
- Giving an incorrect information;
- Violation of air traffic control procedures during the assignment of take-off and landing intervals;
- Abuse of air traffic control regulations when an aircraft goes round;
- Violation of air traffic control procedures during issuing take-off and landing permits;
- Abuse of air traffic control regulations during issuing information about aircraft trail;
- Violation of air traffic control procedures during issuing to aerodrome service information about departing and approaching airplanes;
- Abuse of air traffic control regulations during indicating taxi route, exit from the aerodrome area;
- Untimely execution of air traffic control procedures.

Variant of the Problem Solution

After analysis of the object of research and identification of the problem, it is necessary to propose the variant of its solution. Reduction of the list of the Tower controller mistakes and, accordingly, improvement of his level of a professional preparation, is possible by creation and implementation in the educational process the intelligent training system «Tower Controller», the key element of which is a model of activity that takes into account the problem laws associated with the decision-making process in issuing permits for an aircraft take off and landing (Pilipyonok, 2017). Such approach will complement simulator training in terms of study, supporting and controlling the decision-making process in the management of an aircraft take-off and landing, and thus will ensure the formation of deep knowledge, development of profound skills in a Tower controller (Pilipyonok, 2017).

It is known, there is a large number of scientists from all over the world, who are engaged in the development of various educational systems or work in the field of automation a practical training and professional activity of air traffic controllers. They are De-Jung Hwang, Myung-Suk Jenny Pang, Mark J. Rosenberg, Yatlenko B., Soldatkin V., Bondarenko O., Guseva A., Lisov O., Savelyev O., Tikhomirov V., Ignatova I., Botsul M., Zavisha V., Cillers P., Kryzhanovsky G., Kane V., Boytsov V., Nedilko S., Nedilko V., Obidin D., Sikirda Yu., Chinchenko Yu., Shmelova T., Zhang J., Bayen A., Alem L., Schaefer D., Meckiff C., Magill A., Pirard B., Aligne F., Chone R., Nicolaon J., Mogford R., Guttman J., Morrow S., Kopardekar P., Histon J., Hansman J., Arad B., Schmidt D., Chatterji G., Averty A., Koros A., Della Rocco P., Panjwani G., Ingurgi V., Raar J., Sperandio J., Cannon J. and others. Each scientist offers his own view on the solution of a particular question. Authors of this chapter suggested a conceptual scheme of the intelligent training system «Tower Controller», shown in Figure 1 (Pilipyonok, 2017).

This scheme displays a client-server structure of the system. According to the structure, the knowledge base and the relational database are located on the server. The knowledge base includes:

- Storage of the sample data;
- Fuzzy products of meteorological conditions;

Figure 1. A conceptual scheme of the intelligent training system «Tower Controller»

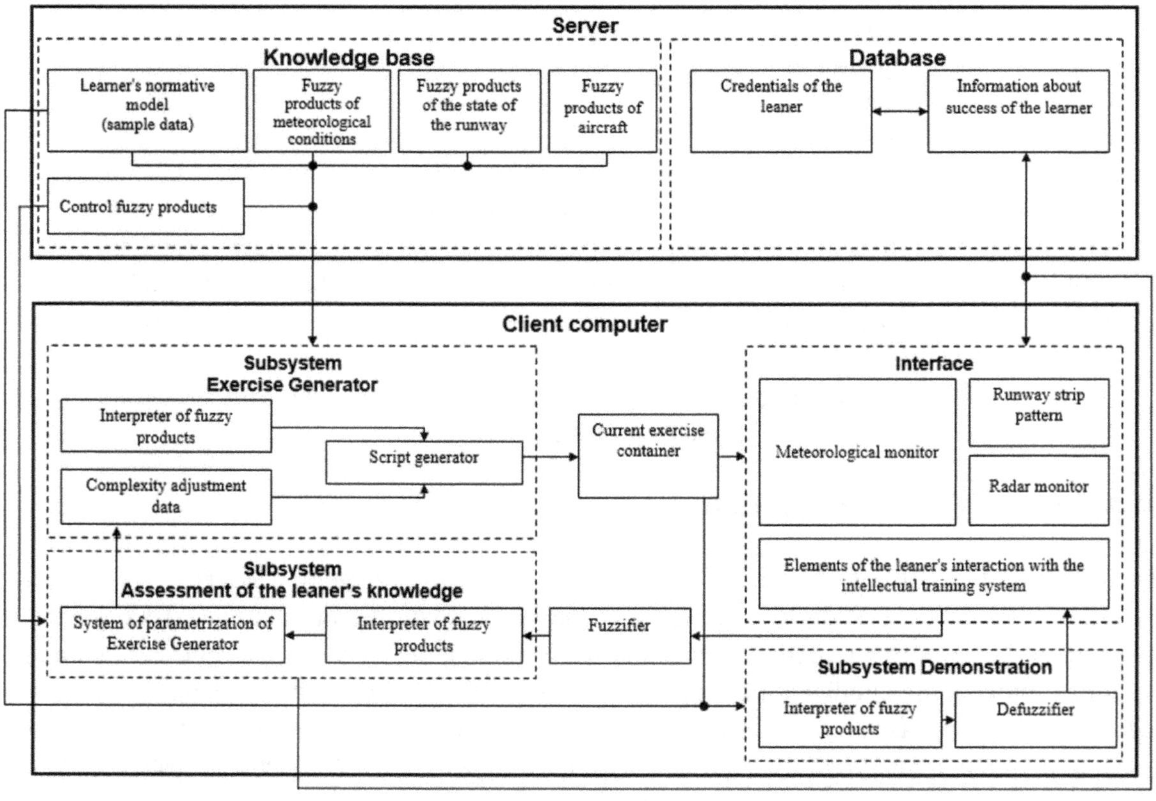

- Fuzzy products of the state of the runway;
- Fuzzy products of aircraft;
- Control fuzzy products.

The relational database contains:

- Repository of credentials of the studied person (surname, name, group number, etc.);
- Information about the success of the learner.

On the client computer, there is the application, which consists of several subsystems. They are:

- Exercise Generator;
- Assessment of the learner's knowledge;
- Demonstration;
- Current exercise container;
- fuzzifier;
- User interface.

The user interface includes:

- Meteorological monitor;
- Runway strip pattern;
- Radar monitor;
- Elements of the learner's interaction with the intellectual training system.

Such subsystems as Exercise Generator, Assessment of the learner's knowledge, Demonstration contain:

- Interpreter of fuzzy products;
- Script generator;
- Complexity adjustment data;
- The system of parametrization of Exercise Generator;
- Defuzzifier.

The main function of interpreter of fuzzy products is the transformation of fuzzy product data into natural data and sequences of logical operators for current computing systems. Script generator creates a script of exercise, which contains information for setting up the user interface, data from storage of the sample data to assess the learner's knowledge and may have control fuzzy products. Complexity adjustment data is responsible for adjusting in the current exercise container the current task according to the learner's execution. The system of parametrization of Exercise Generator defines a set of control fuzzy products for the description of the namespace of fuzzy products, which will directly participate in the script generation process. Defuzzifier converts the system results into clear univocal statements, which are necessary for the generation of exercise scenarios, the assessment of the learner's knowledge, the formation of data for the meteorological monitor, etc. The function of fuzzifier is inverse to the function of defuzzifier.

The proposed structure of the intelligent training system «Tower Controller» provides three modes:

- Demonstration mode;
- Training mode;
- Control mode.

Demonstration mode allows a learner to see how to act in the given situation. Training mode provides an opportunity not only see what to do but also participate in this process for formation skills, which are necessary for the work. Control mode helps to test a learner for the material assimilation.

Implementation of the Variant of the Problem Solution

After the proposal of the variant of the problem solution, the stage of its implementation begins.

Previously, three operating modes of the intelligent training system «Tower Controller» were described. These three modes have a common component, which is a key element of the whole system. This is the learner's normative model. For its development, there was carried out the structural analysis of the decision-making process of a Tower controller in issuing permits for an aircraft take off and landing. The structural analysis allowed identifying the main phases of the decision-making process of a Tower controller under favourable meteorological conditions. They are (Pilipyonok, 2017):

- Selection and indication of the runway working direction;
- Analysis of the category of aircraft by a speed, mode and approach systems;
- Evaluation of the work, which is in progress on the runway;
- Analysis of meteorological data, information about the location and type of departing aircraft, landing aircraft and aircraft under the visual flight rules, which are following the zone by span;
- Assessment of the readiness of departing aircraft to take off immediately;
- Issuance of the instruction to the departing aircraft crew to wait on the holding position, under the condition that landing aircraft will be at the point of entry into the glide path at the time when departing airplane takes off;
- Issuing permits for departing aircraft to come to line up position and take off;
- Issuing a permit to continue landing and transmission of information about the presence of departing aircraft on the runway;
- Monitoring of the moment, in which departing aircraft leaves the runway, and issuing landing permit to landing aircraft.

Described phases allowed presenting the decision-making process of a Tower controller in a form of the algorithmic model (Pilipyonok, 2017). Next step was studying the laws of aircraft flow formation.

Tower controller constantly deals with departing and landing aircraft at his workplace, so he should know the information about the aircraft number, with which he will work on a shift, moments of their entry to his area of responsibility, time intervals between aircraft entries, categories of turbulence in the aircraft trail. That is why the learner's normative model needs also to include this information. For this purpose, there were studied the laws of aircraft flow formation with help of flight plan research of Odesa Regional Branch, UkSATSE, in the busiest period - August and in the least busy period - January. Calculations showed that in August the flight plan was completed approximately by 88%, 12% of

Figure 2. Analysis of flight plan in August

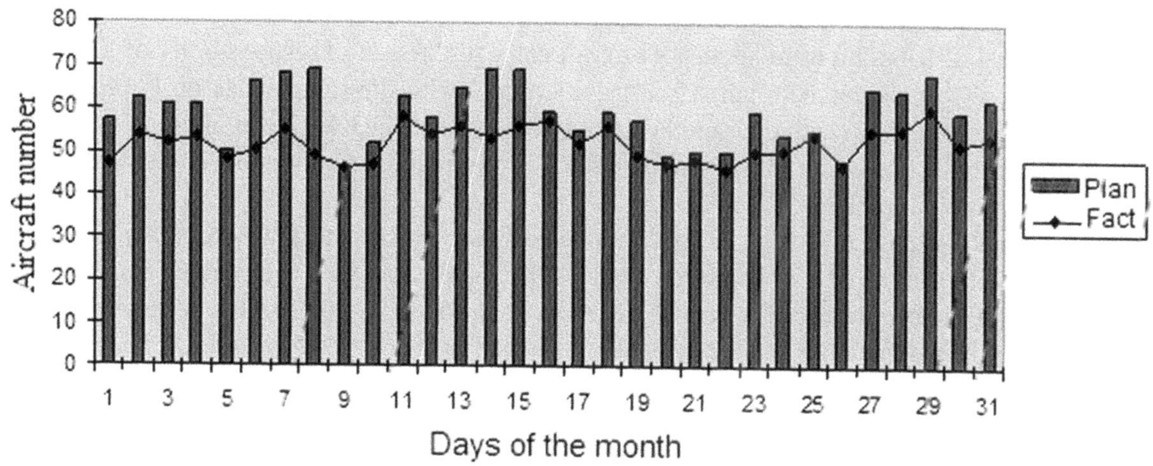

the flights were postponed/cancelled for various reasons (Pilipyonok, 2017). This is shown in Figure 2. In January the flight plan was completed by approximately 74%, and 26% of flights were postponed/ cancelled (Pilipyonok, 2017). This is shown in Figure 3.

The analysis of flight plans allows describing aircraft flow in the area of responsibility of a Tower controller as stochastic (Dzhuma & Pilipyonok, 2015). Consequently, probabilistic modelling methods may be used to present it in the intelligent training system (Dzhuma & Pascal, 2010; Dzhuma & Pilipyonok, 2015).

In the process of handling of data, which describes moments of aircraft entry to the area of responsibility of Tower controller and time intervals between aircraft entries, data empirical distribution was approximated to six continuous distribution laws (Pilipyonok, 2017).

During studying the aircraft number the empirical distribution was approximated to four discrete distribution laws (Pilipyonok, 2017). It should also be stated that usage of 50-60 aircraft during research

Figure 3. Analysis of flight plan in January

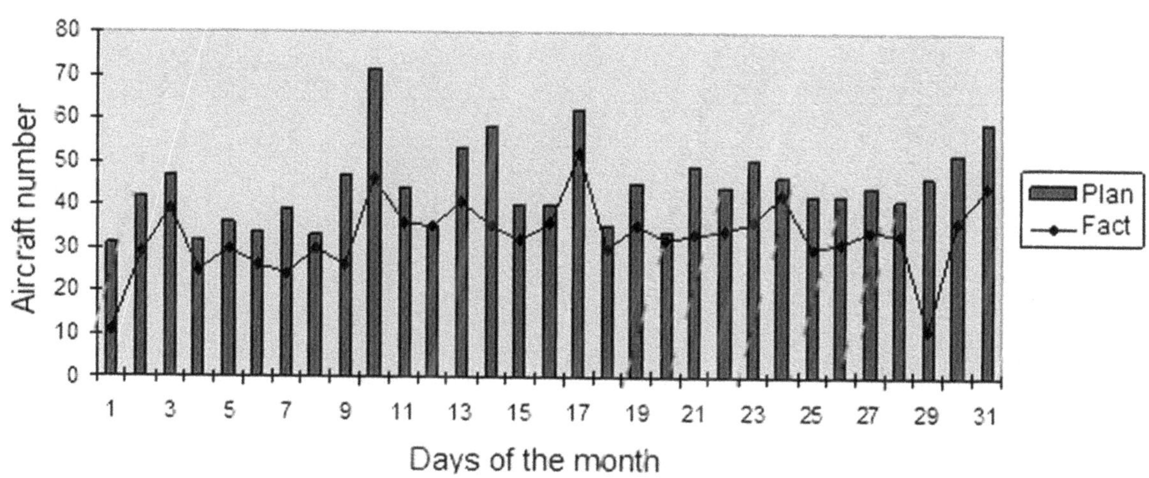

allowed getting the more obvious law. Usage of less than 40 aircraft during research did not allow getting obvious law (Dzhuma & Pilipyonok, 2015; Pilipyonok, 2012a).

The empirical distribution of moments of aircraft entry to the area of responsibility of Tower controller most closely coincides with normal distribution law (with the sample size of 47 aircraft) and rectangular distribution law (with the sample size of 55 aircraft), which is shown in Figure 4 (Dzhuma & Pilipyonok, 2015). According to the results index p for normal distribution law is 0.97592, index p for rectangular distribution is 1.

The empirical distribution of time intervals between aircraft entries most closely coincides with exponential law (with sample size of 47 aircraft) and less accurately, but with acceptable parameters with log-normal (with sample size of 47 aircraft) (Dzhuma & Pilipyonok, 2015). This is shown in Figure 5. According to the results index p for exponential distribution law is 0.93679, index p for log-normal distribution is 0.57718.

Figure 4. The empirical distribution of moments of aircraft entry to the area of responsibility of the Tower controller

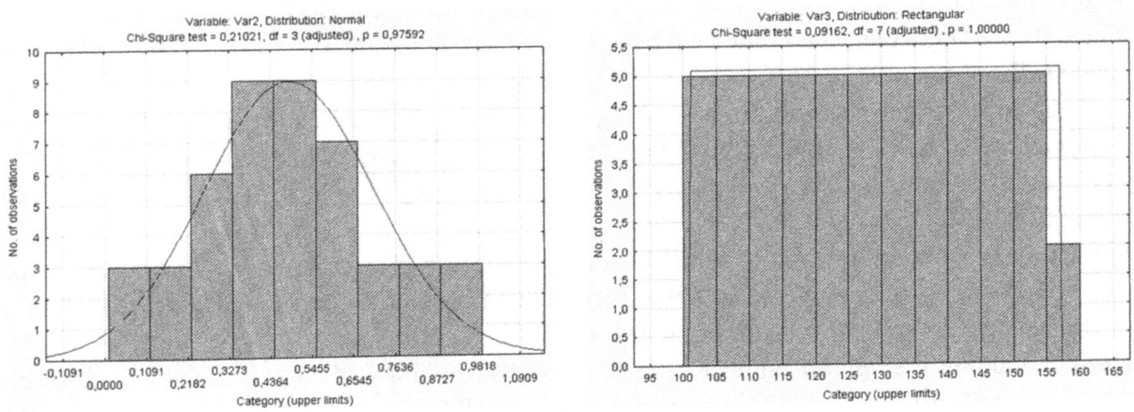

Figure 5. The empirical distribution of time intervals between aircraft entries to the area of responsibility of the Tower controller

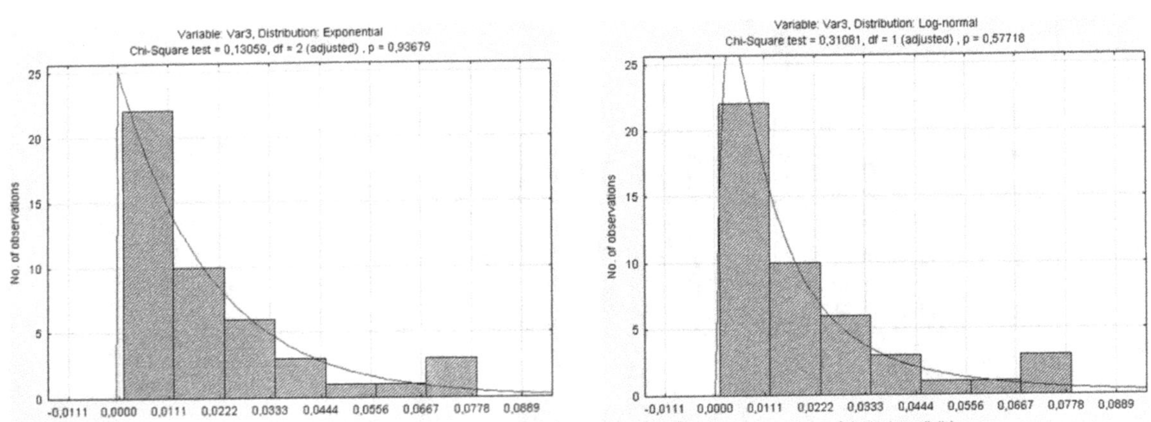

Figure 6. The empirical distributions of the aircraft number in the busiest period and the least busy period of Tower controller work

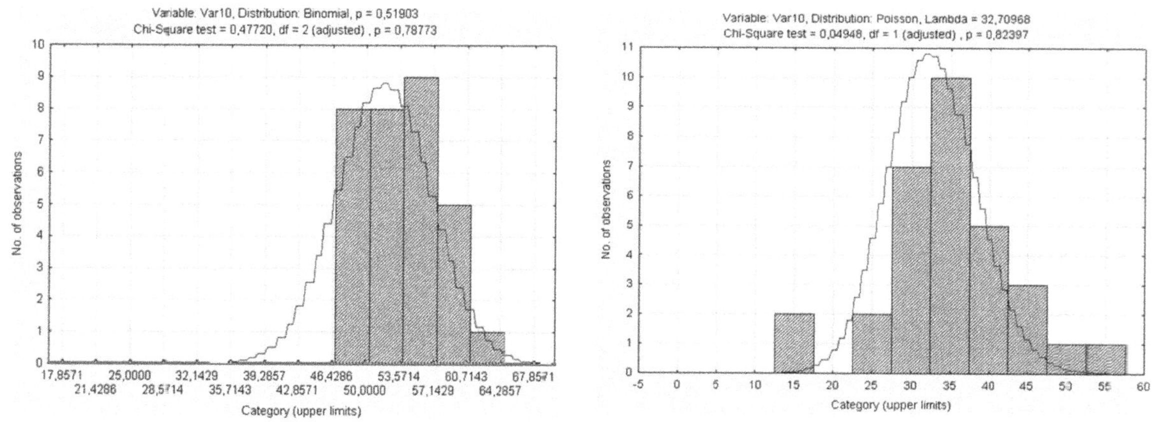

The empirical distributions of the aircraft number in the busiest period and the least busy period of Tower controller work most closely coincide with the binomial distribution (with the sample size of 52 aircraft) and distribution of Poisson (with the sample size of 32 aircraft) (Dzhuma & Pilipyonok, 2015). This is shown in Figure 6. According to the results index *p* for binomial distribution law is 0.78773, index *p* for distribution of Poisson is 0.82397.

Suchwise, obtaining distribution laws of aircraft flow formation are one of development specifics of the Tower controller intelligent training system. They play the role of a factor (among many other factors), which affects the complexity of the learner's exercise and brings the learner closer to real conditions. However, this factor is not enough to develop the complete training system. In connection with this, the next development specific of the Tower controller intelligent training system is studying Tower controller workplace as a queue management system.

According to the queuing theory, there are single-channel and multiple-channel queue management systems. If a Tower controller is responsible for one runway, then Tower controller workplace may be described as a single-channel queue management system. If a Tower controller is responsible for more than one runway, then Tower controller workplace may be described as a multiple-channel queue management system.

A Tower controller may be represented as a service that works with customers (departing and landing aircraft). Customers form the input and output flow in the service system and their quantity is limited by the flight plan. Applications, submitted to the service, may be served immediately or may wait for service in the queue. There is the queue of landing aircraft in the air and the queue of departing aircraft on the ground. The arriving aircraft may not wait for service in the queue but flies to another airdrome. This is a case of refusal to queue. It is also possible for aircraft to leave the queue when the waiting period exceeds permissible.

Such representation of a Tower controller workplace allowed using queuing theory formulas, which helped to calculate the efficiency index of functioning and the main characteristics of the Tower controller workplace as the queue management system. Table 1 summarises indicators, which describe the situation with:

Table 1. Indicators, which describe the situation with 2 places in the queue, an average time intervals between aircraft entries 8 minutes, time for repeated approach 7, 12, 15 minutes

	Indicators	Situation options		
		1	**2**	**3**
Initial data	Time for the repeated approach of aircraft, minutes	7	12	15
	The average time interval between aircraft entries to the area of responsibility of the Tower controller, minutes	8	8	8
	Amount of places in the queue	2	2	2
Calculated data	The intensity of channel load	0.75	0.75	0.75
	The intensity of service by the Tower controller	0.1667	0.1667	0.1667
	The intensity of the flow in the area of responsibility of the Tower controller, aircraft per minute	0.125	0.125	0.125
	Average time, which aircraft spends in the system, minutes	13	18	21
	The average time of Tower controller service, minutes	6	6	6
	Standardized waiting time of aircraft in the system, minutes	1.1667	2	2.5
	The probability of refusal in service by the Tower controller	0.1543	0.1543	0.1543
	Relative capacity	0.8457	0.8457	0.8457
	Absolute capacity	0.1057	0.1057	0.1057
	The average length of the queue in the system	0.5143	0.5143	0.5143
	Average number of applications, associated with the system	1.1486	1.1486	1.1486
	The average waiting time in the queue	4.1143	4.1143	4.1143
	The average time of application stay in the queue management system	9.1886	9.1886	9.1886

- 2 Places in the queue;
- Average time intervals between aircraft entries to the area of responsibility of the Tower controller 8 minutes;
- Time for the repeated approach of aircraft 7, 12, 15 minutes (Pilipyonok, 2017).

Table 2 summarises indicators, which describe the situation with:

- 2, 3, 4 Places in the queue;
- Average time intervals between aircraft entries 8 minutes;
- Time for repeated approach 12 minutes (Pilipyonok, 2017).

Table 3 summarises indicators, which describe the situation with:

- 2 Places in the queue;
- Average time intervals between aircraft entries 1, 2, 3, 8 minutes;
- Time for repeated approach 12 minutes (Pilipyonok, 2017).

The results, given in the tables, indicate that the increase of time for repeated approach leads to the rise of the standardized waiting time of aircraft in the system, which means an increase of fuel consumption.

Table 2. Indicators, which describe the situation with 2, 3, 4 places in the queue, an average time intervals between aircraft entries 8 minutes, time for repeated approach 12 minutes

	Indicators	**Situation options**		
		1	**2**	**3**
Initial data	Time for the repeated approach of aircraft, minutes	12	12	12
	The average time interval between aircraft entries to the area of responsibility of the Tower controller, minutes	8	8	8
	Amount of places in the queue	2	3	4
Calculated data	The intensity of channel load	0.75	0.75	0.75
	The intensity of service by the Tower controller	0.1667	0.1667	0.1667
	The intensity of the flow in the area of responsibility of the Tower controller, aircraft per minute	0.125	0.125	0.125
	Average time, which aircraft spends in the system, minutes	18	18	18
	The average time of Tower controller service, minutes	6	6	6
	Standardized waiting time of aircraft in the system, minutes	2	2	2
	The probability of refusal in service by the Tower controller	0.1543	0.1037	0.0722
	Relative capacity	0.8457	0.8963	0.9278
	Absolute capacity	0.1057	0.112	0.116
	The average length of the queue in the system	0.5143	0.7721	1.005
	Average number of applications, associated with the system	1.1486	1.4443	1.7009
	The average waiting time in the queue	4.1143	6.1767	8.0404
	The average time of application stay in the queue management system	9.1886	11.554	13.607

The rise of the amount of places in the queue leads to decrease of refusal probability. The increase of the time interval between aircraft entries leads to decrease in an average number of applications, associated with the system. The average number of aircraft is equal to the multiplication of input flow intensity by the average time of application stay in the queue management system regardless of the nature of the application flow, the distribution of service time and service discipline.

When the algorithmic model of the decision-making process of a Tower controller is ready, distribution laws of aircraft flow formation are studied, the efficiency index of functioning and the main characteristics of Tower controller workplace as the queue management system are calculated then comes a sequence of the next development specific of the Tower controller intelligent training system. This is the use of object-oriented modelling for getting completed learner's normative model.

In accordance with the object-oriented approach, the decision-making process of a Tower controller may be represented as a behaviour-chart. Generally, a behaviour-chart consists of states, treks, variables, constants, input and output actions, operating conditions. For choosing states, treks, variables, constants, input and output actions, operating conditions of the behaviour-chart of the decision-making process of a Tower controller there were used algorithmic model, distribution laws of aircraft flow formation, some efficiency index of functioning and characteristics of Tower controller workplace (Pilipyonok, 2017). The result is shown in Figure 7.

Running the behaviour-chart opens the visual model, which shows not the only algorithm of the decision-making process of a Tower controller, but also displays prompts, which help to understand a

Table 3. Indicators, which describe the situation with 2 places in the queue, an average time intervals between aircraft entries 1, 2, 3, 8 minutes, time for repeated approach 12 minutes

	Indicators	**Situation options**			
		1	**2**	**3**	**4**
Initial data	Time for the repeated approach of aircraft, minutes	12	12	12	12
Initial data	The average time interval between aircraft entries to the area of responsibility of the Tower controller, minutes	1	2	3	8
Initial data	Amount of places in the queue	2	2	2	2
Calculated data	The intensity of channel load	6	3	2	0,75
Calculated data	The intensity of service by the Tower controller	0.16667	0.1667	0.1667	0.1667
Calculated data	The intensity of the flow in the area of responsibility of the Tower controller, aircraft per minute	1	0.5	0.3333	0.125
Calculated data	Average time, which aircraft spends in the system, minutes	18	18	18	18
Calculated data	The average time of Tower controller service, minutes	6	6	6	6
Calculated data	Standardized waiting time of aircraft in the system, minutes	2	2	2	2
Calculated data	The probability of refusal in service by the Tower controller	0.83398	0.675	0.5333	0.1543
Calculated data	Relative capacity	0.16602	0.325	0.4667	0.8457
Calculated data	Absolute capacity	0.16602	0.1625	0.1556	0.1057
Calculated data	The average length of the queue in the system	1.80695	1.575	1.3333	0.5143
Calculated data	Average number of applications, associated with the system	2.80309	2.55	2.2667	1.1486
Calculated data	The average waiting time in the queue	1.80695	3.15	4	4.1143
Calculated data	The average time of application stay in the queue management system	2.80309	5.1	6.8	9.1886

learner when it is better to issue permits or recommendations and in which form they should be (Figure 8) (Pilipyonok, 2017).

Represented model is quite demonstrative, but, in addition, it should correspond to the actual behaviour of the simulation object. For this purpose, a computational experiment is carried out with the model. Computational experiment with the model of the decision-making process of a Tower controller was carried out using several situations, in which were used different time intervals between aircraft entries to the area of responsibility of a Tower controller, various amount of places in the queue, diverse aircraft number and other characteristics of Tower controller workplace. As a result of this experiment, it was proved that the developed model corresponds to the behaviour of a Tower controller in his workplace.

Thus, the learner's normative model was designed and, accordingly, the demonstration mode was implemented.

For realization training and control modes the model of information support was developed. Here it should be noted the specifics of its development. The specific is the determination of main classes and subclasses of objects, which characterize Tower controller workplace. The first class of objects is an Aircraft with subclasses: Classification of aircraft according to the categories of turbulence, Separation minimums, Stages of take off. The second class of objects is Aerodrome with subclasses: Working area of the aerodrome, Apron. The third class of objects is Monitor with subclasses: Runway working directions, Metar, Additional information. The fourth class is Decision-making process of a Tower controller. The composition of this class varies depending on the tasks, which air traffic controller solves.

Figure 7. A piece of the behaviour-chart of the decision-making process of a Tower controller

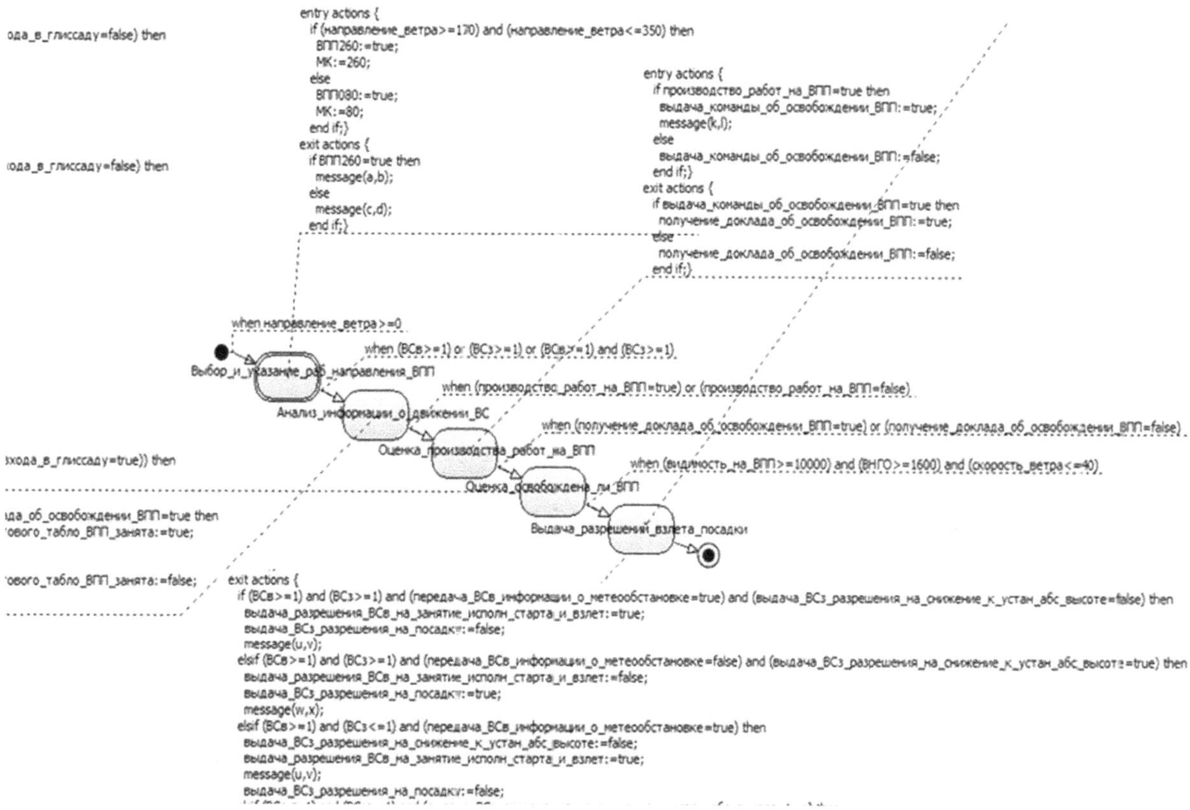

Figure 8. The visual model of the decision-making process of a Tower controller

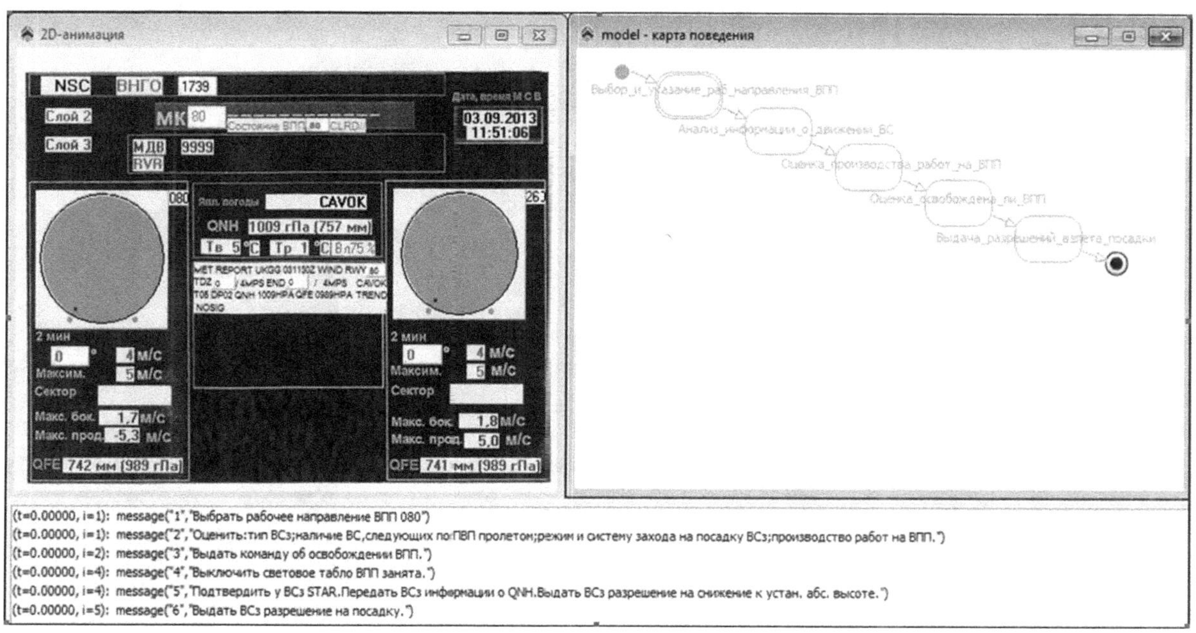

When the model of information support is done, the stage of reproduction of the system prototype begins. This stage has also its specific. This is the creation of a dependency diagram of prototype and prototype testing. Such a diagram makes it possible to synchronize the operation of all system components and get the working prototype of the intelligent training system. The prototype of the intelligent training system «Tower Controller» is shown in Figure 9 (Pilipyonok, 2017). It allows entering decisions, which are made by a learner independently; to show prompts, which help to understand a learner when it is better to issue permits or recommendations and in which form they should be; to display runway strip pattern, radar monitor and meteorological monitor, which are on the real Tower controller workplace.

For testing this prototype, six exercises were developed. According to the scenarios of these exercises, there were three cases, required issuing permission to take off and three cases, in which the correct decision was the return of the aircraft to the apron.

In prototype testing students of Flight Academy of the National Aviation University, Ukraine, participated. They studied in different specialities and had the total or partial lack of knowledge about Tower controller workplace. Analysis of testing results showed that proposed prototype needs improvement in several directions, but even this variant makes it possible to get Tower controller work skills as prototype graphical interface is quite realistic, convenient and easy to read; the completeness of displaying information is sufficient; the complexity of exercises is not high.

Figure 9. The prototype of the intelligent training system «Tower Controller»

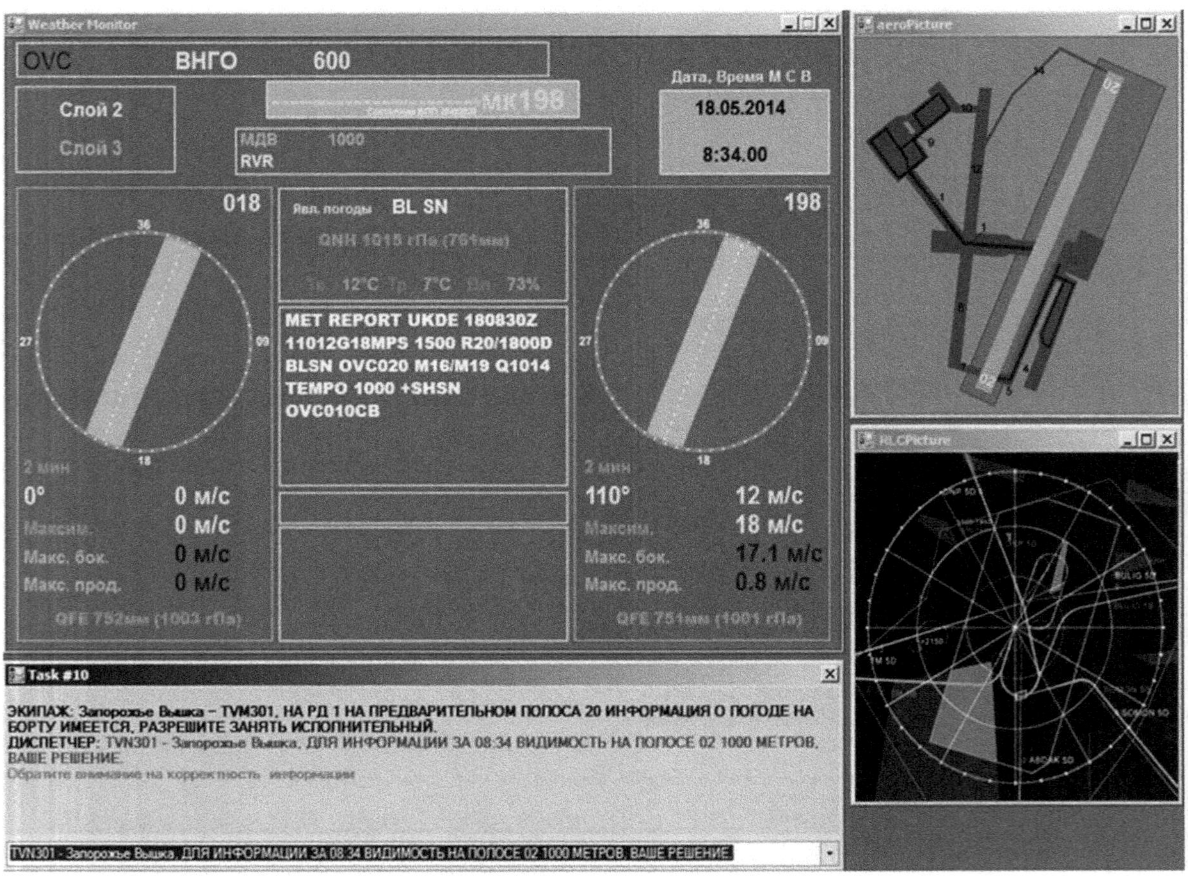

FUTURE RESEARCH DIRECTIONS

This chapter was devoted to development specifics of the Tower controller intelligent training system, which was given on the example of the design of the intelligent training system prototype "Tower Controller". Any prototype needs further development to get a full result. Therefore, future research directions of further development of the intelligent training system "Tower Controller" are:

- Display of aircraft motion on the radar monitor and on the runway strip pattern;
- Automated generation of exercises with various complexity;
- Provision of voice interaction between each particular learner and the intelligent training system;
- Adapting the learning process to peculiarities of a separate user;
- A detailed assessment of the learner's knowledge.

CONCLUSION

The development of an intelligent training system for each sphere of activity has its own specifics. Here were presented development specifics of the Tower controller intelligent training system. They are:

- Analysis of the Tower controller activity;
- Identification of the problem;
- Analysis of the problem cause;
- Highlighting the most vulnerable stages of the decision-making process of a Tower controller with help of interviews, questionnaires and evaluation of statements from the tape recorder of the controller radio exchange;
- The suggestion of the conceptual scheme of the intelligent training system;
- Identifying the main phases of the decision-making process of a Tower controller;
- Presentation of the decision-making process of a Tower controller in a form of the algorithmic model;
- Analysis of flight plans in the busiest period and in the least busy period;
- Finding the empirical distribution of moments of aircraft entry to the area of responsibility of the Tower controller;
- Finding the empirical distribution of time intervals between aircraft entries to area of responsibility of the Tower controller;
- Finding the empirical distributions of the aircraft number in the busiest period and the least busy period of Tower controller work;
- Studying Tower controller workplace as a queue management system and calculation its efficiency index of functioning and the main characteristics;
- Presenting the decision-making process of a Tower controller as a behaviour-chart with states, treks, variables, constants, input and output actions, operating conditions and running the visual model;
- Carrying out the computational experiment with the model of the decision-making process of a Tower controller;

- Determination of main classes and subclasses of objects, which characterize Tower controller workplace and development of the model of information support;
- Creation of a dependency diagram of the prototype of the intelligent training system;
- Testing the prototype of the intelligent training system;
- Analysis of testing results and identifying future research directions.

Described development specifics of the Tower controller intelligent training system may be approximated to design other training complexes for aviation specialists. Moreover, they may be used in different fields, which are connected with the operator's work. Usage of proposed specifics will help to avoid wasting time during creation training system and improve its quality.

REFERENCES

Approach and Landing Accident Reduction (ALAR). (n.d.). Retrieved from http://flightsafety.org/current-safety-initiatives/approach-and-landing-accident-reduction-alar

ATCsimulator. (n.d.). Retrieved from http://www.atcsimulator.com/

ATTower® ATC Tower Simulation. (n.d.). Retrieved from http://www.ufainc.com/products/attower-towsim-tower-simulation/

BAO Tower – New Version. (n.d.). Retrieved from http://www.airliners.net/aviation-forums/aviation_hobby/read.main/73050/

Dnipropetrovsk regional branch, Zaporizhzhya air traffic service system. (2013). Operating instructions of the aerodrome control tower. *Zaporizhzhia, 91.*

Dzhuma, L.N., & Pascal, O.N. (2010). Automation of labor processes in air traffic control system. *Journal of Eastern Ukrainian National University of Vladimir Dal. Scientific Journal, 9*(151), 155-160.

Dzhuma, L.N., Pascal, O.N., & Lishavskaya, V.L. (2011). Application of information technologies for design of Tower controller's workplace. *Journal of Easton Ukrainian National University of Vladimir Dal. Scientific Journal, 7*(161), 241-246.

Dzhuma, L.N., & Pilipyonok, O.N. (2014). Dataware model improvement of the intellectual training system «Tower controller». *Scientific Proceeding of Ukrainian Research Institute of Communication. Scientific and Production Digest, 5*(33), 48-57.

Dzhuma, L.N., & Pilipyonok, O.N. (2015). Modeling of the air traffic flow in the area of Tower controller's responsibility. *Scientific Proceeding of Ukrainian Research Institute of Communication. Scientific and Production Digest, 5*(39), 93-97.

Dzhuma, L. N., Pilipyonok, O. N., & Timoshenko, A. V. (2012). Analysis of the structure of Tower controller workplace and peculiarities of its modeling. *Materials of the first International scientific and practical conference Management of high-speed moving objects and professional training of operators of complex systems*, 32-33.

Gorbunov, A.L., Eliseev, B.P., & Nechaev, E.E. (2010). Innovative approach to the creation of air traffic controller simulations on the basis of mixed reality: the need and the possibility. *Scientific Proceeding of Moscow State Technical University of Civil Aviation, 161*, 5-14.

Multipurpose controller simulator "Expert". (n.d.). Retrieved from http://www.nita.ru/catalog/komplek-snyy-dispetcherskiy-trenazher-ekspert/

National Bureau of Air Accidents Investigation of Ukraine. (2014). *Analysis of flight safety data from the results of the investigation of events with civil aircraft of Ukraine that took place in 2013.* Kiev: Author.

National Bureau of Air Accidents Investigation of Ukraine. (2015). *Analysis of flight safety data from the results of the investigation of events with civil aircraft of Ukraine that took place in 2014.* Kiev: Author.

National Bureau of Air Accidents Investigation of Ukraine. (2016). *Analysis of flight safety data from the results of the investigation of events with civil aircraft of Ukraine that took place in 2015.* Kiev: Author.

NexSim Tower Simulator. (n.d.). Retrieved from https://www.umes.edu/Engineering/Default.aspx?id=26070

Pilipyonok, O. M. (2017). *Method of improving the aircraft management quality of operators of the navigation service systems and traffic control* (Doctoral dissertation). Kirovograd Flight Academy of the National Aviation University.

Pilipyonok, O. N. (2012a). Development of the module of information support air traffic controller activity. *Abstracts of the 12 International scientific and practical conference for young scientists and students Flight. Modern problems of science*, 35.

Pilipyonok, O.N. (2012b). Peculiarities of modeling of the operating conditions of Tower controller. *Materials of the 32 All-Ukrainian scientific and practical conference for young scientists and students Aviation and space: status, achievements and prospects*, 151-152.

Pilipyonok, O.N. (2013). Development of the training system prototype Tower controller. *Materials of the All-Ukrainian scientific and practical conference for young scientists and students Problems of navigation and traffic management*, 32.

Pilipyonok, O.N. (2014). The decision making model as the main component of the intelligent learning system. *Scientific Proceeding of Ukrainian Research Institute of Communication, 4*(32), 54-61.

Software and Solutions for Air Traffic Control and Management. Product Focus – TATIC TWR. (n.d.). Retrieved from http://www.airport-suppliers.com/supplier/Saipher_ATC/press_release/Software_and_Solutions_for_Air_Traffic_Control_and_Management_1/

State Aviation Administration of Ukraine. (2013). *Analysis of flight safety level and identification of potential accident factors with civil aircraft of Ukraine in 2012.* Kiev: Author.

Statistical Summary of Commercial Jet Airplane Accidents. Worldwide Operations 1959-2011. (n.d.). Retrieved from http://www.boeing.com/news/techissues/pdf/statsum.pdf

The FIRSTplusTM Tower Simulator. (n.d.). Retrieved from http://www.network54.com/Forum/248068/thread/1269639662/5/Turkish+HAVELSAN+Simulator

Tower Simulator. (n.d.). Retrieved from http://www.towersimulator.com/

KEY TERMS AND DEFINITIONS

Air Traffic Service System: System that provides air traffic control services, air traffic advisory service, flight information service, and alerting service.

Aviation Simulator: Tool that artificially re-creates working conditions of aviation specialist.

Behavior Chart: Special chart that is formed by states, treks, variables, constants, input and output actions, operating conditions.

Conceptual Scheme: Picture on which the basic idea of development is displayed graphically.

Flight Plan: Document with the details of a particular proposed flight.

Information Flow: Data that is operated by the specialist.

Intelligent Training System: System with intelligent components that allow adapting the learning process to peculiarities of a separate user.

Landing: Process during which an aircraft is going to land on the ground or other surfaces.

Learner's Normative Model: Model that represents the logic of the decision-making process of a learner.

Mode: State of a system in which it works in a special way.

National Bureau of Air Accidents Investigation of Ukraine: Specialized expert institution, the main purpose of which is an investigation of aviation incidents with Ukrainian aircraft and foreign aircraft that occurred on the territory of Ukraine.

Object-Oriented Modeling: Approach in which modelling is executed by the selection of objects.

Personalism: Feature of a particular person.

Professional Competency: The capability of the person to perform the duties of his profession with help of presentation skills.

Prompt: Special text with the help of which a learner may understand what he should do.

Prototype: The basic version of a system that is created for testing and used to obtain a complete system.

Queue Management System: System that consists of service and applications and is used to control queues.

Regional Branch: Unit of Ukrainian State Air Traffic Service Enterprise that performs functions of air traffic management, radio-technical support, air navigation, and meteorological services.

Runway: Special place from which an aircraft may perform takeoff and landing.

Sample: A representative item of something that shows what the rest should be like.

Specific: Something that differs from another; peculiarity.

State Aviation Administration of Ukraine: An executive institution of Ukraine that is responsible for matters relating to civil aviation.

Takeoff: Process during which an aircraft is going to leave the ground or other surfaces.

Tower Controller: Aviation specialist, who is responsible for the safe, orderly, and expeditious flow of air traffic in the zone of takeoff and landing.

UkSATSE: Ukrainian State Air Traffic Service Enterprise, the main mission of which is the provision of an efficient and safe air navigation services in the Ukrainian airspace and in the airspace over the high seas, where the responsibility for providing air traffic services is delegated to Ukraine under the international agreements, subject to the current and future airspace user needs and operational conditions in the air traffic market both in Ukraine and European Region.

Chapter 3
Simulators as an Essential Tool for Shaping the Competence of the Aviation Personnel

Jarosław Kozuba
Silesian University of Technology, Poland

Aleksander Sładkowski
Silesian University of Technology, Poland

ABSTRACT

The implementation of the aviation tasks, including the preparatory ones, is a difficult, complex task requiring from the aviation personnel a high level of general, technical, and specialist knowledge and a wide range of skills, appropriate to the type of technical tools and systems being at the disposal of the aviation personnel as well as the complexity and difficulty of their tasks. Particular importance is currently attached to the development of training devices used in the basic training and in-service training of the aviation personnel. The authors have referred to the role of simulators in achieving the desired level of specialist competence by flight personnel, including, among others, such issues as aircraft simulators development, and their application in the aviation training; aircraft simulators classification in accordance with current aviation regulations; essential functions performed by aviation simulators; flight simulator is an essential tool for basic and in-service training of the aviation personnel.

INTRODUCTION

The results of the analyses conducted by the Author´s of the material clearly point out to the fact that irrespective of the the type of aviation, the aircraft, and its nationality, or the time period taken into consideration, it is the human being – the pilot, mechanic, air traffic controller – that is the underlying factor of almost 70% of undesirable flight-related events (Kozuba, 2011). Z. Błoszczyński, when considering the relationship between the human factor and the undesirable flight-related events, highlights the inadequacy of the actions taken by operators – pilots and other aviation personnel who closely connected with flights, their organization and safety – to the situation that occurred in a certain phase of

DOI: 10.4018/978-1-5225-7709-6.ch003

flight. Such an inadequate action usually leads to an undesirable flight-related event. That situation occurs when threats caused by factors independent from human control have not been removed, or reduced to an acceptable level, despite the real possibilities of doing so. Every action is a result of a particular decision and the related decision-making process. The factor which conditions the emergence of an undesirable flight-related event is usually the occurrence of several consecutive errors in the system of directing (management) an aviation organization, errors in handling the aircraft or in air traffic control, and / or operational errors committed by the air crew. The causes of erroneous decisions made by the pilot-operator are sought for at various stages of investigation whose aim is to discover them, taking into account particular complexity of the aviation system and its environment. Therefore, when discussing the causes of undesirable flight-related events, errors committed by the crew of the aircraft at various stages of the decision-making and implementation process are generally regarded as the key factor resulting in more or less serious consequences. Consequently, in the era when aviation technology is subject to dynamic development, great importance is attached to the quality of flight training, regardless of whether it is basic flight training or professional development training. Modern methods and tools used in the training process are expected to allow for the preparation of highly qualified aviation personnel having a high level of expertise and a broad range of skills which guarantee achieving the desired level of air mission execution at an acceptable level, from the perspective of aviation safety.

Taking into account the wide range of the topic, the authors referred to the main subject of work from the perspective of the pilot - operator of the aircraft. However, the reader should take into account the fact that the advantages and disadvantages of the simulator presented in the material refer both to the simulators used in the training of pilotes, as well as those used in relation to the residual aviation personnel.

BACKGROUND: DOES MODERN AVIATION NEED SIMUALTORS?

Tasks carry out by human in interaction with aviation technologies require the involvement of its basic senses in order to obtain data[1] and information. The relationship between data and information was interestingly presented by Daft, who says that "Information is what changes and supports understanding, while data is the input of a communication channel. The data is tangible and consists of numbers, words, phone calls or computer printouts sent or received. Data will not become information unless people use it to improve their understanding (Daft, 1992). Thus, the pilot to carry out his tasks in a particularly complex and dynamically changing task environment needs information in the sense of a "product of significant data processing" (Clare, 1987). On the other hand, the necessary condition for the process of "significant data processing" is the level of competence (knowledge and skills) and experience presented by the pilot-operator (Figure 1).

Figure 1. The process of data processing into information
Source: Own study

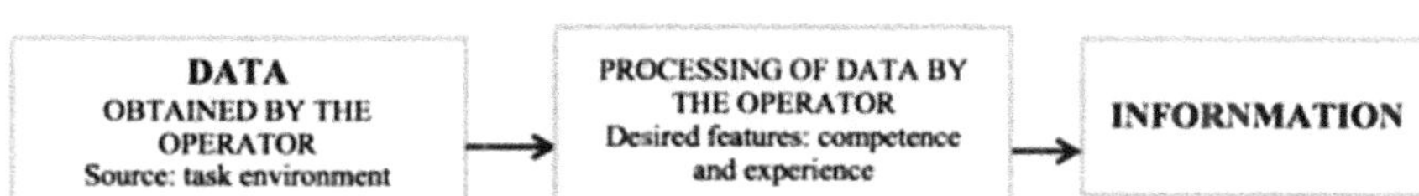

It should be emphasized that the pilot's readiness to act is related to the activity of the complex nervous and muscular systems. On the other hand, human action in the operator - machine relationship is nothing more than a reaction (response) to a specific stimulus. This reaction includes the perception of the stimulus, recognition of the desirability of doing or abandoning the activity (data processing), its selection from the available collection, as the most appropriate in given circumstances. This stimulus is a signal of the operator's need to perform the activity. This stimulus may come from various human senses. This reaction involves the consciousness of the human being in order to evaluate the stimulus and the choice of adequate action.

Depending on the degree of awareness engagement and the need to process data / information, this reaction may be more (data) or less complex (information). In turn, the reaction can be described by the so-called response time[2], particularly important from the point of view of the complexity and state of the environment for the implementation of the aviation task. It should be emphasized that the reflex reaction is the most desirable in the environment of the aviation task implementation. It is related to the level of operator's readiness for the so-called automatic receipt of information and making decisions. It is also important that reaction time is severely limited in this type of reaction. Such a defined response may be a complex action performed by the pilot-operator of the aircraft, including single operations or a combination thereof, as well. In some cases, they may be performed simultaneously or individually, in a specific time sequence, according to circumstances.

Maintaining by the pilot of the desired level of quality of implementation of the above types of processes and activities carried out by him/her, with a certain time limit, they often determine the success of the aviation task, and importantly, they also constitute an acceptable level of aviation safety during the preparation and implementation of aviation tasks.

It should be emphasized that the more stimuli there are to the pilot's operation in a given time interval and the more complex and changeable is the environment of the aviation task implementation (Table 1), the greater the requirements regarding the response to stimuli, their selection, prioritization, the elaboration of decisions and its implementation, will be posed before the pilot - the aircraft operator. The exercises carried out on air simulators are used for this, among others.

The basic role in such stimulated "stimulus-reaction" chains plays the senses and the pilot-operator's ability to conduct correct analytical, decision-making and executive processes.

Regardless of the level of training (basic, advanced, improving, etc.), simulators, next to practical training in airplanes, are an essential tool for building the above stated competencis of pilots - aircraft operators in accordance with the requirements of flight tasks, types of aircraft used, conditions of the task environmental, etc. (Morawski, 2005).

Aerospace technologies present in modern aircraft place ever-increasing requirements on air personnel, particularly on the pilot-operator of the aircraft. Flying is a difficult, complex task requiring from pilots vast general knowledge base and technical expertise and a wide range of skills appropriate to the aircraft type and the tasks performed. The result is that particular attention is currently paid to the development of air training equipment, which is used in the process of basic training and professional development of pilots.

The air training equipment, apart from trainer aircraft is a practical and well tested tool used in assisting training of pilots[3]. This training equipment can be divided in two, particular groups:

1. Theoretical training equipment: content visualization devices, computer software, equipment models, cross-section and cutaway models, training kits, language laboratories, artificial reality laboratories.
2. Practical training equipment: flight / ATC / aviation technical personnel simulator[4], hyper gravity centrifuge (Klukowski, 2005), low pressure test chamber[5], loop (Stelęgowski, 2011), Rhön wheel (Stelęgowski, 2011).

The air training equipment, especially simulators have an important place in the theory of aviation safety. There are a lot of theories concerning flight safety. Mostly, they are related to all the elements related to the preparation and execution of air missions. The most recognized aviation safety theories include among others:

Table 1. Examples of pilot - aircraft operator activity levels, including the involvement of sense functions, analyzers and effectors (Morawski, 2005)

No.	Function of senses (senses)	Function of analyzers (awareness)	Function of effectors (muscles)	The degree of a task complexity
1.	Single stimulus	Detection of the stimulus	Single operation (predetermined)	A
2.	Group of stimuli	Detection of stimuli, selection of the right stimulus, selection of a single activity (their collection) adequate to the stimulus	Single operation (predetermined)	B
3.	Group of stimuli	Identification (selection) of a stimulus from the group, selection of activity or their collection	Single operation (depending on the identified stimulus)	C
4.	Group of stimuli	Identification of the stimuli set, selection of activity adequate to the combination of stimuli	An activity selected from the available set, adequate to the identified stimuli combination	C
5.	Group of stimuli	Identification of the stimuli set, selection of a combination of activities adequate to the combination of stimuli	A combination of activities, according to the stimuli combination	C
6.	Group of stimuli + noise	Separation of stimuli from interference - further activities as in points 2-5	A combination of activities, according to the stimuli combination	D
7.	Group of stimuli + noise + circumstances	Separation of stimuli from interference - further activities as in points 2-5 with regard to circumstances	A combination of activities, according to the stimuli combination and circumstances	E[6]

Singelton Theory. The author holds the view that aviation safety is inextricably linked to the structure and architecture of the machine. He emphasizes the problems of ergonomics. The model related to that theory consists of three major safety optimization strategies, i.e.:

- Operator safety;
- Aviation system (subsystem) safety;
- Aviation organization (appropriate) climate safety.

From the point of view of the above theory, in creating/consolidating desirable characteristics of the pilot-operator, the important role is played by aviation training equipment as a tool for i.a.: qualification and selection of personnel; monitoring the level of training; optimization of instruction and practical training in terms of operator safety; optimization of the selection procedures and training of crews; air accident and incident investigation; research concerning the design and aircraft systems –safety of the system; instilling the right habits of cooperation, and qualities such as trust, truthfulness, loyalty; development and consolidation of desirable characteristics and skills relating to the management and training – safety of the (relevant) climate within the organization.

C.O. Miller and E. Edwards Theory - Model 5M (Man - Machine - Mission - Management - Media). It is based on a system approach to flight safety issues. Also in this case, an important role is played by training equipment, including its role in the proper formation of one of the essential elements, i.e. the pilot. These activities refer to the development and improvement of the conditions for maintaining an acceptable level of interaction between the pilot and the remaining parts of the model, including such areas as:

- Man – recruitment, selection, training and professional development, creation and consolidation of habits, improving multitasking skills, etc.
- Machine – adapting the aircraft cockpit equipment to pilot, developing skills in the use of standby instruments and systems/emergency equipment, etc.
- Mission – acquisition/improvement of skills in the implementation of specific air tasks, understanding the scope and complexity of their execution, air mission execution in extremely different environmental conditions, simulating extremely dangerous mission execution conditions, etc.
- Management – application of mission execution procedures and standards, improvement crew cooperation, optimizing the use of resources at the disposal of the crew, improvement of internal and external communications (e.g. with the ATC services or commanders of other aircraft), improvement of decision-making skills, etc.
- Media – weather, physical conditions of natural environment, flight altitude, terrain, airport conditions, etc.

This model is used among other things in the process of accident investigation, including determination of the causes of undesirable flight-related events and in preventive activities facilitating the avoidance of similar events in the future (Kozuba, 2012).

E. Edwards and Hawkins Theory – $SHEL_1L_2$ Model. The theory of system approach to flight safety in relation to: S – non-material elements of aviation system, H – aircraft, other technical equipment, E – mission environment, L_1 – pilot and other crew members, L_2 – air traffic services personnel, aircraft maintenance personnel, etc. Significant in this theory is the fact that all elements of the system are examined through the prism of human participation in air accidents. When thinking of pilot participation, we mean errors and violations committed by him. In this the theory, aviation training equipment will play a role of a tool used in the prevention of such phenomena, by taking action in the areas of pilot training and development in the following areas:

- S – proper implementation of regulations, procedures and, often mission-specific rules of conduct.
- H – use of technical qualities of the aircraft, including the specifics of the on-board systems use, the use of automation and the improvement of conduct in the event of emergency, etc.
- E – cooperation with aircraft maintenance and air traffic services, proper prevention of dangerous phenomena present in the mission environment, etc.
- L_1, L_2 – selection of aviation personnel, acquiring skills and habits, specialized training, improving health and fitness, multi-crew cooperation, use of resources at the disposal of the crew, interpersonal relationships within aviation organizations, etc.

The letter L_1, in the central point of the model represents the pilot or other crew member, for whom we want to determine interactions with other elements of the model (S,H,E,L_2) after an undesirable flight-related event, e.g. L_1 – crew commander / L_2 – air traffic service personnel, etc. On this basis, we can determine the elements that have failed and had a significant impact on the occurrence of an undesirable flight-related event. The results obtained should be taken into account in the assessment and training, including simulator training, of the aviation organization flight personnel. According to experts[7], it is difficult to imagine the proper application of the findings from this type of analyses, without the use of air training equipment in the process of implementation of the post-accident recommendations. In relation to pilots, such equipment is a flight simulator.

K.K. Platonov and B.F. Lomonov Theory. In this theory the level of flight safety is examined through the prism of the interaction between the four essential elements of the system such as:

- Man – for the needs of the present study considered to be the aircraft operator;
- Aircraft – fixed-wing, rotary-wing, UAV[8];
- Flight management and organization;
- Strategy of the organization with regard to air aviation technology.

According to this theory, the existence of a wide variety of elements in the structure of flight safety system forces a search for such actions that will allow for the existence of specific interactions, both between the individual elements of the system, as well as within each of them. These measures should help to eliminate disturbances with respect to the above-mentioned interactions. On the other hand, they should provide additional support for “strong” air system components, allowing it to function while maintaining the generally accepted and acceptable safety standards. In the opinion of the authors of the presented theory, flight safety problems require systemic analysis, and the flight system components that affect flight safety can be characterized by systemic features (Łomow, 1984). For example, in flight safety system the pilot is described only by his personal and professional qualities, including those resulting from his being an essential element of the aviation system. On the other hand we can say that the level of requirements in relation to the traits that characterize the pilot is highly dependent on the quality of the remaining elements of the aviation system within which he operates.

Systemic approach to the problems of flight safety is necessary, especially when dealing with the "man - machine" co-operation, as the most important factor in the flight safety system is the "pilot - aircraft" system. Adapting the technology to man (e.g. ergonomic aspects of the cockpit) is the condition that guarantees that the cooperation will meet the flight safety requirements (Klich, 2010).

The "air training equipment" element, in the "flight management and organization" subsystem, was clearly articulated in the Lomonov and Platonov model. Taking into account such an approach, it can be concluded that, according to the authors of this theory, air training equipment has a major impact on shaping and improving all elements of the aviation system.

Flight training history dating back over a hundred years suggests that the principles, methods, techniques and tools used in the preparation of pilot candidates developed in parallel with the advances in aviation technology. Changes in aviation technology, and a higher and higher degree of complexity of air operations carried out in highly variable mission environment necessitates constant changes in methodologies of flight training and professional development of pilots. Another factor having a significant impact on the training/professional development of pilots is the need to optimize the cost of air training, depending on the funds at the disposal of a given aviation organization. Taking into account the changes in the approach to air training of pilots, S.K. Kearns has distinguished four basic eras of flight training (Kearns, 2010), whose descriptions have been developed by the author:

1. *Apprenticeship Era (from 1903)* (Kozuba, 2014): learning the skills to perform basic activities related to flying the plane – I know how to do (know-how) and I can use it in flight, without having to understand the steps that are part of the process of handling an aircraft. For example, executing a turn while maintaining certain parameters: altitude +/- 100 m, airspeed +/- 10 km/h and bank angle +/- 5°.
2. *Flight Simulation Era (from 1929):* learning to perform basic operations and the acquisition of certain habits of using simulation techniques (instrument training devices and other flight simulators).
3. *Aviation Safety Era (from 1979):* the need for new elements in air training was clearly articulated and they were introduced in the process of flight training, allowing trainees to gain/enhance their knowledge and skills in aviation safety, which forms the basis for the preparation and execution of air missions at an acceptable level of safety. The beginning of the third era of flight training organization is assumed to be the date of the NASA Conference in 1979, when the representatives of the largest airlines in the world commenced the talks about the alarming increase in the number of aviation accidents caused by human factor.
4. *Dedicated Training Era (from 1990):* in line with the assumptions of its developers, dedicated flight training should be focused on individual abilities and needs of the trainee, in order that he may achieve above-average specialist development – specialist competence.

In summary, the conclusions of the above considerations relating to the position of air training equipment in the selected flight safety theories flight training history suggest that air training equipment is considered to be an essential tool for supporting the system of recruitment, selection, theoretical and practical training of flight crews. Flight simulators take a special position within the above mentioned aspects. That equipment is used as a valuable tool in improving development of the pilot (air traffic controller, aviation technical personnel) skills needed for safely execution / supporting of any air mission. Regardless of the type of aviation and the stage of flight training/professional development, flight simulators are a widely used pilot training tool.

The best proof of utility of the simulators is the history of their development, present state and perspective of development in the future. Authors are going to discuss above mentioned issue in the next chapter.

CHOSEN ASPECT OF THE AVIATION SIMULATORS DEVELOPMENT

The first aerospace designs were imperfect and very unreliable. Therefore, the possibility of applying aviation were limited. Growing expectations set for aerospace designs caused a rapid development of the aviation industry. Airframe and engine designs were refined so as to eliminate the causes of defects, incidents, and accidents. The aim of flying was not just to get off the ground into the air and fly some distance but, above all, to perform some task. And as it often happens, the more difficult task to perform, the greater the demands on the equipment and the skills of pilots.

Aviation pioneers had to deal with a very difficult task, because they attempted to fly without sufficient knowledge of flight aerodynamics and without the appropriate skills necessary in aviation. The first aircraft designs were far from perfect, and controlling them was limited and quite difficult. When learning to fly, many pioneers of aviation had accidents which resulted in their serious injuries or fatalities. Therefore, attempts were made to find ways of teaching the pilots the skills necessary to control the plane, without having to take excessive risk, or at least limiting it. A certain solution of that problem was the use of first trainers that gave pilots opportunity to learn the principles of controlling the aircraft. The design of those trainers was very crude. One of the first ground-based trainers was the "Sanders Teacher" (Figure 2a). That trainer was in fact an aircraft mounted on a universal joint in an exposed position, fixed to the ground. The training consisted in moving aileron, elevator and rudder controls in order to maintain the aircraft at equilibrium using the force of the wind. In order to perform a training, the simulator had to be positioned against the wind, and the wind force had to be appropriate, so as to achieve the efficiency of the control surfaces necessary to maintain the balance. Another solution using the wind to teach the use of controls was the captive monoplane made by the Italian company, Gabardini. In the UK, in 1910 Eardley Billing patented an aircraft fixed to the ground used to give basic flight training (Figure 2b). Another type of trainers were devices whose operation was not based on the use of

Figure 2a. First simulators: Sanders Teacher Trainer Source
Source: (A History of Simulation, 2018)

Figure 2b. First simulators: Eardley Biling Oscillator
Source: (A History of Simulation, 2018)

wind, but were operated by people – Antoinnette Trainer (Figure 2c). These persons simulated roll and pitch of an airplane, and the task of the student was to respond to disturbances introduced.

World War I had a significant impact on the development of aviation. In addition to noticing newer and newer possibilities of using aviation, the losses that were incurred as a result of the aviation equipment imperfections and insufficient skills of the pilots were among the main reasons for the rapid development of aviation. Removing the causes of aircraft failures "perfected" their designs, and the acquisition of the new knowledge on the theory of flight and aviation skills by pilots, resulted in a new approach to flight training. A new discipline of aviation psychology appeared. A variety of tests were developed to verify flying skills in the process of aviator selection. Tests for reaction time and coordination became popular. One of the devices that were used in such tests was Reid & Burton device from z 1924 (Figure 3a). Research studies brought more and more knowledge. The examination determined the influence

Figure 2c. First simulators: Antoinette Trainer
Source: (A History of Simulation, 2018)

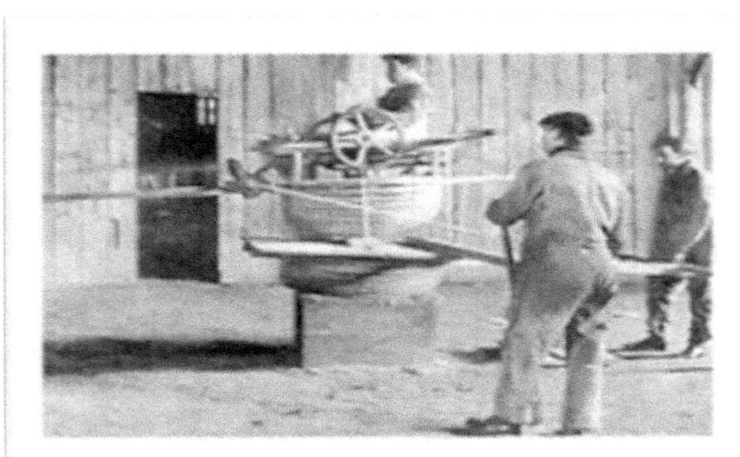

Figure 3a. Simulators – next generation: Reid & Burton device
Source: (Reid, 1924)

of the vestibular apparatus and the sense of sight on the sense of balance. The device simulating the position of the aircraft while performing various aerobatic figures was the Ruggles Orientator (Figure 3b). The device consisted of a seat mounted within a gimbal ring assembly with a student sitting inside, which enabled full rotation in all three axes, simulating pitch, roll, yaw, and in addition provided vertical movement. The task of the trainee was to learn to recognize the position in which he found himself and to correctly recover to a horizontal position.

The end of World War I was the commencement of the air transport era. In order to compete with other modes of transport, airlines had to seek ways of flying in instrument meteorological conditions (IMC). The IFR flight era began. In 1927–29 Edwin Link developed the “Link Trainer” device (Figure 4a), which owing to its capabilities may be called a flight simulator. The simulator cockpit was mounted

Figure 3b. Simulators – next generation: Ruggles Orientator device
Source: (Aerotoons, 2018)

Figure 4a. Simulators to IFR training: Link TrainerSource: a. (Link Flight Simulator, 2018) b,c (Aircrew Training Devices. 2018)

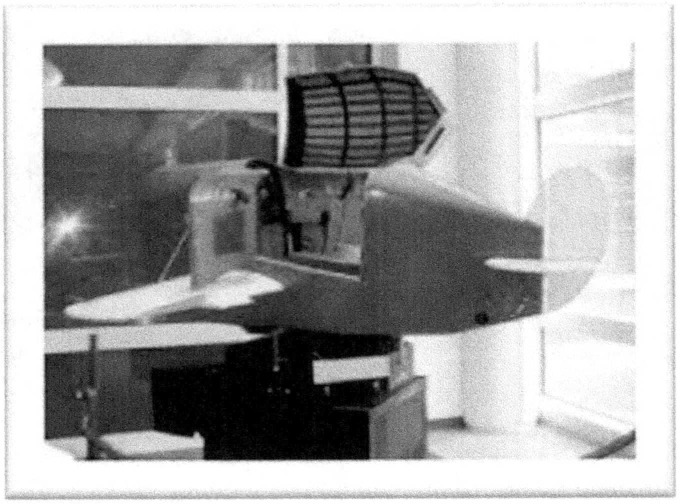

on the fixed base. Pneumatic bellows were used for actuation of the cockpit, which simulated the attitude change of the aircraft in accordance with the deflections of the simulator controls. The first "Link Trainer" did not have any instruments and was used to teach the students to control the simulated airplane. The subsequent versions, however, started to be equipped with the basic instruments for IFR flights, including a compass, turn and slip indicator, airspeed indicator, altimeter, and vertical speed indicator (Figure 4b, 4c). This gave opportunities not only to train and master the rules of controlling the aircraft plane, but also, after closing the cockpit with an opaque canopy, to train IFR flights. Despite more than ten years of work on flight training devices, the existing trainers were regarded as characterized by

Figure 4b. Simulators to IFR training: Controls of a Link Trainer - Link C-3, (approx. 1939)
(Aircrew Training Devices. 2018)

Figure 4c. Simulators to IFR training: "Link Trainer" Link C-8, (approx. 1950)
(Aircrew Training Devices. 2018)

insufficient realism of the flight simulation. In the initial period, the interest in the "Link Trainer" was limited, although its capabilities of simulating the flight conditions were more realistic than those of the pre-existing trainers. The first orders for the "Link trainers" came in 1934 after a series of air accidents. This favored their further development and improvement. Newer versions were produced supplemented with new instruments, adapting the simulator to the requirements of the ordering parties.

World War II gave rise to a renewed interest in flight simulators. The demand for training a large number of pilot candidates led to new orders for simulators. For the purpose of training the pilots operating transatlantic flights, a simulator was built which gave them a possibility to learn to navigate using astronavigation (Figure 5a). New task to be performed on simulators emerged – multi-crew training. Simulators were then used not only for training pilots in IFR flying and navigators in terrestrial and astronavigation, but also for training crews in procedures to be followed in special situations during the flight. Crews were trained in taking actions after the failure of various aircraft systems, and in aircrew evacuation; bombardiers were trained to use weapons sights and bombing instruments, etc. All these challenges contributed to the emergence of increasingly sophisticated simulators. In 1941, "Silloth Trainer" was developed for emergency situation training, i.e. the procedures to be followed in the event of various types of technical failures (Figure 5b). This simulator was a new idea for simulation techniques. Mechanical and pneumatic solutions were no longer adequate for the growing demands concerning the capabilities of simulators. The Americans used the experience of the British. In 1941, a copy of the "Silloth Trainer" was built and evaluated in the USA. The result of this was the decision to outsource the construction of the modernized "electrical" version of "PBM-3" aircraft simulator to Mohler Organ. Curtiss-Wright Corporation also developed an interest in the idea of applying new solutions. Two simulators of the AT-6 aircraft designed by that company were ordered by the US military (Figure 5c).

In 1948, Curtiss-Wright commissioned the simulator of the Boeing 377 Stratocruiser commercial transport for Pan American. It was a simulator, which imitated the entire aircraft cockpit with the simulation of all instruments and equipment operation. This simulator training gave the opportunity to conduct Multi crew cooperation (MCC) training in standard procedures as well as in emergency situ-

Figure 5a. Next generation simulators: Celestial Navigation Trainer
Source: a. (RAAF A63, 2018)

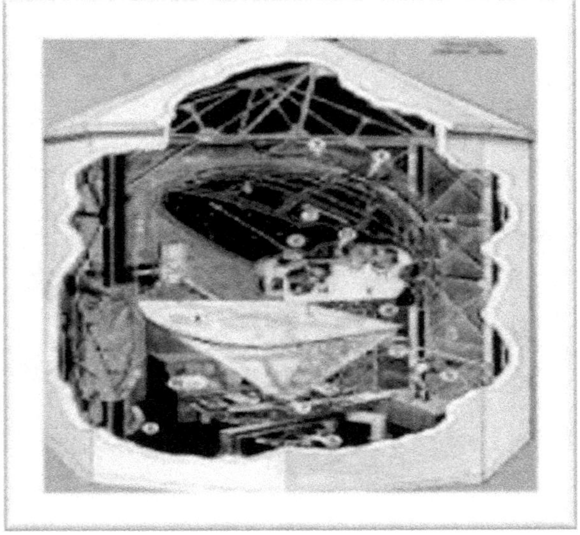

Figure 5b. Next generation simulators: Silloth Trainer
(Silloth Airfield Project, 2018)

ations. Emergency situation scenarios could be easily entered from the control panel by the instructor operating the simulator (Figure 6a). Upon the emergence of jet aircraft, such as the Boeing 707 and DC-8, consideration was given to creating a simulator that would meet the new requirements In the UK under license from Curtiss-Wright company, Rediffusion build the Comet 1 simulator for *British Overseas Airways Corporation* (*BOAC*). Work on the improvement of the simulator continued, which resulted in the creation of the "Comet 4" flight simulator. The cockpit of the Comet 4 aircraft was placed on a platform simulating the aircraft movement in three axes of rotation (Figure 6b). The final element needed to create the realism of the flight was the development of a way of simulating visual cues. Work

Figure 5c. Next generation simulators: Curtiss-Wright Z-1 simulator of the AT-6 aircraft
(Electronic Flight Simulation, 2018)

Figure 6a. Next generation simulators – MCC: Boeing 377 Stratocruise
Source: (Mechaniczne symulatory lotu, 2018)b. (A History of Simulation, 2018)

on this had lasted for a long time. In the 1950s the “Vulcan” bomber simulator introduced the first visual imagery which simulated the atomic bomb explosion. The first tangible achievement in this area was, however, the use of large boards with aerial photographs or models, over which the camera moved, and the image from the camera was displayed on the screen in front of the simulator cockpit. The camera was moving over, e.g. the terrain mock-up responding to the deflection of the simulator controls (Figure 6c). This way of creating visual cues during a simulated flight was used in the construction of new simulators until the end of the 1970s.

Figure 6b. Next generation simulators – MCC: Comet 4, c. Mock-ups applied in flight simulators
Source: b. (A History of Simulation, 2018)

In 1978, Singer-Link company developed the Digital Image Generator (DIG), which started the era of simulators using computer generated imagery (CGI) (Figure 7). Currently built simulators are becoming more and more perfect owing to the use of modern technology because of which they simulate flight conditions increasingly accurately.

Taking into consideration design complexity and the applied degree of fidelity in imitating cockpit instruments, equipment, and systems of the aircraft, as well as the manner of simulating the loads occurring during the aircraft maneuvers, flight simulators (Flight Simulation Training Devices - FSTD) can be divided into four main groups (CS-FSTD, 2012)[9]:

1. Full Flight Simulator (FFS) (Figure 8a), which is used in Polish Air Force Academy (PAFA) - the most technologically advanced simulator type. Complete, full-sized and functional cockpit replica of a certain aircraft type, model or series, combined with the appropriate computer system necessary to imitate the aircraft during ground and air operations. The system of visualization provides a view outside the cockpit, and the system of actuators imitates the physical sensations associated with motion. Devices of this type are used, among others, in training flight crews in dangerous flight conditions, acquiring the appropriate habits, and in maintaining currency.
2. Flight Training Device (FTD) (Figure 8b) - complete, full-sized and functional cockpit replica of instruments, equipment, and control panels of a given aircraft type, combined with the appropriate computer system necessary to replicate the aircraft during ground and air operations. These devices do not have to be equipped with visualization and motion imitation systems.
3. Flight and Navigation Procedures Trainer (FNPT) (Figure 9a) In this type of simulators the cockpit is connected with a suitable computer system necessary to represent a particular type, or group of types, of aircraft during in-flight operations. Devices of this type are used i.a. in training IFR flights and in navigation training.

Figure 7. Flight simulator using computer generated imagery (CGI)
Source: (WidevieW, 2018)

Figure 8a. Full Flight Simulator (Disorientation Simulator)
Source: photos made in PAFA

4. Basic Instrument Training Device (BITD) (Figure 9b): a device imitating the instruments of an aircraft (possibly displaying them on the screen) to be used in practicing at least the procedural aspects of an IFR flight.

Simulators approved for the training of licensed aviation personnel have to meet the requirements described in civil aviation regulations. Military simulators must meet the requirements specified by the

Figure 8b. Flight Training Device (FTD)
Source: photos made in PAFA

Figure 9a. Flight and Navigation Procedures Trainer (FNTP-II)
Source: photos made in PAFA

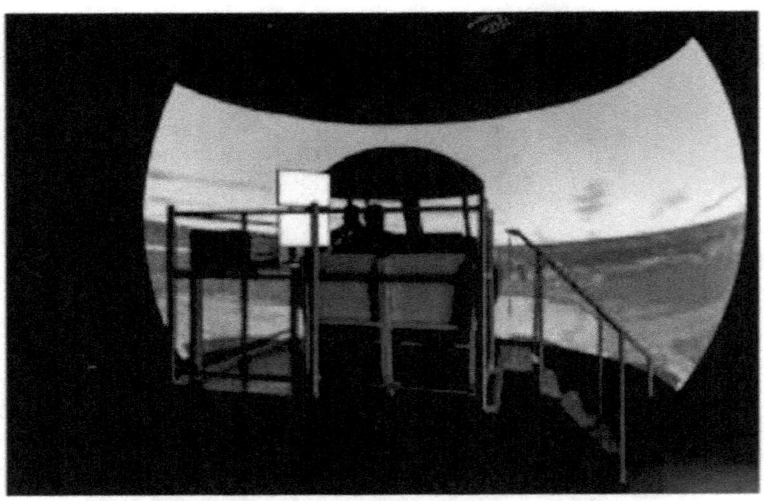

contracting party. These requirements are largely consistent with those applicable in civil aviation. In these requirements great emphasis is put on the fidelity of the cockpit instrument replication, including certain versions of the aircraft, allowing to train the full spectrum of missions executed by a given aircraft.

The use of simulation techniques, especially in the case of complex training systems, such as training of flight personnel for the needs of the armed forces, indicates that there is a need to integrate simulators with other tools and systems to increase training efficiency through greater "approximation" of the training task environment to this real one. This is achieved, among others, by the possibility of com-

Figure 9b. Basic Training Device (BTD)
Source: photos made in PAFA

bining the mission practiced on the simulator with those tasks realistically performed in the air by the crews of aircraft, or the integration of simulators included in the system in a way that enables the joint implementation of the mission by the trainees. An example of such a system can be the one created for the training of the M-346 pilots.

The integrated M-346 aircraft training system consists of the following components (Figure 10):

1. Electronic classroom: is a concept of using computer software to create applications for education and training. With regard to M-346, Electronic Classroom is an application called CBT (Computer Based Training) It contains chronologically and thematically arranged lessons in the following topics: aircraft systems, aerodynamics, navigation, and normal and emergency procedures. Students of the course are responsible for individual study under the specified daily schedule. CBT also includes supervision of lecturers, who, address the emerging questions with their knowledge and experience, and explain potential issues. The technology implemented in CBT uses 2D and 3D interactive models, including audio commentary guiding students through individual topics. Training management and learning progress are recorded in the Learning Management System, which also includes dedicated units to evaluate learning progress.
2. Desktop Trainers / SBT (Simulated Based Training): is a concept of using a personal computer (desktop or laptop) to learn some elements of the knowledge about the aircraft and operating its systems. The most common format is the application that allows the trainee to practice operating the controls in the aircraft cockpit (e.g. operating the UFCP – Up-Front Control Panel and MFD – Multi Function Display). The application is limited to the presentation of the effects of the pilot's

Figure 10. A comprehensive training system for M-346 aircraft pilots
Source: Author's own work, developed on the basis on: (System symulacji, 2018), (Szkolenie, 2018), (Naziemny system szkoleniowy, 2018)

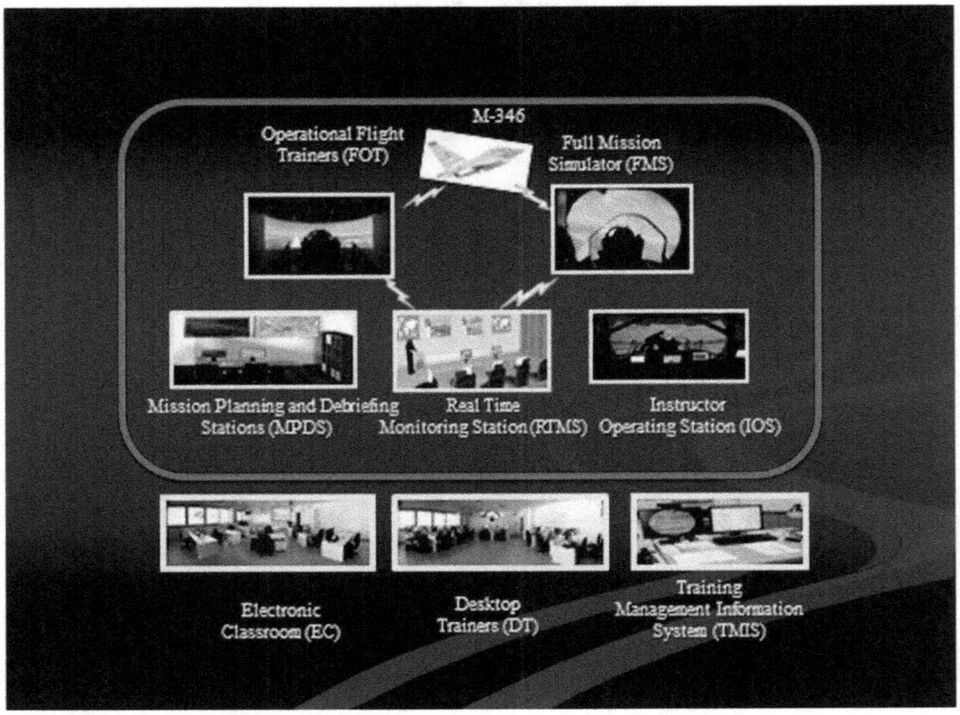

impact on individual systems or aircraft control components (such as e.g . operating the navigation system), with the possibility of performing a flight or executing a mission, but with limited functionality and visualization.

3. Training Management Information System (TMIS): IT system combining the operation of individual IT subsystems managing the M-346 training. Such subsystems include, among others: Integrated Logistic Support (ILS), elements of GBTS (Ground Based Training System), and Learning Management System (LMS). TIMS designed to manage the following: training, the availability of aircraft, the use of simulators, planning training and maintenance activities involving the aircraft and other equipment.
4. Mission Planning and Debriefing Stations (MPDS): common tool used for supporting the M-346 aircraft as well as FMS and FTD simulators. It supports planning, briefing, and uploading the mission planned on a memory card. The data is then uploaded on the aircraft or simulators. They are the basis for setting the aircraft's systems and determine a scenario according to which a flight is performed. MPDS comprise a workstation with a single or multiple screens, wall projection displays, and space planning and debriefing. MPDS is the basic component which manages the planning, debriefing and evaluating the trainee.
5. Real Time Monitoring Station (RTMS): enables real-time monitoring of the task being performed, and its modification from the ground through a Data Link. RTMS technology in conjunction with LVC and Data Link also enables conducting formation flights which combine the use of simulators and real aircraft.

6. Instructor Operating Station (IOS): FMS or FTD simulator workstation which manages training and mission execution. The station is equipped with controls to allow the instructor to take over the control from the student in order to demonstrate and training component. In addition, the station allows the instructor to introduce a target aircraft or an aircraft to fly in a formation, which are piloted by the instructor using the controls located on the console station. Mission management enables to declare the assumed meteorological and tactical conditions, as well as conditions related to the systems of the aircraft.
7. Operational Flight Trainer (OFT)/ Flight Training Device (FTD)/ Part Task Trainer (PTT): part-task simulator used to familiarize the trainee with the aircraft systems, and to train aircraft handling in VFR / IFR navigation flights, emergency procedures, and, in conjunction with the FMS, to serve as a platform for formation flights and the implementation of tactical scenarios. It is characterized by excellent graphics and user-friendly interface but, in comparison with FMS, it has a limited field of vertical and horizontal visualization (120/60 degrees).
8. Full Mission Simulator (FMS) is an example of the highest level of virtual technology implementation in flight training. Using it, it is possible to imitate, in a 1:1 ratio (considering the realness of training), all flight and training phases, including sensor integration (Fire Control Radar, Targeting Pod, Electronic Warfare System, Data Link), in order to create complex tactical scenarios. This helps reduce training costs and increase the availability of aircraft for other tasks. FMS enables to perform, among others, flights in the formation of two and four aircraft, offensive and defensive manoeuvring in dog fight using weapons, to perform tasks using a radar and a targeting pod, perform strike missions using weapons, to fly in NVG (Night Vision Goggles), and using HMD (Helmet Mounted Device). In combination with the FTD, the FMS simulator is able to carry out the formation tasks. Furthermore, the FMS and FTD with LVC technology (Live Virtual Constructive based on Data Link and ETTS) and the use of RTMS (Real Time Monitoring Station) are capable of performing tasks simultaneously in connection with real aircraft executing an air mission[10].

The system described above combines the functions of a basic and an advanced training subsystems, and a subsystem for training pilots who are in the so-called continuous training. Elements of the system fully secure various training stages with regard to academic instruction and practical simulator training.

Currently, flight simulators, in addition to aircraft are considered as one of the essential tools for selection, training, and developing/maintaining currency of flight crews. They are also viewed as a reliable tool for testing pilot knowledge and skills in specific aspects of aircraft operation and air mission execution. Because of the above-mentioned advantages of these devices, simulator training, in addition to theoretical and practical training in the air, plays a crucial role in shaping the competencies needed to achieve the desired level of situational awareness by the pilot at various stages of air mission execution.

FLIGHT SIMULATOR AS AN ESSENTIAL TOOL FOR BASIC AND IN-SERVICE TRAINING OF THE AVIATION PERSONNEL

Flight simulators play a significant role both in pilot training and in sustaining pilot currency. The safety level of modern general, commercial and public aviation depends largely on simulation technologies, used not only for training purposes, but also in air accident investigations, studies of aircraft design or air traffic simulation and on a better understanding of the interactions present in the pilot-aircraft-mission environment system. Conclusions drawn from the analysis of the reference literature (Lintern, 1992) (Hays, 1992) (Leszczyński, 2000) suggest that the role of flight simulators increases with the growing complexity of modern aircraft systems, which means that simulator training is now an integral part of the processes of pilot training and professional development, or conducting the research studies related to adapting aircraft design to human capabilities and limitations. Ever increasing degree of fidelity of representing the pilot natural work environment by the simulator makes flight simulators an invaluable tool used in the process of air accident investigations. How complex has been the development of simulators can be seen from the example of the first flight simulator "Aeronautical Link Trainer", which consisted only of a basic set of cockpit flight instruments and a very basic platform for simulating the aircraft movement. It did not support the function of displaying the external mission environment for the pilot (Bezdek, 2004). Modern flight simulators apart from the faithful reproduction of the internal and external mission environment can also be equipped with motion simulation platform. Modern flight simulator has become the essential type of ground equipment for pilot training and professional development[11]. Currently, it is difficult to imagine modern flight training without the use of flight simulators. The effectiveness of practical training using flight simulators is determined by a number of factors, including, among other things, such as: safety, economic, technical, methodological, etc. Due to the desired level of simulator training efficiency and flight simulator versatility these devices must fulfill a number of functions. The essential, desired functions of a flight simulator designed for pilot training and professional development can include:

- Demonstration Function: it relates to the enrichment of the theoretical training process with demonstrating the trainee the elements of specific procedures, methods and the uses of aircraft instruments, controls and systems.
- Education Function: it relates to such elements as: developing/perfecting trainee's skills and habits with regard to practical implementation of operations in the cockpit, the use of operational procedures, onboard emergency procedures, improving multi-crew cooperation, and the skills in the use of aviation phraseology, developing /perfecting training skills of flight instructors, etc.
- Personality Function: it is associated with the formation and development of the desired personality and professional traits of the pilot (trainee), preparing him for it, among others, for air mission execution under high workload and/or time deficit, and in new situations (onboard and in the mission environment) from the point of view of the current pilot (trainee) experience, etc.
- Adaptive Function: it amounts to improving the pilot (trainee) "recurring" actions on board the aircraft, including actions related to complex and emergency situations. Exercises performed within this function are particularly important with regard to the preparation for the execution of the mission which is new from the point of view of the pilot's current experience.
- Research Function: it is related to checking the behavior of the pilot (trainee) at different stages of the planned mission execution, taking into account the specific solutions relating to the design, ergonomics, and automation of a particular aircraft type. It also amounts to verifying the validity of theoretical assumptions related to the way of solving problems onboard the aircraft by the crew, for example, the validity of assumptions concerning the crew actions in the event of certain onboard emergencies. This function is also applied when it is necessary to confirm the conclusions of the air accident investigation board's proceedings, relating to the causes of undesirable flight-related events.

- Selection Function: it consists in eliminating permanently or temporarily from further aviation training those candidates who have no predispositions (competencies, knowledge, skills) to work as a pilot/perform specific flight tasks.
- Control-Examination Function: it allows to carry out control/examination exercises in relation to the specific license or aviation rating (Leszczyński, 1994).

It should be noted that the effectiveness of the use of flight simulators in the recruitment, selection, training and professional development of pilots, air accident investigations, or research is dependent on an understanding of their capabilities and limitations that apply to specific areas of aviation. On the other hand, in the process of the flight simulator design, it is very important that the designer, apart from applying specific, high-tech aviation technologies, should also strive to create a mission environment model as close to the reality as possible, and take into consideration the knowledge of human capabilities and limitations, including those relating to the ability to perceive and process information, as well as the cognitive factor concerning the operational capabilities of the simulator. Modern flight simulators are also treated as an essential tool for developing the pilot's ability to "automate" the actions in the cockpit and for the efficient reception and selection of signals related to the aircraft status. Many psychologists hold the view (Terelak, 1975), that in relatively simple situations habits can replace thinking – automatic execution of tasks. In complex, unexpected situations – non-routine air incidents – an important role is played by the quick assessment of the situation and the appropriate action programming. From the point of view of situational awareness, the pilot's ability to automatically perform actions on board the aircraft is of primary importance to the perception, selection and analysis of large amounts of information within the time limit and to the construction of the aircraft status mental images. One of the simplest examples of such actions is the cockpit pre-flight check. Research conducted by the author shows that the time differences between a trainee pilot and a pilot in continuous training[12], can range from tens of seconds to several minutes, depending on the aircraft type. Too frequently, especially in the initial stage of flight training, the preparation of a future pilot amounts to learning specific definitions and points of instruction during the theoretical training course, and learning to control certain actions (often without understanding the phenomena of e.g. aerodynamics and flight mechanics) during the practical training.

Meanwhile, with the development of aircraft design, including the more recent introduction of the solutions relating to the areas concerning the aircraft management, pilot's initiative and independent thinking based on vast general and specific knowledge and practical skills, including the skills relating to performing certain actions automatically, become more and more important for the safety and efficiency of air mission execution.

In addition to socialization techniques and teaching techniques, such as brainstorming, decision-making exercises with a limited amount of information and in time deficit, an increasing role is played by flight simulator exercises. A good example of such activities are exercises in acquiring habits relating to pilot conduct in the event of an onboard emergency. By modern flight simulator we mean a flight training device characterized by a faithful simulation of both the internal (aircraft cockpit with instrumentation and indicators) and the external mission environment (everything that the pilot observes during the flight outside the cockpit), sound and air services, events changes that occur during aircraft position in three planes of reference - pitch, roll and yaw, simulation of mechanical and functional control system that allows the aircraft to acquire and consolidate skills to effective decision-making and the efficient and secure management of aircraft, especially in the situation of a heavy workload and with a time limit. Very important in terms of forming correct situational awareness habits is to enable the participation of the pilot in a training covering receiving and processing the data to be used in order to create a status image for particularly complex situations which may be encountered when executing a specific air mission.

Therefore, it should be considered particularly important that the engineer-designer should understand how big an impact on the effectiveness of the simulator is its degree of reality representation in relation to the type of aircraft operated. The simulator designer should then apply the technology which is the most useful from the point of view of the user to re-create the functional and physical characteristics of the aircraft and external mission environment.

Today's flight simulators, regardless of the nature and extent of their use in flight training, prove that their designers are well aware of the fact that a flight simulator is not just a machine equipped with modern computer software. Flight simulator designers also understand that in the design of the flight simulator particular attention should be given to adapting the newly created equipment to the capabilities and limitations of the pilot and his tasks conducted on a specific aircraft type in a particular mission environment.

Until the second half of the 1970s, flight simulators had often been perceived by the majority of the civil aviation community as a kind of attraction. Taking into account the economic factors and impracticality, many airlines and smaller aviation organizations had rejected the use of flight simulators for training purposes. However, with more and more evidence of the simulator training effectiveness, commercial aviation sector became increasingly interested in those devices. The final breakthrough came during the oil crisis in the 1970s, when flight simulators became a widely used training tool in commercial aviation. In fact, flight simulators were regarded as a low-cost training tool, used as an alternative to flight training. Over the years and with the development of aviation technology, flight simulators have become a commonly used training tool which is also used for basic flight training of young aviation adepts. Today, we can say that flight simulators are an essential element of pilot training and professional development, regardless of the kind of aviation and aircraft type taken into consideration.

However, due to the economic aspects related to the purchase and operation of a flight simulator, there are still some restrictions on its use, particularly in relation to *general aviation* where trainer aircraft still remain a major tool used in instruction and training[13]. The exception is the PPL (Private Pilot Licence) training, where simulator training (simulator or FNPT BITD Class)[14] may be substituted for 5 hours of practical training, and procedural training for IFR flights, where 50 hours of flight training may be conducted as: 35 hours of practical Simulator training (FNPT II class simulator) and 15 hours of training on a trainer aircraft[15]. To sum up, in accordance with the current regulations, 20% of the total practical air training time of the pilot trained for CPL licence and IFR rating, can be assigned to simulator training. However, it is not only because of regulations that, despite its number of advantages, simulator training, is rarely used in daily flight training. The second limitation is the fact that the vast majority of training can be conducted using the FNPT-II class simulator, whose purchase price is approx. € 400,000. In addition, the decreasing number of air training candidates each year makes the purchase, operation and maintenance of such equipment in general aviation become increasingly irrational in economic terms. Because of this barrier, only some Flight Training Organizations in Poland possessed FNPT-II class, and few centers BITD class simulators. Bearing in mind that as of 1st July, 2015, there were more than 50 accredited Flight Training organizations, the above-mentioned number of flight simulators seems to be very modest. On the other hand, the results of comparative analysis of the training costs for IFR flights with and without the simulator use shows that the first variant is cheaper by more than 50%[16].

The main argument in favor of a wider use of simulators in flight training are the results of studies related to the transfer of knowledge and skills from the flight simulator to the aircraft with regard to basic flight training. Studies carried out so far clearly confirm the usefulness of flight simulators, including the simplest (BITD), in the basic aviation training process. Lintern et.al. (Linter, 1990) in their research studies analyzed the degree of transferring the landing skills from the simulator training to the trainer aircraft at the beginning of the pilot flight training. The first group of trained pilots held two simulator training sessions covering the landing skills prior to the commencement of practical training in the air. The second group – the controls – did not receive this kind of training before the start of practical training in the air. The investigators demonstrated that the first group of trainees needed approx. 1.5 hours of training flights less before the first solo flight, as compared with the second group, which had not undergone the simulator training. In this case, we refer to only one, particularly important and difficult element of air training – learning to land (Macharella, 2005).

Dennis and Harris (Harris, 1998) examined the impact of the simulator training on the practical flight training using Microsoft Flight Simulator (v4.0) – a PC computer-based game program. Each trainee performed straight and level flight and coordinated turns for 1 hour. Then, the trainees commenced practical training in the air. The results of the comparative analyses concerning the performance demonstrated by the trainees who had received a simulator training session with those who had not explicitly confirmed that it was easier for the first group to assimilate the elements of the exercises which had been carried out before on the flight simulator. Furthermore, these studies confirmed the view that simple PC simulators are an effective tool in the development of flight skills in the initial phase of flight training.

One of the ways to conduct research on how human beings function in a task environment which is dangerous because of its complexity and the rapid, and frequently unpredictable, changes is the use of computer simulation techniques. A wide range of applications of flight simulators which are based on the state-of-the-art computer technology is what causes aviation training organizations to increasingly reach for this class of simulators currently treated as an additional, non-certified training tool. Further development of computer graphics technology or virtual reality means that such devices may be treated in certain flight training areas in the same way as those used today, which are more expensive. Although the main drawback of simulation devices is the fact that while reconstructing the actual operational emotional stress, their level of impact on the emotional side of the human is lower than desired, because of their sufficiently high training effectiveness they could still become a valuable training tool.

The primary objective of the study was to determine whether the PCTAD flight simulator is an effective training tool which allows an airplane/helicopter pilot to achieve the IFR competency at the desired safety level by adhering to the prescribed flight parameters as required.

In order to achieve this goal, quasi-transfer studies were carried out using the "Elitte" air simulator of the PCATD class, and a specially developed simulator training scenario, which included supplementary academic instruction in IFR training modules described in Part-FCL, and practical simulator training. To be eligible for the simulator training the trainee had to pass the theory examination at the conclusion of the academic instruction, and to achieve the test score no lower than 75%. The simulator of the class mentioned above was used both in performing tasks in the subsequent phases of the simulator training – 3 phases were implemented – and in measuring the skill transfer level at the end of each training phase by conducting examination flights.

The study involved 38 pilots, including 16 airplane and 22 helicopter pilots, who participated in its subsequent stages. The participants were from 27 to 53 years old and presented a total flying time of 200-2700 hours. This approach allowed us to determine the suitability of the PCTAD simulator for training, irrespective of the total flying time of the training participants.

The pilots participating in the study were divided into four training groups of up to 10 pilots each and performed the full range of training presented in detail in the next section of the article. Prior to the commencement of the training, all participants received training materials and instructions relating to the course of the research, a description of the simulator, and a detailed description of the exercise performed including the flight parameters. An individual simulator training session lasted 45 minutes. Every day, each pilot-trainee performed a flight along a new trajectory and trained subsequent maneuvers corresponding to those performed during the trial flights and the examination flights. These were performed upon the completion of the first, second and third stages of practical training respectively. All scores were collected and processed. Both academic instruction as well as each simulator flight session concluded with an individual debriefing with the instructors. The evaluation of selected flight parameters at each IFR flight phase was performed using a five-step rating scale, i.e. 1 to 5, where 5 was the maximum score (Table 1).

This approach allowed the assessment of individual trial and examination flights by using the same rating scale for each flight parameter taken into account, thus allowing the comparison of the results obtained by individual trainees and selected groups of them, with regard to e.g. the criterion of having a specific total flying time, as well as the evaluation of the training effectiveness for the entire group.

Using such a method of evaluating the trainees' performance with regard to adhering to predetermined flight parameters, the rate of simulator training efficiency was determined with that respect by formulating the following mathematical relationship:

$$\Delta = \frac{p_{AFE} - p_{APE}}{p_{APE}} \cdot 100\% \tag{1}$$

where:

Δ - percentage coefficient of simulator training efficiency[17];

p_{AFE} - parameter adherence quality index of a pilot (e.g. A – airplane, H – helicopter, F – the entire training group) obtained on the basis of the final practical examination(FA – final exam) on PCATD;

p_{APE} – parameter adherence quality coefficient of a pilot (e.g. A – airplane, H – helicopter, F – the entire training group) obtained on the basis of the first practical trial flight (PE – pre-exam) on PCATD.

Table 2. Rating scale of flight parameter adherence on PCTAD flight simulator with an example of a selected flight parameter – altitude – evaluation

No.	Numerical rating.	Verbal rating	Altitude
1.	1	Fail	± 250ft
2.	2	Sufficient	±200 ft
3.	3	Satisfactory	±150ft
4.	4	Good	±100ft
5.	5	Excellent	±50ft

The theory examination was graded on a percentage basis, i.e. 75% of the correct answers was equivalent to a positive examination result, which with ten open questions and ten single choice questions meant that the respondent was allowed to provide wrong answers to max. 5 questions.

In the initial phase of preparation for the research study, the team of simulator instructors developed a scenario covering theoretical and practical parts, including simulator and examination flights on the PCATD – Elitte simulator (Figure 11). It was assumed that each of the trainees would perform 45 minutes of simulator flight daily. Every day, the pilots improved various parts of the procedure resulting from the scenario, take-off procedures, Standard Instrument Departure, and ILS approach.

The academic part of the training covered six hours of IFR instruction modules according to the EASA regulations and concluded with a 30-minute written examination being a test, which consisted of 15 single choice and five open questions. The practical part of the training consisted of three training phases of IFR flights performed on the simulator. In total, each of the participants conducted 12 hours of simulator flights and took three half-hour examination sessions at the end of each training phase (Table 3).

The following systematic plan of a simulator IFR flight was created for Warsaw Okęcie airport for the purposes of the research presented in this article. (Figure 12) It also be should be stressed that since the pilots involved in the experiment were both helicopter (H) and airplane pilots (A), the prescribed flight parameters partly differed from one another, i.e the en-route altitude while completing the task: H= 4000ft(A)/3000ft(H); bank in the turn ß=25° (A)/15°(H), rate of climb – VVI= 150ft/min.; flight velocity was V=120kt (A)/90kt (H). The flight procedure itself was as follows: After the take-off (1) the pilots climbed on magnetic heading (HDG) 328° to H=4000ft/3000ft. Upon reaching the distance 2Nm DME – (Distance Measuring Equipment (2 DME)) from Okęcie OKC VOR and Hmin=2000ft, they performed a left turn to magnetic heading (HDG) 180° (2). They continued on that heading until reaching the lead point (LP) to intercept the 240° radial from OKC VOR (outbound). Upon reaching the LP they performed a right turn with ß=25(A)/15 (H) to course (CRS) 240° (4), maintaining the prescribed final altitude – H= 4000ft(A)/3000ft(H). Upon reaching the distance of 20 DME (A)/15 DME (H), the pilots made a right turn with ß=25(A)/15(H) – (5) to magnetic heading 90° in order to intercept CRS 60° (inbound OKC VOR). Then, they continued flying CRS 60° (6), and after reaching the distance of

Figure 11. PCATD – Elitte simulator used in the research studies
Source: (PCATD, 2018)

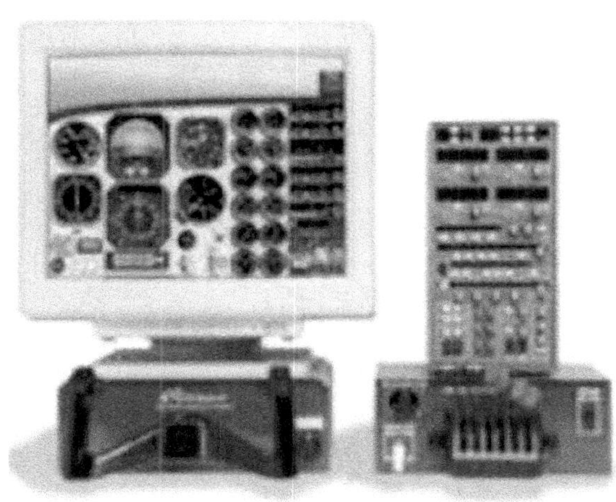

Table 3. Empirical research schedule of IFR flights using the PCATD simulator

No.	Training stages/ training phase	Duration (clock hours)
	I. ACADEMIC INSTRUCTION – 6.5 hours	
1.	1. Refresher academic instruction in selected IFR modules.	6.0
	Theory examination.*	0.5
	II. SIMULATOR TRAINING – 13.5 hours	
2.	1. Simulator training phase – practical familiarization with PCATD and scenario.	4.0
	1a.Practical pre-exam.	0.5
3.	2. Simulator training phase – advanced training	4.0
	2a. Mid-course practical pre-exam.	0.5
4.	3. Simulator training phase – final advanced training	4.0
	3a. Final practical exam.	0.5
Total hours of academic instruction and simulator training.		20

13(A)/11(H) DME, made a right turn to intercept ARC 12(A)/10(H) DME (7). Subsequently they intercepted the ILS to RWY33 (8). Pilots performed the flight under IFR conditions, paying special attention to combined indicator – LOC (localizer – indicates horizontal deviation of the aircraft from the runway heading) and G/S (glideslope indicates the vertical deviation of the aircraft from its optimum path of descent), and, subsequently, they performed the final approach with the predetermined decision height of 200 feet AGL (above ground level).

During the research study, the trainees' adherence to the prescribed simulator flight parameters (Table 4) was evaluated on the basis of the schematic shown in Figure 2.

According to the assumptions of the training scenario, in order to qualify to the subsequent study phase, the participants had to pass a theory examination at minimum 75%, i.e. provide the right answers to 15 out of the total of 20 questions. All participants received the score above the threshold of 75% (Table 5).

Figure 12. Schematic of a planned simulator flight trajectory according to the IFR procedure for Warsaw-Okęcie Airport for research purposes

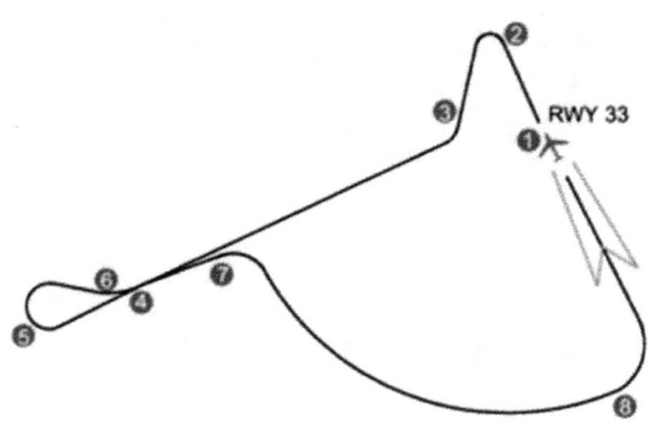

Table 4. Elements evaluated in specific phases of examination flights

Flight phase

CLIMB				LEVEL				DESCEND			ILS APP				ARC
Parameter evaluated															
Dist	V	β	VVI	V	Alt	β	CRS	V	VVI	β	LOC	GS	V	DA	Dist.

Notice: Dist. – distance, V – velocity, ß – bank,
HDG- heading, VVI – wertical speed, LOC – localizer,
GS – glide slope, DA – decision altitude.

Table 5. Participants' scores of the theory examination

No.	Rating scale (%)	Airplane pilots (number)	Helicopter pilots (number)
1.	75-80	6	5
2.	81-90	2	7
3.	91-100	8	9
4.	Average score	88%	89%
5.	A+H average	88.5%	

Analyses of the theory examination results presented above indicate that the academic knowledge presented by the participants was relatively high which is reflected by the average score of 88.5%. This level of knowledge certainly had an impact on the final training results.

The first examination evaluating the correctness of the scenario implementation in practice was conducted after five 45-minute simulation training flight sessions, each of which was part of the procedure described in the previous chapter. Each simulator flight session concluded with an individual debriefing with the instructor. Taking into account the objective of the study, the results achieved by the pilots in subsequent examinations were analyzed from the perspective of the surveyed population of airplane and helicopter pilots, and the test group as a whole (Figure 13).

The mean value of flight parameter adherence quality index was pA = 4,01 for airplane pilots whereas its mean value for helicopter pilots was pH = 3.50. Thus, the difference between the flight parameter adherence quality indices for airplane and helicopter pilots was Δpc = 0.51. The results achieved by the pilots tested during the first trial simulator flight point out to significant differences between the airplane and helicopter pilots in the correctness of the adherence to flight parameters at various stages of the test exercise. These differences are definitely favor of airplane pilots and are in the range of:

- $0.5 > \Delta p_c$ for parameters such as: V, β – level flight, V, – descent;
- $\Delta p_c \geq 0.5$ for adhering to parameters such as: Dist., V – climb; CRS – kl; VVI, β – descent; LOC, GS, V, DA, and Dist – ILS landing approach.

As far as the remaining flight parameters taken into account in the evaluation of the pilots are concerned, the differences of the β and VVI parameter values determined for the climb phase were in the range of one-hundredth of the Δpc parameter.

Figure 13. Mean values of adherence quality indices to individual flight parameters by airplane (blue) and helicopter pilots (red) ˜– first trial flight after 4.0 hours of simulator flights

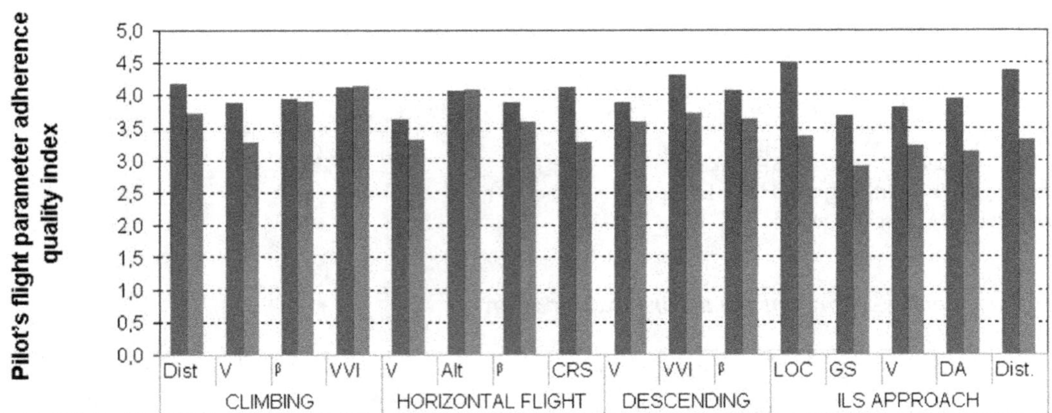

	CLIMBING				HORIZONTAL FLIGHT				DESCENDING			ILS APPROACH				
	Dist	V	β	VVI	V	Alt	β	CRS	V	VVI	β	LOC	GS	V	DA	Dist.
■ Airplanes - A	4,18	3,87	3,93	4,12	3,62	4,06	3,87	4,12	3,87	4,31	4,06	4,50	3,68	3,81	3,93	4,37
■ Helicopters - H	3,72	3,27	3,90	4,13	3,31	4,09	3,59	3,27	3,59	3,72	3,63	3,36	2,90	3,22	3,13	3,31

**For a more accurate representation see the electronic version.*

The reasons for such large differences between airplane and helicopter pilots in the quality of adhering to flight parameters are due, among others to:

- The type of aircraft used by the respondents – airplane or helicopter, and the associated need to operate a given aircraft under IFR conditions. In this case, airplanes are much more likely to perform air tasks under IFR conditions;
- Total flying time characterizing the participants – 100% airplane pilots had the total flying time of more than 200 hours whereas among the helicopter pilots 12 had the total flying time in the 120-200-hour range and the remaining 10 had flown more than 200 hours.

After a further 4 hours of in-flight training, during which the trainees improved their ability to maintain the desired level of flight parameter adherence quality, the next practical test was conducted using the same IFR flight scenario/procedure. The mean numerical value of the flight parameter adherence index presented by the pilots was pA = 4.52, i.e. 0.51 higher compared to the first test, whereas the same index for helicopter pilots was pH = 4.36 so in that case the increase was 0.86 in comparison to the first test. Such a significant increase in the flight parameter adherence quality resulted in a significant reduction of the difference of the flight parameter adherence index presented by both groups of pilots, airplane and helicopter. The difference between those indices decreased from $\Delta pc = 0.51$ to $\Delta pc = 0.16$.

Conclusions from the data presented in Figure 4 indicate that, with respect to 16 flight parameters under evaluation, helicopter pilots achieved the following increase in the quality of their adherence to flight parameters, in comparison to the first test (Figure 14): $pH \leq 0.5$ – 1 parameter, $0.5 < pH \leq 1$ – 11 parameters, and for $pH > 1$ – 4 parameters; whereas in the case of airplane pilots, the values of the respective above indices were – $pA \leq 0.5$ – 4 parameters; $0.5 < pA \leq 1$ – 10 parameters, and for $pA > 1$ – 0 parameters. On the other hand, with respect to the VVI-parameter – climb phase – the pA dropped below zero, i.e. to - 0.25, which should be regarded as a particularly unfavorable phenomenon.

Figure 14. Mean values of adherence individual flight parameter quality indices for airplane (blue) and helicopter pilots (red) – second trial flight after 8.5 hours of simulator flights

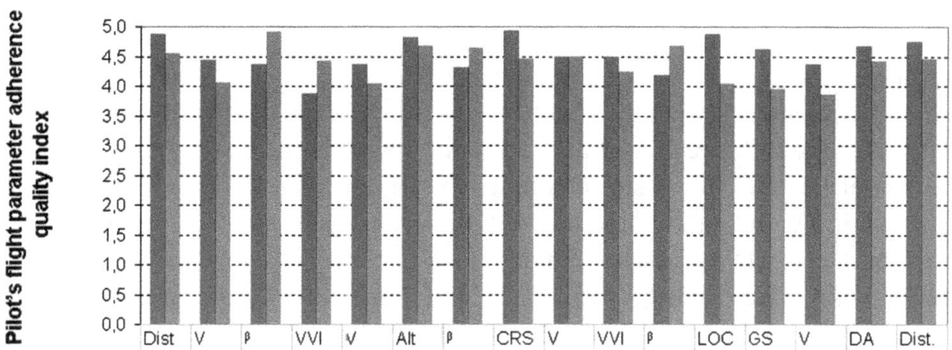

	CLIMBING				HORIZONTAL FLIGHT				DESCENDING			ILS APPROACH				
	Dist	V	β	VVI	V	Alt	β	CRS	V	VVI	β	LOC	GS	V	DA	Dist.
Airplanes - A	4,87	4,44	4,37	3,87	4,37	4,81	4,31	4,93	4,50	4,50	4,18	4,87	4,62	4,37	4,68	4,75
Helicopters - H	4,55	4,05	4,91	4,41	4,04	4,68	4,64	4,45	4,50	4,23	4,68	4,04	3,95	3,86	4,41	4,45

**For a more accurate representation see the electronic version.*

Figure 15. Mean values of adherence individual flight parameter quality indices for airplane (blue) and helicopter pilots (red) – second trial flight after 13 hours of simulator flights

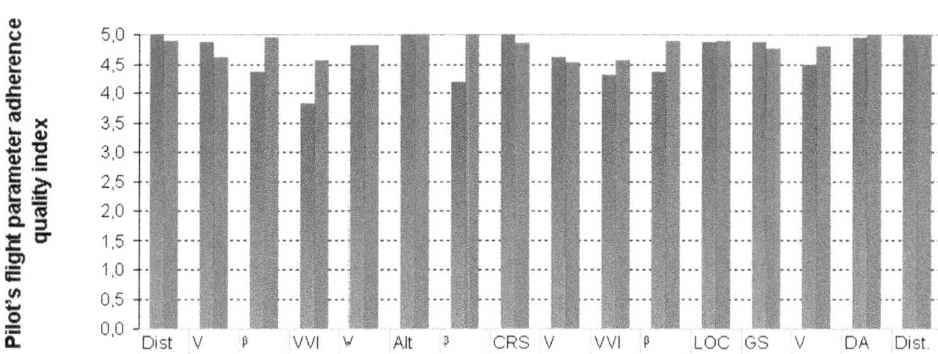

	CLIMBING				HORIZONTAL FLIGHT				DESCENDING			ILS APPROACH				
	Dist	V	β	VVI	V	Alt	β	CRS	V	VVI	β	LOC	GS	V	DA	Dist.
Airplanes - A	5,00	4,87	4,37	3,81	4,81	5,00	4,19	5,00	4,62	4,31	4,37	4,87	4,87	4,50	4,94	5,00
Helicopters - H	4,90	4,62	4,95	4,57	4,81	5,00	5,00	4,86	4,52	4,57	4,90	4,90	4,76	4,80	5,00	5,00

**For a more accurate representation see the electronic version.*

To sum up, despite the fact that airplane pilots had previously executed a relatively higher number of air tasks, which is reflected by their greater total flying time, helicopter pilots relatively quickly began to "catch up" on their competence deficiencies with regard to the quality of adherence to individual flight parameters.

Upon the completion the last stage of the training flights – 4 hours of simulator flights – the participants were given the final test of the course, following the same scenario that was used in the first two tests. The analysis of the exam data in individual groups – airplane pilots and helicopter pilots – indicates the continuation of the trend observed in the analysis of the second exam results i.e. a further increase in the parameter adherence quality indices for helicopter pilots (Figure 15).

The mean numerical value of the flight parameter adherence quality index for airplane pilots was pA = 4.65, i.e. it was by 0.64 better in comparison with the first examination. The same index for helicopter pilots, however, was pH = 4,82 which points out to the increase by 1.32 comparing with the first examination. Such a significant increase of the flight parameter adherence quality index in helicopter pilots caused the reversal of the trend and the reaching of the index value Δpc = 0.17 in favor of helicopter pilots.

Using the formula for calculating the simulator training efficiency ratio, the relevant analyzes were made taking the data from Figure 14 and Figure 16. As a result of these activities, the average PCATD simulator training efficiency ratios were determined for the airplane pilot group, helicopter pilot group, and the entire training group, taking into account the mean values of the flight parameter adherence quality indices for the respective groups (Figure 6).

The values of those indices were ΔA=15.8%, ΔH= 38%, Δc=26.9% respectively. Training efficiency of 15% for airplane pilots or 38% of helicopter pilots fully confirms the usefulness of PCATD simulators in daily air training activities.

The results of the research study have allowed its assumed objective to be achieved. The obtained training efficiency indices of ΔA = 15.8% for airplane pilots, ΔH = 38% for helicopter pilots, and the average efficiency index for the entire group ΔA = 15.8%, ΔH = 38%, Δc = 26.9% definitely confirmed that a PCTAD class flight simulator is an effective training tool to be used by airplane/helicopter pilots for achieving and maintaining an appropriate level of competence necessary for conducting IFR flights at the desired safety level by adhering to the prescribed flight parameters at the required degree. Moreover,

Figure 16. Average percentage increase (decrease) of the flight parameter adherence quality indices of airplane pilots (blue line), helicopter pilots (red line), and the entire experimental group (brown line), based on the comparative analysis of the above mentioned indices achieved in the first and the final examinations

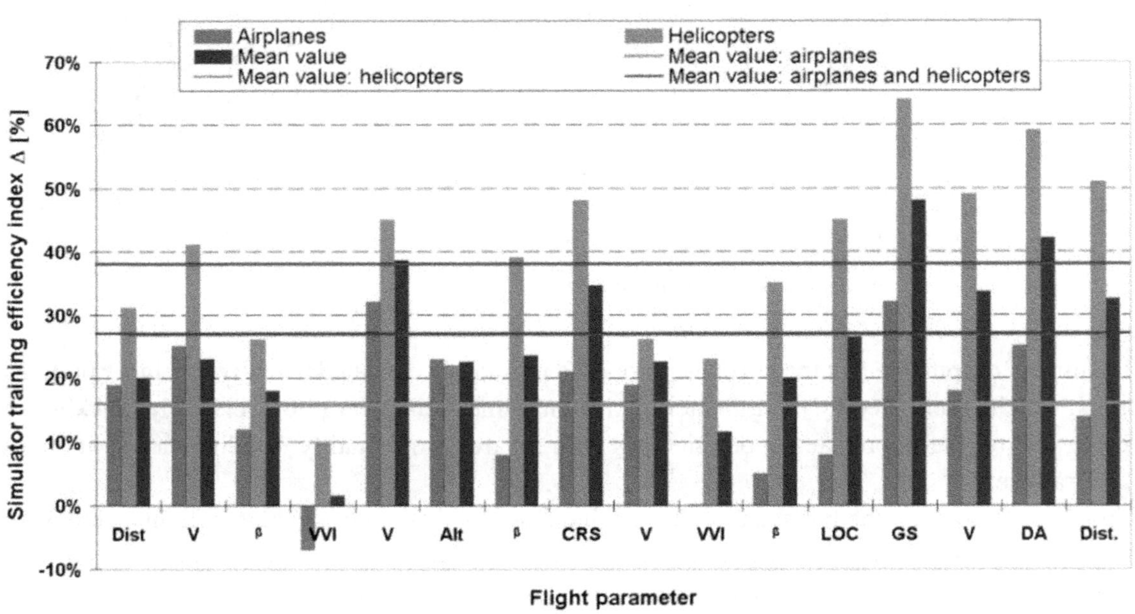

**For a more accurate representation see the electronic version.*

the results achieved by the groups tested – airplane and helicopter pilots – show that this tool is effective not only for pilots having the total flying time of more than 200 hours, but also for those, whose total flying time is significantly lower, in the 130-hour range. To sum up, PCTAD simulators can be a viable alternative for small aviation training centers that want to implement flight simulator training but cannot afford to use the FTD class for financial reasons.

Undoubtedly, PCTAD class simulators can meet the needs of small centers of basic aviation training, with regard to procedural training. This view is supported by both the low purchase costs (in Poland it is approximately € 5,000), as well as operating costs which are similar to the operating costs of a high-end computer. The essential requirement, however, for the admission of this class of simulators to training within the framework of the IFR program is the certification of this type of equipment by national or international aviation authorities. Another condition that would have to be fulfilled is a change in international and, what it involves, national regulations concerning the allowing this simulator category to be used in the pilot training process.

The results of the research studies prompted the authors to continue their work in the years 2017-2018 with further groups of airplane and helicopter pilots. The objective of these studies will be to identify changes to the scenario that would improve the pilot's ability to adhere to some of the qualitative values of flight parameters such as e.g.VVI.

On the other hand, the results of the research study conducted by Vaden and Hall (Vaden, 2005) pointed to the fact that the platform imitating the motion of an aircraft does not serve an important role in the acquisition of skills by the pilot. Using the method of meta-analysis they examined the impact of the aircraft's plane of motion, which is an integral part of the aviation simulator, on the development of the pilot's skills. The study showed little effect of the use of such devices on increasing the flight training effects in cargo aircraft pilots. They also pointed to the fact that the cost of upgrading the simulator with this feature, as well as additional physical load experienced by the pilot during the exercises on a Full Flight type simulator, do not have a significant impact on reducing the number of hours of practical training carried out by the crew in the air. The authors do not comment on the impact of this training on enhancing the crew's skills in the use of the available resources during the flight.

The interviews conducted by the author with a group of instructors from flight training organizations in Poland indicate that FNPT-2 simulators provide highly effective training and selection, particularly in relation to the exercises concerning: IFR flights, emergency procedures, maneuvers performed by the aircraft during mission execution, and cockpit pre-flight check.

Given the above findings, it is possible to formulate a thesis that modern flight simulators ensure a high degree of basic flight training efficiency[18].

Commercial aviation organizations following the current regulations conduct the simulator training within the scope of training for certain aircraft type, MCC, CRM, and mandatory periodic training. The main document regulating the air operator's duties regarding the preparation and professional development training of the crews belonging to commercial aviation organizations is OPS-1, and Part-FCL1 in relation to the training for particular aircraft type. As is the case with general aviation, each flight simulator or air training equipment item replacing the aircraft for training and testing purposes must be certified in accordance with the requirements applicable to air training equipment[19]. Under the current regulations, any member of the flight crew prior to commencing flight operations below the standard set for Cat. I, operations outside the standard of Cat. II and operations within Category II and III (Compa, 2011) [20] must complete the training courses and tests which use flight simulators. These courses include reduced visibility operations, including multi crew cooperation. It is emphasized that MCC and CRM trainings ought to be conducted with the maximum use of flight simulators. In the case of aircraft for which the flight simulator has not been produced, the operator must ensure that the phase of simulator training specific to the visual scenarios of Category II operations is conducted in a specifically approved flight simulator. Moreover, each crew member shall undergo periodic training and testing appropriate to the type or variant of aircraft in which he operates flights. Those trainings are usually conducted using the flight simulator which corresponds to a given aircraft type. As part of ongoing operations the operator must ensure that:

- The pilot is not assigned to operate an aircraft as a member of the minimum approved crew, either as pilot flying or pilot non-flying, if during the past 90 days he has not performed three take-offs and three landings as pilot flying in an aircraft or in a flight simulator of the same type/class;
- The pilot who does not have a valid IFR rating is not assigned to perform night flights as a crew commander, if during the past 90 days he has not performed at least one landing at night, as pilot flying in an aircraft or in a flight simulator of the same type/class.
- Moreover, each operator is required to provide transition training involving such elements as:
- Ground training and tests concerning aircraft systems, routine, non-routine and emergency procedures;
- Trainings and tests in order to familiarize trainees with emergency and safety equipment, which must be completed before the commencement of the practical air training. It should be noted that an important element of these courses are exercises and tests performed with the use of flight simulators.

Importance attached to the simulator training of commercial aviation aircrews in the above-cited regulations as well as the results of analyzes relating to the opinions of pilots from commercial aviation organizations clearly show that simulator training is an essential element of professional development training in this type of organizations. Flight simulators used in the training described above facilitate, on the one hand, a continuous improvement of the flight crew skills, and, on the other hand, they are a valuable tool to determine the level of flight crew preparedness for the tasks facing them. Moreover, taking into consideration the scale and frequency of training it is difficult to imagine a commercial aviation organization capable of affording to conduct the training in full using only the aircraft.

In Military aviation, which is less dependent on the financial factor, flight simulators are widely used in daily operations and training relating to the acquiring/maintaining pilot currency. Flight simulators used in military aviation can be divided into two main groups, according to the tasks performed with them:

1. Simulators for schools and military pilot training organizations. Their use is associated with familiarization of beginner pilots with the aircraft, its handling, navigation and basic combat missions. They are used for the selection and training of pilot candidates, advanced flight training, training of all flight phases, assessing the implementation of specific air missions, and crew training and coordination within the scope of MCC and CRM training. The syllabus of pilot training for PAFA officer cadets includes 120 hours of specialized simulator training during the four years of study. The situation is somewhat different in the case of training/professional development of jet pilots in the units of the Polish Armed Forces. Simulators of this aircraft type alongside handling and navigation functions support the functions related to performing combat missions. The number of simulator training hours depends on the intensity and the type of tasks performed by the pilot.

Apart from the above-mentioned applications, flight simulators are treated in air force units as a valuable tool for preparing to resume pilot currency and skills in the event of long time break between flights.

2. Simulators of the Polish Armed Forces units, where the main emphasis is placed on improving aircraft handling skills associated with aerial combat techniques (Szczepański C. Symulatory lotnicze, stan i perspektywy – report, Warszawa, 1998). Individual training tasks carried out in the air are generally preceded by the training on a flight simulator which is an exact copy of the aircraft operated. As a rule, the trainee is allowed to conduct the tasks in the air after successfully conducting them on a simulator. Simulators are used for a variety of reasons, including those related to the low unit cost of simulator training hours compared to the cost of flight hours. U.S. Air Force estimates that the cost of one hour of training on the C-5 aircraft is approximately $ 10,000, while one hour of training on the C-5 simulator is $ 500 (Moorman, 2002). Especially with regard to military pilots and airline pilots the key advantage of simulators is the fact that they allow pilots to perform simulated flight tasks and train the behavior in non-routine situations, which if performed on the trainer aircraft would involve too high risk of undesirable flight-related events.

Another reason for the use of simulators in the Air Force is the knowledge transfer degree, which is comparable with that acquired during practical training in the air. The results of research concerning the aspects related to the effectiveness of simulator training clearly show that the knowledge, skills and habits acquired during such training exert a direct influence on the pilot's readiness to perform air missions in the air. R. Moorman et al. conducted research relating to knowledge transfer involving flight simulator. Their study covered the period 1957–1986 (Hays, 1992). The authors conducted a study using the meta-analysis method (Marlinussen, 2010) and the results allowed them to conclude that simulator training contributed to a permanent increase of the training effects in jet pilots, which are equal to those of practical training carried out on the specific aircraft type. However, this regularity was not confirmed for helicopter pilots.

In another study, Corveta and Dunlap (Carreta, 1997) focused the research on the effectiveness of simulator training involving jet pilots, with respect to such items as landing maneuver, IFR flights and tasks performed at the firing ground (aerial bombardment). In all of the above areas of research, the results clearly indicate a high degree of usefulness of flight simulators for pilots' skill development.

From the situational awareness perspective, flight simulator plays a particularly important role in military aviation. Despite the high level of automation of modern combat aircraft, because of the complex character of air mission environment – unknown flight zone, limited possibility of using radio navigation instruments, low and very high altitudes, high speeds and G-loads, unpredictable enemy, conducting missions in radio silence, often as a member or head of the combat group, the pilot conducts air missions under heavy psychophysical load. Therefore, the degree of pilot mastery concerning cockpit actions, or the degree of "automation" of these actions has a major impact on the pilot's ability to maintain the desired level of situational awareness at various stages of air mission execution. The results of the analysis of serious air incidents which concerned combat aircraft of the Polish Armed Forces in the years 1970–2004 show that the vast majority of them was caused by insufficient situational awareness of the pilot – approx. 70%.

The advantages of using flight simulators in daily professional development training of military pilots which have been presented in this chapter point to the fact that flight simulators are an effective tool for the development of the essential elements that affect the level of air mission execution safety tasks, including the desired degree of situational awareness. The confirmation of this thesis may be made by the recommendations of the UK Royal Air Force (RAF) following a large number of air accidents which resulted from the introduction of jet aircraft in 1960's. One of the key corrective actions aimed at preventing the similar events in the future was the decision to purchase flight simulators and intensify simulator training of jet pilots.

Regardless of the type of aviation and the stage of flight training/professional development, flight simulators are a widely used pilot training tool. The results of reference literature analysis (Hays, 1992), expert opinions, flight simulation studies, and the author's own air permit the conclusion that the essential benefits of the flight simulator, including those relevant to the pilot's situational awareness can include the following:

1. High training effectiveness. The tests related to measuring the effectiveness of flight simulation training demonstrated that the trainees develop knowledge and skills at a level similar to that achieved in a real flight. During simulator training, due to lack of necessity to physically control the aircraft, the instructor can fully focus on the student and the activities that the student performs. Modern simulators support the function of recording and collecting data from simulated air operations carried out during the training. This facilitates multiple analyses of the compatibility of flight parameters as defined in the exercise scenario with the values registered during the training flight.
2. Maintaining high standards of training safety. Due to the necessity of maintaining a certain level of safety during training, simulators sometimes provide the only way to learn some maneuvers, the elements of air operations performed in the event of dangerous weather conditions (wind shear, turbulence, icing, jet streams, etc.). Performing the above-mentioned maneuvers in a real flight would involve putting the trained crew at high risk. A similar situation occurs in the event of critical aircraft component failure, including, among other things, failures concerning the engine, flight instruments, or control systems, etc.
3. Availability. The fact that the use of the flight simulator in flight training is not dependent on the current ambient weather conditions, the state of the airport, or the good working condition of ground navigation equipment allows for a more methodical approach to the training from the human factor perspective. Moreover, it is possible to simulate non-routine in-flight situations without the necessity to wait for their occurrence in the real air mission execution environment.
4. Repeatability. The simulator does not require the implementation of the full cycle of a given air operation (pre-flight check, take-off, mission execution, landing, psot-flight check) to discuss and repeat a specific part of the exercise by the trainee. It facilitates the repetition of each flight element and making breaks for an exchange of views between the trainee and the instructor.
5. Predictability. Simulator training prevents the occurrence of such dangerous phenomena as air traffic collision, wind shear, icing, weather deterioration, turbulence, closure of airports, etc. Of course, all of these situations are possible to simulate, but the exercise of this kind are carried out by the trainee at the time specified by the instructor, according to the trainee's progress made in the training.
6. "Learning from mistakes" – with unlimited possibilities of applying a number of ways to solve an in-flight operational problem with "zero" risk level, the simulator allows the student to select the solution which is the most optimal from his perspective and to test it. This capability allows the trainee to physically "prove to himself" that some solutions considered by him to be more useful

and safer than those recommended by the instructor or the instruction manuals are not a good option in operational reality and, consequently, they are eliminated by the trainee once and for all.

7. Maintaining pilot currency and proper attention allocation. Systematic training exercises on the flight simulator allows the trainee to consolidate the desired habits (maintain pilot currency) and, what is equally important, allows him to permanently master the optimal attention allocation in various phases of the air mission execution. It allows the beginner pilot, on the other hand, to adequately master the basic elements of cockpit preflight check and such actions as starting up the engine and other aircraft equipment. This allows both to save the time and reduce the costs of practical training.
8. Credibility as a tool applied in air accident investigation. Flight simulators ensure accurate reconstruction of flight conditions, the situation onboard, and the evaluation of the actions taken by the crew of the aircraft in the event of an undesirable flight-related event. The conclusions of these analyses and corrective actions taken in the aspects related to the event – the pilot-operator, the aviation organization, the environment, the aircraft allow to avoid or safely respond to undesirable flight-related events in the future.
9. Simulators can be used for training for the prototype aircraft under design, or aircraft employing new solutions (systems). Performing this task in practice, without the prior simulation and practice of selected flight maneuvers could be associated with a high level of risk and, in extreme situations, with high probability of undesirable flight-related event.
10. High comfort of performing the training tasks by the instructor. An undoubted advantage of the flight simulator in comparison with the aircraft is the fact that the instructor can focus his attention fully on the trainee and the task performed. Equipping the simulator with a module recording the operating parameters, including the correspondence between the instructor and the trainee, allows for deep analysis of multiple elements of the flight in order to draw the appropriate conclusions and facilitates methodological development of corrective actions to be taken in relation to the trainee[21]. In contrast to a training flight conducted in the real mission environment, the instructor may but does not have to immediately correct to the trainee's errors. The instructor's actions may only amount to monitoring the situation on board the aircraft in a particular mission execution environment, and assessing the trainee's response time to committed errors. Such a situation is favorable in that the trainee can fully see the effects of his improper actions on board the aircraft. In real flights allowing such a situation could lead to an excessive risk of an undesirable flight-related event.
11. Optimizing the use of financial resources. The use of the simulator can significantly reduce the training costs as compared to those incurred for the use of the aircraft. Furthermore, simulator training implemented instead of the training on the aircraft allows for significant savings resulting from the extended service life of the aircraft. Taking into account the detailed analysis of the cost vs. effectiveness relationship, Orlansky and String proved that the costs of applying flight simulators in the training of military flight personnel account for 5%–20% of the cost of the same operation performed on trainer aircraft. This allows for the reduction of the flight training cost on average by approximately 12%. They also found that commercial aviation organizations can expect a return of the total purchase price of the simulator within approximately 9 months, and the return of the costs of the necessary training facilities within about two years.
12. The use of simulators significantly reduces the number of hours flown by the trainee, which is beneficial for aircraft operating costs, and the costs resulting from the use of airspace and airports, both controlled and uncontrolled.

Despite the advantages mentioned above, no simulator can currently be defined as a device that can replace the hands-on training in the air. It is still considered as a very important form of preparing or supplementing practical training in the air.

The use of flight training simulator, like any other training device, has a number of disadvantages, which can include:

1. Purchase price of the simulator. The most important barrier resulting in a limited number of simulators available in our country is their purchase price.
2. Insufficient levels of environmental impact on the trainee which is present in real air operations. In a simulated emergency (dangerous weather phenomenon, onboard system failure), the trainee does not experience the same degree of stress as in the case of the same situation enocuntered in the air. However, mastering the proper execution of actions related to a specific emergency situation using the simulator makes it easier to take appropriate action in the event of encountering a similar phenomenon in the air. On the other hand, because of too strict adherence to procedures and not allowing the trainee to experiment and learn from mistakes in preventing dangerous situations during simulator training, the trainee is not able to move beyond the learned schema, and thus is often not able to properly respond to a non-routine situation in the air. This usually leads to an undesirable flight-related event. The proof of that is provided by the results of studies conducted by a NASA research team who used the data from the Aviation Safety Reporting System (ASRS) (Kozuba, 2014).
3. Shorter flight duration in comparison with real flights. Too frequently, selective treatment of certain flight elements causes the effect of mental fatigue and weariness to occur during simulator training with less severity than in a real flight. As a result, it is difficult to assess the resistance of the trainee to weariness or mental fatigue usually present in long lasting flights.
4. No faithful reproduction of "radio traffic reality" of the real mission environment. While conducting simulator training, instructors often place less importance on the so-called "radio traffic reality". Because of that, the trainees with less operational experience have difficulty in understanding the radiotelephony communication and performing air operations in the "congested" areas of controlled aerodromes.

Currently, flight simulators, in addition to aircraft are considered as one of the essential tools for selection, training, and developing/maintaining currency of flight crews. They are also viewed as a reliable tool for testing pilot knowledge and skills in specific aspects of aircraft operation and air mission execution. Because of the above-mentioned advantages of these devices, simulator training, in addition to theoretical and practical training in the air, plays a crucial role in shaping the competencies needed to achieve the desired level of situational awareness by the pilot at various stages of air mission execution. Therefore, the use of these devices for forming and consolidating the pilot capabilities necessary, among others, for the efficient execution of the situational awareness process should be considered as justified.

CONCLUSION AND FUTURE RESEARCH DIRECTIONS

Flight simulators in the space of aviation development have become a valuable tool for the implementation of a number of functions related to aviation activities, with a special emphasis on flight training.

Moreover, systematic introduction of modern aviation technology has led to the rapid development of training tools commonly used in pilot training and in both simulator and practical air training. At present, it is difficult to imagine the training for aviation personnel without this type of training tools, with e special usage of simulators. Despite the fact that simulators are usually expensive, their advantages have caused them to become common training tools. These devices are particularly useful in enhancing the pilot (aircraft crew)-machine (aircraft) cooperation. The relatively low cost of reconfiguring simulation devices causes that they faithfully imitate the cockpits of aircraft used in air operations, and thus allow the pilot, aircraft mechanic, or air traffic controller to achieve a high level of adaptation to acting in a new mission environment task, which is particularly important from the perspective of the complexity and high dynamics of the mission environment variation affecting the above-mentioned air personnel. These activities are additionally supported by the utilization of modern technologies (e.g. control automation, glass-cockpits, etc.) in all aircraft types used at all stages of flight training. Because of all these advantages, aircraft manufacturers, more and more frequently offer comprehensive training systems for pilots (M-346), aircraft mechanics or Air Traffic Controlers, at the same time not forgetting the simplest training devices from today's perspective – PCATD.

Moreover, it is worth to underline that simulators done in technology of an artificial reality are going to be the new way for relatively low cost tool of training for aviation personel. Despite the very strict aviation regulation in the area of using simulators, there is no any restriction in using non certified simulator as a tool for improving the quality of basic and annual training of the in-service training of the aviation personel. Because of that, there is no any obsatcles to state that simulators are an essential tool for shaping the needed competence by the aviation personnel.

REFERENCES

A History of Simulation. (2018). Available at: https://www.historyofsimulation.com/flight-simulators-and-the-beginning-of-the-jet-age/

Aerotoons. (2018). Available at: http://aerotoons.com

Aircrew Training Devices. (2018). Available at: http://www.wildfire-productions.com/atd/

Akceptowalne sposoby potwierdzania spełnienia wymagań (AMC) oraz materiały zawierające wytyczne (GM). *(2012).* Available at: http://edziennik.ulc.gov.pl/api/DU_ULC/2013/92/oryginal/Zalacznik2.pdf (in Polish)

Bezdek, W., Powell, R., & Mays, D. (2004). The history and future of military flights simulators, AIAA modeling and simualation technologies conference and exhibit, *Providence, RI. AIAA Paper, 2004*, 5148.

Carreta, T. R. (1997). Transfer of training effectiveness in flight simulation: 1986 –1997 (technical report AFRL-HE-AZ-TR-1998-0078). Mesa, AZ: U.S. Air Force Research Laboratory, Human Effectiveness Directorate.

Clare, C., & Loucopoulos, P. (1987). *Business Information Systems. Pardigm.* London: Alfred Waller Ltd.

Compa, T., & Rajchel, J. (2012). *Podstawy nawigacji lotniczej, WSOSP.* Dęblin. (in Polish)

CS-FSTD (As) - Certification Specifications for Aeroplane Flight Simulation Training Devices – Annex to ED Decision N° 2012/010/Directorater of the Executive Director of the Aency of 4 th july 2012 on the Certification specifications for aeroplane flight simulation training devices.

Daft, R. L. (1992). *Organization Theory and Design*. St.Paul: West Publishing Company.

Electronic Flight Simulation. (2018). Available at: https://www.simulationinformation.com/education/electronic-flight-simulation

Ground based training system (GBTS). (2018). Available at: http://www.leonardocompany.com/en/-/gbts

Harris, K. A., & Harris, D. (1998). Computer-based simulation as an adjunct to ab initio flight training. *The International Journal of Aviation Psychology*, *8*(3), 261–276. doi:10.120715327108ijap0803_6

Hays, R. T., Jacobs, J., Prince, C., & Salas, E. (1992). Flight simulator training effectiveness. A-meta-analysis. *Military Psychology*, *4*(2), 63–74. doi:10.120715327876mp0402_1

Jones, E. R. (1967). *Simulation applied to education. Mc Donnell-Douglas Corporation paper*. St. Louis, MO: Mc Donnell-Douglas Corporation.

Kearns, S. K. (2010). *E-learning in Aviation*. Farnham, MA: ASHGATE.

Klich, E., & Szczygieł, J. (2010). *Bezpieczeństwo lotów w transporcie lotniczym*. Radom: PIB. (in Polish)

Klukowski K. (Ed.). (2005). *Medycyna wypadków w transporcie*. Wyd. Lek. PZWL. (in Polish)

Kozuba, J. (2011). Impact of human factor on likehood of aircraft accident. *Transport Systems Telematics, TST-11*, 29–36.

Kozuba, J. (2014). *Selected aspect of the pilot situational awareness*. PAFA, Dęblin.

Kozuba, J., & Kowalczyk, P. (2012). Risk management – a crucial safety element in aviation. *Vojenske Reflexie Rocnik*, *7*, 2.

Lee, A. T. (2005). *Flight Simulation*. Burlington: ASHGATE.

Leszczyński, R. (2000). *System selekcji kandydatów do Sił Powietrznych, Praca badawcza*. WSOSP, Dęblin. (in Polish)

Link Flight Simulator. (2018). Available at: http://www.ctandi.org/link-flight-simulator/

Lintern, G., Roscoe, S. N., Koncoe, J. M., & Segal, L. D. (1990). Transfer of landing skills i beginning flight training. *Human Factors*, *32*(3), 319–327. doi:10.1177/001872089003200305

Łomow, B. F., & Płatonow, K. K. (1984). *Eksperymentalna psychologia lotnicza*. Warszawa: PWN. (in Polish)

Macharella, N. D., Arban, P. K., & Doherty, S. M. (2005). Transfer of training from flight training devices to flight for Ab-Initio pilots. *International Journal of Applied Aviation Studies*, *6*(2), 299–314.

Marlinussen, M., & Hunter, D. R. (2010). *Aviation psychology and human factors*. New York: CRC Press.

Mechaniczne symulatory lotu. Jak uczono pilotażu, gdy nie było komputerów? (2018). Available at: https://gadzetomania.pl/3742,mechaniczne-symulatory-lotu-jak-uczono-pilotazu-gdy-nie-bylo-komputerow (in Polish)

Moorman, T.R. (2002). The civilian looping military. *MS&T Magazine, 6.*

Morawski, J. M. (2005). *Human and technology*. Pułtusk: OW Aspra-JR.

Naziemny system szkoleniowy (GBTS). (2018). Available at: http://www.m-346.com/m-346/szkolenie-pilotow/system-gbts (in Polish)

Orlansky, J., & String, J. (1977). *Cost-effectiveness of flight simulator for military training (Rep. No. IDA NO. HQ 77-19470).* Arlington, VA: Institute for Defense Analysis.

Part – FCL – Załącznik nr I do - Rozporządzenie Komisji (UE) NR 1178/2011 z dnia 3 listopada 2011 r. ustanawiające wymagania techniczne i procedury administracyjne odnoszące się do załóg w lotnictwie cywilnym zgodnie z rozporządzeniem Parlamentu Europejskiego Rady (WE) nr 216/2008

PCATD On Top PCATD. (2018). Available at: https://www.indiamart.com/proddetail/pcatd-on-top-pcatd-2527426997.html

Pielacha, R. (1994). *Imitatory kierowania samolotami w kształtowaniu nawigatorów wojskowych, Research Study*. PAFA, Dęblin. (in Polish)

RAAF A63 Link Celestial Navigation Trainer. (2018). Available at: http://www.adf-serials.com.au/2a63.htm

Reid, G. H., & Burton, H. L. (1924). Psychomotor Responses in Relation to Flying. *Proc R Soc Med.*, *17*, 43–53. Available at: https://www.ncbi.nlm.nih.gov/pmc/articles/PMC2201251/?page=6

Silloth Airfield Project. (2018). Available at: https://sillothairfield.wordpress.com

Stelęgowski, A., Kłossowski, M., & Wochyński, Z. (2011). Wpływ ćwiczeń na lotniczych gimnastycznych przyrządach specjalnych (LGPS) na sprawność motoryczną i skład ciała, WIML. *Polski Przegląd Medycyny i Psychologii Lotniczej*, *1*(17), 7–19.

System symulacji szkolenia taktycznego (ETTS). (2018). Available at: http://www.m-346.com/m-346/szkolenie-pilotow/system-etts

Szkolenie M-346 dla polskiego wojska. (2018). Available at: http://www.m-346.com/dla-mediow/galeria-zdjec/m-346-szkolenie/category/5-szkolenie

Terelak, J. (1975). *Higiena psychiczna i pilot.* Warszawa: MON.

Vaden, E. A., & Hall, S. (2005). The effect of simulation platform motion on pilot training transfer: A meta-analysis. *The International Journal of Aviation Psychology*, *15*(4), 375–393. doi:10.120715327108ijap1504_5

WidevieW. (2018). Available at: http://www.wideview.it/my_cockpit.htm

KEY TERMS AND DEFINITIONS

CBT: Computer-based training.

Flight Simulator: A device or a computer program which simulates the aircraft operation in real flight conditions.

FMS: Full mission simulator.

FNPT: Flight and navigation procedures trainer.

FTD: Flight training device.

GBTS: Ground-based training system.

HMD: Helmet mounted device.

ILS: 1) Instrumental landing system, 2) integrated logistic support.

IMC: Instrument meteorological conditions.

IOS: Instructor operating station.

LMS: Learning management system.

LVC Technology: Live virtual constructive technology.

MPDS: Mission planning and debriefing stations.

NVG: Night vision goggles.

OFT: Operational flight trainer.

PTT: Part task trainer.

RTMS: Real-time monitoring station.

SBT: Simulated-based training.

TMIS: Training management information system.

ENDNOTES

1 The data is everything that is or can be processed mentally or computer. This "processability", i.e. the possibility of further use of data is a feature without which data could not exist at all - http://wiadomosci.ngo.pl/wiadomosc/1508583.html - dn. 24.03.2018 r.

2 Reaction - time from the appearance of an activating stimulus to the reaction time.

3 Air training devices are also an effective tool used, among others, in the process of selection of pilot candidates, and as a research tool used in air accident investigations and aviation research.

4 Flight simulator - it is a device or a computer program which simulates the aircraft operation in real flight conditions (real space flow situation / any aircraft or its part).

5 http://www.wiml.waw.pl/node/90.

6 The scale of the complexity of the task: A - a simple task, B - a complex task / the need to choose a stimulus from the group / reaction, C - more complex task - necessity of choosing a stimulus from the group / necessity of a specific reaction, D - highly complex task - necessity of choosing stimuli from the group after removing disturbances / necessity of choosing a specific reaction compilation, E - extremely complex task - necessity of selecting stimuli from the group after removal disturbances / the need to choose a specific compilation of reactions taking into account the circumstances (usually emergency situations in flight).

7 Experts – experienced pilots, general, commercial and national aviation instructors whom the author interviewed in order to collect data for the present work.

8 UAV – Unmanned Aerial Vehicle

9 CS-FSTD (As) - Certification Specifications for Aeroplane Flight Simulation Training Devices – Annex to ED Decision N° 2012/010/Directorater of the Executive Director of the Aency of 4 th july 2012 on the Certification specifications for aeroplane flight simulation training devices, p. 9.

10 Developed based on: (Ground based, 2018), (System symulacji, 2018).

11 In the present work, the author refers mainly to simulators applied in the process of training and professional development of pilots. It should also be emphasized that at present simulators are an essential tool applied in the process of training and professional development of the remaining air personnel (aircraft technicians etc.).

12 Continuous training – a term applied to fully trained air force pilots participating annually in a training program aimed at maintaining/improving pilot currency with regard to handling a particular aircraft type.

13 In this chapter, the author has not addressed the training of helicopter pilots, due to the fact that in Poland, as of 1st July, 2012, only Polish Medical Air Rescue had a certified FNPT II helicopter simulator. The remaining helicopter pilot training organizations in Poland did not use flight simulators. This is primarily due to the relatively high purchase price (approx. € 650,000) and maintenance costs of the simulator with a small number of trainees – several people each year. Such a small number of trainees is primarily due the fact that the cost of training one candidate for professional license CPL(H) is approximately € 125,000–170,000, depending on the training organization and the helicopter type used in the training.

14 Part-FCL – FCL.201A PPL

15 Part-FCL – Annex 6. C (As) modula training - point 7.

16 Assuming the cost of one hour of the Cessna-172 flight to be € 225 and one hour of the FNPT II simulator use to be € 60, at 40 hrs. of simulator training and 15 hrs. of practical training conducted on the Cessna-172, the total cost of the training would be € 5,775. If the whole training were conducted on The Cessna-172, the total cost of the training would be € 12.375. The analysis was based on the average price of 1 hour of the Cessna-172 aircraft flight and 1 hour of the FNPT-II simulator use as of 1st July, 2012 – author's own analysis.

17 This formula can also be used to determine the efficiency of simulator training for a single pilot, as well as for a group of pilots subjected to a study/training – as in the research studies covered by the article.

18 Training up to CPL (A) license and IFR (A) rating.

19 In relation to simulators, these requirements are set by EASA (Akceptowalne, 2012).

20 These categories refer to the categories of ILS (Instrumental Landing System) applicable to certain airports. The basis of category allocation is the possibility of making a safe landing at a minimum cloud ceiling and visibility: Cat. I – DH (Decision Height) no less than 200 ft (60 m) and RVR (Runway Visual Range) no less than 2040 ft (800 m). Cat. II – DH no less than 100 ft (30 m) and RVR no less than 1,050 ft (350 m). Cat. III has been subdivided into three subcategories: - IIIa – DH = 0-100 ft, RVR no less than 600 ft (200 m); IIIb – DH= 0–50 ft, RVR no less than 150 ft (50 m); IIIc – DH = 0, RVR = 0 – Developed on the basis of: T. Compa, J. Rajchel, *Podstawy nawigacji lotniczej*, WSOSP, Dęblin 2011, p. 297.

21 The simulator facilitates collecting the flight data related to the level of compliance. As a part of the diagnostic process, the instructor-pilot can compare the established criteria of conducting a given exercise with those defined by the recorded flight parameters. In addition, the instructor can

make a comparative analysis of the quality of performing the tasks by trainees who are at the same stage of training (the diagnosis of the state). Having the results of progress made by the trainee during the training, the instructor can understand more deeply the causes of mistakes made by the trainee, and offer him a new approach to an element of the exercise, which will allow him to achieve acceptable results. The instructor may also evaluate the feasibility – measurability of the quality of performing various elements of the exercise can be used to enhance the effectiveness of training by changing the approach to the training of selected elements of the exercise.

Chapter 4
Environments Diagnosis by Means of Computer Vision System of Autonomous Flying Robots

Konstantin Dergachov
National Aerospace University – Kharkiv Aviation Institute, Ukraine

Anatolii Kulik
National Aerospace University – Kharkiv Aviation Institute, Ukraine

Anatolii Zymovin
National Aerospace University – Kharkiv Aviation Institute, Ukraine

ABSTRACT

In this chapter, the authors present an approach to the extrinsic environs diagnostics based on using visual information collected by autonomous robots. The possibility of utilizing a computer vision for the purpose of rational control implementation in the condition of the full or partial uncertainty is investigated. In the study, the combined hardware and software computer vision tools were verified. The models, algorithms, and codes for solving the local tasks of obstacle identification and mutual location kinematic parameters estimation have been developed. A series of computational and in-kind experiments that illustrate a practical possibility of implementing the navigational environment diagnosis is carried out with the aim to select a rational flight path.

BACKGROUND

At present time, there is a trend of reducing the number of remotely piloted unmanned vehicles (UAVs) in sphere of aviation appliances. The more topical becomes the UAVs that are capable not only to fly autonomous fully but also perform monitoring, aerial photography, agricultural work, military missions, etc. Therefore, a UAV becomes an autonomous flying robot (AFR), which executes the basic functions without human being participation.

DOI: 10.4018/978-1-5225-7709-6.ch004

The progressive trend in AFRs development is to extend the sphere of application by improving the functionality of the automatic control system.

As a rule, the autonomous robots work in conditions that are partially or completely unknown. The path adaptation to a situational uncertainty is possible (Odarchenko, R. S., Gnatyuk, S. O., Zhmurko, T. O., & Tkalich, O. P. (2015, October)) by appropriate processing the visual information obtained by the on-board video camera. The efficiency of AFR adaptation process depends greatly upon the quality of extrinsic conditions diagnosis.

The environment in which the robot's movement is going on can be characterized by a situational uncertainty caused by a number of factors (Isermann, R. (2006)).

AFRs operate in the condition of certain objectively existing causes which destabilize the navigation task fulfillment (Gulevich, S. P., Veselov, Yu. G., Pryadkin, S. P., & Tyrnov, S. D. (2012)). The destabilizing factors for the AFR could be events like given below

1. The condition of the atmosphere in terms of atmospheric phenomena like the presence or absence of the wind, precipitation, solar activity, cloudiness;
2. The state of the terrestrial and celestial visual orientators (their shift, changes of the reflecting surface);
3. The obstacle occurrence on the flight path (different flying vehicles, other interfering things).

So, the destabilizing factors are the real, objectively existing causes that need to be identified and evaluated in the course of performing a relevant mission (Kulik, A. S. (2014)). The destabilizing impacts produce a situational uncertainty of the extrinsic environs.

Such an uncertainty can be reduced using the procedures intended to perform the operative diagnosis of vicinities of the AFR route. Diagnostic procedures would allow, for example, discovering the interference, identifying and localizing them, i.e. providing for a complete ambiguous events analysis. This information is required in real time to rapidly generate on board the autonomous robot such the obstacle bypassing procedure that would allow rational solving the navigation task in the condition of naturally limited time and energy resources. Similar diagnosis procedure functions should be applied when other destabilizing factors occurred.

For AFR, a task of navigating the flying vehicle from the point of departure to destination involves the need to identify obstacles in the path and bypass them.

Obstacles that arise on the path of AFR movement can be shared into two large groups: the obstacles that can be overcome by the AFR without changing its motion trajectory and those ones that can be avoided by changing the trajectory only. Hereafter, we investigate the second option when you need neccessarily to change the trajectory of AFR movement to avoid the obstacle (Kulik, A. S., & Radomskyi, O. M. (2017)). The following components associated with an intelligent diagnosis can be emphasized. You need providing

1. Obstacle detection;
2. Determination of the distance to the obstacle;
3. Determination of the own speed relative to the obstacle;
4. Estimation of obstacle size;
5. Determination of AFR angular attitude relative to the obstacle.

To solve these AFR tasks, the appropriate complex of hardware and software tools is required. Currently, a wide range of sufficiently high-quality and low-cost video sensors are available to developers; therefore it is advisable to place computer vision complexes directly on board the AFR with the aim to solve tasks of extrinsic environs diagnosis. In autonomous computer vision systems of mobile carriers, the single-board platform Raspberry Pi with the manufacturer's recommended operating system Raspbian is oftentimes used. The basic programming language will be Python here.

Main computer vision tasks for images and video data processing when programming in Python are being solved with usage of the Pillow library (Python Imaging Library), which provides a sufficiently complete set of functions and methods for processing images and video. However, a joint usage of the Pillow library and Open CV library (Open Source Computer Vision Library) resources is the most productive.

OpenCV usage in conjunction with Python is convenient for creating simple and understandable applications, also performing experiments and different prototypes synthesis. In Python, almost all the functionality of the OpenCV library is available. When using a vector-matrix ideology which is similar to that adopted in Matlab, it is possible to generate sufficient rapid codes providing the task solution in real time.

Such combination of hardware and software tools is optimal in terms of price and operation quality (performance) for solving the sort of issues mentioned (Dergachov, K. Yu., Krasnov, L. O., & Radomskyi O.M. (2017)).

A development of the algorithmic software to provide the process of diagnosing the extrinsic environment constitutes a topical problem. One of solutions of the problem is suggested in the work.

Research Objectives on Diagnosing the Extrinsic Environment by an Autonomous Flying Robot

Presently, an actual practical task is a local navigation of autonomous robots. The appropriate information supply as regard the exact location of the transport robot and its spatial position extends the number of possible applications essentially, and also provides for reliable accomplishing the task of autonomous movement control and planning the trajectory based on the principles of rational management (Kulik, A. S. (2014)). Typically, the AFR location is defined in either global or local coordinates.

The satellite navigation systems (SNS) like GPS and GLONASS are widely used for solving positioning problems noway. The main SNS advantages are the global coverage and the sufficient accuracy of outdoor location determination if the differential positioning mode is in use. As major disadvantages of the SNS practice, the following reasons can be named: the dependence on weather conditions and the fundamental impossibility to be used in indoor cases like indoor warehouses, underground parking lots, tunnels, under bridges, in road junctions. The drawbacks limit the range of SNS applicability for local navigation assignments. The essential drawbacks which don't let the SNS usage for the local navigation are the noticable inaccuracy while solving positioning tasks within the ordinary mode of SNS operation and a principal impossibility to take into consideration the geometric dimensions and obstacle placement topology when automatically planning AFR motion. It should be noted that the use of sonars in navigation systems leads to insufficient accuracy, limited operating distance range, and allows only determining the distance to the obstacle, but not identifying it.

The drawbacks mentioned could be corrected by the local positioning system (LPS) having a computer vision system (Kolodziej, K. W., & Hjelm, J. (2006)).

The AFR motion control is usually executed under conditions of uncertainty resulted fromcaused by the robot positioning errors and its spatial orientation (angular attitude) errors, on one hand, and the environment condition changes, e.g. due to steady-state and dynamical interference to the motion (standing obstacles or other moving vehicles), on the other. The most amount of information about the environment in which the movement is carrying out is contained in visual images. The latter circumstance suggests making a collection of images shooted by the video camera installed on the AFR and processing the shots in the computer vision system so that provide the AFR rational control that would be based on the video data obtained.

Thus, the use of the computer vision in AFR control systems makes it possible to significantly reduce the robot flight conditions uncertainty. These may concern a better acquire of its local position and surrounding interferences, as well as the possibility of the angular attitude accurate specifying, hence choosing the safe route and related parameters of motion. Such a usage constitutes the elements of the rational control.

The objective of the mobile robot visual navigation that would use the specified visual orientators can be stated mathematically.

In the space region Ω of $V = a \cdot b \cdot c$ size, some visual orientators are placed and a few AFRs are moving in that space.

Let $w_1, w_2, \ldots, w_n$ – be the exterior visual orientator located in space area Ω;

$AP_1, AP_2, \ldots, AP_l$ are the mobile robots performing the flight; *n* is the number of exterior visual orientators, and ***l*** is the number of robots.

Each i-th visual orientator at the moment of time t_k is specified by a set of parameters:

$$w_i = w_i\left(\bar{r}_i, \bar{g}_i, \bar{c}_i, A_i, t_k\right).$$

where

$\bar{r}_i$ is the vector of coordinate values for characteristic point of the visual reference. In the case of three-dimensional space Ω we have $\bar{r}_i = \bar{r}_i^{3} = \left(x_i, y_i, z_i\right)^T$ in this vector, x_i, y_i, z_i represent rectangular coordinates. For the two-dimensional space, we'd have $\bar{r}_i = \bar{r}_i^{2} = \left(x_i, y_i\right)^T$;

$\bar{g}_i$ is the vector of characteristic geometric dimensions of the ***i-th*** visual orientator;

A_i represents an algorithm of generating a three-dimensional figure for the visual orientator on the basis of $\bar{r}_i$ value;

$\bar{c}_i$ is the vector of color features of the orientator in the color model accepted.

The most popular color model to date is RGB model (Red, Green, and Blue). The three colors are of equal rights and autonomous in the model. If all three RGB components have a zero value, the image constitutes the black background, and if the items accept a maximum value then the white (blank) image would be the case. The RGB model is used in most cases to present photos and videos. OpenCV program utilizes a different, BGR color model by default; here, each pixel is characterized by three values again,

B, G, and R figures (for the blue, green and red component appropriately). The distinction from the classical RGB model is that the green channel conserves its place, while the blue and red channels change places. If you need to discover the object on the image and reliably detect (recognize) the object motion in the video frames sequence then one often utilizes a technique of objects distinction by color features (primitives are in use). This method provides for well noise immunity indices due to low sensitivity of the "color filtering" algorithms to a considerable change in illumination of the local element of the scene.

Any color space is specified as a data representation model based on certain color coordinates. Usually the signals of conventional web cameras present a sequence of RGB color model images the use of which for images distinction is unproductive since the search for areas of the desired color involves an analysis of all three, R, G, and B, components.

At the same time, for implementing the search for the objects picture by the color and brightness feature the HSV color model is usually used, where H, S and V stand for parameters that specify the hue, saturation and value (brightness) of the image. In the model, the items are normalized as follows:

hue: 0 – 360°; saturation: (0 – 100) %; value: (0 – 100) %.

For the HSV model the vector of color features $\overline{c_i}$ contains a set of attributes identifying the visual orientator.

The autonomous robot ARj in the area of space Ω at time instant t_k is specified by parameters represented in

$$AP_j = AP_j\left(\overline{f_j}, \overline{q_j}, R_j, \overline{S}_j, t_k\right),$$

In the record, $\overline{f_j}$ is the vector that defines the coordinates of a characteristic point of the mobile robot; usually, the center of mass of the robot is taken as such a point. In the case of three-dimensional space Ω and using the Cartesian coordinates, we have

$$\overline{f_j} = \overline{f_j}^3 = \left(x_j^{mp}, y_j^{mp}, z_j^{mp}\right)^T;$$

$\overline{q_j}$ – is the vector of the mobile robot angular orientation; in general, $\overline{q_j} = \overline{q_j}^3 = \left(\vartheta_j^{mp}, \varphi_j^{mp}, \psi_j^{mp}\right)^T$;

$\overline{S}_j$ is the purpose vector of motion; in general, $\overline{S_j} = \overline{S_j}^3 = \left(x_j^S, y_j^S, z_j^S\right)^T$;

x_j^S, y_j^S, z_j^S are destination coordinates which is the goal of the movement;

R_j represents the algorithm for AFR spatial attitude restore and dimension marks recovery.

The problem of rational visual navigation of a moving object, e.g. the j-th AFR of the foregoing case study, can be stated as a sequential solution of the three local tasks below.

1. Determination of the AFR location.
2. Determination of the AFR angular attitude relative to the center of mass.

Figure 1. Modelling a space area Ω

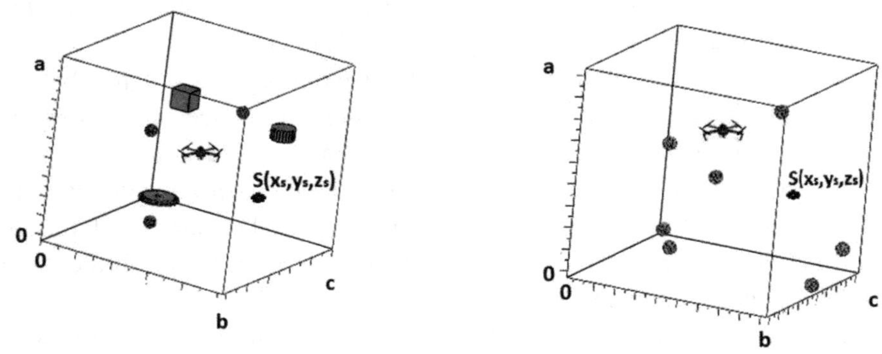

a: $(j=1, i=6, \overline{r_i}^3, \overline{f_1}^3, \overline{q_1}^3, \overline{c_i} = \text{const})$ b: $(j=1, i=7, \overline{r_i}^3, \overline{f_1}^3, \overline{q_1}^3, \overline{g_i} = \text{const}, \overline{c_i} = \text{const})$

3. Determination of the AFR rational flight path.

Let us specify each of these problems.

1. To determine the *AFR location* you need calculating appropriate coordinates, namely the components of the vector $\overline{f_j}$, . This requests the knowledge of methods and algorithms dedicated to handling the composite function (the functional Q) that meets the expression

$$m_{\text{МП}}^{0} = Q\left(F\left(w_1, w_2, \ldots, w_n\right), \overline{f_j}\right),\ m_{\text{МП}}^{0} \in M_{\text{МП}};\ j = 1,2\ ..l,$$

where $m_{\text{МП}}^{0}$ is the selected method of location determining;

$M_{\text{МП}}$ is the array of methods and algorithms for determining location of the AFR;
$F\left(w_1, w_2, \ldots, w_n\right)$ is the functional of detection, identification, and analysis of the visual orientator.

2. To determine the *angular orientation* of the AFR, we need resolving the components of the vector $\overline{q_j}$ representing the appropriate angles. This implies using the knowledge of methods and algorithms related to handling the functional Q that meets the expression

$$m_{\text{уП}}^{0} = Q\left(F\left(w_1, w_2, \ldots, w_n\right), \overline{q_j}\right),\ m_{\text{уП}}^{0} \in M_{\text{уП}};\ j = 1,2\ ..l,$$

where $m_{\text{уП}}^{0}$ is the chosen method of determining the angular position;

$M_{\text{уП}}$ is the set of related methods and algorithms of AFR spatial attitude determination;
$F\left(w_1, w_2, \ldots, w_n\right)$ is one more functional related to special signs of the exterior visual orientators.

3. To propose a trajectory for the robot AP_j. to travel from the current position toward the destination S, a variety of interference to AFR motion should be included into the base; each obstacle is specified by given dimensions and placement in the space area Ω. In general, an objective can be stated as a development of methods and algorithms for defining a functional Q that meets the notation

$$m^0_{\tau p} = Q\left(F\left(p_1, p_2, \ldots, p_g\right), Z\left(\overline{f_j}, \overline{q_j}, S\right), T_j\right), m^C_{y\Pi} \in M_{y\Pi};\ j = 1,2\ ..l,$$

where $m^0_{\tau p}$ is the chosen method of determining the AFR trajectory;

$M_{\tau p}$ is the set of methods and algorithms for determining the trajectory of AFR;

g is the obstacles quantity which is considered known in the case of deterministic problem solution and unknown if the AFR flies at uncertain conditions;

$F\left(p_1, p_2, \ldots, p_g\right)$ represents the functional of the search, recognition and analysis of the interference.

T_j – is the AFR flight trajectory, given by the set of points representing coordinates of the center of mass of the AFR at the time instant t_k.

Autonomous Robot With the Computer Vision System

To solve the task of determining the location of a mobile robot, a contemporary navigation technology is proposed. The technique implies extended usage of the video information containing in shots of the exterior video orientators shooted by the robot on-board camera. Among possible and favorable approaches to resolve the AFR coordinates, the expedient ones are the following:

- Pseudorange method based on finding AFR distance to the exterior visual orientator with given coordinates of the characteristic point, and with primitives' known color features and dimensions (primitives' parameters) (Guevara, J., Jiménez, A. R., Prieto, J. C., & Seco, F. (2012));
- method of visual odometry implementing a consecutive reading of image frames from the video sequence; the video reflects the AFR movement from the origin point with known coordinates and angular position (Huang, A. S., Bachrach, A., Henry, P., Krainin, M., Maturana, D., Fox, D., & Roy, N. (2017));
- Combinations of both methods.

Let we analyze the existing algorithms of local navigation and estimate the possibility of their usage in the system of local navigation of the AFR (Retscher, G. (2016)). We take that the vehicle has been equipped with a vision system purposed to perform the diagnosis of the environment. To complete the analysis, we have examined a few algorithms known as Proximity (the closest visual orientator), Centroid (center of gravity), Weighted Centroid (center of mass), Lateration (lateratsiya), Deduced Reckoning (dead reckoning), Hough Transformation. The analysis outcomes are presented in Table 1.

A computational algirithm of the visual odometry method advertised in Deduced Reckoning techniques presentation (Table 1) is illustrated by the chart in Figure 2.

Table 1. Overview of local positioning algorithms

Algorithm	Contents	Advantages	Disadvantages
Proximity	AFR will be prescribed the coordinates of a point that is the closest to the exterior visual orientator	Implementation simplicity, low computational complexity and a need to get information on the exterior visual orientator's placement only	Low evaluation accuracy, which is commensurate with the distance to the exterior visual orientator
Centroid	The algorithm is based on computing the geometric center of a plane figure involving a few exterior visual orientator	Implementation simplicity, low computational complexity and a need to get information on the exterior visual orientator's placement only	Low evaluation accuracy, which is commensurate with half a distance to the exterior visual orientator
Weighted Centroid	A version of the Centroid algorithm improved due to taking into account the magnitude of the distance to the visual orientator. The coordinates are calculated as a linear combination of exterior visual orientator coordinates computations when taking into account the distance to the exterior visual orientator	Implementation simplicity, low computational complexity	Strong response to the number of exterior visual orientator that are in the field of sight
Lateration	A geometric approach based on computing distances between a point and at least three visual reference points followed by subsequent solution of the system of N nonlinear equations and determining a spheres intersection region in which AFR is staying	High accuracy, resulting from precise optical sensors applied	There must be at least three exterior visual orientator in the field of view
Deduced Reckoning	Calculation or extrapolation of mobile robot current position, based on data of previous position, direction, speed or traveled distance estimates	Ability to determine the location autonomously	A need of getting the initial location and a complexity of the visual odometry algorithm
Hough Transformation Techniques	A geometric approach based on techniques of finding lines, circles, and other simple shapes (primitives) on the image. The location can be found by the proportional comparison of the object actual size and the corresponding object dimensions evaluated from the image of the vision system	Robustness in the case of visual orientator fuzzy localization	Large computational complexity

By investigating changes and transformormation of video fields of the frame and estimating the rate of image elements alteration in the frame, you can gather information about video camera movements and, consequently, on the AFR motion and its approaching to obstacles. Decision making may include the following steps:

- Accumulating differences between images captured;
- Computing the optical stream characterizations;
- Estimating parameters of the object motion using the accumulated optical stream data.

The visual odometry algorithm, along with assigning a few exterior visual orientators, supposes evaluation of the relative movement of the AFR concerning the specified reference points. Existing difference in coordinates resolving accuracy specifies the class of applicability of each algorithm to the AFR computer vision system in use.

Thus, the analysis of algorithms for local positioning the object shows the possibility of their implementation in AFR navigation and control systems. The algorithms are different in terms of computational complexity and achievable solutions accuracy, which depends largely upon the character of primary processing the video information in the AFR computer vision system and the amount of computations required. In a most simple case, when the Proximity algorithm is utilized, it will be necessary to allocate a single known sign in the video frame received. For the other techniques (Centroid, Weighted Centroid, Lateration algorithms), a primary appointment of a few visual orientators, the assessment of the distance to them and further their positioning are required.

When using pseudo-range methods based on the principles of lateration, in order to solve the navigation problem it is necessary to resolve the system of equations like

$$\left\| \overline{r_i} - \overline{f_j} \right\| = \overline{d_{ji}} + \delta;$$

in order to solve the navigation problem . In the expression, parameters meaning are the following:

$i = 1, 2, .., n$ imply the quantity of exterior visual orientators;
$j = 1, 2, .., m$ suppose specifying the number of mobile robots in work;
$\| ..... \|$ sign presents a symbolic designation of the measure of distance between the points specified;
$\overline{d_{ji}}$ is the meaning of the numerical measure of distance between the specific point of the AFR and the visual orientator;
δ is the distance evaluation error between objects.

When using the Cartesian measure of distance and $\overline{f_j}^3$ description of the AFR location, the distance between points of interest can be defined as

$$\overline{r_i}^3 - \overline{f_j}^3 = \left| \sqrt{\left(\left(x_i - x_j^{mp} \right)^2 + \left(y_i - y_j^{mp} \right)^2 - \left(z_i - z_j^{mp} \right)^2 \right)} \right| + \delta;$$

$i = 1, 2, .., n$; $j = 1, 2, .., m$.

On the other hand, the distance between the characteristic point of the AFR and the characteristic point of the visual reference can be derived from geometric relationships resulted from the measurements associated with the video sequence frames examination following the frames preliminary processing (equalization, allocation of specified areas having appropriate to the selected space color characteristics c_i, further localization of the visual orientators $\widehat{w_i}$ adopted for decision making).

The procedure for estimating by the video image the distance from AFR's focus to the exterior visual orientator $\widehat{d_{ji}}$ at time instant *t* consists of measuring the geometric dimensions of the visual orientator shot and comparing the obtained outcomes with the prescribed physical dimensions g_i using of geometric similarity rules:

Table 2. Implementation features of local positioning algorithms

Algorithm	Implementation	Computational complexity
Proximity	$(X_0, Y_0, Z_0) = \arg\min_{i=1..N} \{d_i(x, y, z)\}$	O (N)
Centroid	$X_0 = \frac{1}{N}\sum_{i=1}^{N} x_i; Y_0 = \frac{1}{N}\sum_{i=1}^{N} y_i\ ; Z_0 = \frac{1}{N}\sum_{i=1}^{N} z_i.$	O (N)
Weighted centroid	$X_0 = \sum_{i=1}^{N} \mu_i x_i;\ Y_0 = \sum_{i=1}^{N} \mu_i y_i;\ Z_0 = \sum_{i=1}^{N} \mu_i z_i;$ $\mu_i = \left(P_i^2 \sum_{j=1}^{N} \frac{1}{P_j^2}\right)^{-1}$	O (N)
Lateration	$d_{ij}^2 = r_i^2 - r_j^2,$ $d_{ij}^2 = \begin{vmatrix} (X_i - X_0)^2 + (Y_i - Y_0)^2 + (Z_i - Z_0)^2 - \\ -(X_j - X_0)^2 - (Y_j - Y_0)^2 - (Z_j - Z_0)^2 \end{vmatrix},$ где $i, j = 3 \dots N$.	O (N^2)
Deduced reckoning	the procedure is illustrated by the chart of Figure 2.	Depends on image frame size
Method lies on identifying reference points with the help of Hough's Transform	Determines distance to visual orientator of given dimensions. Computational procedure supposes identification of circle primitives in image frames	O(k log k)

Notes for Table 3: (X_0, Y_0, Z_0) .– AFR coordinates; x_i, y_i, z_i., $i = 1 \dots N$.– visual orientators coordinates; P_i^2 . distance squared to i-th orientator; μ_i .– normalization factor; k is the quantity of the points of circle.

$$\widehat{d_{ji}} = F\left(\overline{r_i}, \overline{c_i}, \overline{g_i}, I_k^i, t\right),$$

where I_k^i is the video frame instance, and k$= 1, 2, .., q$ specifies the image frame increment.

To determine the value $\widehat{d_{11}}$, it is necessary to carry out preliminary processing of the video image in order to allocate an exterior visual orientator on the image. If you need searching for a visual orientator of the circular shape (round primitive) the technique of Hough transformation can be applied (Ballard, D. H. (1981)).

The Hough transform is conventionally used for searching lines on a raster image; hereafter, we understand a circumference (periphery of a circle) to be a locus of points that are equally distant from the point meaning the center of the circle. The equation of the circle on a plane has the form

$$(x \cdot a)^2 + (y \cdot b)^2 = R^2,$$

Figure 2. Visual odometry algorithm flowchart

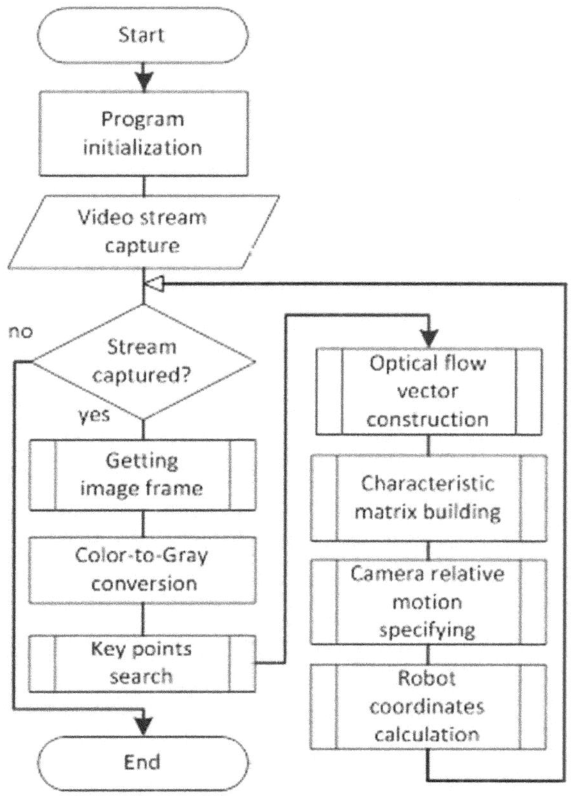

where (x, y) are coordinates of a point of the circumference line; (a, b) are coordinates of the figure's center; R is the circle radius.

Thus, any circular figure can be specified by the three parameters: one for the radius (R) and two others for coordinates of the center – (a, b). Proceeding from this, the accumulating Hough space will be three-dimensional with R, a, b coordinate axes. If the circle radius is known when you are looking for a primitive, then the accumulating space becomes two-dimensional with unknown parameters for coordinates of the center, (a, b), only.

The detection algorithm for finding circles of a given radius on gray images, which utilizes an evaluation of the normal's orientation in planimetric points of voting, is as follows. At the first step, the algorithm identifies the pixels-boundaries of the object. The Sobel operator (Jin-Yu, Z., Yan, C., & Xian-Xiang, H. (2009, April).) can be used for this, for instance, which allows one to obtain an estimate of the value (magnitude) and direction of the gradient vector. The planimetric points of voting are the places with a large value of the gradient modulus. For each supposed extreme pixel, an estimate of the position and orientation of the contour is calculated by passing in the direction of the normal to the contour (in gradient vector direction) the distance R from the extreme pixel; the aim is to determine the center of a circular object of radius R. If the procedure has been repeated for each extreme pixel, a set of predicted center points would be found. The individual estimates can be averaged to determine the exact location of the center. The sketches in Figure 3 interpret the procedure.

Figure 3. Illustration of the principle of finding a circle of unknown radius

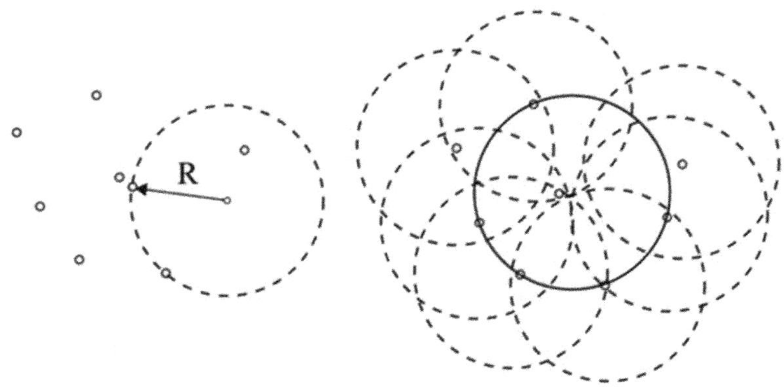

If one needs to determine the circumference of an unknown radius on an image the principle of the vote may be as follows. When the radius of the circle is unknown or varying one can add a variable R onto the accumulating parametric space; the search procedure will now involve a finding of the maximum radius value and the center position by taking into account the third dimension of the parametric space.

Case Study of the Visual Orientator Detection and Localization

Consider the problem of visual navigation of AFR with the computer vision system. The system has to navigate the robot toward the visual orientator, the color and shape features of which are known. The next objective is to lead the AFR to destination by a given distance. Let a red ball be that visual orientation since such a choice simplifies the experiment arrangement and allows Hough transform application. In order to avoid a collision of the robot with the ballobject it is necessary to provide determination of the distance to the orientator. The AFR management algorithm will use the red ball physical parameters required for the mathematical model and used by the method algorithm when comparing ball actual dimensions with measured ones. After detecting and recognizing the visual orientator, it is necessary to generate a control signal required to target the AFR to the central point of the visual orientator which is in the video frame.

The experiment was carried out with AFR and the red circle being the visual landmark for experimental verifying the algorithm workability. A photo of the experiment arrangement is shown in Figure 4. As working instrument for identifying the orientator by the robot, a sequence of video images shooted by the AFR computer vision system was used.

A multi-rotor quad-copter drone equipped by the computer vision system was chosen for the case study. For practical system implementation, a variant of the hardware based on the Raspberry Pi microcomputer and a low resolution USB camera was chosen (Figure 5).

Specifications of the Raspberry Pi 2 (модель B) computer platform the exterior view of which is shown in Figure 6 can be seen in Table 3.

The specifications illustrate the fact that the platform is the contemporary single-board computer means for profitable employment in the computer vision systems of mobile vehicles like AFRs

Figure 4. Robot-in-flight and orientator-primitives

Figure 5. Motherboard of the prototype AFR computer vision system

Figure 6. Exterior of Raspberry Pi 2 board (model B)

Table 3. Microcomputer Raspberry Pi 2 (model B) specifications

	Parameter	Characteristic
1	Processor (CPU)	Broadcom BCM2836 ARM Cortex-A7 Quad Core 900 MHz
2	Number of processor cores	4
3	RAM	1 GB
4	External storage	MicroSD
5	Networking	Ethernet 10/100
6	Video Output	HDMI
7	USB Ports	4
8	Audio output	3,5 Jack
9	GPIO	40 pin
10	Power supply	5V, 2A
11	Options	2 diodes
12	Dimensions, weight	85,6 x 53,98 x 17mm, 45 gram

Thus, for solving the case study tasks declared, the identification of the color visual orientators will be carried out on the base of Raspberry Pi 2 platform and using the Hough transform technique. A flowchart of the corresponding algorithm is presented in Figure 7.

Let us focus on the implementation of the algorithm aimed to search the reference landmark and specifying its geometric dimensions. At the algorithm initialization stage, the following data are input:

1. The area boundaries of the visual orientator in the frame (specified in the screen coordinate system);
2. The maximum radius value of the visual orientator in the video image frame.

Figure 7. Flow chart of Hough transformation algorithm to search circles

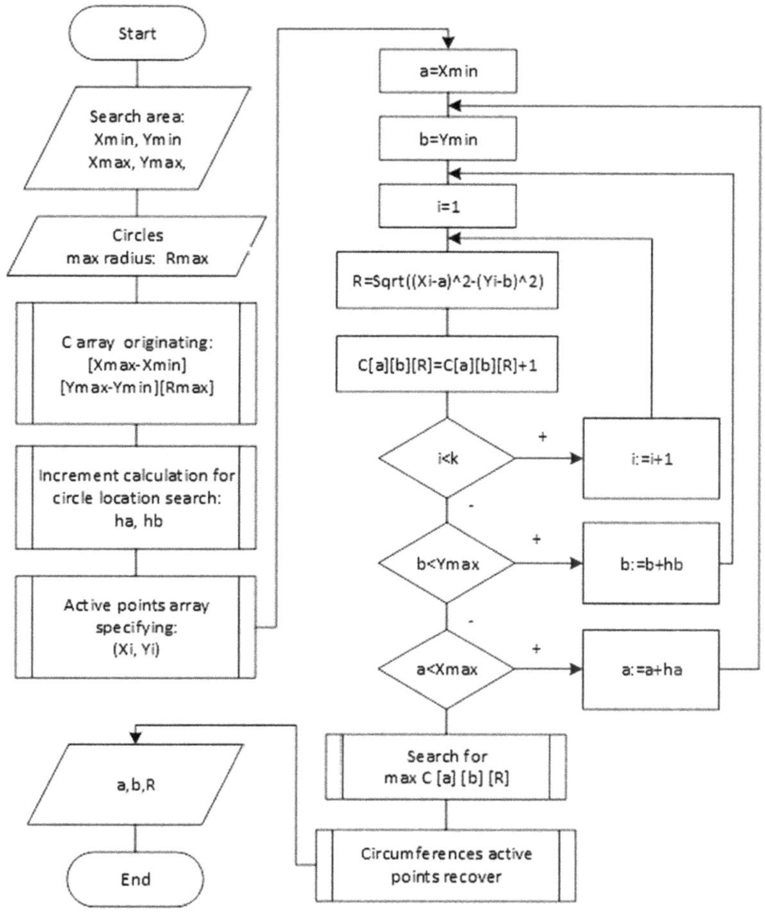

After the input of initial data, the structures of the data are initialized to provide necessary storing and processing of the image digital information contained in video frames. The basic such structure is the three-dimensional data array C, i.e. [Xmax ... Xmin] [Ymax ... Ymin] [Rmax], which is appointed to accumulate current values of the Hough transform operating points.

The initiation data entering activates computing of the extra variables associated with, first, the increment value of the Hough transform operational parameters while running the code and second, the algorithm execution speed index. The former reason concerns the need to match the increment step with the size of the visual orientator area.

At the next step, a choice of active points in the image and filling in the data array is performed; the instances selected will take part in the vote calculation procedure by Hough.

After formulating the algorithm initial conditions, a triple nested loop is realized in which a meaning is calculated for every active point in the limited frame area:

$$R = \left| \sqrt{\left(x_i - a\right)^2 + \left(y_i - b\right)^2} \right|.$$

where (x_i, y_i) is the location of current active point of the video frame; (a,b) is the assembly of current Hough transformation parameters for the visual orientator of circular shape.

Next, meanings of all the active points are accumulated in the array C [a] [b][R].

After all the cycles have been completed, the maximum meaning in the array C[a] [b][R] is determined. For this specific value, the active points will be reconstructed and meanings a, b, R saved.

In OpenCV program library, the Hough transform procedure designed to finding circles is implemented by the function *cvHoughCircles*() (Bradski, G., & Kaehler, A. (2008)). Resulted from the transform application, the geometric dimensions of the visual primitive's shot in the image frame are determined. To find the distance from the characteristic point of the AFR to the visual orientator, one must use appropriate geometric relations.

The limitation of this algorithm is that only one orientator can be identified (detected) and localized by such a procedure.

An algorithm the flow chart of which is shown in Figure 8 provides for the AFR autonomous motion towards the visual orientator. The prerequisites to supply the drone with necessary control when approaching the target will be the flight path map, also the color model and radius of the orientator primitive being the point of destination.

Figure 8. Flowchart of visual orientator identification algorithm

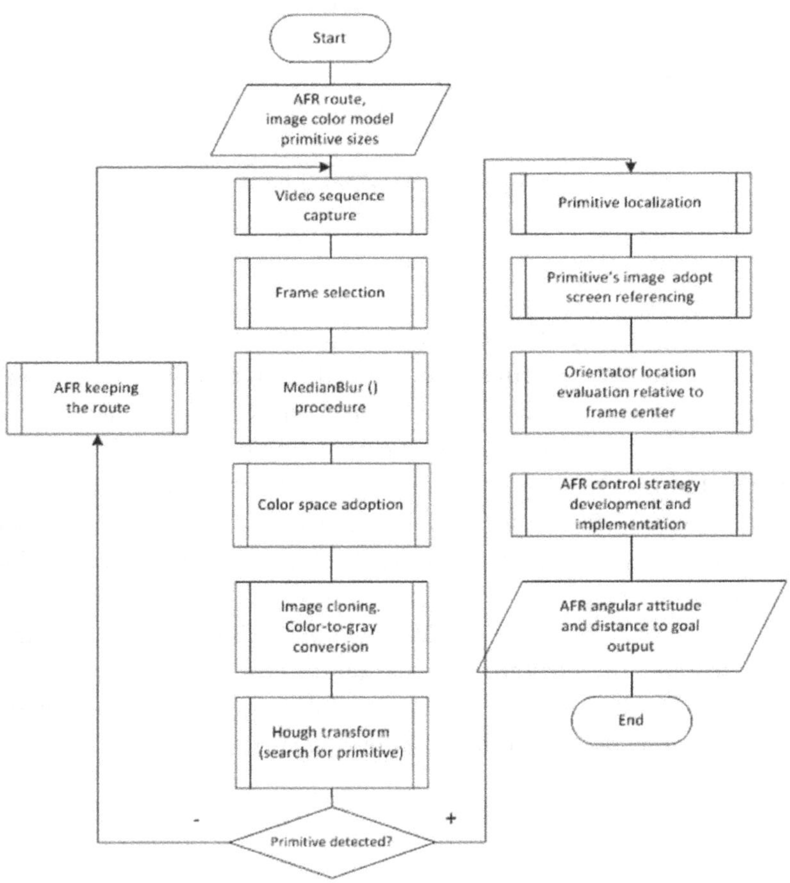

A few comments on the AFR control stages organization when using the computer vision systems

1. Because of the fact that the processing of the video sequence is carried out on board the AFR in real time or nearly so the initialization of the video sequence received isn't a plain task.
2. Video frames are extracted from video sequence in a discrete time interval.
3. To suppress background noise, after color filtering you must perform image binarization with the brightness threshold cut-off.
4. After noise filtering, the image is usually converted to a hardware-independent digital space, e.g. HSV space.
5. Conventionally, a color image is cloned and transferred to the gray (intensity-based) format to save resources of the computer.
6. Using a cloned image, the system tries detecting the visual orientator on the base of Hough transform technology.
7. In the event of orientator wrong detection, the AFR keeps following the predetermined flight path.
8. When the program discovers a certain orientator in the video frame the algorithm will continue the procedure of its allocation followed by the orientator image dimensions estimation.The latter aids computing a distance to the target. The displacement of the visual reference relative to the center of the image frame makes it possible to determine the angular position of the AFR relative to the visual reference.
9. Based on identified values of the distance to the visual orientator and the angular position of the reference image with respect to the center of the frame, a proper control by the robot may be realized to avoid uncertain collisions and safely approach the visual orientator environs.

The output of the algorithm is the current value of the distance from the AFR to the visual reference point, and also the angular position of the visual reference relative to the center of the image frame.

In conducting experimental studies, the position of the AFR is characterized by the viewing angle which appears a displacement of the AFR video camera along the horizontal axis (for the flight in a single plane). One can evaluate this parameter by the location of the visual orientator shot in the image frame.

Figures 9 -10 display the results of experiments on recognizing a visual orientator by the AFR computer vision system. To pick up the experimentation data, Raspberry Pi 2 microcomputer has been linked to a virtual desktop installed at 0.64 m height over visual orientator primitives. Figure 10 demonstrates results of detecting the primitive using the Hough transform.

Figures 11-12 show similar results dedicated to a different experiment condition – the 0.32 m altitude of the desktop.

The outcomes of the investigation are autonomous robot flight parameters evaluation that became a result of measurements picked up during a specific number of real test flights. The outcomes relate to the robot to orientator distance estimation and also the calculation of the AFR viewing angle relative to the orientator center point located in the frame plane.

The obtained results, namely the error in estimation of the distance, which constituted 1.5 to 2 cm, and attitude determination errors of 2 to 2.5 degrees let concluding the efficiency of the proposed approach on diagnosing environs of flying robots during their either controlled or autonomous flight.

Figure 9. The shot of camera matrix with coordinates of the visual orientator center calculated

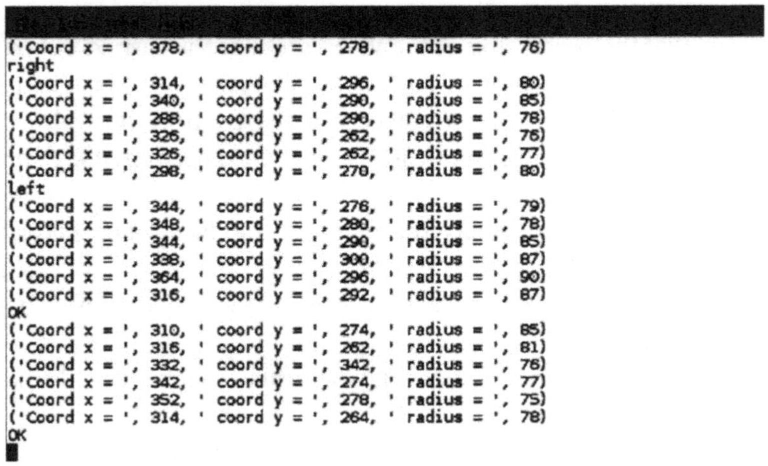

Figure 10. Window shot with the visual orientator at distance 63 cm

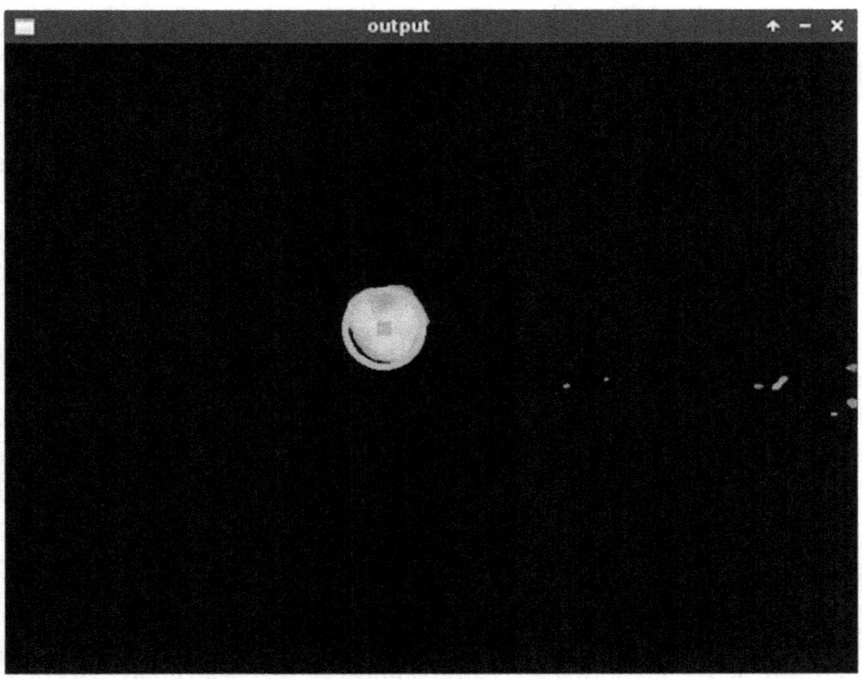

FUTURE RESEARCH DIRECTIONS

In future, it is supposed to push forward the research so that to provide for such an algorithmic support of the AFR computer vision system that would be capable specifying several visual orientators contained in video frames. To realize such algorithms, it is necessary to have information a priori about geometric features and color models of multiple attribute primitives. It is the progress in developing adaptive

Figure 11. Window shot with the visual orientator at distance 32 cm

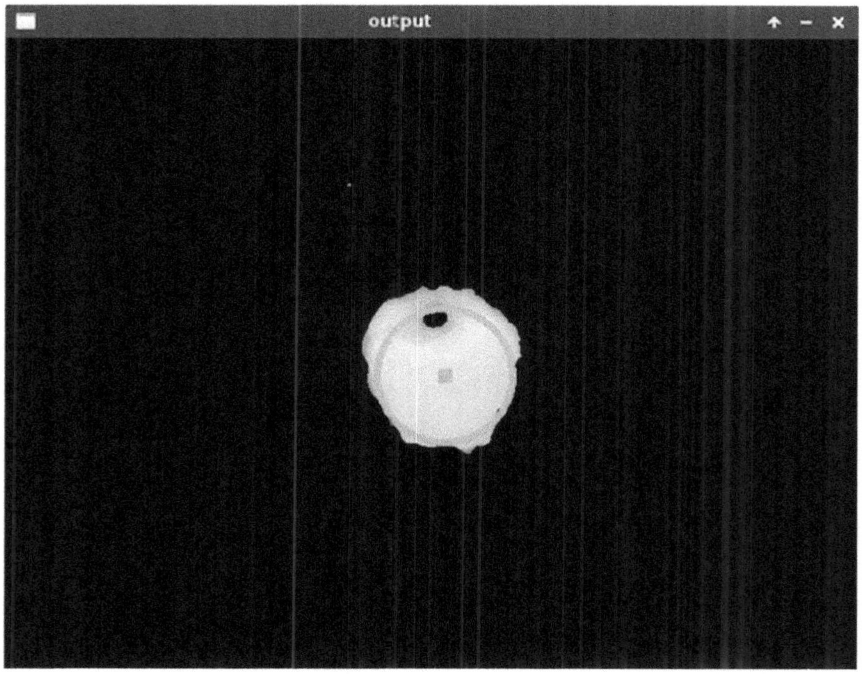

Figure 12. Command-line window after the program termination

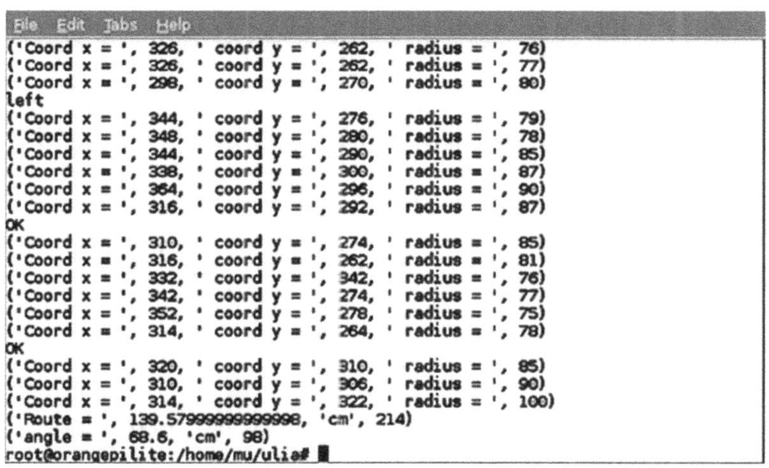

aids of video images averaging in terms of the state of illumination level that can raise visual attributes identification performance significantly.

The next research direction may concern the task of developing the rational algorithms for an autonomous vehicle control. These would offer the robot ability sidestepping obstacles and other uncertain circumstance when knowing the special visual orientators.

Figure 13. Experimental study results

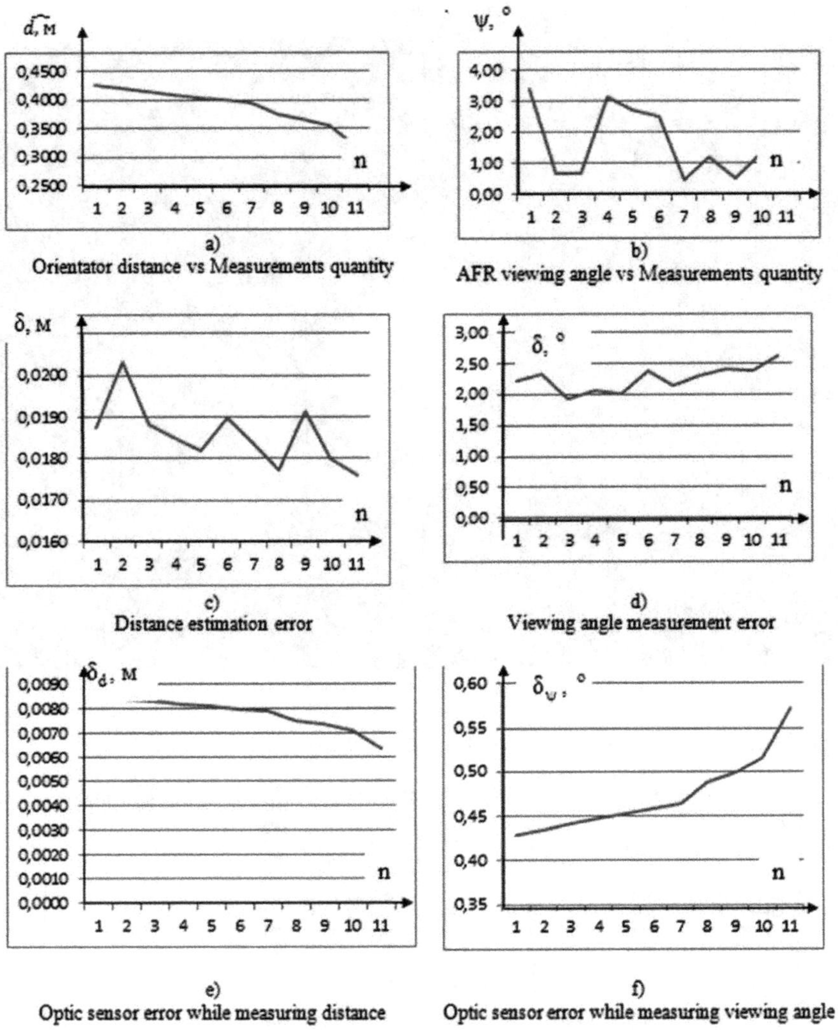

CONCLUSION

The environment in which AFR moves is characterized by situational uncertainty, which is caused by a number of destabilizing factors. Such situational uncertainty during the AFR flying must be reduced through procedures of operational diagnosis of the vehicle extrinsic environs.

Diagnostic procedures starting from detection of obstacles and special visual references assist identifying and localizing them, i.e. providing for a complete diagnosis of the flight circumstances what allows the mission to be worked in safer conditions.

It is the usage of the computer vision in the AFR control system that offers opportunity to significantly reduce a situational uncertainty. For a computer vision system implementation in a present-day autonomous robot, it is advisable to use the Raspberry Pi hardware platform with the Raspbian operat-

Figure 14. Interval meanings of experimentally determined values

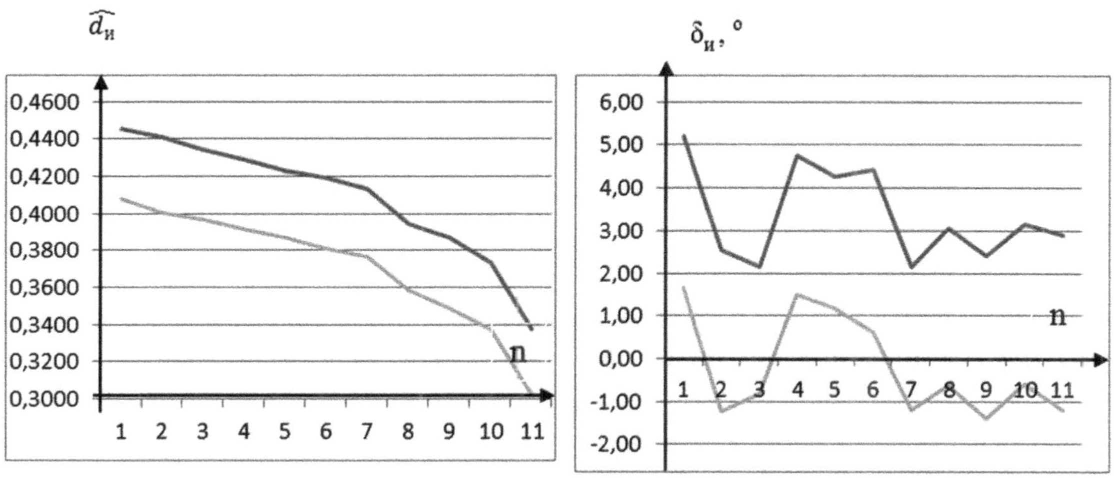

a)
Distance-to-orientator interval estimate

b)
Interval estimate of viewing angle relative to camera frame horizontal axis

ing system recommended by a manufacturer. In this case, the basic programming language will be the Python which is adapted to share the Pillow and OpenCV libraries resource.

Analysis of algorithms for solving navigation problems based on video information processing shows that most of them have the comparable computational fidelity, but differ significantly in terms of the accuracy of final outputs. One of efficient algorithms dedicated to searching for visual orientators in the system of computer vision is the technique of Hough transform.

In the course of research, the experimental setup for testing an autonomous flying robot equipped by the computer vision system and the schedule of appropriate experiments were created. To provide the robot flight toward a special target (visual orientator), the movement control algorithm has been developed. As a result, the experimental estimates of the robot to orientator distance and of the measurement of the viewing angle during a flight have been gotten.

REFERENCES

Ballard, D. H. (1981). Generalizing the Hough transform to detect arbitrary shapes. *Pattern Recognition*, *13*(2), 111–122. doi:10.1016/0031-3203(81)90009-1

Bradski, G., & Kaehler, A. (2008). *Learning OpenCV: Computer vision with the OpenCV library*. O'Reilly Media, Inc.

Dergachov, K. Y., Krasnov, L.O., & Radomskyi, O.M (2017). Methods of local navigation of mobile robots based on of technical vision. *Electrotechnic and Computer Systems*, (25), 295-300.

Guevara, J., Jiménez, A. R., Prieto, J. C., & Seco, F. (2012). Auto-localization algorithm for local positioning systems. *Ad Hoc Networks*, *10*(6), 1090–1100. doi:10.1016/j.adhoc.2012.02.003

Gulevitch, S. P., Veselov, Y. G., Pryadkin, S. P., & Tyrnov, S. D. (2012). Analysis of factors affecting safety of flight of unmanned aerial vehicles (UAV). Causes of accidents of UAVs and methods of preventing them. Science and education Russia, Bauman Moscow State Technical University.

Huang, A. S., Bachrach, A., Henry, P., Krainin, M., Maturana, D., Fox, D., & Roy, N. (2017). Visual odometry and mapping for autonomous flight using an RGB-D camera. In *Robotics Research* (pp. 235–252). Cham: Springer. doi:10.1007/978-3-319-29363-9_14

Isermann, R. (2006). *Fault-diagnosis systems: an introduction from fault detection to fault tolerance*. Springer Science & Business Media. doi:10.1007/3-540-30368-5

Jin-Yu, Z., Yan, C., & Xian-Xiang, H. (2009, April). Edge detection of images based on improved Sobel operator and genetic algorithms. In *Image Analysis and Signal Processing, 2009. IASP 2009. International Conference on* (pp. 31-35). IEEE.

Kolodziej, K. W., & Hjelm, J. (2006). *Local positioning systems: LBS applications and services*. CRC Press. doi:10.1201/9781420005004

Kulik, A. S. (1991, September). Fault diagnosis in dynamic Systems via signal-parametric approach. In *IFAC/IMACS Symp. Baden-Baden* (*Vol. 1*, pp. 157-162). Academic Press.

Kulik, A. S. (2014). Rational control of aerospace objects operability under the influence of the destabilizing actions. *Aerospace Engineering and Technology*, (1), 31-38.

Kulik, A.S., & Radomskyi, O.M. (2017). Usage of a technical vision unit in a mobile walking robot control system. *Electrotechnic and Computer Systems*, (25), 347-356.

Odarchenko, R. S., Gnatyuk, S. O., Zhmurko, T. O., & Tkalich, O. P. (2015, October). Improved method of routing in UAV network. In *Actual Problems of Unmanned Aerial Vehicles Developments (APUAVD), 2015 IEEE International Conference* (pp. 294-297). IEEE. 10.1109/APUAVD.2015.7346624

Retscher, G. (2016). Indoor navigation. Encyclopedia of Geodesy, 1-7. doi:10.1007/978-3-319-02370-0_9-1

Shmelova, T. F., & Bondarev, D. I. (2015, October). Graph theory applying for quantitative estimation of UAV's group flight. In *Actual Problems of Unmanned Aerial Vehicles Developments (APUAVD), 2015 IEEE International Conference* (pp. 328-331). IEEE.

Sładkowski, A. (Ed.). (2013). *Some Actual Issues of Traffic and Vehicle Safety. Faculty of Transport*. Silesian University of Technology.

ADDITIONAL READING

Mojsilovic, A. (2005). A computational model for color naming and describing color composition of images. *IEEE Transactions on Image Processing*, *14*(5), 690–699. doi:10.1109/TIP.2004.841201 PMID:15887562

KEY TERMS AND DEFINITIONS

Computer Vision: Implements a complex process of extracting, identifying, and converting video information, which includes six basic steps: 1) obtaining (perceiving) information, or sensing; 2) preconditioning, or preprocessing; 3) segmentation; 4) description; 5) recognition; and 6) interpretation.

Computer Vision Systems: Sensory assemblies that provide shooting of the work scenes and objects, and images converting, processing, and interpretation using an UAV on-board computer and further the transfer of results to the management device.

Destabilizing Factors: Such objectively existing interfering impacts that are needed to be identified and evaluated in the course of carrying out the respective mission.

Environs Diagnosis: A set of activities aimed at reducing situational uncertainty in the operation of autonomous flying vehicles, for example, the measures include procedures for interference detecting, identifying, and localizing to ensure the complete knowledge of outdoor circumstances.

Hough Transform: An algorithm used to extract image elements. The technique supplies the image analysis and digital processing in a variety of computer vision facilities; is designed to search for primitives that belong to a certain shape types with utilizing a voting procedure.

Off-Nominal Situations (Contingency Events): Uncertain events with respect to both a moment of their occurrence and to an obscure feature of the reason causing the situation.

Visual Reference, Visual Orientator: An object perceived by the eye of a person or an optical device as a quest visual form.

Chapter 5
Decision Support Systems in Aeronautics and Aerospace Industries

Ramgopal Kashyap
https://orcid.org/0000-0002-5352-1286
Amity University Chhattisgarh, India

ABSTRACT

The aim of this chapter is to utilize the upkeep hypothesis of on-condition, condition checking. It proposes support ideas: for example, advanced upkeep, proactive upkeep which is bolstered by solid checking and finding strategies to enhance the effectiveness of upkeep extraordinarily; so there must be decision support system framework to develop a present-day flying upkeep framework. Maintenance DSS can give powerful choice help to aeronautics support. It additionally makes strides in upkeep effectiveness and controls support costs. The flying MDSS still needs further research on the accompanying subjects. Since current airplanes have numerous frameworks and complex structures, numerous flight parameters need to be screened. Step-by-step instructions to assemble an effective database and research criteria for judging the adequacy of ongoing information should direct further research. Since continuous flight status information has cost issues, how to set up the data download criteria should direct further research.

INTRODUCTION

The essential reason for flying support is to keep the diminished levels of unique flying machine airworthiness, security and unwavering quality outline at the least cost, which is of extraordinary hugeness to the running of air transport endeavors. As per insights, flight upkeep represents roughly 20% to 30% of direct working expenses, excluding roundabout expenses caused because of support, for example, flight deferrals, material and extra parts, corporate picture. Lately, as the quick improvement of common avionics industry, a substantial number of new advancements have been utilized as a part of planes, which increment the multifaceted nature of the common flying support, particularly the use of computerized data innovation. In light of present-day avionics support of the new circumstance, in order to adjust to

DOI: 10.4018/978-1-5225-7709-6.ch005

the future improvement pattern of computerized upkeep and enhance the productivity and nature of current flying machine support, building up the vital current air transport flying machine upkeep choice emotionally supportive network turns out to be progressively dire. Therefore, this chapter tops to bottom investigation the issue of how to construct the flight Maintenance Decision Support System (MDSS). The current correspondence innovation, fault conclusion innovation, and present-day control innovation, flight upkeep choice emotionally supportive network can accomplish constant information trade amongst air and ground, and it can give powerful choice help to fast fault finding and advanced support. This framework ought to be made out of data securing and handling, flying machine upkeep bolster, material administration, upkeep administration, support occasion assessment, and so forth.

Information Obtaining and Preparing

Aircraft Decision Support System (DSS) needs an assortment of continuous information and chronicled information for choice help judgments, and every one of this information are gathered by data obtaining and preparing module with the vast number of data innovation connected in common airship, present-day airplanes are outfitted with focal support PC, airship condition checking framework and other continuous airship state parameters securing and screen frameworks, which make it conceivable to screen airplane state parameters amid the entire flight. In particular, the gathered airplane state parameters are inputted to the airship online fault determination framework, which analyzes the fault of the airship by the built up online fault finding model (Novák, Kameníková & Podhadský, 2016). The result ought to be constantly transmitted to the ground framework through the air-ground information interface, for example, Aircraft Communications Addressing and Reporting System (ACARS), in order to give choice help to ground support and accomplish quick repair of airship disappointment, consequently enhancing productivity in the utilization of airship as appeared in Figure 1.

Historical Information Obtaining and Handling

Flight data is typically firmly connected to airship disappointment recorded by Quick Accesses Recorder (QAR), through the inside and out an investigation of recorded flight status parameters; we can take

Figure 1. Flying upkeep choice framework graph

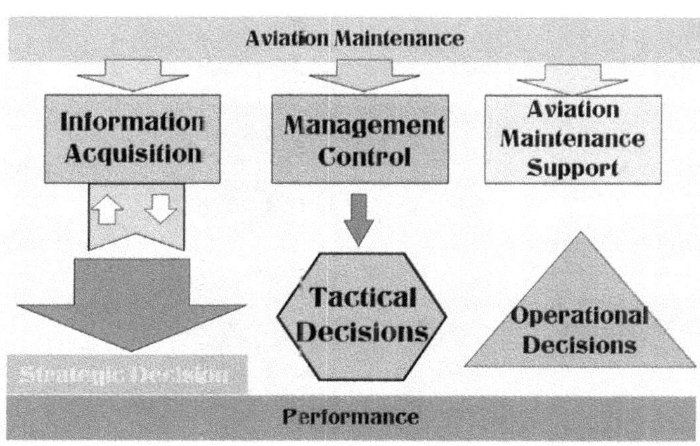

in the state of the plane in flight. Utilizing the causal connection between the framework disappointment manifestations and the reason for the glitch, we can make online flying machine fault finding model to give choice bolster for online flying machine fault determination, and in view of which build up an on-load up fault analysis framework. Aircraft support records are the delineation of the general development of the airplane and frameworks (Dalkilic, 2017). Flying machine data can be utilized to acquire a profound relationship between the framework disappointment manifestations and the reason for the breakdown, it moves toward becoming especially valuable when comes to take in the connection amongst framework and condition, and the cooperation chain connections amongst framework and framework, that makes it conceivable to direct more extensive and inside and out fault finding on the premise of online fault determination, to give firm choice help to the ground support. The procedure of flying upkeep will aggregate a lot of verifiable information. This information can be partitioned into three sorts, as appeared in Figure 2.

Air material is the material of aeronautics support, to accomplish the many-sided quality capacities and superior of the current flying machine, there are numerous sorts of flying machine frameworks, while diverse sorts of air material have distinctive cost, diverse loss of consistency and diverse stockpiling conditions the required temperature, stickiness, stockpiling time (Тамаргазін, 2016). Hence the despicable choice of air material control methodology may prompt inadequacy or overabundance of air material, then again, if make great utilization of air material chronicled utilization information, at that point we can comprehend the status of air material misfortune and build up a fitting obtainment technique to keep up a sensible level of air material save, to enhance aeronautics material use and the effectiveness of undertakings.

Figure 2. Authentic information securing and handling outline

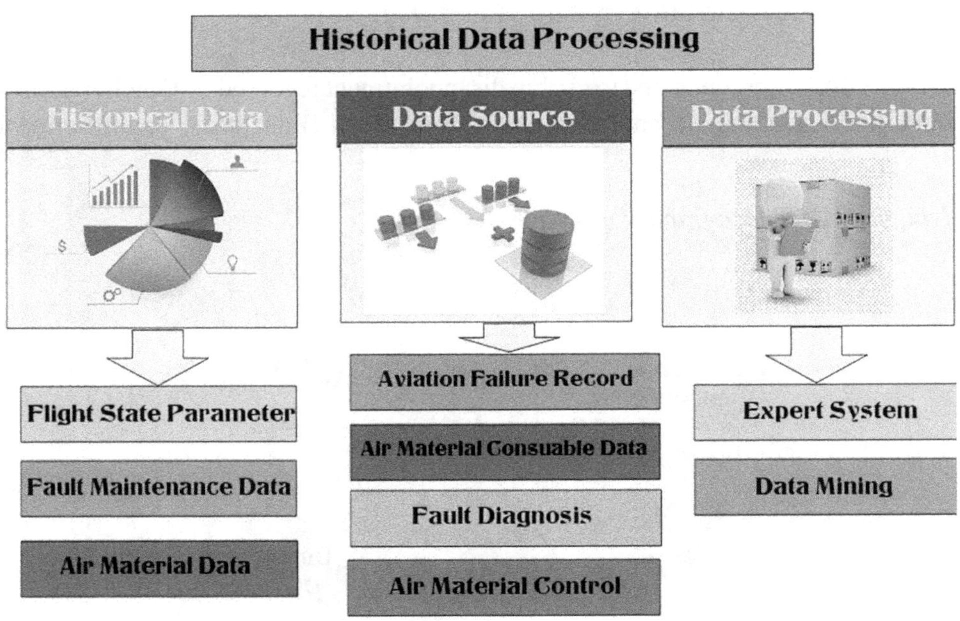

Aviation Upkeep Specialized Approach

Aeronautics upkeep specialized help modules depend on the gathered and prepared information and the current upkeep assets. In particular, it can be partitioned into two angles:

The Fault Diagnosis Model and Technique

Fault diagnosis demonstrates is partitioned into online and disconnected analysis display. On-line finding model gets history airship state parameters from information securing and handling module, through information securing and preparing module for the airship state parameters of the procedure history information, utilizing more advanced master framework, fluffy distinguishing proof and other customary fault analysis, set up to recognize diverse frameworks furthermore, segments of the on-line fault conclusion demonstrate in light of the instance of genuine activity, can be utilized time of determination, symptomatic precision and different parameters on-line fault conclusion model to assess the outcomes in view of assessment of demonstrative model of the online database to be refreshed (Sayed-Mouchaweh & Lughofer, 2015). The plane landed utilizing Wireless Fidelity/Personal Digital Assistant (WiFi/PDA) and other specialized methods for transmission fault code list, lessening the stream of ground information interface and enhance the proficient utilization of ground information connect. Disconnected fault diagnosis model in light of data securing and preparing module of the airship support record data acquired through inside and out examination of the framework fault diagnosis (Oravec & Jadlovská, 2017), especially between the framework and nature, frameworks and cross-linking between the laws of the framework, the foundation of fault manifestations, and the causal connection between the fault kinds, the utilization of counterfeit neural system, Petri net fault analysis innovation of present-day flying machine far-reaching symptomatic information. Determination entered into the flying upkeep specialized documentation for the intellectual framework, coordinate access to a suitable upkeep program. Likewise, as per the framework with authentic fault information can likewise utilize drift investigation, bend fitting, time arrangement and different strategies for various sorts of issues anticipate the pattern anticipated outcomes utilizing the fault determination can be evaluated for disappointment symptomatic help.

Maintenance of Specialized Documentation

The flying upkeep specialized documentation is another critical part of avionics upkeep specialized help, specifically, support of avionics intuitive electronic archives. Present day flying machine is a coordinated assemblage of cutting-edge fabricating innovation, process innovation, and data innovation. It delivers and uses heaps of specialized data in the outline, creation, utilizes and support process, for instance, information from the producer including Aircraft Maintenance Manual (AMM), Fault Reporting Manual (FRM), Fault Isolation Manual (FIM). Generally, a large number of these specialized archives utilize paper for the capacity medium, or then again to some extent, have been digitized, yet an expansive number of specialized records have issues of questions troublesome, difficult to convey and protection and hard to refresh (Jin, Chen, Jia & Zeng, 2016).

Intuitive flight upkeep specialized manual coordinates favorable circumstances of innovation of application PC, interactive media, database and arrange, sort out and oversee complex content working manuals and support data as indicated by significant principles, execute exact execution of required data

Figure 3. Graph of air-ground fault finding

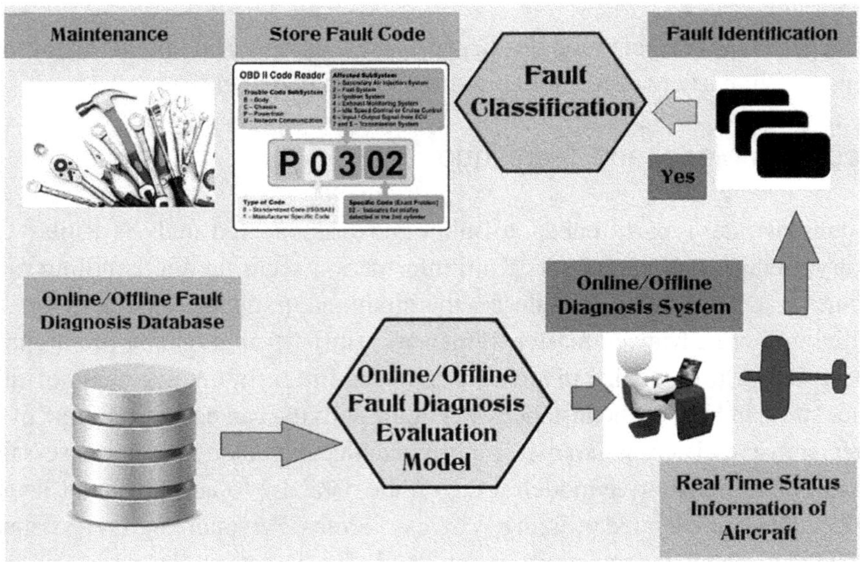

Figure 4. The types of support innovation archives

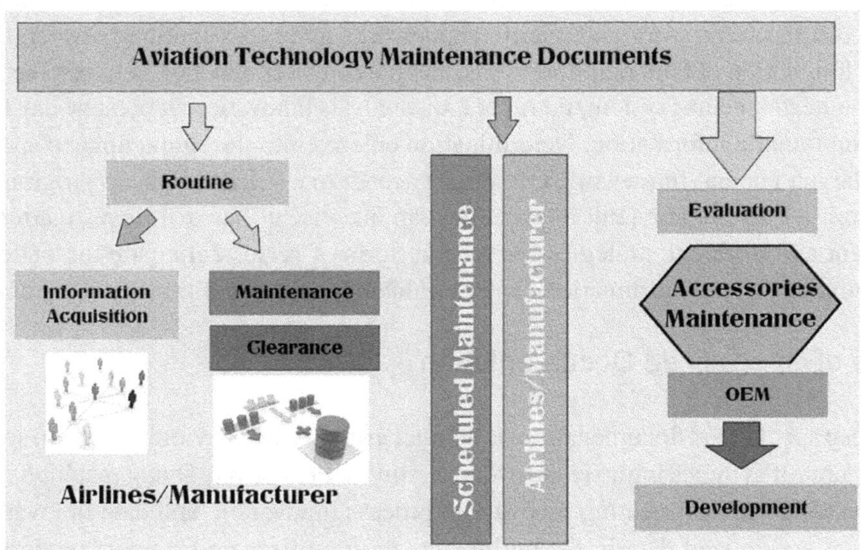

of support staff with streamlined way. The reason for it is to quicken the advance of aeronautics upkeep, improve the viability of support.

MANAGEMENT AND CONTROL OF AIR MATERIAL

Air material stock, which is the most elevated cost in flying support process, is an imperative part of aeronautics support process for the multifaceted nature of present-day airship segments, an assortment of airplane composes and materials, and the utilization of various guidelines of air material, distinctive capacity prerequisites, and flight wellbeing. In this manner, control of air material has an extraordinary effect on the effectiveness of ventures. Through we utilize recorded data for data obtaining and handling module to get air material, it can pick up the normality of misfortune and purchase of air materials utilizing systems of factual investigation and information mining, for example, the misfortune consistency of repairable air materials and unrepeatable air materials, the kept up normality of kept up air materials, and so forth, at that point through the levelheaded control procedure, it sets up the relating misfortune mode, executes streamlining of air material stock and enhance the nature of upkeep (Alexander & Merkert, 2017).

What's more, air material stock control ought to think about the technique for shared provincial flying support units, particularly with high esteem, and there might be more genuine issues with the absence of basic flight hardware, for example, airship motors, and so on. Through displaying and investigation of chronicled information, it can build a sensible model for local air material sharing, and afterward, add to improve the effectiveness of the business for the entirety district.

Aviation Support Administration

Flying support administration is isolated into upkeep program advancement, designation of support undertakings, and upkeep assignments planning under the preface of security, the fundamental objective of flying support is monetary. A quick, exceptionally solid support can build the quantity of airship flight hours, accordingly upgrading financial proficiency. Flying Maintenance Management Module oversees and controls the support venture in view of the rundown of upkeep errands from specialized bolster module. Upkeep design is the center of support assignments. It decides the usage of upkeep strategies in light of the level of support undertakings, the current upkeep assets labor, gear, offices, material, and so on and the airplane status. Upkeep design incorporates numerous angles, for example, doing well ordered

Figure 5. Administration and control of air material demonstrating

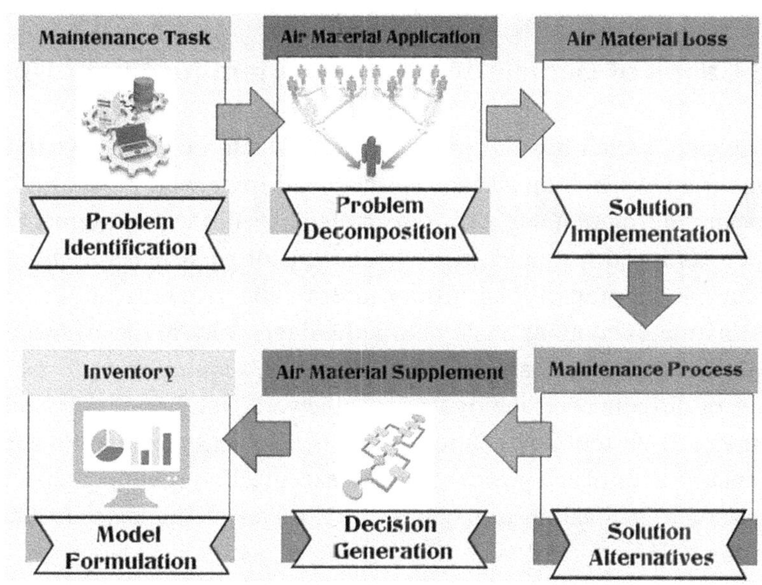

or in parallel ways, the name of upkeep, repair content, support specialists, and materials required for the upkeep, and so on (Le Grange & Buys, 2012). Support design makes sensible plans for the future condition of repair to guarantee the smooth execution of flight support. Solid repair work is finished by mechanical, electronic, basic, adornments upkeep workshop. The portion of support errands is to keep up a sensible circulation of work to the proper workshop, to empower the repair shops doing the upkeep work well ordered or in a parallel way and guarantee the smooth advance of upkeep ventures. In the real upkeep ventures, support projects will definitely be balanced, since the calendar may be preposterous, or something comes up as a crisis or mischance. Support booking is to utilize sensible planning technique and fitting break down or join support errands or simply change the grouping to guarantee the smooth usage of upkeep works as per the advance of the arrangement.

Maintenance About Assessment

The avionics upkeep choice emotionally supportive network is an open framework to accomplish a goal what's more, sensible evaluation, there ought to be a sensible upkeep occasion parameters assessment framework, consequently accomplishing ideal and refresh upkeep choice help framework. Assessment of upkeep occasions ought to contain no less than three viewpoints: (1) Pertinence of on-line fault finding model. On-line fault finding framework is an onboard framework, so its analysis ought to be continuous and needs clear disappointment depiction while its relative analysis precision isn't high. (2) The relevance of disconnected fault diagnosis (Weckman, Marvel & Shell, 2006). Disconnected fault analysis framework has adequate assets and time to judge the on-line determination framework's outcomes in detail, along these lines, disconnected fault finding framework isn't basic on the finding time, yet the precision of conclusion must be high, subsequently sensible support errands can be resolved, subsequently, the successful support can be executed and upkeep expenses will be lessened. (3) Reasonable air materiel control show. Air material includes different viewpoints, for example, mechanical parts, electronic segments, and there are repair materials and support consumables. Sensible air materiel setup meets the upkeep necessities, as well as decreases the cost from the acquisition, stockpiling, transportation, and so on and generally the carrier may have an exceptionally negative effect on tasks.

Hypothesis and Cases of Decision Support System for Flight Upkeep

The DSS of flight upkeep, which has the properties of discrete occasion dynamic frameworks and nonstop time dynamic framework, is an average half and half framework. The arrangement of securing and handling of constant information ought to contain elements of the procurement of parameters data, separating, change et cetera, and it should utilize hypothesis of persistent variable unique frameworks to develop while securing and preparing of history information generally utilize pertinent hypotheses of insights and investigation. It contains a parcel of individuals work in the arrangement of administration and control of air materials, for example, upkeep process, acquirement process, and so forth., and has evident properties of discrete occasion dynamic framework, yet through sensible transformation, it can utilize an assortment of present-day control techniques for least fluctuation control, ideal control, stochastic control, versatile control and power control to execute. Flying support specialized records may give careful consideration to innovation and hypothesis of mixed media, database/information distribution center (Yu & Wang, 2013).

At the same time, since frameworks utilizing discrete occasion dynamic framework property cannot be communicated utilizing the customary issue of parsing, it can utilize knowledge hypothesis to explain, for instance, fault analysis can utilize hypotheses of the neural system, Petri nets, fluffy thinking, specialist, et cetera. With the advancement of hypothesis and innovation of PC recreation, it can utilize more approaches for the exploration of the model and strategy for flying machine fault conclusion, the model of air materiel utilize and the air material stock control, choice emotionally supportive network of aeronautics support, and utilize full-reproduction examination of the framework to build up a sane furthermore, proficient choice emotionally supportive network for flight upkeep. The accompanying is a sure kind of flying turbine stream motor fault conclusion model of airplane tire building process and stock control process for instance.

Aircraft Fault Analyze in View of Versatile Fuzzy Petri Nets

Air motor, which has auxiliary multifaceted nature, has various segments, works in the high temperature and cruel condition, so it might be fault whenever. At the point when there is a fault, on account of the quantity of comparative fault, it makes challenges for air motor investigating. For instance, a specific kind of air bearer turbojet has the fault of a fuel utilization of vast slip; it can utilize versatile Fuzzy Petri Nets (FPN) to set up fault analysis model to position the reason for such fault (Amin & Shebl, 2014). The procedure of model set up as takes after:

1. The meaning of model information factors and yield factors, the model information factors is the rationale condition of fault side effects, characterizing as

$X_1 X_2 X_3 X_4 X_5 X_6 X_7$

The significance as appeared in table 1; show yield factors are level of certainty of fault causes, characterizing as

$Y_1 Y_2 Y_3 Y_4 Y_5$

The significance is as appeared in table 2.

Table 1. Definition of fault symptom and input variables

Variable name	X1	X2	X3	X4
Symptom of Phenomenon	Oil in spout at the base of the turbine	Oil at blower entrance, fore tank base skin	Oil at plane tail cone	White smoking at toward the back weight reducer
Variable name	**X5**	**X6**	**X7**	
Symptom of Phenomenon	Oil dabs at radial ventilator's vent	Ejector tube radiating purple smoke	Trip noticeable all around, oil utilization overwhelming	

Table 2. Meaning of fault reason and yield factors

Variable Name	Y1	Y2	Y3	Y4	Y5
Fault Cause	Oil pump's one-way valve not fixing	One-path valve of radiator interface not fixing	Outward valve layer break	In bearing hole, weight as well high	Three class rummaging pump wasteful

2. Initializing the model parameters, enrollment parameters of information fault indications and fault reason, this is given by the master framework and repair industrial facility's measurable information, separately set to as appeared in table 3 and table 4.
3. According to the fluffy surmising principles of FPN, it characterizes organize calculations on each floor, utilizes the most extreme level of enrollment and the guideline of participation review limit ($\gamma>0.8$) and consolidates with the meaning of fluffy calculations to develop the fault conclusion show FPN.
4. According to the real upkeep condition, correct participation between fault side effect and fault causes in beginning model parameters and the correction equation is as takes after:

$$d_{ij} = \frac{\text{times of side effect which have a place with j reason}}{\text{the aggregate circumstances of I manifestation}}$$

The parameters mean. It can influence FPN to correct model parameters online as per fault event.

Table 3. The participation review relationship measurements between fault indications and fault causes in the repair plant

Fault Symptom xi							Membership Grade	Fault Cause yi				
X1	X2	X3	X4	X5	X6	X7	d_{ij}	Y5	Y4	Y3	Y2	Y1
0	0	0	0	0	0	1	d_{01}	0.2	0.4	0.3	.1	0
0	0	0	0	0	1	0	d_{02}	0.15	0.5	0.35	0	0
0	0	0	0	1	0	0	d_{03}	0	0	0.6	0.3	0.1
0	0	0	1	0	0	0	d_{04}	0.25	0.5	0	0	.25
0	0	1	0	0	0	0	d_{05}	0.75	0	0	25	0
0	1	0	0	0	0	0	d_{06}	0.8	0	0.15	.5	0
1	0	0	0	0	0	0	d_{07}	0	0.2	0	0.8	0

Table 4. Expert database of membership grade relations between fault symptoms and faultcauses

Fault Symptom xi							Membership Grade	Fault Cause yi				
X1	X2	X3	X4	X5	X6	X7	d_{ij}	Y5	Y4	Y3	Y2	Y1
0	0	0	0	0	0	1	d_{11}	0.45	0.3	0.25	0	0
0	0	0	0	0	1	0	d_{12}	0.35	0.6	0.05	0	0
0	0	0	0	1	0	0	d_{13}	0	0	0.4	0.5	0.1
0	0	0	1	0	0	0	d_{14}	0.50	0.4	0	0	0.1
0	0	1	0	0	0	0	d_{15}	0.85	0	0	0.15	0
0	1	0	0	0	0	0	d_{16}	0.7	0	0.15	0.15	0
1	0	0	0	0	0	0	d_{17}	0	0.3	0	0.7	0

As per the above stream, versatile FPN fault finding model is set up as appeared in Figure 6. fault indication state vectors, reproduction is done, reenactment comes about and the genuine reason affirmed in support (Peng Zhang at el 2011). This demonstrates the dependability of model analysis is generally high there is some deviation between yield of fault model finding and aftereffect of real support, since taking technique for limit participation review to dispense with fluffy, demonstrative proficiency little reductions, yet at the same time to maintain a strategic distance from effect of irregular variables, which could bring about the deviation of symptomatic comes about.

TIRE STOCK ADMINISTRATION AND CONTROL SHOW

Assume that a similar kind of airships in an armada of aircraft "A" devour and supplant tires regular, measure of which is X1 (k) k is a date serial number, and X1 (k) is the irregular process. Considering the normality of flights, we can imagine that there is no relationship between loss of factual laws of tires and dates from the designing purpose of large-scale see. In the meantime, it is viewed as that the measure of tire loss of any day does not influence the measurement of tire loss of any day, later on, that is, X1 (k) and X1 (k+1) (h $\neq$ 0) are independent. In view of above-given conditions, clearly, tires required changing regularly is taken after Poisson appropriation, so it is a Poisson procedure. For controlling air material tires stock needs new tires to supplement at a specific period, or invigorate utilized tires. Set the period is M days, we ought to think about the measurable character of the process

XM(k)= X1(k)+X1(k+1)+……+ X1 (k + M -1). X1 (k), X1 (k+1), ……, X1 (k + M -1)

is free of each other, took after a similar conveyance, and has restricted numerical desire and fluctuation. As indicated by the focal point of confinement hypothesis, when M watches out for limitlessness really, as long as M is sufficient huge XM (k) complies with the ordinary conveyance. It is that XM (k) is taken after a typical irregular process. It ought to be noticed that the diverse armadas, distinctive airplanes have diverse X1 (k). As distinctive aircrafts with various courses, taking off and landing recurrence of flights are very diverse consistently from two times per day to dozen daily, and distinctive flying machines may have an enormous distinction in tire wear, consequently, the tire's misfortune properties measurable of various armada are extraordinary (Svenmarckt, P., & Dekker, S.,2003).

After the obtaining of air material stock and support information from various carriers, it utilizes Statistical Product and Service Solutions (SPSS), MATLAB and another programming for information measurable preparing. An instance of aircraft "An" is investigated and utilized the tire misfortune information of an armada which incorporates ten same flying machines from March 1, 2008, to November 1, 2008. After information examination, there are tables as takes after the consequences of fitting investigation with the ordinary process, even process, poisson process, and exponential process. Examination underneath appears:

1. Tires misfortune process is an arbitrary procedure and has a high fitting degree with a Poisson process.
2. The expansion in figuring cycles, the fitting degree with the typical procedure is expanding.
3. From the autocorrelation examination, we discover that when the deferral is 0, the relationship coefficient is 1; the pertinent degree is under 0.2 under the rest delay. We can get the conclusion from beneath examination as takes after:
    a. Taking the date serial number as an arbitrary factor, tires misfortune time-arrangement can be dealt with as a Poisson procedure.
    b. Tire misfortune information relationship is feeble between various days. In this way from a designing perspective, the loss of tire number of various dates is autonomous.
    c. With the expansion of figuring cycles, arbitrary conveyance watches out for ordinary circulation. Use of building control hypothesis to take care of the issue of ideal stock appears to be essential.

Prospects and Difficulties

As of late, in light of generally utilizing of the supported hypothesis of on-condition, condition observing, and so on., it proposes support ideas, for example, advanced upkeep, proactive upkeep et cetera, which bolstered by solid checking and finding systems to enhance the proficiency of upkeep extraordinarily, so there must be solid choice help framework to develop current flying support framework. MDSS can give powerful choice help to flight support. It likewise makes strides upkeep proficiency and controls support costs (Ursavas, 2014). In any case, the flying MDSS still needs further research on the accompanying points:

1. Since present-day airplane has numerous frameworks and complex structure, numerous flight parameters need to screen. Step by step instructions to assemble an effective database and research criteria for judging the viability of ongoing information should direct further research.

2. Since ongoing flight status information downloaded identify with cost issues, how to build up the fault data download criteria should lead additionally look into.
3. The framework has a property of cross breed framework, and there are numerous courses for DEDS and CVDS displaying; however hypothesis and strategy for demonstrating the half and half framework is as yet required to be enhanced, so how to natural join CVDS framework with DEDS framework and displaying the half breed framework should lead additionally look into.
4. MDSS is firmly related to the carrier working condition, as there are self-evident contrasts between carriers, how to choose suitable parameters, and how to beat the terrible materialness should direct further research. On present-day warrior airship, an ever increasing number of frameworks are actualized keeping in mind the end goal to enhance the pilot's mindfulness about the present circumstance of the airship and the world encompassing it. As the number and intricacy of these frameworks increment so do the number of dangers to the flying machine and suitable countermeasures for the pilot to browse. To enable the pilot to settle on a legitimate equivocal activity at the point when a danger happens a Decision Support System (DSS) can be executed on board the airship. This part depicts the requirement for a DSS in a warrior airplane, the necessities such a framework must conform to, and the stream of information on which choices from the framework must be made. Existing trial and operational frameworks are additionally portrayed.

Problem Description

A DSS is portrayed as "a gathering of PC based intuitive applications, which in light of area particular learning and data bolster a chief in at least one periods of the choice procedure." A DSS might be in view of a specialist framework which is a PC program that expands upon area particular information from at least one specialist. A DSS for military pilots will expand upon master learning inside the field of EW. Rockets might be let go from the ground or ocean-based platforms, or they might be let go from adversary contender airship in an aerial commitment. In the last case, the workload on the pilot will be higher than it will be for land or ocean based dangers since the position, height, and speed of the adversary flying machine must be taken into thought too. In this work, the subject is to explore intends to plan a DSS for discovering reactions to ground-based dangers as it were at the point when connected by ground-based dangers at least one rockets might be propelled towards the flying machine (Noh & Choi, 2012). The pilot may utilize countermeasures to stay away from the effect of a moving toward rocket. The countermeasures to pick depend on e.g. the sort of rocket, the separation and course between the airship and the rocket, the antagonistic vibe of the earth, elevation of the airship, and the accessibility of countermeasures. Information about dangers that might be met in the not so distant future may likewise impact the decision of appropriate countermeasures and the grouping in which they are utilized (Mahdavi & Shirazi, 2010). On the off chance that the airship is outfitted with a constrained measure of disposable countermeasures it may lessen the survivability of the flying machine for the whole mission if at any given time more expendables were utilized than should be expected, along these lines leaving none for a later need. Models for the DSS are outlined utilizing procedures from the fields of Artificial Intelligence (AI) and Operations Research (OR) from the field of AI the likelihood of utilizing the Prolog programming dialect is analyzed. This is picked since the strategies for reactions to ground-based dangers can be figured as a set of principles that can be actualized utilizing Prolog. At DDRE a theory has been composed regarding the matter of utilizing Bayesian System (BS) for choice help for military pilots. BN may likewise be considered as an AI innovation, and it is developed the encounters picked up by that work by looking at additional utilization of BN. The choices proposed to the military pilot may enhance if the DSS is composed to deal with transient viewpoints. These angles may portray confines on the utilization of countermeasures amid a mission. For this, it is portrayed the issue utilizing OR methods. The issue is portrayed by a scientific model that can be understood to optimality. A heuristic is likewise connected, and here an exchange off between the nature of arrangements and the time it takes to discover them is made.

Survivability

The point of this work is to outline a DSS that may expand the survivability of the warrior flying machine when flying a mission over adversary domain. In this survivability is named Aircraft Combat Survivability (ACS) and it is characterized as: "The ability of an airship to stay away from or withstand a man-made antagonistic condition." The survivability is identified with the terms: powerlessness, helplessness, and killability as depicted beneath (N.G. Buromensky, 2016).

Vulnerability

The weakness of an airplane depicts the failure to keep away from rockets, radars, firearms, and different components of the unfriendly condition. Defenselessness at the point when the components of the unfriendly condition cannot be kept away from the defenselessness depicts the failure of the airplane to withstand the condition.

Killability

The nullability depicts the likelihood of the airship being "slaughtered" because of adversary activities. This relies upon both the helplessness (the airplane must be hit) and the defenselessness (this hit must reason adequate harm to slaughter the airplane) of the flying machine.

The survivability is the inverse of the killability having a high likelihood of being slaughtered will bring about a low likelihood of surviving, and the other way around. All through the writing on flying machine survivability (Noh & Choi, 2012). The survivability is regularly alluded to as PS, while the killability is alluded to as PK. The connection between PS and PK is PS = 1−PK. For a few dangers, the likelihood of having an effect on the flying machine can be built up. In the event that the likelihood for e.g. a vicinity melded rocket being intertwined by the flying machine is known as PF and the likelihood of the airship being executed by a nearness combined rocket is known as PK|F the survivability of such a rocket assault is given by:

$$PS = 1 - PK = 1 - PF \bullet PK|F.$$

The likelihood PK|F relies upon e.g. the development of the airplane. While PS might be figured for a given rocket assault, discovering it for a whole battle mission is more unpredictable. Here the probabilities of e.g. the flying machine being recognized by the foe, and the foe participating in an assault of an identified airship should likewise be built up.

Design Requirements

A DSS for military pilots must satisfy various prerequisites to be appropriate amid a mission underneath these prerequisites are portrayed. In planning a strategy for proposing activities to the pilot these necessities must be taken into thought (Alfredson & Ohlander, 2016).

The framework needs to discover answers for happening dangers in close real time it might take as meager as 2-3 seconds from a rocket has been propelled until the point when it achieves the flying machine. Before moves can be made to maintain a strategic distance from the approaching rocket, sensors on-board the flying machine must distinguish the rocket, the framework must locate a suitable arrangement of activities, and these activities should be exhibited to the pilot. To give the pilot sufficient time to perform sly activities it is evaluated that the framework has roughly 200 milliseconds from an adjustment in the danger situation happens until an arrangement of activities must be proposed to the pilot. Since the last execution of the framework must be kept running in a warrior airship, the equipment required for building up the framework must match the prerequisites given to equipment in the flying machine. The outcomes from a DSS relies upon information contribution from different gadgets onboard the airship and thus it must be anything but difficult to coordinate the DSS with these frameworks.

The portrayals of dangers and direction frameworks are continually advancing as are new countermeasure frameworks. In this manner, the framework created must be effectively refreshed and looked after. To guarantee this the framework should ideally be information-driven, and refreshing the framework will at that point involve refreshing information on rocket frameworks, direction strategies, and so forth. The calculations utilized must have none or minor updates as it were. Any arrangement from the framework must appear to be sensible to the pilot. Generally, the framework won't be trusted and henceforth not utilized. This necessity can be separated into two sub-prerequisites: the framework must propose a sensibly answer for any adjustments in the situation, e.g. at the point when another risk happens, and when no dangers are impending no recommendations ought to be given. Both in battle and amid tests in the advancement stage the client of the framework will be a military pilot. In the event that the pilot does not see how the arrangements proposed by the framework can be induced, he may not trust their legitimacy. Accordingly, the idea of the surmising parts of the framework must be moderately straightforward. The handiness of a DSS is its capacity to enhance the survivability of the pilot. On the off chance that the framework is restricted to just propose activities to the pilot inside a small amount of the considerable number of circumstances he may get himself and the airplane in it is of no utilization. The contrast between this prerequisite and the past is that a framework might be dependable just inside given cutoff points without being helpful to the pilot. An illustration about is a framework that proposes activities to alleviate e.g. RF dangers as it were.

Results from the framework must be introduced to the pilot in such a path, to the point that they are simple and quick to translate. The introduction can be visual, utilize sound, or be a mix of both. Amid the assessment of conceivable strategies for building up the framework the introduction of the proposed activity is of minor significance. In the improvement of a framework that will be handled this prerequisite must take a high need. In more issues are specified as basic to the plan of a DSS. Among these issues are the utilization of information pressure procedures, an easy to use database interface for refreshing the information utilized by the framework, and powerful memory administration. None of these issues are viewed as significant in this work and the dealing with of these is past the extent of the work. Information from various sensors, or from a similar sensor over some stretch of time, might be combined to give the pilot a superior situational mindfulness. The learning of performing information combination is its very own point, and it too is considered past the extent of the work.

Mission Data Flow

Prior to a mission is started data about conceivable places of adversary radars and platforms are gathered (Bernardini, 2017). This is finished amid a period of the arrangements known as the Intelligence Preparation of Battlefield (IPB) from this the Electronic Order of Battle (EOB) is set up. The EOB depicts the front line in subtle elements utilized by the pilot and by various frameworks onboard the airship. Amid the mission, the pilot and on-board frameworks will persistently recover data about the front line from onboard sensors and perhaps likewise from different sources, for example, ground workforce or another flying machine (Skalický & Palasiewicz, 2017). After the mission, the military pilot will be questioned, and any perceptions amid the mission will be recorded for later utilize. The information stream appeared in Figure 6 portrays the gathering of information previously, amid, and after a mission is flown.

Intelligence Preparation of Battlefield

The IPB is depicted as a military technique for gathering and handling knowledge about the war zone. The strategy depicts how exact and pertinent knowledge might be sorted out and given to a military chief in an opportune manner. Hotspots for insight may differ from onlooker articulations and reports from military staff to radio blocks and satellite photographs (Skalický & Palasiewicz, 2017). The IPB depicts how insight is to be utilized to pick up learning about the front line. It is a rehashed process comprising of the four stages portrayed beneath:

Figure 6. Processing of data stream

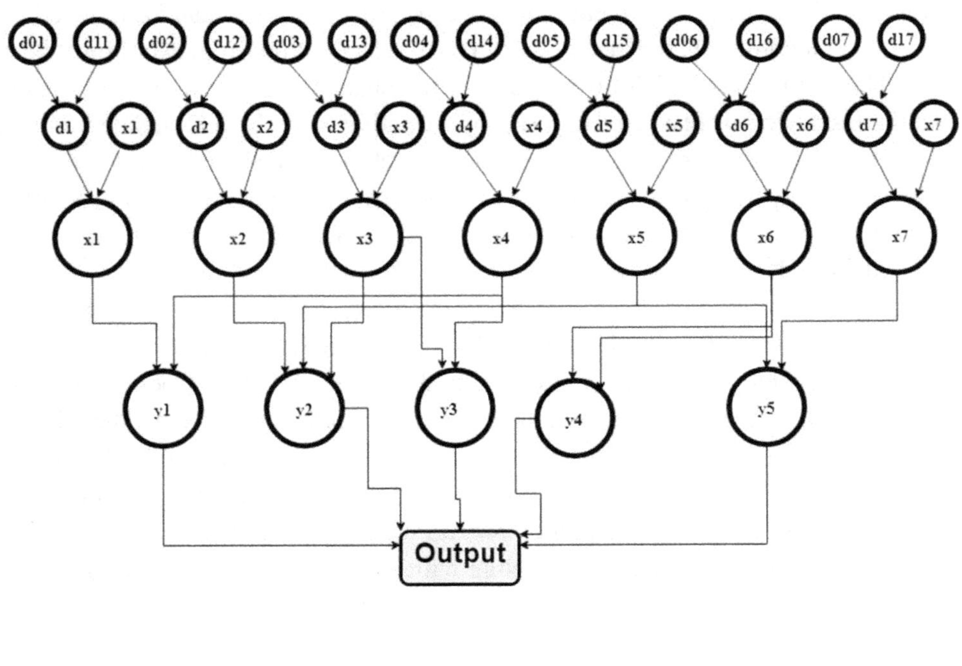

1. Characterize the war zone the war zone is a land territory and the airspace above it. The leader will focus on choices in regards to powers inside this zone. The span of the region may fluctuate amid battle as per the information of the combat zone accessible.
2. Portray the impacts of the war zone here the effect on the mission by e.g. the landscape or the present climate is built up. This progression may discover passageways in where the airship can fly without being recognized by for radar, or courses in where the flying machine can remain relatively undetectable due to haze.
3. Assess the risk a profile of the adversary's capacities inside the combat zone is made. This may e.g. incorporate the number and places of adversary radar framework.
4. Decide adversary approaches with the learning picked up from the initial three stages of the IPB the plausible courses of adversary activities are decided.

Electronic Order of Battle

The EOB is characterized as a rundown of the areas, recognizable pieces of proof, capacities, and capacities of electronic gear utilized by a military power. This data is made accessible to the pilot amid the pre-flight planning of the airship. Data about the arranged course and current hardware and ammo on the airship may likewise be stacked electronically to a flying machine PC. The arranged course will regularly be depicted utilizing a settled number of areas. Such an area is known as an Intermediate Point (IP). The stock comprising of countermeasures and weapons depends on the EOB. Since countermeasures and weapons will regularly need to have similar stations set under the airship the more countermeasures that are considered fundamental the fewer weapons might be conveyed by and large the evaluation of the war zone will exist in one of the classifications given underneath (Li, Liu & Ke, 2013). While war zones inside a few of these classifications are assessed as being improbable, they are said here for fulfillment.

- The front line contains no known dangers and the airship isn't prepared with ECM.
- The adversary has aloof rockets just and the airplane will be furnished with flares.
- The foe has dynamic rockets just and the flying machine will be furnished with waste. This circumstance is improbable.
- The adversary has dynamic rockets as it were. These rockets might be stuck and the flying machine is outfitted with both debris and jammer. This circumstance is impossible also.
- The adversary has various kinds of weapon or the piece of weapons is obscure. The flying machine is furnished with waste, flares, and a jammer.
- The adversary has uninvolved rockets just or rockets that might be stuck. The flying machine is furnished with jammer and flares. This circumstance too is impossible.

System Data Flow

The essential information stream for the workplace of the DSS various frameworks onboard the airship. With the guise of an information base and the built SA, the DSS finds an answer. This arrangement would then to be displayed to the pilot, as well as programmed reactions from other onboard frameworks can be started. Information is originating from sensors on the left-hand side of the figure. Choices from the DSS impact the subsystems on the correct hand side. The pilot is persistently assembling data from the on-board frameworks to enhance and overhaul his Situational Awareness (SA). With a specific end goal to help the pilot the DSS needs to manufacture and keep up its own particular SA. At the point when an answer is found by the DSS, it must be executed. One approach to do this is to display the answer for the pilot and let him execute the activities recommended. Since one reason for acquainting a DSS is with diminishing the workload of the pilot the DSS might be connected specifically to important subsystems for naturally conveying countermeasures, potentially in conjunction with appropriate equivocal moves (Damangir & Abedini, 2014). Basic leadership in military spaces is frequently depicted by the four activities: Watch, Orient, Decide, Act. These activities are performed over and over in what is known as the OODA circle at the point when connected by a rocket the military pilot will initially watch the rocket; he will decide whether the rocket is representing a prompt danger; on the off chance that it is he will choose legitimate equivocal activities; lastly he will act to stay away from the effect of the rocket. On the off chance that the military pilot is to be helped by a DSS it will itself go through the initial three periods of the OODA circle before a choice is exhibited to the pilot. In the event that a legitimate reaction must be discovered no later than 200 milliseconds after a danger happens.

Acquiring Data

The specific end goal to settle on activities to recommend to the pilot the framework needs contribution from different sources. In most contender's flying machine, these sources will be frameworks associated with an information transport onboard the airplane. In numerous flying machines, this transport will agree to the MIL-STD-1553B standard. On a MIL-STD-1553B transport, the correspondence is controlled by a transport controller. Information can originate from various remote terminals, and it can be perused by a number of transport screens. The transport controller fills in as a judge that permits the remote terminals to utilize the transport each one in turn. Information on the transport is transmitted as datagram's from one remote terminal to at least one transport screens. The configuration of a datagram is portrayed in an Interface Control Document (ICD), and since each remote terminal may utilize a particular arrangement for each transport screen getting information from the terminal, countless might be expected to portray correspondence on the transport. Since a DSS may require a contribution from numerous remote terminals, and these may shift starting with one airplane then onto the next, the DSS must be planned so it can be effortlessly adjusted to consent to new arrangements of ICDs. Most datagrams are transmitted with a generally low recurrence (Ji, Li & Chen, 2013). In the event that permitted by the transport controller, a run of the mill datagram will have a transmission recurrence in the request of 1 to 20 every second, contingent upon the surveyed significance of the datagram. Joined with a generally low clock recurrence controlling the transport, a slack time in the size of 0.1 to 0.5 seconds may well show up between the times a sensor framework has distinguished a danger till a transport screen gets a notice about it. This slack leaves just a brief timeframe for the DSS to discover and recommend an action to the pilot. Other than onboard sensor frameworks information to the DSS might be provided through strategic information connect. The connection is a standard for such a strategic information connection, and it is utilized for sharing data e.g. distinguishing proof and voice charges between partnered units flying machine, ships, and so on in the front line. A DSS may profit by information given through Link-16 to keep up a model of dangers, their tracks, sizes, numbers, and positions. The sensors and hotspots for identifying dangers onboard the airplane vary in a few angles. They work in various groups of the electromagnetic range, utilize diverse strategies to decide the dangers, and since they may decipher their simple information in an unexpected way, they may not concede to the dangers found. To give the DSS a solitary "ground truth" to work with, it might be fundamental that information from the distinctive sort of sources is combined before given to the framework.

Threat Evaluation

Not all progressions to the risk situation will require the DSS to propose new activities. All through a mission one of the objectives of the DSS is to limit the workload of the pilot. Along these lines, when the airplane flies over the cordial domain, and no dangers are inescapable, no activities must be recommended by the DSS. When the MWS issues a notice, or the RWR recognizes radiation from perhaps antagonistic radar, the domain can never again be viewed as agreeable, and the framework will recommend activities to the pilot. Indeed, even while flying over for region the DSS should just recommend sly activities whenever fundamental. To decide when e.g. admonitions from the RWR come from beforehand undetected radar framework, or if the radar has quite recently returned on the RWR in the wake of being lost for a minute, a danger evaluator subsystem might be included to the DSS. Such a subsystem should construct its assessment with respect to an enrollment of ongoing admonitions (Shibl, Lawley & Debuse, 2013). The RWR may free track of radar frameworks if the flying machine is situated with the end goal that either itself or the landscape underneath it keeps the radar from being perceptible by the RWR receiving wires. For the MWS a few admonitions can be overlooked as well. In the event that a notice is issued for a small amount of a second just, it is likely to be caused by e.g. a reflection that is obvious just for a minute, rather than by a rocket following the airplane. On the off chance that the risk evaluator needs to look at each cautioning to past alerts, the main admonitions of a risk will be overlooked. At the point when alerts are never again disregarded even less time is left for the DSS to discover an arrangement.

Executing Decisions

As a rule, a DSS will bolster a leader in choosing appropriate activities at the point when the recommendations from a warrior airplane DSS is introduced to the pilot, the pilot should settle on the activities to perform before completing them. Since a constrained time is accessible for the pilot to do this the DSS can play out the activities itself; even without the assent of the pilot. Most warrior flying machine utilizes fly-by-wire control, i.e. all controls are altogether electronic. This implies the DSS can take control of the airplane and play out the correct moves while apportioning expendables. While this might be specialized possible it isn't really a circumstance needed by the pilot, the flying machine taking control was an abomination to them. Having the DSS working in collaboration with a portion of the ECM framework on-board the airplane, without taking complete control, may, in any case, enhance the survivability of the airplane. The DSS may e.g. inform the jammer about RF dangers not distinguished by the jammer itself, or it might discharge flares or refuse when the flying machine is flying the best possible way. While the control of the airplane stays in the hands of the pilot, the DSS will build his survivability and part of his workload will be taken from him.

Models and Systems

The ways to deal with the advancement of a warrior airship DSS depicted in this work are not the main endeavors on bringing a DSS into the field of flight. This area depicts various test models that might be utilized as a part of a DSS, and a couple of existing frameworks. A portion of the models utilize procedures from the fields of AI and additionally, for example, fluffy sets, impersonate nets, and hereditary programming these methods are not portrayed here.

Experimental Models

The Countermeasure Association Technique (CMAT) naturally suggests countermeasures and moves to a pilot under rocket assault. The method chooses reactions that will expand the survivability in connection to both the present and future dangers. It is finished by combining potentially inadequate sensor input also, bolstering the outcomes to a copy net. An arrangement of battle situations is displayed to EW specialists and the copy net is prepared to imitate the reactions from these specialists. Arrangements found by the CMAT are called procedures and given a danger situation, various procedures are found. The best of these procedures, i.e. the procedure that best imitates the reactions from EW specialists, is picked. In 1986 the Pilot's Associate undertaking was started by the United States Defense Advanced Research Projects Agency. The idea of the Pilot's Associate is an arrangement of information based subsystems: two assessors, two arranging subsystems, and a Pilot-Vehicle Interface (PVI). The Situation Assessment subsystem decides the present condition of the outside world, while the Framework Status subsystem gains information about the condition of onboard frameworks. The Tactics Planner subsystem underpins the pilot in picking reactions from conceivable choices when quick dangers happen. If there should arise an occurrence of deviations from the mission design the Mission Planner will adjust the arrangement to the current circumstance. The PVI adjusts the data appeared to the pilot to the current circumstance. At the point when for example the pilot needs to react to a fast approaching risk, admonitions that can be postponed, e.g. of breaking down subsystems, won't be appeared.

In the Pilot's Associate is contrasted with a French framework named Copilote Electronique this framework depends on a top to bottom examination of the pilot's subjective handling, and the point is to discover approaches to help the pilot in light of his current exercises and an estimation of his current mental workload. The idea of General Aviation Pilot Advisory and Training System (GAPS) is to run a flight administration framework on an economical PC. It is propelled by the Pilot's Associate framework, and it comprises of two fundamental parts, the Flight Mode Mediator (FMI), and the Pilot Advisor (PA) (Коростелев, 2016). In view of e.g. flying machine state factors, for example, elevation, velocity, and rate of climb, the FMI decide the current flight mode. This is finished utilizing fluffy sets furthermore, the outcome is consistently nourished to the PA. In view of the present flight mode and an arrangement of standards, the PA will utilize the Heads-Up Display (HUD) and the Heads-Down Show (HDD) to show important data to the pilot. The Missile Countermeasure Optimization (MCO) issue is the subject of a number of papers, the point is to consolidate airship moves what's more, countermeasures to survive an assault from a solitary surface-propelled antiaircraft rocket. The best blend of countermeasures and moves is discovered utilizing hereditary programming. The strategy envelops vulnerabilities about both the sort and the condition of the moving toward rocket.

The Automated Threat Response utilizing Intelligent Agents (ATRIA) framework has advantage operator speaks to a military resource, for example, a rocket as well as airplane following framework, a radar framework, or a rocket. The operators in the framework share a developing portrayal of the battle situation and keeping in mind that discovering reactions for themselves they convey these to other cordial resources in the combat zone (Debar, Thomas, Cuppens & Cuppens-Boulahia, 2007). In the utilization of fleeting thinking during the time spent basic leadership is depicted. Here the situation is Beyond

Visual Range (BVR) battle with different for the airship, and the extension is to give the pilot a visual elucidation of the circumstance. For every one of them for the airship, decency esteem is computed, showing which airship will have the upside in a nearby experience. The integrity esteem depends on the time it will take for every flying machine to start suitable activities. BNs are proposed as methods for assessing the properties of an adversary warrior airship. In waste and flare, programs are upgraded utilizing a factual model. This display expresses the survivability as an element of different parameters including rocket assault rate, false alert rate, rocket identification likelihood, and the length of a mission. The objective of the work portrayed isn't to improve the survivability in a particular situation, but instead to think about how extraordinary parameters influence the refuse and flare programs.

Existing Systems

An Integrated Defensive Aids System (IDAS) is a framework that gathers input from various sources and consolidates these to e.g. naturally apportion expendables or use a jammer to counter adversary frameworks. Info may originate from distinctive sensors onboard the airplane or it might be handed-off from sensors on different stages, for example, another flying machine, or ground-or ocean-based sensor frameworks. Truly EW gear on an airship has been considered as extra units what's more, not as incorporated parts of the airship, it is specified that an IDAS must be considered as an incorporated piece of the flying machine frameworks in much the same route as the motor, and it is contended that the part of both is to get the airplane to the objective and back. Every one of the two frameworks depicted in this segment can be viewed as an IDAS. The Electronic Warfare Management System (EWMS) to diminish the pilot's workload and to guarantee facilitated and viable utilize of onboard EW frameworks. It can be worked in three modes: In manual mode, the pilot actuates the countermeasure program to utilize. At the point when in self-loader mode the Electronic Combat Adaptive Processor (ECAP) will dissect the nearness of dangers and picked a successful mix of countermeasures. This mix is introduced to the pilot who can give his assent by enacting a switch. After assent is given content for the mix of countermeasures is started. The pilot does not have to give his assent when the framework is in programmed mode. A content identified with the best utilization of countermeasures found will be started.

The Threat Response Processor (TRP) is produced at Georgia Tech Research Establishment (GTRI), it is a mix of direct enhancements of the survivability and a specialist framework, and it has experienced broad flight testing over seven years. At first contributions from different frameworks locally available the airplane are related, ambiguities are taken care of, and inputs portraying e.g. diverse dangers are organized after this various contents are generated, contingent upon the character of the information. In addition to other things, these contents select appropriate countermeasures to apply and messages to show to the pilot after each brought forth content is executed the ideal reaction is found from the outcomes of the contents. This reaction is then encouraged to the EWMS to deal with e.g. administering of expendables.

Decision Support System and Responsibilities

This section depicts the requirement for a warrior airplane DSS to help the pilot in finding the best reactions while being locked in by ground-based dangers. Since arrangements from a DSS are planned to enhance the survivability of the pilot, the idea of survivability is presented. The survivability relies upon the weakness, helplessness, and killability of the flying machine. A DSS must agree to various plan necessities. Prerequisites to reaction time, capacity to keep running on flying machine equipment, capacity to be effortlessly refreshed, what's more, the framework being both dependable and valuable are depicted. Notwithstanding this, a DSS must have a simple to utilize UI/PVI. The plan of the client interface is outside the extent of the work. A DSS will chip away at mission information in deciding ideal reactions. The mission information stream is portrayed, and the ideas of Intelligence Preparation of Battlefield (IPB) and Electronic Order of Battle (EOB) are presented. At the point when airborne the DSS depends on information from onboard sources. Information may frequently be gained by means of an airplane information transport. MIL-STD-1553B is a standard for such a transport, and some of the attributes of this standard are said. Exploratory models and existing frameworks are portrayed. The tests models apply methods from the fields of AI as well as while frameworks that have as of now been actualized in airplane appear to utilize more logical methodologies.

Take-Off Time Prediction

The choice emotionally supportive network must anticipate a set aside off time for every airship in a take-off succession. A cost can't be resolved until the point when this has been finished. Keeping in mind the end goal to have the capacity to foresee take-off times for flying machine, a supposition is made that airplane will take-off as ahead of schedule as would be prudent, given all of the requirements upon how early the assume off can take a position. An airplane can't as a rule takes off sooner than the computed most punctual take-off time except if decreased partitions are being utilized and there is no favorable position to be picked up from removing any later than the most punctual take-off time (Peng, Dan & Qu, 2011). As opposed to this, in the landings issue flying machine can regularly fly speedier or take an alternate way with a specific end goal to arrive sooner than the perfect arrival time, at the cost of expanded fuel consume or workload. The elements that compel how early a flying machine can take-off are recorded underneath:

Taxi Time to the Runway

An airship can't take off before it can achieve the runway the most punctual runway entry time can be ascertained by including a holding zone traversal time which can be founded on the way taken through the holding territory to the landing time of the airplane at the holding zone this implies the traversal ways should be resolved before take-off time forecast.

Moving Time

There might be times when an airship must sit tight for some other development to occur before it can move to the runway. For this situation, it is imperative to take into consideration this prerequisite while foreseeing take-off circumstances. The most punctual take-off time is adjusted amid the plausibility test, to take into consideration any defer sitting tight for other airship to pass and is a motivation behind why the attainability test must be performed preceding anticipating take-off circumstances.

Wake Vortex Partition Rules

A flying machine can't take off before the wake vortices from the past take-off have had time to scatter. Takeoff course and speed partition rules with a specific end goal to control the workload for downstream controllers and accomplish in-flight partition separations, a division time is implemented at take-off between airplane flying along comparable courses.

Planning Time

A portion of the pre-flight checks might be made after push-again from the stand, while an airship is navigating to the runway. On the off chance that taxi times are short it might be vital not to plan a flying machine to take off not long after it achieves the holding zone. For instance, if a huge airship pushes once more from a stand near the runway it can without much of a stretch achieve the runway before it has had sufficient energy to play out the pre-flight checks.

Take-Off Schedule Opening

Any airplane with a take-off schedule opening may not take off before they begin time of the vacancy. The first and second take-off time limitations in the rundown above are taken care of in this detailing of the issue by the plausibility test, depicted in section 6. Notwithstanding deciding the achievability of the re-sequencing, the attainability test additionally yields a most punctual take-off time, fi, for every airship (Atkin, Burke & Greenwood, 2011), which mulls over the time at which the flying machine touches base at the holding zone, the measure of time the airplane should cross the holding region and any postponement caused by waiting for other airship to move before the flying machine can take off.

Separation Rules

The detachment rules for the third and fourth take-off time confinements are managed so a required partition can be figured in light of the weight classes, flight courses and speed gatherings of each match of airplane for any flying machine, I, let vi be the weight class of the airship, ri be the takeoff course distributed to the airplane and si be the speed gathering of the flying machine. Let the situation of the airplane I in the take-off arrangement be given by ci, so ci = 1 for the main flying machine to take off. For any match of the airship, I and j, where flying machine I take off before airship j, values can be resolved for the required partitions between the airships, as per current directions. Give Vij a chance to be characterized as the required wake vortex partition between driving flying machine I and following airship j, decided from the weight classes vi and vj of the airship. Let R ij be characterized as the required detachment in light of the SID courses, ri and rj, and the speed gatherings, si, and sj, of the airship I and j. A base required division, Mij, would then be able to be figured for each arranged combine of flying machine I then j, has appeared.

Ready Time

The impacts of the fifth take-off time imperative are approximated by guaranteeing that there is the least time between push-back and take-off for all airship. This base time is resolved in view of the weight class of the flying machine, as bigger airship require longer getting ready. The capacity Pi is characterized for airship I to give the most punctual time the airship will be prepared for take-off. This is characterized as the time at which the flying machine left it's remaining in addition to a base time contingent on the weight class of the flying machine. In the reenactment comes about introduced in this proposition, Pi was dealt with as a steady for every airship, paying little mind to the coveted take-off arrangement. A live framework would rather at first estimated this esteem yet allow a controller to supersede this with another time as indicated by data from the pilot. For instance, expanding the esteem if a pilot pronounces there are a few troubles and the flying machine won't be prepared for take-off at a predefined time. The recreation displayed in this proposition keeps on utilizing the assessed an incentive all through the examinations as no refreshed esteem will ever be accessible to the hunt. The capacity utilized as a part of the analyses introduced in this proposition is a straightforward one which restores an estimation of one moment, three minutes or five minutes after push-back for light, medium and overwhelming airplane separately. With a two-minute holding region traversal time, this is just typically compelling upon the substantial flying machine which pushes once again from stands near the runway.

Take-Off Schedule Opening

Numerous airplanes have a Calculated Time of Take-off (CTOT), On the off chance that a flying machine, a CTOT allotted, at that point let bi be the begin time of the resulting take-off window what's more, given li a chance to be the finish of the take-off time window. The fifteen moment CTOT schedule vacancy is at that point meant when interim [bi, li]. On the off chance that airship I don't have a CTOT then bi is set to a low esteem (Atkin, Burke & Greenwood, 2011) sooner than the begin time of the dataset and li is set to a substantial esteem, sufficiently vast that it is past the last take-off time of the last airship in the dataset. This adequately gives any airship that does not have a CTOT, a take-off window which traverses the whole scope of take-off circumstances. As a result, no time window limitation is connected to this flying machine and an extraordinary case to manage the nonattendance of CTOTs is superfluous.

Predicted Take-Off Time

Given the above meanings of bi, fi, Pi and ei for flying machine I, the condition for the anticipated take-off time, di, can be resolved for every flying machine, I, in take-off request, utilizing condition di = max(bi, fi, Pi,ei).

OBJECTIVE FUNCTION

The point of this model is to empower the choice emotionally supportive network to discover groupings with low deferrals and high CTOT take-off schedule opening consistency while controlling the imbalance of postponement in the succession and guaranteeing that the re-sequencing would be worthy to a controller. Each of the targets is considered underneath, trailed by the portrayal of the general target work.

Definitions

Give hello there a chance to signify the time at which airplane I arrived, or will touch base at the holding territory. Let di signify the anticipated take-off time of airship I. Give li and bi a chance to mean the points of confinement of the take-off schedule opening, as characterized prior. By communicating all circumstances like a flash, in ordinary activity li = bi + 900 for flying machine which has CTOTs. Where airship doesn't have CTOTs, li and bi are set to exceptionally little and huge qualities individually, so di dependably lies between them. Give ai a chance to mean the situation of flying machine I in the landing arrangement at the holding region and ci indicate the situation of flying machine I in the planned take-off succession. Where there is a formerly known arranged take-off grouping, let oi signify the situation of flying machine I in this beforehand take-off succession. On the off chance that there is no past grouping containing flying machine I, at that point, let oi be set to the same esteem as ai, so the primary start things out served take-off succession is accepted.

Compliance With CTOT Availabilities

Flying machine must stick to their CTOT openings. It is constantly conceivable to defer flying machine to move them into the CTOT space, however, it isn't generally workable for flying machine to take-off sufficiently early to meet the CTOT confinements. As depicted a CTOT opening by more than five minutes is much more regrettable than missing it by up to five minutes. All misses ought to stay away from however in the event that a miss is unavoidable then it is smarter to miss by less than five minutes.

The Holding Zone Delay

The significance of keeping postpone low has just been set up and identifies with fuel consumption, contamination and traveler disappointment and additionally keeping up a high runway throughput. The time spent in the holding territory is here utilized as a measure of the postponement for every flying machine. Given the above meanings of di and hello, the time spent via flying machine I in the holding region can be seen to approach greetings.

Schedule Equity

It is critical to guarantee a level of value in the take-off arrangement it is smarter to evade superfluous surpassing or re-sequencing of the airship. Flying machine can be thought to be positional postponed when their situation in the take-off arrangement is later than their situation in the landing arrangement at the holding region. At the point when the flying machine is positionally postponed in the take-off arrangement, the influenced pilots can feel disappointment with the booking because of the apparent shamefulness. Vast positional postponements can likewise confine the re-sequencing open doors for another flying machine, since a postponed airship must be held someplace, in this manner obstructing a piece of the holding region structure and lessening the re-sequencing adaptability (Akdeniz, 2001). Little positional deferrals are not an issue but rather bigger postponements turn into an expanding issue, both regarding injustice and compelling later re-sequencing. A negative positional postponement (or positional progression) is an issue for neither the reasonableness nor the adaptability of the holding zone. The positional deferral for flying machine is given by ci-ai if positive, or then again zero generally. A cost has been incorporated into the target work that is relative to the entirety of the squares of the positional postponements for every flying machine.

Schedule Stability

There are regularly situations where there are numerous groupings with fundamentally the same as expenses. For this situation, it is smarter to support a formerly utilized succession as opposed to permitting grouping changes that will have little advantage. This is particularly essential if there is some vulnerability in the information used to settle on the choices as little saw advantages might be simply down to information mistakes. To give for this, a factor of abs(ci -oi) is incorporated into the goal work because of the meaning of oi over, this factor will support the main start things out the served arrangement if no past take-off succession is known. Penalizing deviation from past arrangements is certifiably not another one. As depicted a punishment has been connected in past landing sequencing research into the dynamic issue with a specific end goal to limit or punish deviation from a past entry arrangement. For the take-off sequencing issue, given that a controller would not have the capacity to order a take-off grouping until the point that the airship achieve the holding territory given the presumption that the runway controller is doing the sequencing, and that, giving adequate data is accessible about navigating airship, the sequencing is normally performed while the airplane is still on the runways, it is less critical to extremely punish deviations from past plans than it was in the landing sequencing research.

Avoiding Punishing Particular Airplane Composes

The capacity Z(ci, ai, vi, si) was added to stay away from superfluous penalization of the certain flying machine. The goal to lessen deferral will move tricky divisions to a later place in the succession. Sometimes, postponing airship does not help and only compels the holding zone re-sequencing by blocking the holding region pointlessly. For instance, there are so few light flying machines flying from Heathrow that it is profoundly improbable that deferring one light flying machine will make it conceivable to aggregate it together with another light airplane to maintain a strategic distance from the high wake vortex partition that will go before a light flying machine. In the event that a flying machine which will dependably have an expansive partition related with it, for example, a light or then again moderate airship is postponed to be sequenced close to the finish of the take-off arrangement, at that point this capacity will apply a punishment to punish the succession. Without this factor, there is a propensity for the framework to succession the airplane later and later, to maintain a strategic distance from the partition. The nearness of this capacity keeps this. In the examinations performed for this postulation, this capacity was executed to apply a punishment to any succession which positional postpones light or moderate airship. The cost was parameterized however the parameters were chosen with the goal that any positional deferral of in excess of one position in the arrangement connected a cost equivalent to an expanded aggregate postponement of two minutes in the succession. Perception of the outcomes demonstrated that this little factor, in the mix with the punishment for over the top positional defer that is connected to any positional deferred flying machine, was adequate to keep the positional postponement of such airship.

CONCLUSION

In this chapter, we have examined the utilization of case-based thinking strategies and model-based fault identification for the flight line support and fault determination of air motors. Specifically, the chapter has focused on specific parts of this work featuring how there is a move from nearby help for diagnostics to diagnostics from an incorporated virtual condition working crosswise over-appropriated assets. The diagnostics will dependably be performed on the motor, the appropriation of a unified administration which upkeep workforce at different levels can get to is exceptionally profitable as new advancements end up accessible normally with the blast of the AI. Aero engine specialists who are viewed as high-esteem assets can be activated to examine complex issues all around from remote areas. The show frameworks displayed can be utilized to recognize deficiencies at aircraft and furthermore anticipate approaching flaws crosswise over armadas of motors through inclining. The business advantages of this open, adaptable, proactive way to deal with motor checking and support are improved fault finding execution, as well as reusable administration gatherings, better versatility, better viability, higher accessibility, diminishment in unscheduled upkeep and coming about airplane downtime.

REFERENCES

Akdeniz, A. (2001). The impact of mandated aging airplane programs on jet transport airplane scheduled structural inspection programs. *Aircraft Engineering and Aerospace Technology*, *73*(1), 4–16. doi:10.1108/00022660110366827

Alexander, D., & Merkert, R. (2017). Challenges to domestic air freight in Australia: Evaluating air traffic markets with gravity modeling. *Journal of Air Transport Management*, *61*, 41–52. doi:10.1016/j.jairtraman.2016.11.008

Alfredson, J., & Ohlander, U. (2016). System Characteristics and Contextual Constraints for Future Fighter Decision Support. *International Journal of Information System Modeling and Design*, *7*(1), 1–17. doi:10.4018/IJISMD.2016010101

Amin, M., & Shebl, D. (2014). Reasoning dynamic fuzzy systems based on adaptive fuzzy higher order Petri nets. *Information Sciences*, *286*, 161–172. doi:10.1016/j.ins.2014.07.011

Atkin, J., Burke, E., & Greenwood, J. (2011). A comparison of two methods for reducing take-off delay at London Heathrow airport. *Journal of Scheduling*, *14*(5), 409–421. doi:10.100710951-011-0228-y

Bernardini, P. (2017). First data from the DAMPE space mission. *Nuclear And Particle Physics Proceedings*, *291-293*, 59–65. doi:10.1016/j.nuclphysbps.2017.06.014

Buromensky, N. G. (2016). Survivability of Military Communication Systems. *Military Thought*, *25*(3), 117–121. doi:10.21557/MTH.47912151

Dalkilic, S. (2017). Improving aircraft safety and reliability by aircraft maintenance technician training. *Engineering Failure Analysis*, *82*, 687–694. doi:10.1016/j.engfailanal.2017.06.008

Damangir, H., & Abedini, M. (2014). System identification and subsequent discharge estimation based on level data alone gradually varied flow condition. *Flow Measurement and Instrumentation*, *36*, 24–31. doi:10.1016/j.flowmeasinst.2014.01.002

Debar, H., Thomas, Y., Cuppens, F., & Cuppens-Boulahia, N. (2007). Enabling automated threat response through the use of a dynamic security policy. *Journal In Computer Virology*, *3*(3), 195–210. doi:10.100711416-007-0039-z

Fioriti, M., Vercella, V., & Viola, N. (2018). Cost-Estimating Model for Aircraft Maintenance. *Journal of Aircraft*, 1–12. doi:10.2514/1.c034664

Jao, C. (2011). *Efficient decision support systems*. Rijeka, Croatia: InTech.

Ji, W., Li, X., & Chen, J. (2013). Remote Update on Flash-Based Hydrologic Remote Terminal through GPRS Technology. *Advanced Materials Research, 846-847*, 1711-1715. Retrieved from www.scientific.net/amr.846-847.1711

Jin, X., Chen, S., Jia, Z., & Zeng, Z. (2016). The Research of Compose Method for Removal/Install Part in Aircraft Maintenance Manual. *Applied Mechanics and Materials, 835*, 828-834. Retrieved from www.scientific.net/amm.835.828

Коростелев, A. (2016). Five-step preparation of the pilot of civil aviation in a system of continuous training. *Proceedings Of The National Aviation University*, *6*(1-2). doi:10.18372/2306-1472.6.9721

Le Grange, L., & Buys, A. (2012). The management of maintenance technology transfer in the South African aviation industry. *South African Journal of Industrial Engineering*, *13*(2). doi:10.7166/13-2-312

Li, T., Liu, P., & Ke, Y. (2013). Battlefield Awareness Network Research Based on Intelligence Role Division and Wireless Sensor Network. *Applied Mechanics and Materials, 300-301*, 580-584. Retrieved from www.scientific.net/amm.300-301.580

Mahdavi, I., & Shirazi, B. (2010). A Review of Simulation-based Intelligent Decision Support System Architecture for the Adaptive Control of Flexible Manufacturing Systems. *Journal Of Artificial Intelligence*, *3*(4), 201–219. doi:10.3923/jai.2010.201.219

Noh, S., & Choi, C. (2012). Predicting the Operational Effectiveness of Aircraft Survivability Equipment Suite. *IACSIT International Journal of Engineering and Technology*, *4*(4), 372–375. doi:10.7763/IJET.2012.V4.387

Novák, M., Kameníková, I., & Podhadský, J. (2016). Modern Communication Devices in the General Aviation Aircraft Maintenance. *MAD - Magazine of Aviation Development, 4*(17), 5. doi:10.14311/mad.2016.17.01

Oravec, M., & Jadlovská, A. (2017). Sensors Fault Diagnosis Algorithm Design of a Hydraulic System. *Acta Electrotechnica Et Informatica*, *17*(2), 30–37. doi:10.15546/aeei-2017-0013

Peng, H., Dan, M., & Qu, H. (2011). Research of Aero-Engine Life Prediction Based on Take-Off EGTM. *Applied Mechanics and Materials, 99-100*, 286-292. Retrieved from www.scientific.net/amm.99-100.286

Sayed-Mouchaweh, M., & Lughofer, E. (2015). Decentralized fault diagnosis approach without a global model for fault diagnosis of discrete event systems. *International Journal of Control*, *88*(11), 2228–2241. doi:10.1080/00207179.2015.1039594

Shibl, R., Lawley, M., & Debuse, J. (2013). Factors influencing decision support system acceptance. *Decision Support Systems*, *54*(2), 953–961. doi:10.1016/j.dss.2012.09.018

Skalický, P., & Palasiewicz, T. (2017). Intelligence Preparation of the Battlefield as a Part of Knowledge Development. *International Conference Knowledge-Based Organization, 23*(1), 276-280. 10.1515/kbo-2017-0045

Svenmarckt, P., & Dekker, S. (2003). Decision support in fighter aircraft: From expert systems to cognitive modeling. *Behaviour & Information Technology*, *22*(3), 175–184. doi:10.1080/0144929031000109755

Тамарґазін, О. (2016). The technique of periodicity choice of maintenance control of aviation engineering products. *Proceedings Of The National Aviation University*, *6*(1-2). doi:10.18372/2306-1472.6.9716

Ursavas, E. (2014). A decision support system for quayside operations in a container terminal. *Decision Support Systems*, *59*, 312–324. doi:10.1016/j.dss.2014.01.003

Weckman, G., Marvel, J., & Shell, R. (2006). Decision Support Approach to Fleet Maintenance Requirements in the Aviation Industry. *Journal of Aircraft*, *43*(5), 1352–1360. doi:10.2514/1.19749

Yu, K., & Wang, F. (2013). Research and Design on Intelligent Decision Support System Frame for Vehicle Maintenance. *Advanced Materials Research, 765-767*, 3-7. Retrieved from www.scientific.net/amr.765-767.3

KEY TERMS AND DEFINITIONS

DSS: Decision support systems (DSS) are a class of modernized data framework that helps basic leadership exercises. DSS are intuitive PC based frameworks and subsystems expected to help leaders utilize interchanges advancements, information, archives, learning and additionally models to finish choice process errands. A DSS may display data graphically and may incorporate a specialist framework or computerized reasoning. It might be gone for business administrators or some other gathering of information specialists. Common data that a choice help application may assemble and present would be (1) accessing all data resources, including heritage and social information sources; (3) comparative information figures; (3) projected figures in view of new information or suspicions; (4) consequences of various choice choices, given past involvement in a particular setting.

IPB: Intelligence preparation of battlefield (IPB) is a deliberate, persistent procedure of examining the risk and condition in a particular geographic region. It is intended to help staff appraisals and military basic leadership. Applying the IPB procedure helps the leader specifically apply and boost his battle control at basic focuses in time and space on the war zone by deciding the danger's presumable COA. Portraying the condition your unit is working inside and the impacts of nature on your unit. IPB is a consistent procedure which comprises of four stages which you play out each time you direct IPB: Characterize the combat zone condition. Portray the war zone's belongings. Assess the danger. Decide danger COAs. The IPB procedure is persistent. You lead IPB before and amid the summons underlying making arrangements for a task; however, you likewise keep on performing IPB amid the direct of the activity. Each capacity in the process is performed constantly to guarantee that the results of IPB stay finish and substantial.

Chapter 6
AF 447 as a Paradigmatic Accident:
The Role of Automation on a Modern Airplane

Antonio Chialastri
STASA, Italy

ABSTRACT

In this chapter, the author presents a human factors problem for automation: why, when, and how automation has been introduced in the aviation domain; what problems arise from different ways of operating; and the possible countermeasures to limit faulty interaction between humans and machines. This chapter is divided into parts: definition of automation, its advantages in ensuring safety in complex systems such as aviation; reasons for the introduction of on-board automation, with a quick glance at the history of accidents in aviation and the related safety paradigms; ergonomics: displays, tools, human-machine interaction emphasizing the cognitive demands in high tempo and complex flight situations; illustration of the AF 447 case, a crash happened in 2009, which causes are linked to faulty human-machine interaction.

INTRODUCTION

Human error is considered the first cause of accident. Actually, as many safety scholars affirm, the human error is only an epiphenomenon. The real cause of accident is what induces the human error, such as human performances and limitations, poor teamwork, organizational pressures on crews to obtain unreasonable performances, faulty human-machine interaction and lately some psychological upset, where the pilots posed an intentional threat to the safety of flight.

Although humans could be considered somehow a threat to safety, they are also the main resource to cope with unexpected events, unruly technology, changing environment and uncodified system failures. System designers often conceive the human being as a superman or a robot. Actually, we have psychological limitations (constant or transient), physical constraints (due to ageing, fatigue, etc.), big differences among individuals, vices and emotions. All these elements are powerful sources of variability.

DOI: 10.4018/978-1-5225-7709-6.ch006

This brings pros and cons. Variability is often considered something negative, because aviation is based on standard procedures. On the other hand, variability provides the needed flexibility to make the entire system resilient. Resilient systems are characterized by flexibility and robustness. Automation provides the latter characteristic. It is important to point out that we cannot achieve one alone, be it flexibility or robustness. Both are needed. The problem is how making this two necessary elements work together. The aim of this chapter is to show why, when and how automation has been introduced in the aviation domain, what problems arise from different ways of operating, and the possible countermeasures to limit faulty interaction between humans and machines. This chapter is divided into four main parts:

1. Definition of automation, its advantages in ensuring safety in complex systems such as aviation;
2. Reasons for the introduction of on-board automation, with a quick glance at the history of accidents in aviation and the related safety paradigms;
3. **Ergonomics:** Displays, tools, human-machine interaction emphasizing the cognitive demands in high tempo and complex flight situations;
4. Illustration of the AF 447 case, a crash happened in 2009, which causes are linked to faulty human-machine interaction.

WHAT IS AUTOMATION

According to a shared definition of automation, it may be defined in the following way: "Automation is the use of control systems and information technologies to reduce the need for human work in the production of goods and services". Another plausible definition, well-suited the aviation domain, could be: "The technique of controlling an apparatus, a process or a system by means of electronic and/or mechanical devices that replaces the human organism in the sensing, decision-making and deliberate output" (Webster, 1981). The Oxford English Dictionary (1989) defines automation as:

1. Automatic control of the manufacture of a product through a number of successive stages;
2. The application of automatic control to any branch of industry or science;
3. By extension, the use of electronic or mechanical devices to replace human labour.

According to Parasumaran and Sheridan, "Automation can be applied to four classes of functions:

1. Information acquisition;
2. Information analysis;
3. Decision and action selection;
4. Action implementation.

Information acquisition is related to the sensing and registration of flight data. These operations are equivalent to the human perception. Let's imagine a video camera. It helps to replace continuous, boring, monotonous human observation with reliable, objective and detailed data on the environment. Automation may handle easily and reliably these functions, as it is more efficient than humans in detecting.

At the same time automation offers the possibility of positioning and orienting the sensory receptors, sensory processing, initial data preprocessing prior to full perception, and selective attention (e.g.: the focus function in a camera).

Information analysis follows the perception. Unlike the human cognitive functions, that could show some biases (Kahneman, Tversky, 1994), automation assures reliable working memory and inferential processes. Working memory allows for quick retrieval of information, thanks to a data-base with stored elements allows to match the incoming information with some known pattern of analysis.

After the data processing come the *decision and action* **selec**tion. Even here automation is useful because it involves different levels of augmentation or can replace the human decision-making with machine decision making. It is generally acknowledged that human decision-making processes are subject to several flaws, because of cognitive limitations, flawed logic and high variability. Moreover, machines are not affected by emotions or illusions, fatigue or boredom.

The fourth stage involves the *implementation o*f a response or action consistent with the decision taken. Generally, it this stage automation replaces the human hands or voice. Certain features in the cockpit allow automation to act as a substitute for pilots. For instance, this occurs when – following an alert and warning for windshear conditions – the automation system detects an imminent danger from a power setting beyond a pre-set threshold. In this case, the autopilot automatically performs a go-around procedure, which avoids a further decrement in the aircraft's performances.

Finally, the automation *informs* the pilots on which function is actually in use. This complete the feedback loop to keep the pilots aware of what is constantly going on. This information should be given either via visual, aural or haptic means. Usually, when an information is particularly relevant all means are used in a redundant way. For instance, since the stall is the "worst case situation" for a pilot, manufacturers chose to reinforce the "Stall" information via a red light (STALL), an aural warning, a voice call (STALL) and a haptic device (stick-shaker) that make the cloche to vibrate. All these indications together make almost impossible that such an important information could be left undetected.

HOW TO DESIGN THE AUTOMATION?

Besides being applicable for the abovementioned functions, automation has different levels corresponding to different AF 447: a paradigmatic accident? The role of automation uses and interactions with technology, enabling the operator to choose the optimum level to be implemented, based on the operational context (Parasumaran, Sheridan, 2000). When designing the automation, the manufacturer may opt for different levels, corresponding to the actual requirements of the operational scenario. These levels are:

1. The computer offers no assistance; the human operator must perform all the tasks.
2. The computer suggests alternative ways of performing the task.
3. The computer selects one way to perform the task and
4. Executes that suggestion if the human operator approves, or
5. Allows the human operator a limited time to veto before automatic execution, or
6. Executes the suggestion automatically then necessarily informs the human operator, or
7. Executes the suggestion automatically then informs the human operator only if asked.
8. The computer selects the method, executes the task and ignores the human operator.

Based on these level of automation, the manufacturer offers a series of options to pilots, who have to choose when and how to use it. Usually, the higher the workload, the higher the level of automation used, in order to displace the mental energy from doing (hand flying) to monitoring.

Accidents Trend

Let's look at the graph below, showing the accident curve along the years, from Sixties to today. The vertical axis corresponds to the number of accidents per million take-offs, while the horizontal axis corresponds to the relative decades. We can clearly notice some trends in flight safety. First of all, the safety level improved along the years. This is a good news. Unfortunately, there is also a bad news: the accident curve never reached zero. The third news is puzzling: from time to time the accident curve rises up, surges, according to a change in the nature of accidents (Figure 1).

During the Sixties accidents' causes were linked to a series of factors; among them human performances, capabilities and limitations on one side, and poor ergonomics on the other. We should first investigate the reasons leading to the introduction of onboard automation. The main causes of aviation accidents were believed to be related to the human factor, such as "active failures" leading to loss of control of the aircraft. In those cases, the pilots failed to keep the aircraft under control, reaching over-speed limits, stalling, excessive bank angles, etc. The root cause for that kind of accidents was a flawed performance that eventually caused the loss of control (the effect). Factors related to human performance, e.g. the impact of fatigue, attention, high workload sustainability, stress mismanagement, etc. were consequently addressed.

Technological solutions were sought to help pilots manage these factors. Innovation at that 20 10 60 70 80 90 00 00 Dashed line: 0,5 per million / take off time eventually led to the introduction on massive scale of the auto-pilot, auto-throttle, flight director, etc. After the mid-Sixties, as a result of these innovations, the accident curve dramatically dropped. As we can observe, after a significant improvement the accident curve rose again during the mid-Seventies. Aviation safety experts were faced with accidents

Figure 1. Source: ICAO doc. 9683/950. Accident rate over the years

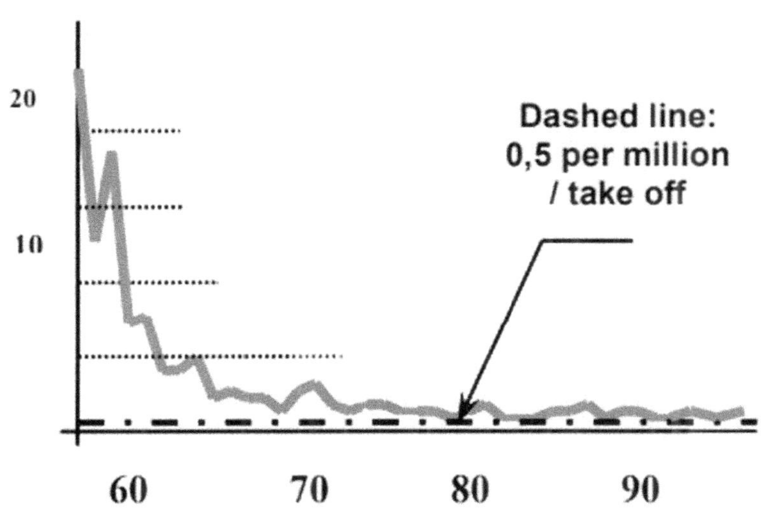

involving a perfectly functioning aircraft, with no evidence whatsoever of malfunctions. In these cases (known as "Controlled Flight Into Terrain" - CFIT), the aircrafts were hitting obstacles with the pilots in full control. The accidents were caused by loss of situational awareness, either on the horizontal or on the vertical path. The evidence showed that an improper interaction between pilots was the main cause behind the accident, so this time the solution came from psychology. Human factor experts developed techniques and procedures to enhance cooperation between pilots, and specific nontechnical training aspects became mandatory for pilots. For instance, a Cockpit Resource Management (CRM) course is nowadays mandatory in a pilot's curriculum: its topics may include leadership, communication, cross-checking and criticizing fellow colleagues, assertiveness, resolution of conflicts, etc. The faulty interaction between pilots was not an unexpected phenomenon since the human factors issue emerged at the IATA Istanbul Conference in 1975. In 1977 the worst ever accident in the commercial aviation happened in Tenerife, where 583 lost their lives.

This accident could be considered as paradigmatic, because after that a series of reforms in the approach to safety where implemented. The focus, according to SHELL (Software, Hardware, Environment, Liveware-Liveware) paradigm shifted from automation to relational skills, even if it is true that some technological solutions were very important to avoid the CFIT phenomenon, such as the Ground Proximity Warning System. This device informs and warns the pilots about a dangerous proximity to terrain, providing also some safety features such as the aural "Go-around" prompt to pilots in order to climb immediately.

During the Eighties, some new typology of accidents emerged. In 1978 the de-regulation act was issued in the United States. This introduced initially a fierce competition between carriers that in most cases cut on expenses and investments to stay afloat on the aviation market. The conflict between protection and production led in some cases the crews to bend on profit-oriented decisions rather than on the safety-led judgement. Wrong or flawed decisions were taken in order to maximize the profit. Pressing organizational climate led the crews to erode the usual safety margins, such as landing under a heavy thunderstorm, limiting the uplift fuel, or accepting stressing and challenging duty time. The accident curve rose again. This time, to cope with this new kind of threats, the solution arrived from a normative point of view: stricter regulations, more severe checks by the State on airlines regarding training, maintenance and crew rostering. It was not only the commercial aviation field to be impacted by the new managerial approach, profit-oriented. The NASA itself experienced this new approach, accompanied by the "Cheaper, faster, better" mantra, obviously in a very challenging, demanding and extreme environment you cannot be at the same time cheaper, faster and better; if you are cheaper and better, you hardly be faster. If faster and cheaper, maybe you lose in quality, so any better. If faster and better, it is unlikely to be cheaper. The solution to the managerial approach was normative, with emphasis on the checks by the regulators, in order to make organizations compliant with norms, rules and procedures. NASA itself had two major mishaps (Challenger and Columbia) that showed several flaws at the organizational level. At the end of the decade, James Reason put forward a new safety paradigm, highlighting the impact of organizational factors in causing most of the accidents. His masterpiece "Human error" is published in 1990, just at the end of the deregulation decade. His conception of safety is that to avoid error we set a series of defensive barriers at every level (political, managerial, professional, technological). Since every barrier contains some flaws, induced by the variability of human performance, the accident dynamics appears when all holes present at every stage line-up to let the initial perturbation pass through all the defensive barriers. That's why we know the Reason's model as the "Swiss cheese model".

During the Nineties the pendulum has swung back leading to loss of control as a major cause of aviation accidents. However, compared to the characteristic accidents experienced during the Sixties, the factors leading to loss of control appear to be different. Whereas in the beginnings of aviation, human performance was impaired by "under-redundancy", or insufficient aids available to pilots for avoiding the psychophysiological effects of fatigue, workload and stress which reduced the pilots' performance, nowadays we may quote another phenomenon: "over-redundancy". This means that increasing automation may drive the pilot out-of-the-loop, thus causing reduced situational awareness, automation complacency or over-confidence and loss of skills, due to lack of practice in manually flying the aircraft. As a result, pilots may not be able to regain control once automation has failed, or may be incapable of effectively monitoring the performance of automated systems (and questioning it when required). Susan Bainbridge call such a situation "ironies of automation" to address a paradoxical approach to safety, namely the engineers' approach. This point of view sees the human contribution as a threats to safety that has to be replaced, assisted or guided by automation. The paradox comes out when the designers allocate the human intervention where the automation fails or is not adequate to some tasks. In that particular moment, during an emergency or another critical situation, a complacent pilot, accustomed to rely on the automation, under-skilled because of lack of training, should do something that automation can't; from main threat to main resource in a few seconds.

The safety philosophy behind the adoption of increasing on board automation is based on the assumption that human error is the main cause of accidents. Therefore, since the human (Liveware) component of the system is the flawed link in the accident chain, we ought to look for a substitute capable of handling the tasks once performed by pilots. This is partially true, as we'll see later on. It is first necessary to understand what are the pros and cons of human contribution to safety, at what levels of operation does automation offer undoubted advantages and where the latter should end to leave room for pilots' decisions. Pilots and machines are not a consumption, cheaper maintenance and flexible pilot training. Concern for the adaptability of pilots to these new solutions only came at a later stage and only after some severe mishaps. To achieve this balance, we'll briefly analyze what levels of operation are involved in flight and where automation – primarily conceived to replace certain human operator tasks – should give way to the pilot's intervention.

According to a paradigm proposed by Jan Rasmussen and also developed by James Reason, human activity can be grouped into three main modes: skill-based action, rule-based action and knowledge-based action. The first field is based on the human capability to accomplish physical tasks, such as providing correct input to flight control (so-called "stick and rudder training"), responding to external stimuli in a quick and consistent way, and coordinating the body in order to obtain a desired result. It is mainly an area in which the psycho-physiological aspect is paramount. Unfortunately, the human brain developed in the savannah where it was designed by evolution to move on two axis: back/forth, left/right. The entire perceptual, emotional and cognitive systems are adapt to this movement. We tend to forget that we fly since the beginning of 20th century. In about 100 years we cannot modify our main natural computer. So perception is altered inflight because of visual illusion, the secretion of adrenalin interferes emotionally with our decisions and the cognitive biases make us prone to errors in our judgment.

At the first level, the perceptual one, we use automation for monitoring tasks, such as detection, identification and response to external signals stemming from habits (body automatism or conditional reaction). Automation has played an important role at this level by replacing human performance rather well. We may now rely on a source of help when flying in clouds, immersed in thick fog, or at night. As a result, autopilots, auto-throttle (and later on, auto-thrust computers) have come to gradually replace pilots in "hand-flying". Generally speaking, an autopilot can tolerate workloads that are hardly sustainable by a pilot. Let's imagine an oceanic flight during the night; an autopilot is able to maintain (with no effort at all) altitude, speed, track and so forth, whereas pilots are subject to tiredness, attention lapses, distraction, etc. On the other hand, the systematic replacement of basic flying skills has led to the erosion of competence, because, as Germans put it: "*Die übung macht den meister*" ("Use makes master"). In the U.S. the FAA (Federal Aviation Authority) has suggested adopting "back to basics training", in which pilots are taught how to fly without the help of automatisms and how to retrieve the elementary notions of aerodynamics, in order to avoid grossly misreading altitude, speed and power. Unfortunately, the new conception of fly-by-wire is based on the so-called "flight laws". Based on the configuration, the electrical sources available plus countless parameters, a middle-sized aircraft such as the Airbus A-320 (and similar) responds with different logic when the pilot gives an input to the flight control. The problem is that when everything works as per design and no failure is present the aircraft works in "normal law". It means that almost everything is managed automatically, the flight envelope has numerous protections (against stall, excessive bank, excessive speed and so on). When the first failure shows up there is an initial degradation in "alternate law" and eventually in "direct law" where most of the protections disappear. It is to be noticed that other intermediate configurations are possible. Basically, the more a pilot is in trouble, the less aids he/she receives from automation. Moreover, it is difficult to have a big picture about strange configurations since they depend on several parameters (flight phase, speed, electrical interactions, etc.). It is evident that a pilot can hardly keep him/herself well trained, because he/she cannot simulate different configuration while in a normal commercial flight. The pilot can see and experience different states of the aircraft only in a simulator session. Unfortunately, training costs are less and less affordable in a competitive market, so airlines tend to reduce to a minimum (imposed by the international rules) the simulator sessions available to pilots. Basically, the pilots fly normally an airplane that during an emergency turns out to be different from what they expect.

The second level in the Rasmussen taxonomy – the rule base action – is the conceptual layer. It indicates compliance with the rules, norms, laws, and everything laid down in the official documentation. It is unproblematic to apply a given rule whenever conditions warrant it. This is the case of a limit set for a device, e.g. the maximum temperature for operating an engine (EGT). When the upper limit is exceeded, something happens: red indications on instruments, alarms, flashing light on to attract the pilot's attention, automatic exclusion of the failed system and so forth. A machine can easily detect whether the operating conditions are normal or abnormal, by matching the real values with an operating envelope. Since the pilot may forget some rules, apply them incorrectly, or fail to apply valid norms, certain functions (especially those relating to the monitoring activity that induces boredom and complacency) are assigned to automation. It is a consequence of automation, therefore, that the flight engineer is no longer required in the cockpit. Some problems were initially detected in the normal flying activity of newly designed cockpits, since two pilots were required to manage a three-pilot cockpit, with automation playing the role of a "silent crewmember".

The third level – the knowledgebase – includes the sound judgement of pilots in deciding if and when the given rules are applicable. This implies the notion of a complex system. Complexity is evident at every level of reality, from physics to biology, from thermodynamics to meteorology (Morin, 2001). Different conceptions of complexity emerge in the current scientific debate, but generally speaking, we may highlight some commonalities between the different theories: refusal of reductionism, different level of system description (be it physical or biological or even man-made) according to the level of observation, emergent proprieties, etc. Since aviation is a complex system made up of complex subsystems such as humans, technology and the environment, it is almost impossible to govern everything in advance through rules and norms. There will always be a mismatch between the required task and the final outcome (Hollnagel, 2006). The resilience engineering approach to safety is aware of this complexity and focuses on the ambivalent role – in such a system – of man, who simultaneously constitutes a threat and resource in coping with unexpected events, unforeseeable situations and flawed procedures. Much of this activity, which continuously and strategically adapts the means to the goal, is undetected either by the top management, or by the front-line operators themselves (pilots). These micro corrections are so pervasive that the person involved in accomplishing a task fails to even realize how much he/she deviates from a given rule. The front-end operator should always seek to compromise between efficiency and thoroughness. Hollnagel calls this compromise the ETTO (Efficiency, Thoroughness Trade Off) principle. The paradox emerging from the blind application of rules – the so-called "white strike" or "working by the rule" – is that it leads to a paralysis of the entire production line (Hollnagel, 2009).

The effects of such shortcuts during normal operations are another area of concern affecting flight safety, due to the systems' opacity, the operator's superficial knowledge, uncertainties and ambiguities of the operational scenario. If we turn to the initial introduction of aircraft automation, we may detect some commonalities regarding the way pilots have coped with the innovations. The third level of Rasmussen, the knowledge level, is the ability to cope with complex situation, to invent from the scratch a solution, or to be creative. It's the work of an intelligent operator and at the moment could not be replace by computers, or not with the same reliability. At this stage we should introduce the artificial intelligence, that has a series of objections hard to overcome. In fact, given the undoubted progress of Artificial intelligence, we are focusing on what could made artificial. Actually, philosophers and scientists are striving from the beginning of ancient Greek civilization to understand what intelligence is. It is here that the human contribution enter the scene at the third level of Rasmussen's conception. Computers can act upon what a pre-designed program foresees. Reality is beyond our present ability to model it. The mismatch between designers' idea of the world and what actually really happen in the operational scenario, the gap between imagination and reality is continuously bridged by professionals. Alas, they don't realize how many times they make the system flexible with their sound judgment.

Automation Should Follow an Evolutionary Path Rather Than a Revolutionary Approach

Its adoption on board aircrafts does not respond to the planned purpose of enhancing safety "from scratch" in a consistent way, but rather resembles a biological organism trying to continuously adapt to the challenges posed by its environment (fly-fix-fly). This trial-and-response approach can be observed regardless of the fact that innovation introduced on board generally lags a step behind the overall level achieved by the industry. In fact, one of the requirements for a certain technology to be implementable in the aviation domain is its reliability; it is preferable to have a slower yet reliable system rather than a high-speed one that is not completely tested or tried in an operational environment. We may identify three main generations of aircraft automation systems: mechanical, electrical and electronic. In the beginnings of commercial flight there were no instrumental aids to help pilots to fly. A piece of string was attached to the wing to indicate whether the airflow over it was sufficient to sustain flight. Later on, the first anemometers and altimeters were introduced to indicate pilots the airspeed and altitude, respectively. These were the first steps toward the "virtualization of the environment" (Ralli, 1993).

The invention of the pneumatic gyroscope (replaced, shortly after, by the electric gyroscope), used to stabilize an artificial horizon, helped pilots to understand their situation even in meteorological conditions characterized by extremely poor visibility, while at the same time preventing dangerous vestibular illusion (a false sense of equilibrium stemming from the inner ear). These simple instruments were merely capable of providing basic indications. Early signs of automation were introduced on board aircrafts during the decade from 1920 to 1930, in the form of an autopilot based on a mechanical engineering concept that was designed to keep the aircraft flying straight: a very basic input to control the flight at a "skill" level. Moreover, as airplanes became bigger and bigger, it became necessary to apply some form of amplification of the pilots' physical force, because of the airflow over large aerodynamic surfaces. Servomechanisms were introduced on board, alongside certain devices aimed at facilitating perception of the force acting on such surfaces (artificial feel load, Mach trim compensator) and absorbing the effects of the so-called Dutch-roll, an abnormal behaviour whereby the airplane yaws and oscillates in an uncoordinated manner (yaw damper). This (mechanical) innovation was the first of multiple steps that began widening the gap between the pilot's input (action on the yoke) and the final outcome (aerodynamic movement). Instead of direct control, with the yoke mechanically attached to the ailerons, airplanes began to be constructed with a series of mechanisms intervening between the pilot's input and the expected output. In this case, the virtualization of flight controls accompanied the parallel virtualization of flight instruments introduced by the artificial horizon (Attitude Display Indicator). At this stage, automation aided pilots mainly in their skill-based activity.

The second generation of automation included electric devices replaced the old mechanisms. Electric gyroscopes instead of pneumatic ones, new instruments such as the VOR (Very High Frequency Omnidirectional Range) to follow a track based on ground aids, the ILS (Instrumental Landing System) to follow a horizontal and a vertical path till the runway threshold, and so forth. The 1960s saw plenty of innovations introduced on board aircrafts that enhanced safety: electric autopilots, auto-throttle (to manage the power setting in order to maintain a selected speed, or a vertical speed), flight directors (used to show pilots how to maneuver to achieve a pre-selected target such as speed, path-tracking and so forth), airborne weather radars, navigation instruments, inertial platforms, but also improved alarming and warning systems capable of detecting several parameters of engines and other equipment. Whereas the first generation of automation (mechanical) managed the pilot's skill-based level, the second generation managed the skill-based and rule-based levels previously assigned to pilots. The airplane systems were monitored through a growing number of parameters and this gave rise to a new concern: the inflation of information with hundreds of additional gauges and indicators inside the cockpit, reaching almost 600 pieces (Boy, 2011). At this stage, pilots used the technology in a tactical manner. In other words, their inputs to automation were immediately accessible, controllable and monitored in the space of a few seconds. For example, if the pilot wanted to follow a new heading, he would use a function provided by the autopilot. The desired heading value was selected in the glare-shield placed in front of the pilot's eye and the intended outcome would be visible within a few seconds: the airplane banked to the left or right to follow the new heading. End of the task. At this stage, automation helped pilots also at a rule-based level, since monitoring of thousands of parameters required efficient alarming and warning systems, as well as recovery tools. The third generation of innovation involved electronics, and was mainly driven by the availability of cheap, accessible, reliable and usable technology that invaded the market, bringing the personal computer into almost every home. The electronic revolution occurring from the mid-80s also helped to shape the new generation of pilots, who were accustomed to dealing with the pervasive presence of technology since the early years of their life. Electronics significantly helped to diminish the clutter of instruments on board and allowed for replacing old indicators - gauges in the form of round-dial, black and white mechanical indicators for every monitored parameter - with integrated colored displays (e.g.: CRT: Cathode Ray Tube, LCD: Liquid Crystal Display) capable of providing a synthetic and analytic view of multiple parameters in a limited area of the cockpit. It is worth mentioning that the type of operations implied by the Flight Management System shifted from tactical to strategic. In fact, whilst in the previous electrical automation stage, pilots were accustomed to receiving immediate feedback visible shortly after the entered input, in the new version, a series of data entered by the operator would show their effects at a distance of hours. The data was no longer immediately accessible and visible, therefore this new way of operating placed greater emphasis on crew co-ordination, mutual cross-checking, operational discipline, not only in the flying tasks but also in the monitoring activity. The Flight Management System database contains an impressive amount of data, from navigational routes, to performance capabilities and plenty of useful information that can be retrieved from pilots. Further on we analyze the traps hidden behind this kind of automation.

However, it is important to point out that the actual discontinuity introduced by this generation of automation: was the notion of electronic echo-system. Compared to the past, when pilots were acquainted with the inner logic of the systems they used, their basic components, and normal or abnormal procedures for coping with operational events, in the new cockpit pilots are sometimes “out of the loop”. This occurrence forces them to change their attitude towards the job. On an airplane such as the A-320 there are almost 190 computers located in almost every area of the fuselage. They interact with each other without the pilot being aware of this interaction. Every time the pilot enters an input to obtain a desired goal (e.g. activating the hydraulic system), he/she starts a sequence in which not only the selected system is activated, but also a number (unknown to the pilot) of interactions between systems depending on the flight phase, operational demands, the airplane’s conditions, etc. The unmanageable complexity of the “electronic echo-system” is a genuine epistemological barrier for the pilot. Whereas before, the pilot had thorough knowledge of the entire airplane and could strategically operate in a new and creative manner whenever circumstances required, the evolution of the cockpit design and architecture has brought about a new approach to flight management that is procedural and sequential. Only actions performed in accordance with computer logic and with the given sequence are accepted by the system. In acting as a programmer who cannot perform tasks that are not pre-planned by the computer configuration, the pilot has lost most of his expertise concerning the hardware part of the system (the aircraft). He too is constrained by the inner logic which dictates the timing of operations, even in high-tempo situations. Consequently, in the international debate on automation and the role of pilots, the latter are often referred to as “system operators”. This holds true up to a certain point but, generally speaking, it is an incorrect assessment. Recalling what has been said about the levels of operation, we may say that at the skill-based level, pilot have become system operators because flying skills are now oriented to flight management system programming. As a flight instructor once said: “We now fly with our fingers, rather than with a hands-on method”. What he meant was that the pilot now appears to be victim of the “push-the-button” syndrome, that leads them to look for switches and devices to operate rather than govern the yoke and throttle. At the skill-based level, automatisms may be in charge for the entire flight, relegating the pilot to a monitoring role. However, at a rule-based level, computers manage several tasks once accomplished by pilots, including monitoring the pressurization system, air conditioning system, pneumatic system and so forth. This statement is no longer valid when referred to the knowledge-based level. Here, the pilot cannot be replaced by any computer, no matter how sophisticated the latter is. Sound judgment of an expert pilot is the result of a series of experiences in which he/she fills the gap between procedures and reality. Since this paper discusses automation rather than the human factor in aviation, anyone wanting to investigate concepts such as flexibility, dealing with the unexpected, robustness, etc. may refer to authors adopting a Resilience Engineering approach (Hollnagel, 2008) (Woods, Dekker, 2010). The fact that pilots are no longer so acquainted with the airplane as in the past, has led them to only adopt the procedural way to interact with the airplane. This is time-consuming, cognitively demanding, and above all, in some cases it may lead to miss the "big picture", or situation awareness.

This new situation introduces two major consequences: automation intimidation (ICAO, 1998) and a restriction of the available tools for coping with unexpected events. In this sense, we can compare the natural echo-system – made up of a complex network of interactions, integrations, retro-actions that make it mostly unforeseeable and unpredictable – with the electronic system. On a modern aircraft, the electronic echo-system is mostly not known by its user. Pilots only know which button they have to press, what the probable outcome is, and ignore what lies is in between. It is a new way of operating that has pros and cons. To assess the real impact on safety, we ought to analyze the relationship between man and technology, by shifting from HMI (Human Machine Interaction) to HCI (Human Computer Interaction). A further step will lead to Human Machine Engineering and/or Human Machine Design (Boy, 2011). The core of this approach is that the focus of the research should be on human-centered design; in other words, the final user – with his/her mental patterns activated in real scenarios – should be regarded as the core of the entire project for a new form of automation. From an engineering perspective, it is very strange that concepts routinely used to describe the role of aircraft automation miss a basic focal point: the final user. In fact, when we talk about technology, we refer to bolt-on versus built-in systems. These expressions indicate the different pattern of integration of aircraft technology. Bolt-on indicates the introduction of a new technology on board an airplane conceived without automation. It is a reactive mode that strives to combine the old engineering philosophy with new devices. It is a kind of "patchwork" which requires several local adjustments to combine new requirements with old capabilities. The expression "built-in" indicates the development of a new technology incorporated in the original project. Every function is integrated with the airplane's systems, making every action consistent with the original philosophy of use. The paradox is that the final user has been removed and left out of the original project, when new types of aircrafts were conceived during the Eighties. Nowadays, after several avoidable accidents, pilots are involved in the early stage of design in order to produce a user-friendly airplane. It is worthy to mention another aspect that could be interesting related to this topic. Not only pilots are unaware of the aircraft philosophies and design, but designers themselves are. After the Concorde accident in Paris in 2002, the prosecutor called the project manager of the aircraft, the one who conceived, plan and execute the project that led to make the Concorde flying. He knew every single part, he knew every system and knew who, where and when subsystems were designed. At the time, it was possible to master a project. Today, most of the airplanes are conceived, designed, built and certified in different places, by different designers, with multiple working groups that have a restricted view on the overall project. It implies as well side-effect problems regarding the liability in case of malfunction or accident. Who is responsible, accountable and liable for an accident induced by automation?

Why Automation?

Two main reasons led to the decision to adopt aircraft automation: the elimination of human error and economic aspects. The first element stems from the general view whereby human performance is regarded as a threat to safety. Such a topic would require a paper in itself, so it is more appropriate to briefly mention some references for students eager to investigate the topic thoroughly. The second element is easier to tackle since we can even quantify the real savings related to, say, lower fuel consumption. According to IATA estimates, "a one percent reduction in fuel consumption translates into annual savings amounting to 100,000,000 dollars a year for IATA carriers of a particular State". (ICAO, 1998). Aside from fuel, the evolution of aircraft technology over the years has led to a dramatic improvement in safety, operational costs, workload reduction, job satisfaction, and so forth. The introduction of the glass cockpit concept allows airlines to reduce maintenance and overhauling costs, improve operational capabilities and ensure higher flexibility in pilot training.

1. *Fuel consumption.* A crucial item in an airline's balance sheet is the fuel cost. Saving on fuel is vital to remain competitive on the market. The introduction of the "fly-by-wire" concept helps to reduce fuel consumption in at least three areas: weight, balance and data predictions.
    a. The *fly-by-wire* concept has brought a tangible innovation. Inputs coming from the pilots' control stick are no longer conveyed via cables and rods directly to the aerodynamic surfaces. In fact, the side-stick (or other devices designed to meet pilots' demands regarding a conventional yoke) provides input to a computer which – via optic fibers – sends a message to another computer placed near the ailerons or stabilizer. This computer provides input to a servo-mechanism to move the surfaces. Therefore, there is no longer any need for steel cables running through the fuselage and other weighty devices such as rods, wheels, etc. This also significantly reduces the aircraft's weight and improves fuel consumption, since less power is required to generate the required lift.
    b. The second area that contributes to saving fuel is *aircraft balance*. The aircraft must be balanced to maintain longitudinal stability (pitch axis in equilibrium). This equilibrium may be stable, unstable or neutral. In a stable aircraft, the weight is concentrated in front of the mean aerodynamic chord. Basically, this means that the stabilizer (the tail) should "push-down" (or, technically, induce de-lifting) to compensate for the wing movements. In an unstable equilibrium, the balance of weight is shifted sensibly backwards compared to a stable aircraft. In other words, the stabilizer should generate lift to compensate for the wing movements. Where does the problem lie with unstable aircrafts? A stable aircraft tends to return to its original state of equilibrium after it deviates from the latter, but is less maneuverable since the excursion of the stabilizer is narrower. On the contrary, unstable equilibrium causes an increasingly greater magnitude of oscillations as it deviates from the initial point. It makes the aircraft more maneuverable, but unstable. In practical terms, in these kinds of airplanes pilots are required to make continuous corrections in order to keep the aircraft steady. This is why the computer was introduced to stabilize the airplane with continuous micro-corrections. This significantly reduces the pilots' workload for flying smoothly. Moreover, due to the distribution of weight concentrated on the mean aerodynamic chord, an unstable aircraft consumes less fuel.
    c. The third factor helping pilots is a database capable of computing in real time any variation to the *flight plan* either on the horizontal path (alternative routes, shortcut, mileage calculations, etc.) or on the vertical profile (optimum altitude, top of descent to manage a low drag approach, best consumption speed, and so forth). This enhances the crew's decision-making task in choosing the best option in order to save fuel.
2. *Maintenance costs.* The glass cockpit concept enables airlines to reduce maintenance and overhauling costs. In conventional airplanes, every instrument had its box and spare part in the hangar.

Whenever a malfunction was reported by the crew, maintenance personnel on the ground fixed it by replacing the apparatus or swapping the devices. All these actions required a new component for every instrument. If we consider that an airliner has roughly one million spare parts, we can easily understand the economic breakthrough offered by the glass cockpit concept. In these airplanes, a single computer gives inputs to several displays or instruments. The maintenance approach is to change a single computer rather than every component or actuator. Based to this operating method, few spare parts are required in the hangar: no more altimeters, no more speed indicators, no more navigation displays (often supplied different manufacturers). Moreover, training of maintenance personnel is simplified as it focuses on a few items only which, in turn, allows for increased personnel specialization.

3. *Selection and Training* **costs.** The fast pace in growth of the airline industry over the last decades has generated concern about the replacement of older pilots, since training centers cannot provide the necessary output for airline requirements. Hiring pilots from a limited base of skilled workers creates a bottleneck in the industrial supply of such an essential organizational factor as are pilots for an airline. Automation has facilitated the hiring of new pilots, since the basic skills are no longer a crucial item to be verified in the initial phase of a pilot's career. If, barely thirty years ago, it would not have been sufficient "to have walked on the moon to be hired by a major airline", as an expert pilot ironically put it, nowadays the number of would-be pilots has increased exponentially. In the current industrial philosophy, almost everyone would be able to fly a large airliner safely with a short amount of training. This phenomenon gives rise to new and urgent problems, as we'll now see. Besides the selection advantages, broadening the potential pilot base enables airlines to save money for the recurrent training of pilots or to reduce transition costs. Indeed, once the manufacturer sets up a standardized cockpit display, the latter is then applied to a series of aircraft. If we look at the Airbus series comprising A-319, A-320, A-321, A-330, A340, etc., we realize how easy is to switch from one airplane to another. In this case, the transition course costs are significantly lower since pilots require fewer lessons; indeed, aside from certain specific details, the only difference involves the performance (take-off weight, cruise speed, landing distances, etc.). Since pilot training costs make up a considerable portion of an airline's budget, it is financially convenient to purchase a uniform fleet made up of same "family" of airplanes. Often, the regulations enable airlines to use a pilot on more than one airplane belonging to a "family". Consequently, this leads to shorter transition courses, operational flexibility as pilots get to fly several aircrafts at a time, and in the long run, better standardization among pilots. Some authors have pointed out how automation has redefined the need for different training processes and crew interaction (Dekker, 2000).
4. *Operational flexibility.* A pilot flying with no aids at all, be they mechanical, electrical or electronic, is limited in many ways. He/she must fly at low altitude, because of his/her physiological limits (hypoxia), he/she cannot fly too fast since the effort on the yoke exceeds his/her physical power, he/she cannot even fly in bad weather (clouds or poor visibility) since he/she must maintain visual contact with the ground. Automation allows to overcome such limitations. Higher flight levels also mean lower fuel consumption and the possibility of flying out of clouds. Faster speeds allow for reaching the destination earlier and completing multiple flights a day. Aircraft instruments enable pilots to achieve better performance. Let's imagine an approach in low visibility conditions. In the beginnings, pilots would rely on a Non-Directional Beacon (NDB) as an aid to find the final track to land on the runway. Safety measures were implemented such as the operating minima. These implied that during a final landing approach, the pilot had to identify the runway before reaching a certain

altitude (landing minima). Obviously, as instruments became more reliable, the landing minima were lowered. Subsequently, the introduction of the VOR (Very High Frequency Omnidirectional Range), which provided a more accurate signal, allowed for performing an approach closer to the runway and at a lower "decision altitude", as the safety margins were assured. When the ILS (Instrumental Landing System) was introduced on board, pilots could also rely on vertical profile indications. This implied higher safety margins that led the regulatory bodies to once again lower the landing minima. As the ILS became increasingly precise and accurate, the landing minima were lowered till they reached the value of zero. This means that an airplane can touch down with such low visibility that pilots must rely on autopilot to perform such an approach. This is due to human limitations such as visual and vestibular illusions (white-out, wall of fog, duck under, etc.) that may impair the pilot's performance. Let's imagine an approach during the '50s: with a visibility of 1,000 metres at the destination airport, the airplane should have diverted to another airport because the pilots would not have been able to attain the airport visual reference at the decision altitude (landing minima). Nowadays, the same airport could operate with 100 metres visibility due to improved transmitting apparatus on the ground and flight automation that enhances pilot performance. A problem evidenced by several authors concerns the shift in responsibility of the people managing the automation system. With the performances brought about by automation that exceed human capabilities, a failure in the automatic system places pilots in a situation in which they have no resources available for coping with an unexpected event. In one case, where the pilots were performing a low visibility approach, the automatic system went out of control at very low altitude, causing the airplane to pitch down and hit hard on the runway. Experts acknowledge that the pilots' recovery in that case was beyond reasonable intervention, that is to say "impossible" (Dismukes, Berman, Loukopoulos, 2008). Indeed, the time available for detecting, understanding and intervening was so short that it was virtually impossible to solve the airplane's faulty behaviour. Who is responsible in such a case? Is it correct to refer to the pilots' mismanagement, poor skill, or untimely reaction when they are operating outside their safety boundaries? If the operations are conducted beyond human capabilities we should also review the concept of responsibility when coping with automation. As Boy and Grote put it: "Human control over technical systems, including transparency, predictability, and sufficient means of influencing the systems, is considered to be the main prerequisite for responsibility and accountability" (Boy, 2011).

Different Cockpit Presentation

In recent years, cockpit design has undergone somewhat of a revolution. Old aircraft were designed to satisfy pilots' needs and relied upon slight modifications in the previous pilots' panel scanning pattern. A common standardized cockpit display emerged in the early '50s: the T-model. It included the basic instruments such as artificial horizon (in the top centre), anemometer (speed indicator) on the left, the altimeter on the right and compass (indicating the heading and track) in the bottom-centre position. Two further instruments, the side-slip indicator and vertical speed indicator were added later on. The adjacent picture (Figure 2) shows the classic T-model. The pilot was accustomed to perceive configuration, disregarding the precise indications.

With the introduction of the glass cockpit, the traditional flight instrument display was replaced by a different presentation encompassing more information, greater flexibility, color coding and marking, but at the same time, it could lead to information overload, as several parameters are displayed in a

Figure 2. Classic T-model of cockpit display

compact area. In a single display, known as the PFD (Primary Flight Display), multiple information is included not only for basic parameters, but also for navigation functions, approach facilities, automatic flight feedback, flight mode awareness, and so forth. Since the pilot needs to cross-check a great number of instruments at a glance, the design of traditional instruments was developed according to a pattern of attention, from the more important information to the more marginal information. Each instrument had its own case and its functional dynamics was perfectly known by the pilot; it was easy to detect, easy to use during normal operations and easy to handle in case of failure. The switch positions and shapes were positioned on board according to a pattern of use and were instantly recognizable at the touch. It was common for pilots to perform the checklist according to the "touch and feel" principle, by detecting whether a switch was in the required configuration by confirming its ON/OFF position. Almost every switch had a peculiar shape: i.e., rough and big for operative ones, smooth and small for less important switches. A form of training implemented in most flying schools was the "blind panel" exercise, which consisted in covering the trainee's eye with a bandage and asking him/her to activate the switch prompted by the instructor. Being familiar with the physical ergonomics of the cockpit proved very helpful. The pilot knew how much strength was required to activate a command, how much the arm had to be stretched to reach a knob and so forth. This exercise enhanced the pilot's skill, or the skill level of operation (according to Rasmussen's Skill-Rule-Knowledge paradigms). With the new cockpit configuration (dark panel) there is another kind of complacency and lack of skills regarding the activation/deactivation of systems. Since during normal operations the overhead panel (containing most of the system's control panels) the pilots don't touch something that works automatically. So they are not familiar with particular use that may arise during an abnormal or emergency situations. For example, in the older aircraft the pressurization system was handled by the pilot not flying (who talked with the ATC, read the checklist and managed the systems). The management of this system was quite simple, but required continuous attention and practice. In highly automated airplanes, the pressurization system is not even touched during normal operations. When something happens to the pressurization system, the pilot should monitor, act and cross-check not only the cabin altitude indications, but should move the switches, trimming the equivalent altitude. It is an extremely remote hypothesis, but nevertheless it absorb completely one pilot attention during the remaining of flight, causing a major threat to crew integration.

Regarding the input to the flight controls, it was assured via flight control wheels and sticks, cables and rods, in order to enable the pilot to intervene directly on the aerodynamic surfaces such as the ailerons, rudder and stabilizer. With regard to displays, the traditional instrument configuration – made up of drum pointers, single instrument boxes and "touch and feel" overhead panel – was replaced by a new approach towards information available to the pilot. The "touch and feel" philosophy was replaced by the "dark cockpit" approach. A dark, flat, overhead panel was adopted to show the pilot that everything was OK. Failures or anomalies were detected by an illuminated button, recalled by a master caution/warning light just in front of the pilots' eyes. In the unlikely (but not impossible) situation of smoke in the cockpit severely impairing the pilots' sight, (due to the concentration of a thick layer of dense, white fog), pilots of conventional airplanes could retrieve the intended system's configuration "by heart", by detecting the switch position with a fingertip. Vice-versa, let's imagine a dark panel with smoke on board. It is truly challenging to spot where to place your hand and, above all, what the system's feedback is, since the system configuration is not given by the switch position but by an ON/OFF light, which is often undetectable.

Ergonomics

The problem with innovation, especially in a critical context for safety as is aviation, lies in the interaction between a community of practice and a new concept, conceived and implemented by engineers. This implies a relationship between automation and ergonomics. Ergonomics is a word deriving from the Greek words ergon (work) and nomos (law), and it is a field of study aimed at improving work conditions so as to guarantee optimal adaptation of the worker to his/her environment. We may identify three main types of ergonomics: physical, cognitive and social (or organizational). Initially, with the onset of the first generation of automation (mechanical), ergonomists studied physical ergonomics, namely: how to reach a control, how much force is required to operate a lever, visibility of displayed information, seat position design, and so forth. For example, several accidents occurred due to misuse of the flap lever and landing gear lever. Indeed, these were positioned near one another and had similar shapes, leading the pilots to mistake automation them (Koonce, 2002). The ergonomists found a straightforward and brilliant solution: they attached a little round-shaped rubber wheel to the lever tip to indicate the landing gear, and a wing-shaped plastic cover indicating the flaps. Moreover, the two levers were separated in order to avoid any misuse. In the next generation of automation, namely electrical automation, the ergonomists aimed to improve the cockpit design standardization. A pilot flying on an airplane had difficulty in forming a mental image of the various levers, knobs and displays, as every manufacturer arranged the cockpit according to different criteria. Dekker has clearly illustrated a case of airplane standardization: the position of the propeller, engine thrust and carburetor in a cockpit in different airplanes, even if they were built by the same manufacturer (Dekker, 2006).

Pilots transitioning to other aircrafts struggled to remember the position of every lever since they were adjacent to each other, and the possibility of mistaking them was very high. Moreover, during the training process there was a risk of the so-called "negative transfer", that is, the incorrect application of a procedure no longer suited to the new context. For example, the switches on Boeing aircrafts have top-down activation, while Airbus has chosen the swept-on system (bottom-up). A pilot transitioning from a Boeing 737 to an Airbus A-320 may quickly learn how to switch on the systems, but in conditions of stress, fast-paced rhythm and high workload, he/she may return to old habits and implement them incorrectly. Social ergonomics tries to eliminate such difficulties. Cognitive ergonomics studies the adaptability of the technology to the mental patterns of the operator. The third generation of automation introduced on board aircrafts gave rise to many concerns about whether the instrument logic was suitable for being correctly used by pilots. From a cognitive perspective, the new system should be designed according to the user's need, while bearing in mind that pilot-friendly is not equivalent to user-friendly (Chialastri, 2010). A generic user is not supposed to fly, as some engineers erroneously tend to believe. Pilots belong to a professional community that share a mental pattern when coping with critical operational situations. The modus operandi is the result of a life-long inflight experience. A professional community such as the pilot community is target-oriented, and aims at the final goal rather than observing all the required procedural tasks step-by-step. This is why we talk about "shortcuts", "heuristics", and so forth. For example purposes, let's discuss the introduction of a new airplane with variable wings: the F-111. It was conceived by the designers with a lever in the cockpit to modulate the wings from straight (useful at low speed) to swept (to fly at high speed). In the mind of the designer (who is not a pilot), it seemed perfectly sensible to associate the forward movement of the lever in the cockpit with the forward movement of the wings: lever forward – straight wings, lever back – swept wings. When the airplane was introduced in flight operations, something "strange" occurred. Pilots associated the forward movement with the concept of speed. So, when they wanted high speed, they pushed the thrust levers forward, the yoke down to dive and, consequently, the wing lever forward (incorrectly so). For a thorough analysis of this case, see Dekker (2006). All this occurred due to the confusion between generic users and specialized users, as are pilots. As Parasumaran puts it: "Automation does not simply supplant human activity but rather changes it, often in ways unintended and unanticipated by the designers of automation and, as a result, poses new coordination demands on the human operator" (Parasumaran, Sheridan, 2000).

Automation Surprise

The interaction between pilots and technologies on board the aircraft raises some concerns regarding an acknowledged problem: the automation surprise. This occurs when pilots no longer know what the system is doing, why it is doing and what it will do next. It is an awful sensation for a pilot to feel to be lagging behind the airplane, as – ever since the early stages of flight school – every pilot is taught that he/she should be "five minutes ahead of the airplane" in order to manage the rapid changes occurring during the final part of the flight. When approaching the runway, the airplane's configuration demands a higher workload, the communication flow with air traffic controllers increases and the proximity to the ground absorbs much of the pilot's attention. Once all the fast-pace activities are handled with the aid of automation, the pilot may shed some of the workload. When automation fails or behaves in a "strange" manner, the workload increases exponentially. This is only a matter of the effort required to cope with automation, and could be solved with additional training. A more complicated issue – arising from investigations on some accidents – is the difference between two concepts, namely: "situational surprise" and "fundamental surprise" (Wears, 2011).

To summarize our explanation, we may refer to the following example on the everyday life. We are driving in the highway and suddenly we hear a loud bang from left bottom part of our car. We are surprised, but after three seconds we realize that a tyre is burst, so we steer towards the edge of the highway, braking to slow down our speed and looking for the behaviour of other cars behind us. This is a situation surprise. Let's now think to the same situation, with a single difference: while driving, our car start flying. For the first three seconds we are astonished. We don't speak, trying to figure out what the hell is going on. We literally don't know what to do, which resources we have to cope with this incredible situation. We cannot even plan an alternative, and even if we can, we cannot communicate it. Obviously, the reader may object "Car don't fly". On the other hand we may say that "Planes don't stall" (actually, they do, as we'll soon show, but many pilots assume they don't). Therefore, translating these concepts in the pilot's operational life, we may say that a situation surprise arises when something occurs in a pre-defined context, in which the pilots know how to manage the situation. Such a situation is unexpected, being an abnormal condition, but nonetheless remains within the realm of the pilot's knowledge. Moreover, it returns to being a manageable condition once the pilot acknowledges the difference between a normal and an abnormal condition, by applying the relevant abnormal procedure. After the initial surprise due to the unexpected event, the pilot returns to a familiar pattern of operation acquired with training. A so-called fundamental surprise is a very different circumstance: it implies a thorough re-evaluation of the situation starting from the basic assumptions. It is an entirely new situation for which there is no quick-fix "recipe" and is difficult to assess, since it is unexampled. The pilots literally do not know what to do because they lack any cognitive map of the given situation. This concept can be well summarized by the findings following an accident involving a new generation airplane. When analyzing the black box after an accident, it is not rare to hear comments such as: "I don't know what's happening" – this is a something of a pilot's nightmare. It should be noted that some failures occurring on board highly automated aircrafts are transient and are very difficult to reproduce or even diagnose once on the ground. Even the manufacturer's statements regarding electrical failures are self-explaining, as a failure may produce different effects in each case or appear in different configurations depending on the flight phase, airplane configuration, speed and so forth. Sometimes, an electrical transient (also influenced by Portable Electrical Devices activated by unaware passengers) could trigger certain airplane reactions that leave no memory of the root cause.

Automation Issues

Although automation has improved the overall safety level in aviation, some problems tend to emerge from a new way of operating. A special commission was set up in 1985 by the Society of Automotive Engineers to determine the pros and cons arising from flight deck automation. Nine categories were identified: situation awareness, automation complacency, automation intimidation, captain's command authority, crew interface design, pilot selection, training and procedures, the role of pilots in automated aircrafts (ICAO, 1998). Sometimes, pilots lose situation awareness because they lag behind the automation logic. A pilot should always know what the automation system is going to do in the next five minutes, to either detect any anomalies or take control of the airplane. In fast-paced situations – such as crowded skies, proximity with the terrain and instrumental procedures carried out in poor visibility – a high number of parameters must be monitored. As the workload increases, the pilot tends to delegate to the automation system a series of functions in order to minimize the workload. It is important to specify that the physical workload (the number of actions performed within a given time frame) differs from the cognitive workload in that the latter implies a thorough monitoring, understanding and evaluation of the data coming from the automation system. The paradox involving airplane automation is that it works as an amplifier: with low workloads, it could lead to complacency ("let the automation system do it") or boredom that reduce alertness and awareness. Plainly said: "When good, better; when bad, worse". Poor user interface design is another issue. Norman, Billings et al. have studied the way humans interact with automation and have developed cognitive ergonomics.

There are several studies concerning the optimal presentation (displays, alarms, indicators, etc.) for pilot use. For example, old displays were designed in such a way that each single instrument provided data for a specific domain only: the anemometer for speed, the altimeter for altitude and flight levels, and so forth. With the integrated instruments, all the required data is available at a glance in a single instrument, which not only provides color coded information but also on-demand information by means of popup features, so that it can be concealed during normal operations and recalled when needed. Several issues are related to displays, but only two are addressed here: cognitive mapping of instruments and human limitations in applying the color-code to perception. We have to consider that the replacement of round-dial black and white instruments, such as the old anemometer, with colored multifunction CRT (Cathodic Ray Tube) instruments, has determined a new approach by pilots in mentally analyzing speed data. Indeed, with round-dial indicators, pilots knew at a glance in which speed area they were flying: danger area, safe operations or limit speed. Although it is not important to know the exact speed, according to a Gestalt theory concept we need to see the forest rather than every single tree.

Furthermore, from a cognitive perspective, we tend to perceive the configuration as a multiple instrument: for example, during the approach phase, the anemometer needle will point to 6 o'clock, the vertical speed indicator to 8 o'clock and so forth. In the new speed indicator, the values appear on a vertical speed tape rolling up and down, with the actual speed magnified for greater clarity. Nevertheless, the real perception of the speed range which the pilots are flying is somehow impaired by this kind of indication. For example, a dangerous situation in flight may arise whenever the airplane approaches its maximum operating altitude. As the altitude increases the relevant stall speeds tend to close in to a single value (coffin corner). Since the speed window is limited to 20 kts above and 20 kts below the actual speed, the pilots may think that they have a greater margin on stall than they actually have. Look at the following picture (Figure 3).

The acceptable speed is the black band, while the amber and the red represent the no-fly zone (high speed buffet and low speed buffet). If we compare the acceptable area with the no-fly zone, apparently

Figure 3. The acceptable speed on devise in dangerous situation in flight

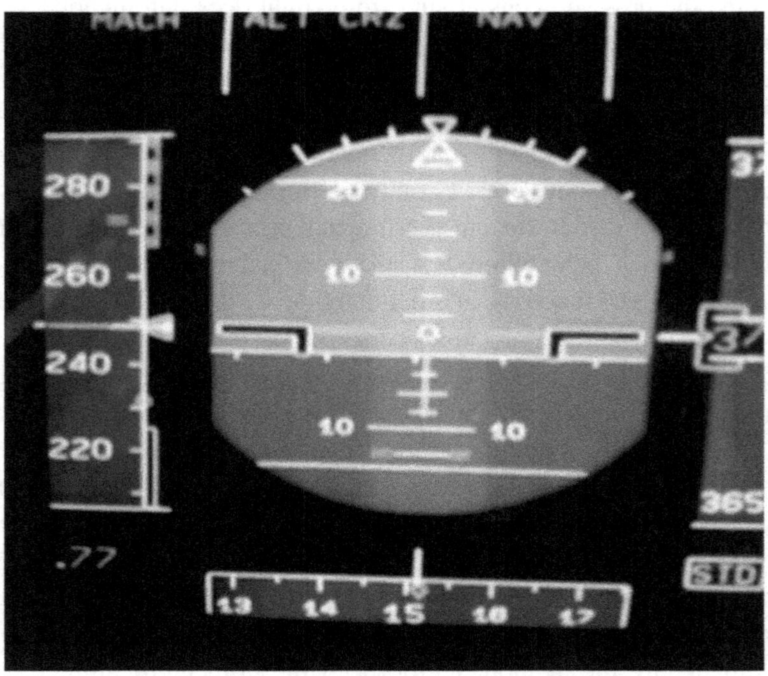

we have 50% of the field indicating that we have margin over the flight boundaries. Actually, if we translate the forty knots in a rounded up dial, we realize that the acceptable speed range reduces to 10% of the selectable speed. A safety concern may arise whenever the pilots accept to fly at an altitude where the low speed buffet is very close to the high speed buffet (the so called coffin corner). In this case an increase in the wind, an unexpected moderate turbulence or some thunderstorm activity below which induces a convective movement may cause an oscillation in speed triggering some consequences. In fact, in the highly automated aircrafts some protections (such as high speed or low speed protection) automatically activate, with unpleasant consequences. For example, the airplane may dive down to avoid a loss of speed or it can climb to limit the increase of speed. In any case, these are upsetting conditions for a pilot flying at cruising altitude. The speed tape shows also additional indicators are displayed on the speed tape, such as minimum speed, flap retraction speed and maximum speed for every configuration. In many ways, this system facilitates the pilot's task. Problems arise whenever failures occur (electrical failure, unreliable indication etc.), as the markings disappear.

The situation worsens compared to the old indicators, as pilots needs to create a mental image of the speed field in which they are flying. For example, it is unimportant for us to know whether we are flying at 245 or 247 knots, but it becomes important to know that we are flying at 145 knots rather than 245 knots. This means that the mental mapping of the speed indication involves a higher workload, is time-consuming and also energy consuming. Let's take a look at these two different speed indicators to grasp the concept. The second issue is related to human performance and limitations. The human eye may detect colors, shape, light and movement with foveal vision. There are two kinds of receptors in the fovea: the cones and the rods. The cones are responsible for detecting shapes and colors, in an area roughly large 1 cent, and account for focused vision. Instead, rods are responsible for peripheral vision and are able to detect movement and light. This means that color-coded information should appear in an area just in front of the pilot's head, as only the cones may such information. In fact, since an input appearing to the side can be detected by the rods, which are insensitive to color and shape, the pilot will not notice it unless he stares at it directly. Other relevant indicators for pilots are symmetry and context. The needle in a round-dial instrument satisfies both these requirements, since we may easily detect which field of operation we are currently in (near the upper limit, in the middle, in the lower margin), in addition to the trend (fast-moving, erratic, gradual) and symmetry with nearby parameters. A digital indicator shows this information in an alternative way, by arranging numbers on a digital display, as can be clearly seen in the above picture. The configuration on the left can be grasped at first glance, while the second set of information requires a considerable effort to detect differences.

Disrupted Symmetry of Digital Information

Another area worthy of attention is the greater magnitude of the errors made by pilots when flying highly automated airplanes. Pilots flying older aircrafts normally relied on their ability to obtain the required data using a heuristic and "rule of thumb" approach. This method was not extremely accurate but roughly precise. Nowadays, with the all-round presence of computers on board, flight data can be processed very precisely but with the risk of gross errors due to inaccurate entries made by pilots. An example taken from everyday life may help to explain the situation: when setting older-version alarm clocks, the user's error might have been limited to within the range of a few minutes, whereas the new digital alarm clocks are very accurate but subject to gross errors – for example 8 PM could be mistaken for 8 AM. Lower situation awareness means that pilots no longer have the big picture of the available data. The aviation truism, "trash in-trash out", is applicable to this new technology as well. Having lost the habit of looking for the "frame", namely the context in which the automation system's computed data should appear, pilots risk losing the big picture and, eventually, situation awareness. Unsurprisingly, there have been cases of accidents caused by inconsistent data provided by the computer and uncritically accepted by pilots. This is neither a strange nor uncommon occurrence. Fatigue, distractions, and heavy workload may cause pilots to lower their attention threshold. In one case, a pilot on a long haul flight entered the available parameters into the computer to obtain the take-off performance (maximum weight allowed by the runway tables, speed and flap setting). He mistook the take-off weight with other data and entered the parameters into the computer.

The end result was a wrong take-off weight, wrong speed and wrong flap setting. The aircraft eventually overrun the runway and crashed a few hundred meters beyond the airport fences. Another issue linked the introduction of aircraft automation is a weakening of the hierarchy implied by the different way of using the automation system. The captain is indeed the person with the greatest responsibility for the flight and this implies a hierarchical order to establish who has the final word on board. This hierarchical order is also accompanied by functional task sharing, whereby on every flight there is a pilot flying and a pilot not flying (or monitoring pilot). Airplanes with a high level of automation require task sharing, in which the pilot flying has a great degree of autonomy in programming the Flight Management System, in deciding the intended flight path and type of approach. Above all, the pilot flying also determines the timing of the collaboration offered by the pilot not flying, even in an emergency situation. In fact, since the crew must act in a procedural way in fulfilling the demands of the automation system, both pilots must cooperate in a more horizontal way compared to the past. The hierarchical relationship between captain and co-pilot is known as the "trans-authority gradient" (Hawkins, 1987) and, from a human factor perspective, should not be too flat nor too steep.

Case Studies

Currently, several case studies could be cited to illustrate the relationship between pilots and automation. In the recent past, accidents have occurred due to lack of mode awareness (A-320 on Mount St. Odile), automation misuse (Delhi, 1999), loss of braking leading to loss of control (S. Paulo Garulhos, 2006), loss of control during approach (B-737 in Amsterdam, 2008), and several other cases. From these accidents, we chose a couple of peculiar cases. The first case occurred in 1991 and involved an Airbus flying at cruise altitude. While the captain was in the passenger's cabin, the co-pilot tried to "play" with the FMS in order to learn hands-on. He deselected some radio-aids (VOR: Very High Frequency Omnidirectional Range) from the planned route. After the tenth VOR was deselected, the airplane depressurized, forcing the pilot to perform an emergency descent. From an engineering perspective there was a bug in the system but, moreover, it was inconceivable for the pilot to link the VOR deselection (a navigation function) to the pressurization system. The second case is related to a B-727 performing a low-visibility approach to Denver. The flight was uneventful till the final phase. The sky was clear but the city was covered by a layer of fog that reduced the horizontal visibility to 350 ft. and the vertical visibility to 500 ft. The captain was the pilot flying, in accordance with the company regulations, and used the autopilot as specified in the operating manual. The chosen type of approach was an ILS Cat II. The ILS is a ground based navigation aid that emits a horizontal signal to guide the airplane to the centre of the runway together with a vertical signal to guide the aircraft along a certain slope (usually 3°) in order to cross the runway threshold at a height of 50 ft. The regulations state that once the aircraft is at 100 ft., the crew must acquire the visual references to positively identify the runway lights (or markings). If not, a go-around is mandatory. Once the captain has identified the runway lights he must proceed to land manually, by switching off the autopilot at very low altitude. There are certain risks associated with this maneuver, since – with impaired visibility – the pilot loses the horizon line, depth perception and the vertical speed sensation, and could be subject to spatial disorientation due to the consequences of the so-called "white-out" effect. In this case, however, nothing similar occurred. The autopilot duly followed the ILS signals, but due to a random signal emitted in the last 200 ft., the aircraft pitched down abruptly. The captain tried to take over the controls, though unsuccessfully. He later reported – during the investigation – that the window was suddenly "full of lights", meaning that the aircraft assumed a very nose-down pitch attitude. The high rate of descent, coupled with the surprise factor and low visibility, did not allow the captain to recover such a degraded situation: the airplane touched down so hard that it veered off the runway, irreversibly damaging the fuselage. This case emphasizes the relationship between pilots and liability. Operating beyond AF 447: a paradigmatic accident? The role of automation human capabilities, the pilot finds himself/herself in a no-man's land where he/she is held responsible even when he/she cannot sensibly regain control of the airplane. As stated by Dismukes and Berman, it is hardly possible to recover from such a situation.

The Air France 447 Crash

Few accidents may lead us to reflect on the automation role than the AF 447 case. This flight, carried out on June 2009 from Rio de Janeiro to Paris, disappeared from the radar screens while it was over the Atlantic Ocean. Different causes were identified to explain why the pilots lost the control for about three and a half minutes, diving from an altitude of 33000 ft. to the Ocean surface. Before making any hypothesis, it was necessary to recover the wreckage and read the Flight Data Recorder in conjunction with the cockpit voice recorder. In the history of accidents losing control from cruise altitude is very rare. Investigators wondered how it happened, trying to figure out the real dynamics of the accident. Reading the flight data recorder and the cockpit voice recorder it emerged that after reaching the cruise level the Captain left the cockpit, leaving two first officer in the flight deck. Before leaving his seat he recommended the first officers not to deviate excessively from the assigned route, in order to save fuel. The ITCZ (Inter Tropical Converging Zone) a very challenging meteorological phenomenon, with thunderstorms, lightning, turbulence and ice formation, was lying ahead and many airplanes on the same route deviated significantly to the North. The problem with the fuel was due by the high airplane weight at take-off. In order to avoid a commercial setback, offloading cargo and/or passengers, the Captain decided to uplift the minimum (also the maximum, given the actual conditions) fuel to reach Paris. On the other hand, flying with minimum fuel allows few options to deviate from the programmed route. Large deviations would have meant an extra landing in Bordeaux to refuel and reach Paris. On a commercial point of view this is a "sacrificing decision". On the safety point of view, and with hindsight, this was a trade-off that eroded the safety margins. Following the Captain instructions, the first officers in the flight deck did not deviate as other planes did. They encountered some phenomena such as the St. Elmo Fire (electrostatic discharges on the windshield) that made the younger copilot upset. Another phenomenon, unknown to pilots, was the ice accumulation on the pitot tubes, the probes that allow the speed computer to calculate the difference between dynamics and static air pressure.

When ice accumulate on these probes, the speed indication becomes unreliable. Several problems arise from this unreliable indication. First, there is no warning flag, no alerting light nor any chime recalling the pilots that what they see is not valid. It is the same difference, in the everyday life, between accepting a 30 euro banknote (which doesn't exist) and a 500 euro banknote (it exists, but could be false). Pilots must rely upon they knowledge and expertise to detect an unreliable indication, be it altitude, speed or attitude. As soon as they detected a "strange" speed the pilot flying pull back his side-stick. The aircraft entered a steep climb at high altitude, gaining almost 3000 ft. In the meantime the actual speed decreased till the stall speed. The Airbus A-330 has different flight laws, that normally protect against excessive speed or insufficient speed along the entire flight envelope. When a failure is detected the flight laws reconfigure in degraded modes, such as the so called alternate and direct law. In these conditions the airplane may stall. That's what happened. With missing speed sources, the airplane AF 447: a paradigmatic accident? The role of automation degraded in alternate law, that lack any protection against low speed stall. When the co-pilot pulled back his side-stick, making the airplane lose energy, the speed decreased below the minimum to sustain the flight and the angle of attack increased beyond the maximum allowable. From then on, the pilots lost the situation awareness, entering in a dive that last three minutes and forty seconds, with no way to sort out. The startle effect is a phenomenon well known to the human factors experts. It is a fundamental surprise, that makes the pilots astonished, unable to understand, plan and check what is going on. The airplane behaved in a way incomprehensible to them. The "STALL" alarm sounded at least seventy times during the fatal descent. Nobody in the cockpit acknowledged the they were in a low speed situation leading to a stall. Or, better, they had two different situation awareness but didn't verbalize they perceptions. The more experience first officer, seated on the left side, thought to be in a low speed situation, while the younger co-pilot thought to be in a high speed situation. In fact, he pulled back his side-stick in order to reduce the speed. Or, at least, this is what we may infer reading the final report published by BEA. Having two different, incompatible, perception of the flight progress, they didn't speak out to state the problem (unreliable speed), nor to explain which was their main concern (too slow, too fast). The airplane alarm system saturated the cockpit with aural warnings (a chime, recalling they were leaving the assigned altitude), voice calls ("ALTITUDE", "STALL" "DUAL INPUT"). Pilots themselves were saying something, plus the aerodynamic noise, the thunderstorm that gave also an additional buffet to the physical perception of the pilots. The first officer called the cabin attendant to wake up the Captain. When he came back into the cockpit found a situation already degraded. The aircraft was diving down with an attitude of 5°, but with an angle of attack of about 40°. The engines were at full thrust, while the vertical speed was at 10000 ft per minute (almost four times the usual rate of descent). The Captain was probably in the sleep inertia phase, following an abrupt awakening. The situation he had to cope with was extremely challenging because he didn't follow the evolution of the airplane state from the beginning. He entered the cockpit and didn't figure out what was the real problem. The first officer reported several times "I don't know what's happening", adding confusion to the Captain effort to understand the nature of the problem. In the meantime the younger copilot was acting continuously backward on his side-stick. Unfortunately, the flight controls design of the A-330 makes almost impossible to detect which are the pilot input on the sidestick, because the side-sticks are not coordinated; when a pilot moves his control the other doesn't follow. The only hint is when both pilots act on their respective sidestick. In this case, the aural warning "DUAL INPUT" activates. It is to be mentioned that some features of the A-330 led the entire crew in confusion. For example, the STALL warning sounds only if the speed is above 60 kts. With the unreliable speed below that threshold the warning stopped. The consequence was that when the pilot pushed down his side-stick the STALL warning came on and when he pulled back his side-stick (trespassing the activation threshold) the warning went off. The recommendations following this accident were aimed at some design issue. First, the non-coordinated side-stick movements. Second, the STALL warning must be reinforced with a visual indication. Third, the crew coordination has been recognized to be ineffective, but most of this issue is due to the fundamental surprise that impaired their communication skills. Some other recommendations were issued, but the focus of this paper is to show how an improper design could lead (even decades later) to accident because the human factors dynamics were not addressed when the airplane was conceived. It is worth to mention that some years later another Airbus A-320 crashed after the pilots lost the control at cruise altitude. Even in that case, there was no way to sort out the upset attitude at high altitude.

CONCLUSION

This brief paper highlights certain aspects concerning the introduction of automation on board aircrafts. Automation has undeniably led to an improvement in flight safety. Nevertheless, to enhance its ability to assure due and consistent help to pilots, automation itself should be investigated more thoroughly to determine whether it is suitable in terms of human capabilities and limitations, ergonomics, cognitive suitability and instrument standardization, in order to gradually improve performance.

REFERENCES

Air Pilot Manual. (2011). *Human Factors and pilot performance*. Pooleys, England Alderson.

Amalberti, R. (2000, September). The paradox of the ultra-safe systems. *Flight Safety Australia*.

Anderson, E. W. (1994). Cross-category variation in customer satisfaction and retention. *Marketing Letters*, *5*(1), 19–30. doi:10.1007/BF00993955

Anderson, E. W., & Sullivan, M. (1993). The antecedents and consequences of customer satisfaction for firms. *Science*, *12*(2), 125–143.

Antonio, C. (2010). *Virtual Room: a case study in the training of pilots*. Cape Canaveral: HCI Aero-conference.

Antonio, C. (2011). Human-centred design in aviation. *Proceedings of the Fourth Workshop on Human Centered Processes*.

Antonio, C. (2011). Resilience and Ergonomics in Aviation. In *Proceedings of the fourth Resilience Engineering Symposium*. Mines ParisTech.

Bagnara, S., & Pozzi, S. (2008). Fondamenti, Storia e Tendenze dell'HCI. In A. Soro (Ed.), *Human Computer Interaction. Fondamenti e Prospettive*. Monza, Italy: Polimetrica International Scientific Publisher.

Bainbridge. (1987). The Ironies of Automation. New Technology and Human Error.

Barnes, C., Elliott, L.R., Coovert, M.D., & Harville, D. (2004). Effects of Fatigue on Simulation-based Team Decision Making Performance. Ergometrika, 4.

Boy, G. (2011). The Handbook of human Machine Interface – A Humancentered design approach. Ashgate.

Chialastri, A. (2015). *Human factor – Il rapporto uomo-macchina*. Roma: IBN.

Cooper, G. E., White, M. D., & Lauber, J. K. (Eds.). (1980). Resource management on the flightdeck. *Proceedings of a NASA/Industry Workshop (NASA CP-2120).*

Dekker, S. (2003). *Human Factor in aviation: a natural history*. Technical Report 02. Lund University School of Aviation AF 447.

Dekker, S. (2000). Sharing the Burden of Flight Deck Automation Training. *The International Journal of Aviation Psychology, 10*(4), 317–326.

Dekker, S. (2005). *Why We need new accident models*. Lund University School of Aviation, Technical Report 2005-02.

Doyle, J.C. (2010). Contrasting view of Complexity and Their Implications For Network-Centric Infrastructures. *IEEE Transactions on Systems, Man, and Cybernetics. Part A, Systems and Humans*, *40*(4).

Erik, H. (2009). *The ETTO Principle – Efficiency-Thoroughness Trade-Off*. Ashgate.

Ferlazzo, F. (2005). *Metodi in ergonomia cognitiva*. Roma: Carocci.

Flight Safety Foundation. (2003). The Human Factors Implications for Flight Safety of Recent Development In the Airline Industry. In *Flight Safety Digest, March April Hawkins Frank (1987), Human factor in Flight*. Aldershot, IL: Ashgate.

Hollnagel, E. (2008). Critical Information Infrastructures: should models represent structures or functions. In *Computer Safety, Reliability and Security*. Heidelberg, Germany: Springer.

Hollnagel, E., Woods, D., & Leveson, N. (2006). *Resilience Engineering – Concepts and Precepts*. Ashgate.

Hutchins, E. (1995). How a cockpit remembers its speeds. *Cognitive Science*, (19): 265–288.

IATA. (1994). *Aircraft Automation Report, Safety Advisory Sub-Committee and Maintenance Advisory Sub-committee*. ICAO - Human Factors Digest No. 5, Operational Implications of Automation in Advanced Technology Flight Decks (Circular 234).

ICAO. (1998). Doc. 9683-AN/950. Montreal, Canada: ICAO.

James, R. (1990). *Human error*. Cambridge, UK: Cambridge University Press.

James, R. (2008). *The human contribution*. Farnham, UK: Ashgate.

Koonce, J. M. (2002). *Human Factors in the Training of Pilots*. London: Taylor & Francis.

Loukopoulos. (2008). The limits of expertise. Ashgate.

Maurino, D., & Salas, E. (2010). *Human Factor in aviation*. Academic Press.

Morin, E. (1989). Il Metodo – La conoscenza della conoscenza. Feltrinelli editore.

Morin, E. (2001). Il Metodo – La natura della natura. Raffaello Cortina editore.

Morin, E. (2004). Il Metodo – La vita della vita. Raffaello Cortina editore.

Morin, E. (2008). *Il Metodo – Le idee: habitat, vita organizzazione, usi e costumi*. Raffaello Cortina editore.

Navon, D. (1977). Forest before trees: The precedence of global features in visual perception. *Cognitive Psychology*, *9*(9), 353–383. doi:10.1016/0010-0285(77)90012-3

Norman, D. (1988). *The psychology of everyday things*. New York, NY: Basic Books.

Parasumaran, R., & Wickens, C. (2008). Humans: Still Vital After All These Years of Automation. *Human Factors*, *50*(3), 511–520. doi:10.1518/001872008X312198 PMID:18689061

Parasumaran, S. (2000). A Model for Types and Levels of Human Interaction with Automation. *IEEE Transactions on Systems, Man, and Cybernetics. Part A, Systems and Humans*, *30*(3).

Peter, W., Steven, P., & Bob, F. (2002). *The Prescription and Practice of Work on the Flight Deck*. Department of Computer Science University of York.

Ralli, M. (1993). Fattore umano ed operazioni di volo. Libreria dell'orologio.

Rasmussen, J. (1983). Skills, Rules, Knowledge: Signals, Sign and Symbol and Other Distinctions in Human Performance Models, AF 447: a paradigmatic accident? The role of automation. *IEEE Transactions on Systems, Man, and Cybernetics*, SMC-13.

Robert, W., & Kendall, W. L. (2011). Fundamental on Situational Surprise: a case study with Implications for Resilience. In *Proceedings of the fourth Symposium on Resilience Engineering*. Mines-Tech.

Salas, E., & Maurino, D. (2010). *Human factors in aviation*. Burlington, MA: Elsevier. doi:10.1016/B978-0-12-374518-7.00001-8

Wolfgang, K. (1967). Gestalt psychology. *Psychological Research*, *31*(1), 1830.

Woods, D., Dekker, S., Cook, R., Johannesen, L., & Sarter, N. (2010). *Behind human error*. Aldershot, UK: Ashgate publishing.

Woods, D. D. (1990). Modeling and predicting human error. In Human performance models for computer aided engineering (pp. 248-274). Academic Press.

KEY TERMS AND DEFINITIONS

Attribute Evaluation: A data mining procedure which estimates the utility of attributes for given task (usually prediction). Attribute evaluation is used in many data mining tasks, for example in feature subset selection, feature weighting, feature ranking, feature construction, decision and regression tree building, data discretization, visualization, and comprehension.

Context of Attributes: In a given problem the related attributes, which interact in the description of the problem. Only together these attributes contain sufficient information for classification of instances. The relevant context may not be the same for all the instances in given problem.

Evaluation of Ordered Attributes: For ordered attributes the evaluation procedure should take into account their double nature: they are nominal, but also behave as numeric attributes. So each value may have its distinct behavior, but values are also ordered and may have increasing impact.

Feature Subset Selection: Procedure for reduction of data dimensionality with a goal to select the most relevant set of features for a given task trying not to sacrifice the performance.

Feature Weighting: Under the assumption that not all attributes (dimensions) are equally important feature weighting assigns different weights to them and thereby transforms the problem space. This is used in data mining tasks where the distance between instances is explicitly taken into account.

Non-Myopic Attribute Evaluation: An attribute evaluation procedure which does not assume conditional independence of attributes but takes context into account. This allows proper evaluation of attributes which take part in strong interactions.

Ordered Attribute: An attribute with nominal, but ordered values, for example, increasing levels of satisfaction: low, medium, and high.

Chapter 7
Synthesis of the Laws of Motion Control of a UAV Group With Natural Obstacles

Dmytro Kucherov
National Aviation University, Ukraine

Minglei Fu
Zhejiang University of Technology, China

Andrei Kozub
National University of Defense of Ukraine, Ukraine

ABSTRACT

A solution is proposed for the task of controlling a group of UAVs moving along a given route. The group is considered as a limited-size formation consisting of n-agents moving relative to the leader, which allows us to treat the group as some aggregate with the center of motion. The quantitative composition of a group can change while maintaining the integrity of the group. The chapter proposes the use of smooth laws governing the motion of a group. The safety of motion is ensured by introducing into the law of control the components equivalent to the creation of attractive and repulsive fields.

INTRODUCTION

Group or collective management is one of the priority areas for the development of technical systems. This is due to an increase in the number of technical objects endowed with a certain share of artificial intelligence, which in turn requires the resolution of technical issues related to the organization of their joint actions, such as coherence, mutual assistance, and security.

The first among them is consistency, which is manifested in synchronous movements with an observance of distances, intervals, and exceedances when considering these objects in 3D space. This requires traffic control, for example, when driving along a route. For a UAV of various configurations, en-route movement is a task of varying complexity, for example, when flying in free space where there

DOI: 10.4018/978-1-5225-7709-6.ch007

are no obstacles, at low and very low altitudes, where obstacles are surface structures, or, for example, in confined spaces need different maneuverability characteristics are required.

Complex maneuvers are capable of performing UAVs of special designs. A good maneuvering is distinguished quadrotors – 4-engine aircraft. They are able to hang over the subject and sharply change the direction of motion. Currently, most of the technical issues associated with ensuring their flight, namely: power supply, the execution of hulls, hangers are to some extent resolved. Nevertheless, the development and programming of controllers that implement effective control, and especially the implementation of a group or swarm control remains an actual task.

Advantages of collective management lie in the possibility of obtaining better characteristics of time, accuracy, the probability of accomplishing the final task assigned to a group of air platforms. Recently, a group of objects that provide a joint task is called a multi-agent system, a separate element of the system is an agent capable of performing a number of the simplest activities. The actions of the system's agents must be safe for other agents. In this case, the group should be given an allowable area within which the actions of an agent do not interfere with other agents of the system. From the experience of group flights of manned air vehicles, there are known minimum distances equivalent to 2 – 3 hulls of the vehicles and ensuring safe maneuvering.

Existing approaches to the solution of the problem of swarm management are borrowed from observations of the behavior of bio-systems such as an anthill, a swarm of bees, a flock of birds. To achieve such mutual understanding and coordination of movements what have this listed bio-systems, it is necessary to develop new algorithms, which this section is devoted to. The proposed algorithms are based on modern achievements of science in the field of control of UAV, and also offers a number of solutions for traditional tasks of traffic control along a given route, and also considers possible reconstructions that are of interest for the study of movements in a confined space.

It is believed that a useful task performed by a quadrotor group is monitoring the earth's surface. To accomplish this task, the robotics platform is equipped with a target and auxiliary equipment that provides the acquisition and transfer of information to the point of reception and analysis. Therefore, the main efforts are directed to the development and synthesis of control laws that ensure the action of the UAV agent under standard flight conditions, and also the properties of control laws that ensure a stable and controlled movement whole the group are established.

The section is structured as follows: the next subsection contains an overview of the works that allows us to evaluate the state of the problem. The problem statement and the description of the UAV group are proposed in Section 3. In Section 4 the analysis of the stability of various types of movements is given. The procedure for synthesizing the control law is presented in Section 5. Simulation of the movement of a group's element and the group as a whole is given in Section 6. Conclusions on the work are presented in the last section.

RELATED WORKS

There are many studies on multi-agent systems. The most general idea and concepts of multi-agent systems are given in fundamental publications (Hoek & Wooldridge, 2008 & Murray, 2007). Fundamental research in this area is devoted to the formalization of knowledge representations, for example, as is done by (Hoek & Wooldridge, 2008). Two areas are of interest here. The methods of representing the intelligent state of agents and the methods for determining the rational actions that lead to this state are referred to the first direction. The second direction is to describe the structure of the multi-agent environment and to provide the authorities that agents are assigned to act as part of the group.

The key applications of group control systems and the description of algorithms for their implementation are presented in the papers (Murray, 2007; Shmelova et al, 2016). The areas covered include the use of unmanned vehicles and aircraft, as well as mobile sensor networks. Each control problem consists of a description of the system and their final position by a system of first-order differential equations, and also by the definition of an integral function of quality (productivity) that represents a certain amount of cost, in particular, it can be time and a final state. Control law in the control problems under consideration is a smooth function of the state and the role of the machine.

Many works in this area are aimed at studying the consensus problem in the implementation of joint actions. Among them related this works (Ren et al 2005; Rezaee & Abdollah, 2016; Olfati-Saber et al, 2007; Amelin et al, 2013; Ni & Cheng, 2010; Ma & Zhang, 2010). Thus, the authors (Ren et al 2005) carried out an analysis of the coordination of actions of the multi-agent system to achieve coordinated actions. Consistency is understood as the correspondence of information states of interacting agents. For this, the agents of the system are provided with coordinating information. The conditions for achieving consensus in the conditions of a constancy of topology in the exchange of information and under conditions of dynamically changing topology are determined. The authors note that the analysis is convenient if the process of information exchange is described by integrators not higher than the second order.

In (Ren et al, 2005), a protocol was obtained for coordinating the actions of agents in a multi-agent linear system consisting of underwater autonomous machines having a randomly switched topology. The problem is solved with the help of supermartingales and solving linear matrix inequalities. The problem solution does not depend on the topology of the system and is not computationally expensive.

An overview of the key concepts of the theory of consistency and its applications for network systems within a single structure is presented in (Rezaee & Abdollah, 2016). The work includes basic concepts of consistency in information theory and methods of control of convergence, as well as analysis of the effectiveness of reconciliation protocols based on theories of matrix analysis and spectral graphs. The network system is built on the basic assumption of a local interaction of mobile or static agents, equipped with means of measuring the state, performing calculations and communication. An analysis of the behavior of the flock is shown, and new algorithms and principles for the synthesis of fault-tolerant individual nodes and channels of sensor networks are developed.

Optimization of the flight of small UAVs based on the consensus approach is considered in (Olfati-Saber et al, 2007). To control flight, thermal ascending flows are used. The center of thermal ascending fluxes is detected by the stochastic approximation algorithm. The use of thermal ascending flows saves energy during the flight.

The consensual problem of multi-level high-order agent systems is considered in (Amelin et al, 2013). In the system, the dynamics of each agent and the leader is a linear system. The management of each agent is constructed using local information, as well as a detailed analysis of the consensus on leadership for different topologies, obtained through the description of agents as a linear system, which makes it possible to synthesize the management of linear algebra.

Coherence is a fundamental problem in the field of MAS research. The necessary condition for consistency with respect to a set of admissible consensus protocols for multi-agent systems (MAS) is studied in (Ni & Cheng, 2010) on the basis of the joint influence of a dynamic agent with a linear structure and the coupling topology on consistency. For MAS with fixed topology and stationary agents, necessary and sufficient conditions are shown.

In (Ma & Zhang, 2010) discusses the consensual problem of managing multiple agents with a delay in communication time. Here the dynamics of each agent is described by a system of linear differential equations of the second order, the time delay of the connection is not constant and the interconnections of agents are represented by a directed diagram. The paper also proposes a necessary and sufficient condition for the stability of control for continuous and switching systems.

In (Hu & Hong, 2007; Zhang & Chen, 2014; Ferber et al, 2004; Xiao et al, 2009; Franchi et al, 2011; Pipattanasomporn et all, 2011; Kala & Warwick, 2012; Hao et al, 2018; Flemisch et al, 2017; Olfati-Saber et al, 2003; Koo & Shahruz, 2001), the results of the optimization of multiple-agent systems by different methods are presented. In (Hu & Hong, 2007), the solution is proposed to be a linear multi-agent system based on the construction of the feedback vector in the system. The feedback vector is constructed on the basis of the solution of a linearly-quadratic problem. The authors also show that, in order to achieve a two-sided consensus, a pair of factors are crucial, one of which is the appropriate choice of estimating a consensus error, and the other is an appropriate topology of the graph that requires the existence of a spanning tree and the structural balance of the digraph.

In (Zhang & Chen, 2014), an analysis is proposed for the effectiveness of multi-agent systems based on the minimal description of the model in the form of agent-group-role (AGR). This approach involves eliminating the drawbacks of the model proposed by the MAS community. Among them is the openness of the agent, which causes security problems, modularity, which makes the structure cumbersome in upgrades and excludes the possibility of flexible integration of substructures, as well as the need for an identical structure of agents based on the unification of the programming language, which causes interaction problems between agents.

The main idea of the authors of the article (Ferber et al, 2004) assumes the separation of information in the multi-agent system into global and local. At the same time, global information is provided to the leader and agents responsible for the navigation of the team and also determines the geometry of the desired formation. Local information determines the relative positions of followers in the framework of the education established by leaders. This type of information is only available during the interaction process. The leader is determined by the topology of the information flow, and not by the type of information. The proposed approach allows us to define a new class of nonlinear consensus protocols for coordinating the state of multi-agent systems. The protocols considered have the form of continuous feedback with a finite convergence time, which can be useful for multi-agent systems with many agents.

In the paper (Xiao et al, 2009) proposes a distributed control strategy for remote control of a fleet of unmanned aerial vehicles (UAVs). Control methods are considered that provide flexible UAV behavior for arbitrary separation and integration, ensuring system stability. The general remote control system is considered passive and has stable behavior both in free movement and in interaction with unknown passive obstacles.

Intellectual control of a multi-agent system describing a distributed power circuit is considered in (Franchi et al, 2011). Agents in the multi-agent system work together to detect upstream faults and react accordingly, providing autonomous operation of the microgrid. This capability can be perceived as an alternative to the software of a traditional zone protection system that isolates microgrid. Multi-agent system provides more flexible protection, redefining the zone boundary "on the fly."

Planning the movement of mobile ground robots through the introduction of lateral potentials as measures that allow vehicles to solve their lateral positions on the road is proposed in (Pipattanasomporn et all, 2011).

The authors of (Kala & Warwick, 2012) proposed an optimization strategy for a game with overloads when the objects of the system are representable in a matrix form. A necessary and sufficient condition is formulated to ensure the conversion of the system into a game with overload according to a given performance criterion of the system presented in the form of a potential function based on the development of the proper functions of the value of the object. Using this technology allows you to achieve a global optimization of the dynamic system based on objects in which each agent optimizes its cost function. This approach can be extended to those systems that are partially or almost convertible. The proposed methodology can be applied only to classic games with overloads, where a user with several unit requirements is not allowed.

In (Hao et al, 2018) technologies of group management of aviation and automobile vehicles in automated control systems are presented. The authors note technological and ethical reasons limiting the possibility of realizing a fully autonomous vehicle in the near future. The paper outlines the principles of control based on concepts of separate and cooperative management, as well as the concept of people's control and cooperation and an unmanned vehicle.

Algebraic representation of formations on the basis of triangulated graphs enabled the authors (Flemisch et al, 2017) to determine the cost of the task by a group of vehicles. This presentation together with navigation information allowed the authors to formally solve the problem of tracking the desired trajectory behind a group of six cars and optimizing their actions.

The laws governing the group of unmanned aerial vehicles (UAV) relative to the leader of the group that specifies the required trajectory of flight of driven vehicles is presented in (Olfati-Saber et al, 2003). The paper proposes a solution to the control problem for separate and simultaneous take-off scenarios.

A description of the interaction of the agents by the Laplacian matrix (simply Laplacian) or a pair of independent digraphs (by two Laplacian matrices) is presented in (Koo & Shahruz, 2001), where the estimates of the largest lower bound of the bounded Rayleigh coefficient with respect to the general OC vector and the Laplace EB matrix characterizing the convergence of the agents to the leader.

The practical application of multi-agent systems is given in (Li, 2018; Filaretov et al, 2017; Yu et al, 2017; Baolin et al, 2017; Ögren et al, 2004; Liu et al, 2017; Han et al, 2017; Yun & Jian, 2018; Xie et al, 2018). In (Li, 2018), the development of a joint control system for a group of mobile robots (MR) based on the onboard, universal information management system (ICS) is considered. To implement this cooperative management system, it is proposed to introduce an additional component that ensures the interaction of MR during their transfer to the ICS of each MR. The leader-follower strategy is used to consider the data bandwidth limitation between the MR group. The conducted studies confirm the high efficiency of the proposed system.

The authors of (Filaretov et al, 2017) consider a multi-agent system in the form of a platoon of vehicles following the main line. The law of management is based on the principles of feedback and takes into account the road bias. The individual stability of the vehicle is estimated by the Raus-Hurwitz method.

Replacing the traditional mechanism of deploying a dynamic object with two degrees of freedom by an object with many degrees of freedom similar to the skeleton mechanism of Chrysopelea is proposed in (Yu et al, 2017). A dynamic structure consisting of a combination of serial and parallel connections driven by servo motors is described by some multi-agent matrix. The effectiveness of managing this structure is achieved by introducing an algorithm for optimizing system parameters based on swarm optimization.

The strategy of managing groups of vehicles in moving and reconfiguring in response to a distributed medium response is shown in (Baolin et al, 2017). Each vehicle in the group serves as a mobile sensor and the transport network is a matrix of sensors capable of moving and changing the configuration. The proposed management strategy divides the management of network formation from performing network maneuvers. To achieve the management effect, a coordination structure has been developed that uses virtual bodies and artificial potentials, which allows vehicles to avoid obstacles on the route. The task of searching for a mobile network of sensors with a local maximum in the field of the environment is being constructed. The network can adapt its configuration in response to a sensitive environment in order to optimize its gradient ascent.

In (Ögren et al, 2004), the problem of distributed cooperative management under conditions of reconfiguration of a UAV swarm was investigated. The consensus protocol has a linear form of the law of feedback, sufficient conditions for achieving the required reconfiguration of the UAV formation are obtained on the basis of the Lyapunov theory and the LMI method.

The reconfigurable wireless control system (Liu et al, 2017) is designed for a robotic system with two hands interacting with each other. It appears to be a distributed multi-layer system consisting of a coordinated controller, two motion controllers corresponding to two arms, and ZigBee wireless communication modules.

Development of an intelligent towing system for water transport, where manual interaction has been preserved so far, is presented in (Han, 2017).

The control system for the position and height of a group of space vehicles is presented in (Yun & Jian, 2018). The proposed control system uses the algorithm of adaptive cooperative control, in which the estimates of the coordinates of the spacecraft in the presence of measurement interference are estimated by the method of extended least ones for several spans.

A review of co-management with fixed time in multi-agent systems, ensuring a guaranteed time of establishment, regardless of the initial conditions, is presented in (Xie et al, 2017). These algorithms are able to provide better characteristics of closed overlap and deviation from interference in comparison with asymptotic control algorithms. Unlike finite-time control, co-management with fixed time provides a higher convergence rate and provides a clear estimate of the establishment time, regardless of the initial conditions, which is desirable for multi-agent systems.

In papers (Zuo et al, 2018; Kucherov et al, 2016; 2014; 2015; Kulik & Dergachov, 2016; Kozlovskyy, 2016), problems associated with the environment of systems of a similar kind are considered. The main problem of this chapter is the methodology for synthesizing the management of a multi-agent system when it moves along a certain route, where obstacles are possible, the overcoming of which is not associated with a height maneuver.

PROBLEM STATEMENT

The actions of the UAV group performing the task of airspace monitoring are considered. We denote a separate UAV that is part of the group by the object A_i. We introduce the Euclidean space *XYZ* to describe the current position of Ai in space. Let consider an arbitrary motion of S_i objects in the chosen space. In the interests of the task, the objects start, assemble into a group, perform a group flight, perform some flight mission, and then complete the flight at a given point. The flight mission in question is shown in Figure 1, which is performed by a group of 4 UAVs.

Figure 1 shows the route of the group, on this route, there are obstacles of natural origin or single buildings. When determining the mission, the presence of obstacles was not taken into account. It is

Figure 1. Group of UAVs, N =4

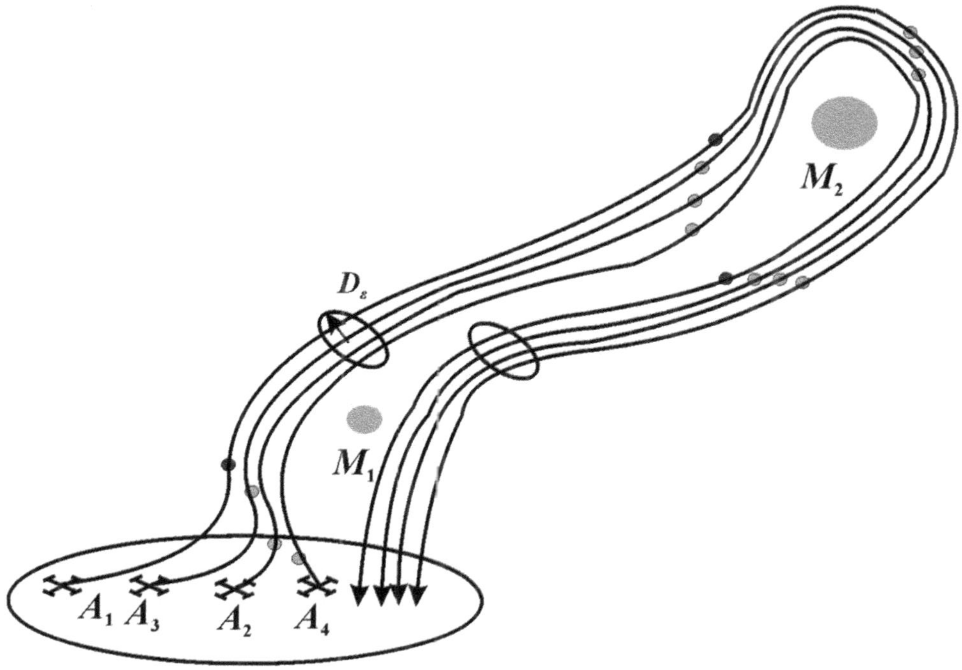

not advisable for the group to maneuver in height, and therefore during the flight, a decision is taken on the variant of the obstacle's bending. The group moves at a constant speed, therefore, the velocities of individual objects of the group are sufficiently close, so the group in the interest of research can be described by some formation called aggregation with a center corresponding to the position of the leader of the group. It is assumed that the agent has the necessary control, navigation, and communication equipment to support the flight. It is also planned to change the number of groups based on the reconciliation protocols.

The obstacle course is on curve trajectory. Thus, the proposed trajectory of motion includes rectilinear and curvilinear parts of the motion. The motion of an agent in a 3-dimensional Euclidean space is considered. The position of the agent is described by some positions pi with coordinates $p_i = (x_i, y_i, z_i)$, and the controlling signal u_i. Considering the management of agents of a multi-agent system, one can use the dynamic model of a second-order integrator, namely

$$\ddot{p}_i(t) = k_i u_i(t). \tag{1}$$

In equation (1), k_i is the coefficient describing the reaction of the ith agent, $t \geq 0$ is the time. We also introduce the parameter $D_\varepsilon > 0$, which determines the size of the multi-agent system, which represents its aggregate for $\forall t \to \infty$, i.e.

$$p_i \to D_\varepsilon(p_c), \tag{2}$$

This suggests that the agent leader is an agent located in the center of the group

$$p_c(t) = \frac{1}{n}\sum_{i=1}^{n} p_i(t), \tag{3}$$

and

$$D_\varepsilon(p_c) = \left\{ p \in \Re^3 : d = || p - p_c ||< \varepsilon \right\}. \tag{4}$$

In (4), ε acts as a constructive parameter that specifies the safe distance between agents. Thus, formula (4) serves as a criterion for the integrity of the group. It should be noted that the configuration of the group can vary, but the inequality

$$D(p_c) \leq D_\varepsilon(p_c) \tag{5}$$

for the group is always executed.

Let us set the task of synthesizing a control law that ensures the movement of a group on a route with stationary obstacles provided that the group integrity condition (5) is ensured.

THE METHOD OF THE POTENTIAL FIELD

As already noted, we assume that the agent management system has measurement tools and computing facilities that allow the measurement results to be converted into signals equivalent to attractive and repulsive forces. The force of attraction ensures the integrity of the group, while the forces of repulsion allow avoiding collisions with each other and with obstacles. The emergence of such forces allows us to affirm the existence of potential energies of attraction and repulsion.

By the definition of (Korn & Korn, 1968), the potential energy of the attractive field is written in accordance with the expression

$$E_{att}(\Delta p) = \frac{1}{2}\mu(\Delta p)^2, \tag{5}$$

where $\mu > 0$ is the scale factor, and $\Delta p_i = || p_i - p_c ||$ is the Euclidean distance between the agent and the leader of the group. The form of the homogeneous potential field in the field of the agent-leader is shown in Figure 2.

The effect of this field is most pronounced at the edges and weakened in the center where the leader is located. The force generated by this field is determined by the expression (Jain, 2009)

$$u_{att}(\Delta p) = -grad\ E_{att}(\Delta p) = -\ \mu(\Delta p). \tag{6}$$

Figure 2. Attractive field acting on agent A_c

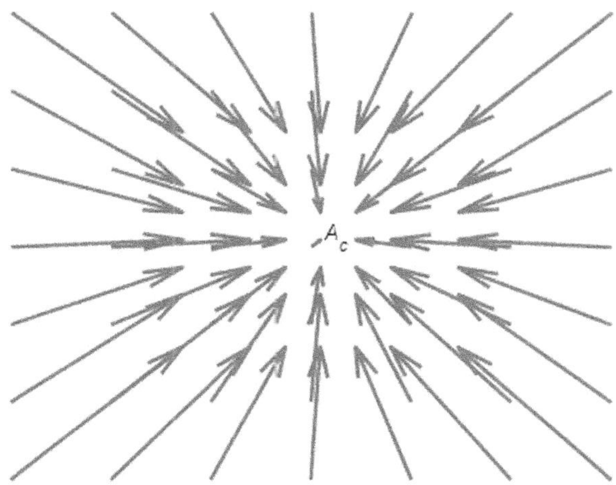

The energy of the repulsive field has the form (Korn & Korn, 1968)

$$E_{rep}(\Delta p_i) = \begin{cases} \frac{1}{2}\nu\left(\frac{1}{\Delta p_i} - \frac{1}{d_0}\right)^2, & if\ \Delta p_i \le d_0. \\ 0, & if\ \Delta p_i > d_0, \end{cases} \quad (7)$$

In this expression, $\nu > 0$ is also a scale factor, and $d_0 > 0$ is a certain constant that determines the minimum distance to the nearest neighboring agent. The form of this field is shown in Figure 3, the action of this field is opposite to attracting, so the strongest repulsion occurs at the center and weak at the edges of the aggregation region.

Figure 3. Repulsive field acting on agent A_c

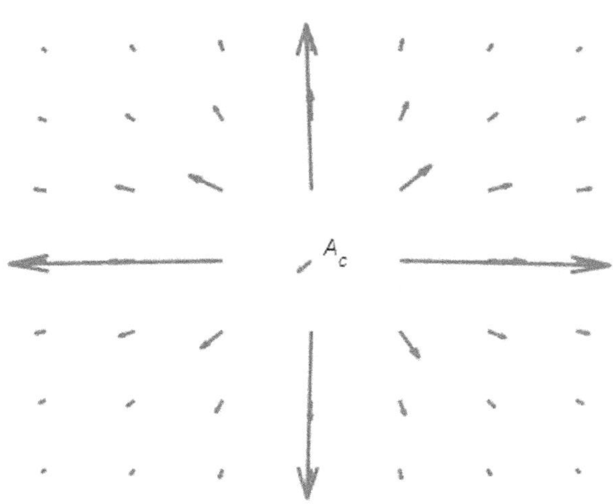

Analogously to (7), we represent the repulsive force by the expression

$$u_{rep}(\Delta p_i) = -grad\ E_{rep}(\Delta p_i) = \begin{cases} \nu\left(\dfrac{1}{\Delta p_i} - \dfrac{1}{d_0}\right)\dfrac{1}{(\Delta p_i)^2}\dfrac{p_i - p_0}{d_0}, & if\ \Delta p_i \le d_0, \\ 0, & if\ \Delta p_i > d_0. \end{cases} \tag{8}$$

where p_i is the closest agent to p_0 or the obstacles p_M, and

$$d_0 = \max_{1 \le i \le N} || p_i - p_0 ||.$$

Proposition 1. The law governing the *i*th agent is representing by the following expression

$$u_i^g = u^i + u_i^{att} + u_i^{rep}, \tag{9}$$

in which u_i has the meaning of the mastering action in aggregation, u_i^{att}, u_i^{rep} is the components of the attractive and repulsive forces, respectively.

Proof. This follows from the accepted idea of the motion of the system in the potential field of attraction and repulsion forces.

The input control u_i^g of the agent-leader differs from the control law of the *i*th agent since the control of the latter is building with respect to the leader. Setting the control law of the leader, we will construct an aggregation control trajectory in accordance with the system of equations

$$\begin{aligned} \dot{p}_1^l &= p_2^l, \\ \dot{p}_2^l &= k^l u^l, \end{aligned} \tag{10}$$

In accordance with (10), the trajectory of the leader starts at the moment $t = 0$ from the point $p(0)$ to the point $p(t_k)$. Next, consider the features of aggregation motion on the rectilinear and curvilinear sections of the trajectory.

Control on a Straight Part of a Trajectory

Considering the assumption that the group moves with approximately equal velocities, i.e. without acceleration, then the system of equations (10) is transformed into the form

$$\dot{p}_1^l = p_2^l = const. \tag{11}$$

The main task of the agent when moving in a group is to maintain his position, avoiding collisions with neighboring agents or obstacles. For this, the following control law is proposed, representing a smooth function with respect to positions with two neighboring agents

$$u^i = \begin{cases} -k_{p_1}(p_{1i} - p_{1j}) - k_{p_2}(p_{2i} - p_{2j}), & if\ i,\ j\ is\ member, \\ 0, & if\ i\ is\ not\ member. \end{cases} \tag{12}$$

In (12), $k > 0$, positive coefficients that set the value of non-zero deviations $\xi_{p1} \neq 0$, $\xi_{p2} \neq 0$ between the positions of neighboring agents $|p_{1i} - p_{1j}| \leq \xi_{p1}$ and $| p_{2i} - p_{2j} | \leq \xi_{p2}$. Assuming the central position of the agent-leader, his position is determined by equations

$$p^l = \frac{1}{N^*}\sum_{i=1}^{N^*} p_i,\ \dot{p}^l = \frac{1}{N^*}\sum_{i=1}^{N^*} \dot{p}_i, \tag{13}$$

where $N^* = N - 1$ number of neighboring agents in the group. This type of motion is shown in Figure 4.

Proposition 1. The movement of each agent of a group that consists of N agents is described by the system of differential equations of second order in the form

$$\begin{aligned} \dot{p}_1 &= p_2, \\ \dot{p}_2 &= \bar{u} = -k_{p_1}(p_{1i} - p_{1j}) - k_{p_2}(p_{2i} - p_{2j}). \end{aligned} \tag{14}$$

where $p = p_i - p_j$ is the appropriate coordinates of i-th and j-th agents, $i \neq j$, $i \leq n$, and $j \leq n$. If $j = l$, is a leader of the group member.

Figure 4. Motion UAVs swarm along line

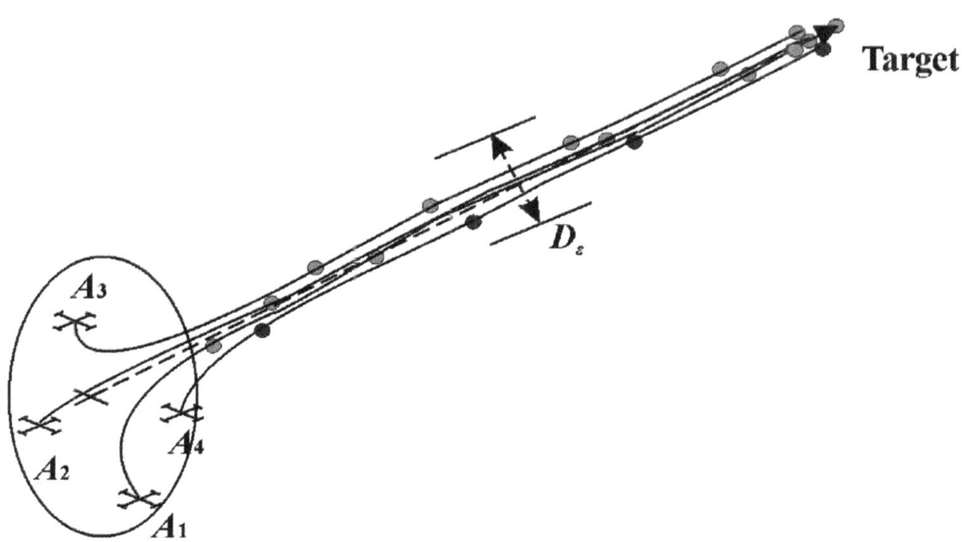

Proof. This statement follows from (9), (11).

Corollary 1. Closed loop dynamic system (14) converges to the equilibrium position

$$p_{1i} - p_{1j} \neq 0, \;\; p_{2i} - p_{2j} \neq 0. \quad (15)$$

Corollary 2. Agent-members' velocity asymptotically converges to the agent-leader velocity.

Corollary 3. In system (14) coordinates p_i is $p_i \neq p_j$ for any $i \neq j$, $t \geq 0$.

Corollary 4. If $d = \| p_i - p_j \| < \varepsilon$ is the maximal size between UAVs in a group, the size of formation is $D(p) < D_\varepsilon(p)$.

Movement on a Curvilinear Part of the Trajectory

In this case, the system of equations (11) is transformed into (10), the aggregation moves with acceleration

$$\begin{aligned} \dot{p}_1 &= p_2, \\ \dot{p}_2 &= \bar{u}. \end{aligned} \quad (16)$$

In (16), the control signal takes values from the condition $| \bar{u} | < U_{\max}$, the group size becomes larger than on the rectilinear part of the trajectory, but the integrity condition of the group (5) is preserved, as shown in Figure 5.

Proposition 2. Let given factors k_y and k_p such as $k_y > 0$, $k_p > 0$ is true inequality $0,5k_y / k_p < 0$, the system (15) has exponential stability.

Figure 5. Motion UAVs along a curve

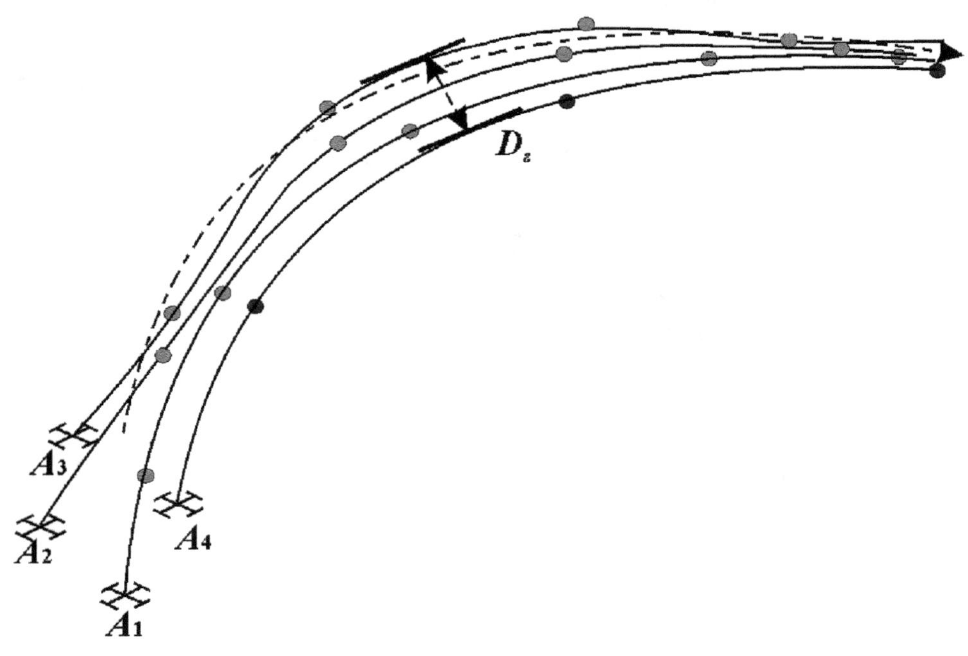

Proof. Let represent the equations system (16) by some differential equation of the second order

$$k_p \ddot{e}_p(t) + k_y \dot{e}_y(t) + k_{\dot{y}} e_{\dot{y}}(t) = 0, \tag{17}$$

in which $e_p = p_i - p_j$, $e_y = y_i - y_j$, $e_{\dot{y}} = \dot{y}_i - \dot{y}_l$. If the factors $k_p \neq 0$, and the (k_p, k_y, $k_{\dot{y}}$) are positive, we can write this equation in the form

$$s^2(t) + k_y / k_p s(t) + k_{\dot{y}} / k_p = 0. \tag{18}$$

As known, the exponential stability of (15) is satisfied, if the roots of equation (18) have a negative real part. It is true if the condition $k_y / (2k_p) < 0$ is satisfied.

Corollary 1. If factor $k_y = 0$, we have oscillation motion around the leader with constant amplitude and its frequency is equal to $\sqrt{k_{\dot{y}} / k_p}$.

Corollary 2. If condition $k_y / \sqrt{4k_{\dot{y}}k_p} > 0$ is true, we have asymptotically stability motion.

Corollary 3. If $0 < k_y / \sqrt{4k_{\dot{y}}k_p} < 1$, the agent performs oscillation motion with an error that is decreasing.

Corollary 4. If $k_y / \sqrt{4k_{\dot{y}}k_p} > 1$, the agent performs a periodic motion.

Features of Motion in Conditions of Obstacles

Movement in obstacle conditions is the most complex movement since it requires the inclusion of additional components in the law of control, based on the forces of attraction and repulsion. The motion occurs along the curvilinear part of the trajectory, therefore, to describe the dynamics of the system, a system of equations (16) with a control law of the form

$$u_i = \begin{cases} u_{rep\ pi} + u_{rep\ p_{1i}} + u_{rep\ p_{2i}}, \\ \quad if\ (d - 2r) \le d_0,\ p_{1i} > 0,\ p_{2i} > 0 \\ u_{rep\ p_i} + u_{rep\ p_{1i}}, \\ \quad if\ (d - 2r) \le d_0,\ p_{1i} > 0,\ p_{2i} \le 0, \\ 0,\ if\ (d - 2r) > d_0,\ or\ p_{2i} \le 0, \end{cases} \tag{19}$$

in which

$$u_{rep\ p_i} = k_1 \left(\frac{1}{d - 2r} - \frac{1}{d_0} \right), \tag{20}$$

$$u_{rep\ p_{1i}} = k_2 p_{1iobs}, \tag{21}$$

$$u_{rep\ p_{2i}} = k_3 p_{2iobs}, \tag{22}$$

parameter r sets the maximum size of the agent in the control plane,

$$p_{1iobs} = p_{1i}^T n_{aobs}, \tag{23}$$

and

$$p_{2iobs} = p_{2iobs}^{\ T} n_{aobs}, \tag{24}$$

where n_{aobs} is a unit vector pointing from i-th agent to obstacle.

The parameter p_{1iobs} sets the direction of the agent's movement to the obstacle, so if $p_{1iobs} > 0$, then the agent moves towards the obstacle, otherwise, i.e. if $p_{1iobs} \leq 0$ the agent moves from the obstacle. One of the options for overcoming obstacles by a group of 4 agents is shown in Figure 6.

Since the obstacles are arranged so that the group is able to overcome this type only if the configuration is changed, i.e. when reconfiguration. In this case, the reconfiguration occurs with the fulfillment of the condition (13) for the agent leader.

MODEL UAV

Main Concepts

Under the conditions of the problem under consideration, each agent of the group is an unmanned aerial vehicle (Bouabdallah & Siegwart, 2005), the traction force of which is created by 4 propellers controlled by motors via autonomous control channels. Propellers are installed in the same plane at equal distances

Figure 6. Motion N=4 UAVs in obstacles

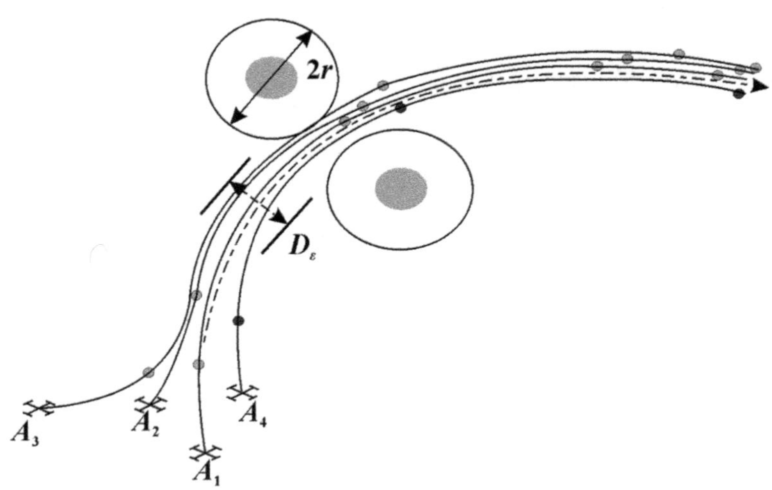

from the center of mass and on mutually perpendicular axes. Then scheme for creating control actions is similar to a helicopter (Olfati-Saber, 2001), which allows vertical takeoff and landing, as well as easy to change the direction of the vehicle, including hanging. In this case, the direction of rotation of a pair of propellers located along one axis is set opposite to the direction of rotation of the second pair located on the perpendicular axis, Figure 7.

Then, to represent the UAV, we use the equations of motion of the helicopter (Olfati-Saber, 2001; Bouabdallah & Siegwart, 2005)

$$\begin{cases} \dot{x} = v, \\ m\dot{v} = RF_b, \\ \dot{R} = R\hat{\omega}, \\ J\dot{\omega} = -\omega \times J\omega + \tau_b. \end{cases} \tag{25}$$

In equation system (25) x is the position vector, and v is the velocity vector of the UAV center of mass relative to the origin, m is the mass of the apparatus, J is the inertia matrix, $J \in \Re^{3\times 3}$, it is given by the formula

$$J = \begin{vmatrix} I_{xx} & I_{xy} & I_{xz} \\ I_{yx} & I_{yy} & I_{yz} \\ I_{zx} & I_{zy} & I_{zz} \end{vmatrix} = -\int_V \rho(r)\hat{r}^2 dV, \tag{26}$$

and R is the rotation matrix of the UAV in the frame O, it has the form

$$R = \exp(\psi e_3)\exp(\theta e_2)\exp(\varphi e_1) \tag{27}$$

where $e_i \in \Re^3$ is the unit vector basis for the XYZ coordinate system, and φ, θ, ψ are the angles associated with rotations about the x, y, z axes, in the Eulerian coordinate system these angles are called roll, pitch,

Figure 7. The configuration of 4-motor UAV with fixed body frame and inertial frame O

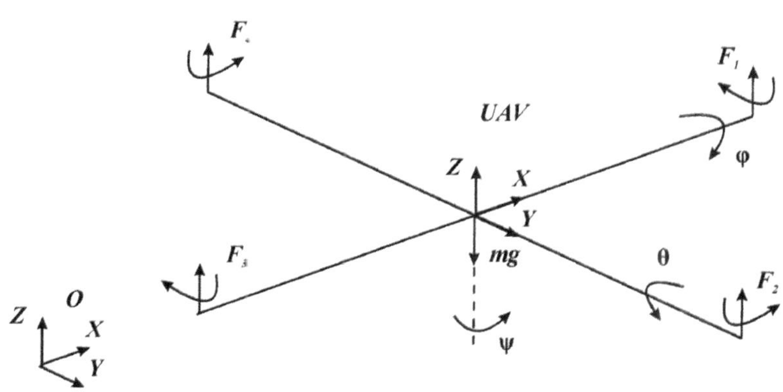

and yaw, respectively. ω denotes the angular velocity vector, $\omega = (\omega_1, \omega_2, \omega_3)^T$, the symbol (×) denotes the vector product. F_b, τ are the forces and moments that are written in the form

$$\begin{aligned} F_b &= -mgR^T e_3 + Te_T, \\ \tau_b &= Q(T)e_T + D\tau, \end{aligned} \tag{28}$$

T is the value characterizing the main thrust of the apparatus, and e_T is the unit vector along the path of the aircraft, which has a small inclination and angles of rotation relative to the central axis,

$$e_T(a,b) = \begin{vmatrix} \sin(a) \\ -\cos(a)\sin(b) \\ \cos(a)\cos(b) \end{vmatrix}. \tag{29}$$

In equation (28), $\tau = (b_T, a_T)$, and $D = \mathrm{diag}(l_b, l_a)$ is the matrix whose elements of the main diagonal are positive numbers. Thrust may vary according to law

$$Q(T) = Q_0 + Q_1 T^{1.5}, \tag{30}$$

with values $Q_0 > 0$, $Q_1 > 0$.

We will assume small values a, b, and $D_\tau = u \in \Re^3$ and use the approximation $Q(T)e_T \approx Q(T)e_3$ then (28) is written in the form

$$\begin{aligned} F_b &= -mgR^T e_3 + Te_3, \\ \tau_b &= Q(T)e_3 + \tau. \end{aligned} \tag{31}$$

Further simplification is connected with the assumption small values (a, b), and $e_T = e_3$, equation (29) can write in the form

$$e_T \approx e_3 + \begin{vmatrix} a \\ -b \\ 0 \end{vmatrix} = \begin{vmatrix} a \\ -b \\ 1 \end{vmatrix}. \tag{32}$$

By setting the values of a, b, essentially small, and giving the values of T, τ one can obtain the system of equations (25) look as follows

$$\begin{cases} \dot{x} = v, \\ m\dot{v} = -mge_3 + (Re_3)T, \\ \dot{R} = R\hat{\omega}, \\ J\dot{\omega} = -\omega \times J\omega + Q(T)e_3 + \tau. \end{cases} \tag{33}$$

Now present Re_3 on the product of the unit normal vector and matrix R, then matrix R we can represent in the form

$$R = \begin{vmatrix} C_\theta C_\psi & S_\phi S_\theta C_\psi - C_\phi S_\psi & C_\phi S_\theta C_\psi + S_\phi S_\psi \\ C_\theta S_\psi & S_\phi S_\theta S_\psi + C_\phi C_\psi & C_\phi S_\theta S_\psi - S_\phi C_\psi \\ -S_\theta & S_\phi C_\theta & C_\phi C_\theta \end{vmatrix}, \tag{34}$$

where $C(\cdot)= \cos(\cdot)$, $S(\cdot) = \sin(\cdot)$. The dynamics of the agent in the system (33) is given by the variable T, which should be associated with the master effect. We replace it by the letter U, which is usually used in the theory of control systems, then we represent the second equation in (33) in the form

$$m\dot{v} = -mge_3 + (Re_3)U, \tag{35}$$

where

$$U = F_1 + F_2 + F_3 + F_4 \tag{36}$$

(see Figure 7), and

$$F_i = \alpha_i \omega_i^2. \tag{37}$$

Here $\alpha_i > 0$ is the some constant, and ω_i is angular velocity i-th motor ($i = 1, \ldots, 4$).

We are going to consider only the longitudinal motion on the axis Y, then the vector τ will have the form

$$\tau = (\tau_\phi,\ \tau_\theta,\ \tau_\psi)^T, \tag{38}$$

with the components

$$\tau_\phi = l(F_4 - F_2), \tag{39}$$

$$\tau_\theta = l(F_3 - F_1), \tag{40}$$

$$\tau_{\psi} = l(F_4 + F_2 - F_3 - F_1). \quad (41)$$

In equations (39) – (41), the variable l denotes the distance from the center of mass of the apparatus to the motor.

We will assume that the input signal in (25) is used to control the quadrotor when flying to a given point. In this case, the yaw pull component $\tau\psi$ is close to zero, and the pitches thrust τ_{θ} and roll τ_{φ} provide stabilization of the angular position. This makes it possible to simplify the right-hand side in (25) for rotational motion and to replace the input signals for individual moments.

Then (25) look like follow

$$\ddot{\psi} = \bar{\tau}_{\psi},. \quad (42)$$

$$\ddot{\theta} = \bar{\tau}_{\theta}, \quad (43)$$

$$\ddot{\phi} = \bar{\tau}_{\phi}, \quad (44)$$

$$m\ddot{x} = (C_{\phi}S_{\theta}C\psi + S_{\phi}S\psi)U, \quad (45)$$

$$m\ddot{y} = (C_{\phi}S_{\theta}S_{\psi} - S_{\phi}S\psi)U, \quad (46)$$

$$m\ddot{z} = -mg + C_{\phi}C_{\theta}U. \quad (47)$$

The equations (43) – (47) coincide with the equations (6) from (Ren et al, 2005) if in right parts (6) from (Kucherov, 2016) enter variables $\bar{\tau}_{\psi}$, $\bar{\tau}_{\theta}$, $\bar{\tau}_{\phi}$, and we can obtain the equations (18) – (23) from (Bouabdallah & Siegwart, 2005) if we take ψ=0, and

$$U = b(F_1 + F_2 + F_3 + F_4). \quad (48)$$

Control Law

In the interests of modeling, equations (42) – (43) can be represented in the form of differential equations with respect to angular velocities $\dot{\varphi}, \dot{\theta}, \dot{\psi}$, taking into account the moments of inertia I_1, I_2, I_3 (26) by axes X, Y, and Z of the apparatus

$$I_1 \ddot{\varphi}(t) = (I_2 - I_3)\dot{\theta}(t)\dot{\psi}(t), \quad (48)$$

$$I_2 \ddot{\theta}(t) = (I_3 - I_1)\dot{\psi}(t)\dot{\varphi}(t), \quad (49)$$

$$I_3 \ddot{\psi}(t) = (I_1 - I_2)\dot{\varphi}(t)\dot{\theta}(t). \quad (50)$$

Then a model of quadrotor is presented by following equations system

$$\begin{cases} \ddot{\varphi} = \dot{\theta}\dot{\psi}\left(\dfrac{I_y - I_z}{I_x}\right) - \dfrac{J_r}{I_x}\dot{\theta}w + \dfrac{l}{I_x}U_2, \\ \ddot{\theta} = \dot{\varphi}\dot{\psi}\left(\dfrac{I_z - I_x}{I_y}\right) - \dfrac{J_r}{I_y}\dot{\varphi}w + \dfrac{l}{I_y}U_3, \\ \ddot{\psi} = \dot{\varphi}\dot{\theta}\left(\dfrac{I_x - I_y}{I_z}\right) + \dfrac{1}{I_z}U_4, \\ \ddot{z} = -g + (\cos\varphi\cos\theta)kU_1, \\ \ddot{x} = (\cos\varphi\cos\theta\cos\psi + \sin\varphi\sin\psi)kU_1, \\ \ddot{y} = (\cos\varphi\cos\theta\cos\psi + \sin\varphi\cos\psi)kU_1 \end{cases} \quad (51)$$

Here (ϕ, θ, ψ) is the roll, pitch and yaw angles, $I_{x, y, z}$ is the inertia UAV, J_r is the inertia rotor, l is the lever in coordinate system with the Quadrotor, $k = 1/m$, m is the mass UAV, w is the rotor speed, and U_1, U_2, U_3, U_4, are calculated as

$$\begin{cases} w = w_2 + w_4 - w_1 - w_3, \\ U_1 = b(w_1^2 + w_2^2 + w_3^2 + w_4^2), \\ U_2 = b(w_4^2 - w_2^2), \\ U_3 = b(w_3^2 - w_1^2), \\ U_4 = d(w_2^2 + w_4^2 - w_1^2 - w_3^2), \end{cases} \quad (52)$$

where b is the thrust factor, and d is the drag factor.

For simulation, the model (51) is transforming to a space-state form $\dot{X} = f(X, U)$, where $X \in \Re^{12}$ is the state vector with the coordinates

$$\dot{X} = \begin{cases} x_2, \\ x_4 x_6 a_1 + x_4 a_2 w + b_1 U_2, \\ x_4, \\ x_2 x_6 a_3 + x_2 a_4 w + b_2 U_3, \\ x_6, \\ x_4 x_2 a_5 + b_3 U_4, \\ x_8, \\ -g + (\cos x_1 \cos x_3) k U_1, \\ x_{10}, \\ u_x k U_1, \\ x_{12}, \\ u_y k U_1, \end{cases} \tag{53}$$

where

$$\begin{aligned} & x_1 = \varphi,\ x_2 = \dot{x}_1 = \dot{\varphi},\ x_3 = \theta,\ x_4 = \dot{x}_3 = \dot{\theta}, \\ & x_5 = \psi,\ x_6 = \dot{x}_5 = \dot{\psi},\ x_7 = z,\ x_8 = \dot{x}_8 = \dot{z}, \\ & x_9 = x,\ x_{10} = \dot{x}_9 = \dot{x},\ x_{11} = y,\ x_{12} = \dot{x}_{11} = \dot{y}. \end{aligned} \tag{54}$$

$$\begin{aligned} & a_1 = \frac{(I_y - I_z)}{I_x},\ a_2 = -\frac{J_r}{I_x},\ a_3 = \frac{(I_z - I_x)}{I_y},\ a_4 = \frac{J_r}{I_y}, \\ & a_5 = \frac{(I_x - I_y)}{I_z},\ b_1 = \frac{l}{I_x},\ b_2 = \frac{l}{I_y},\ b_3 = \frac{l}{I_z}, \end{aligned} \tag{55}$$

$$\begin{aligned} u_x &= \cos x_1 \sin x_3 \cos x_5 + \sin x_1 \sin x_5, \\ u_y &= \cos x_1 \sin x_3 \cos x_5 - \sin x_1 \sin x_5. \end{aligned} \tag{56}$$

The translation model dynamic (51) can be represented in form three-dimensional double-integrators

$$\begin{aligned} & \dot{\eta} = p, \\ & p = U, \end{aligned} \tag{57}$$

then

$$\dot{x} = V_x,\ \dot{V}_x = (C_\phi S_\theta C_\psi + S_\phi S_\psi)U / m, \tag{58}$$

$$\dot{y} = V_y,\ \dot{V}_y = (C_\phi S_\theta C_\psi - S_\phi S_\psi)U / m,, \tag{59}$$

$$\dot{z} = V_z,\ \dot{V}_z = -g + C_\phi C_\theta U / m. \tag{60}$$

The control law of the *i*th agent is synthesized on the basis of the coordinate information regarding the position and speed of the leader, it does not coincide with the management of the leader and other agents not belonging to the group

$$U_i = \begin{cases} -k_y(y_i - y^l) - k_x(x_i - x^l), & if\ i\ is\ agent - member, \\ 0, & else\ . \end{cases} \tag{61}$$

Factor $k > 0$ set safe distance between agents. In contrast to [39], which used to create a sliding mode for control, the control actions (U_1, U_2, U_3, U_4, u_x, u_y) of each agent are smooth functions of time *t*.

SIMULATION

In this section, we consider the motion of a group of quadrotors moving in the potential field of the forces of attraction and repulsion towards the target along a route on which there are obstacles.

We will assume that the group consists of four agents, one of them is the leader. The route of agents is given by two points: the initial and final points. The stratum position of the quadrotors $p_{A1} = (101, 31, 0)^T$, $p_{A2} = (76, 22, 0)^T$, $p_{A3} = (131, 51, 0)^T$, $p_{A41} = (120, 72, 0)^T$, endpoint of the trajectory $p_t = (610, 275, 0)^T$. Coordinates of the stationary obstacles located on the route $p_{O1} = (265, 133)^T$, $p_{O2} = (221, 211)^T$ and $D_\varepsilon(p_A) \leq 45$ are also given. It is assumed that the height obstacles are higher than the altitude of the quadrotors, which leads to maneuvering around the obstacles. The height coordinate is taken equal to $z = 15$. The simulation of the movement of a group of agents is shown in Figure 8.

It can be seen from Figure 8 that a group of agents when passing a route, maneuvers around the obstacle and continues to move to the target along the planned route. If in order to overcome obstacles, it is required to change the size of its aggregation, the group increases distances, and the agent aggregation form remains unchanged if the route has no obstacles.

Figure 9 shows an example of the reconfiguration of a UAV group in the presence of obstacles, the placement of which does not allow the use of an obstacle avoidance maneuver. At the same time, the group is lined up if obstacles along the route are on the way. In this case, the UAV tactics are similar to the column and the UAVs follow in a strictly defined order: A4 → A3 → A2 → A1, where A1 is the first and A4 is the last one.

Figure 8. Bypass movement of agent-leader and agent-members to the target

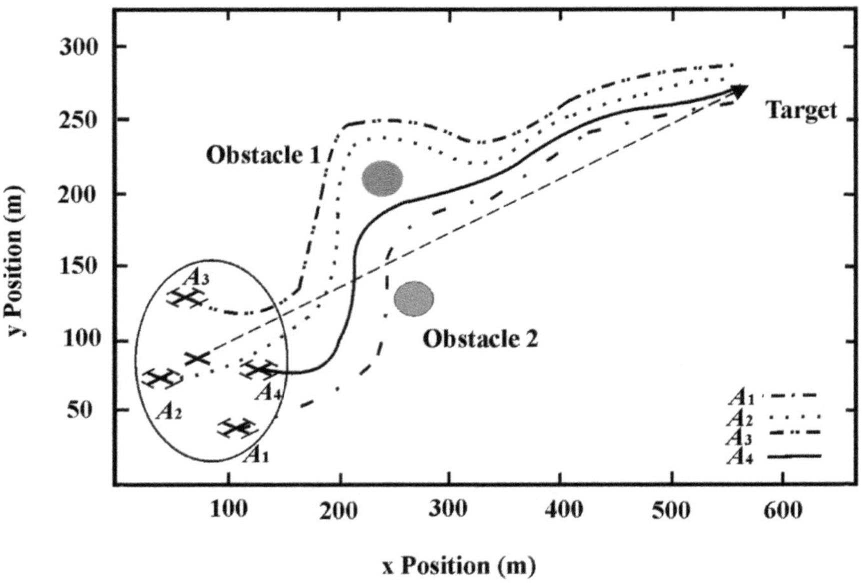

Figure 9. Group of UAVs overcomes side-by-side obstacles by line type motion

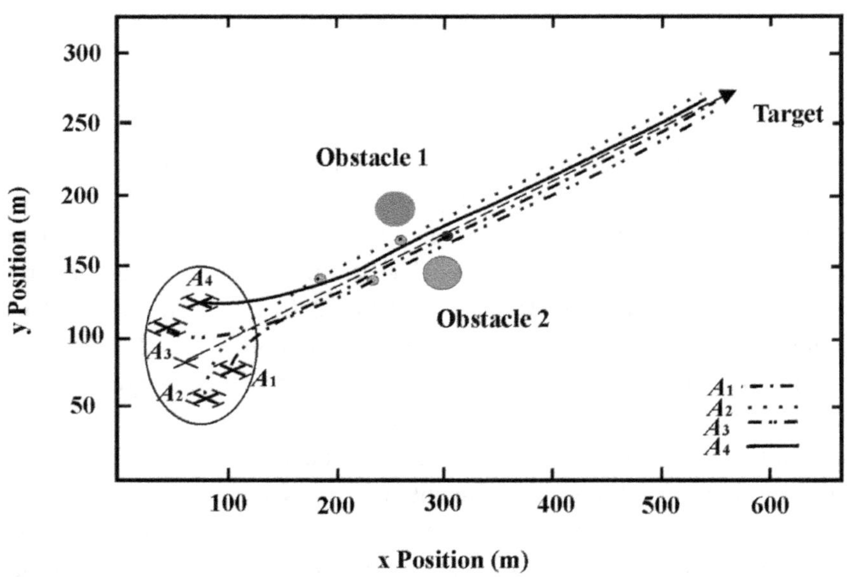

CONCLUSION

The article deals with the problem of synthesis of a smooth motion control law by a group consisting of a limited number of agents following a given route on which obstacles to movement are possible. Obstacles can be overcome in two possible ways: namely to bypass the obstacle without changing the traffic configuration, and the other is based on the reconfiguration of the group, preserving the integrity of the group. In order to avoid the risk of collision of group agents with obstacles and with each other, additional components associated with repulsion forces are introduced into the control law, which does not allow agents to approach closer than a certain predetermined distance and to be removed in order to keep the specified configuration. The need to overcome obstacles leads to a change in the initial configuration of the group.

Based on the idea of the motion of a group of agents in the potential field of attraction and repulsion, the control law of the movement of a group is synthesized, and an algorithm was developed to simulate the agent system under investigation. The properties of the algorithm were studied, which makes it possible to determine certain parameters of the stable motion of a group under given conditions. The presented examples allow talking the efficiency of the algorithm in the considered flying situations. Modeling made it possible to provide the aperiodic character of the trajectory of agents in possible sections of rectilinear and curvilinear motion. The success of the simulation also depended on the choice of parameters, like μ and ν potential functions.

Since the motion along the curve of the trajectory is more general, during its investigation it was found necessary to increase the amplitude of the control signal in order to exclude the increase in the time of the motion. However, the operation with the maximum amplitude control signal can cause "sliding mode" when the input signal is frequently switched. Investigation of the agent system in a sliding mode is one of the future topics of our research.

REFERENCES

Amelin, K., Amelina, N., Granichin, O., Granichina, O., & Andrievsky, B. (2013). Randomized Algorithm for UAVs Group Flight Optimization. In *11th IFAC International Workshop on Adaptation and Learning in Control and Signal Processing*. University of Caen Basse-Normandie. 10.3182/20130703-3-FR-4038.00130

Baolin, F., Lu, L., Lu, X., & Zhu, K. (2017). An Optimized Cooperative Control Algorithm of Multi-Agent Array Based on the Skeletal Mechanism of Chrysopelea. *Proceedings of the 2017 IEEE International Conference on Robotics and Biomimetics*, 2003 – 2008.

Bouabdallah, S., & Siegwart, R. (2005). Backstepping and Sliding-mode Techniques Applied to an Indoor Micro Quadrotor. *IEEE International Conference on Robotics and Automation*, 2259 – 2264. 10.1109/ROBOT.2005.1570447

Ferber, J., Gutknecht, O., & Michel, F. (2004). From Agents to Organizations: An Organizational View of Multi-Agent Systems. *LNCS*, *2935*, 214–230.

Filaretov, V., Yukhimets, D., & Bidyadhar, S. (2017). Information-Control System for Cooperative Control of a Mobile Robots Group. *International Conference on Industrial Engineering, Applications and Manufacturing (ICIEAM)*, 1 – 6. 10.1109/ICIEAM.2017.8076169

Flemisch, F., Canpolat, Y., Altendorf, E., Weßel, G., Flemisch, F., … Schutte, P. (2017). Shared and cooperative control of ground and air vehicles: Introduction and General overview. *IEEE International Conference on Systems, Man, and Cybernetics (SMC)*, 858 – 863. 10.1109/SMC.2017.8122717

Franchi, A., Giordano, P. R., Secchi, C., Son, H. I., & Bulthoff, H. H. (2011). A Passivity-Based Decentralized Approach for the Bilateral Teleoperation of a Group of UAVs with Switching Topology. *IEEE International Conference on Robotics and Automation*, 898 – 905. 10.1109/ICRA.2011.5980368

Han, L., Cheng, X., Xu, W., & Tan, G. (2017). Reconfigurable Wireless Control System for a Dual-arm Cooperative Robotic System. *Proceedings of the 2017 IEEE International Conference on Robotics and Biomimetics*, 202 – 207. Retrieved October 09, 2018, from https://ieeexplore.ieee.org/document/8324418/

Hao, Y., Pan, S., Qiao, Y., & Cheng, D. (2018). Cooperative Control via Congestion Game Approach. *IEEE Transactions on Automatic Control*, *9*. Retrieved from https://ieeexplore.ieee.org/document/8334284/

Hoek, W., & Wooldridge, M. (2008). Multi-Agent Systems. In *Handbook of Knowledge Representation* (pp. 1–41). Amsterdam: Elsevier. doi:10.1016/S1574-6526(07)03024-6

Hu, J., & Hong, Y. (2007). Leader-following Coordination of Multi-Agent Systems with Coupling Time Delays. *Physica A*, *374*(2), 853–863. doi:10.1016/j.physa.2006.08.015

Jain, M. C. (2009). *Textbook of Engineering Physics (Part I)*. Delhi, India: PHI Learning Private Ltd.

Kala, R., & Warwick, K. (2012). Planning Autonomous Vehicles in the Absence of speed Lanes using Lateral Potentials. *Proceedings of the 2012 IEEE Intelligent Vehicles Symposium*, 597-602. 10.1109/IVS.2012.6232148

Koo, T. J., & Shahruz, S. M. (2001). Formation of a Group of Unmanned Aerial Vehicles (UAVs). *Proceedings of the American Control Conference Arlington*, 69-74. Retrieved October 09, 2018, from https://ieeexplore.ieee.org/document/945515/

Korn, G. A., & Korn, T. M. (1968). *Mathematical Handbook for scientist and engineers*. McGraw-Hill Company.

Kozlovskyy, V., Parkhomey, I., Odarchenko, R., Gnatyuk, S., & Zhmurko, T. (2016). Method for UAV trajectory parameters estimation using additional radar data. *Proceedings IEEE 4th International Conference Methods and Systems of Navigation and Motion Control (MSNMC)*, 39 – 42. 10.1109/MSNMC.2016.7783101

Kucherov, D. P. (2014). Agent-based approach to the problem movements by the route, *Visnik of the Volodymyr Dahl East Urk. National University*, *4*(211), 168–171.

Kucherov, D. P., & Kozub, A. N. (2015). Control System Objects with Multiple Stream of Information. *Proceedings IEEE 3rd International Conference "Actual Problems of Unmanned Aerial Vehicles Developments (APUAVD)"*, 290 – 293. 10.1109/APUAVD.2015.7346623

Kucherov, D. P., Kozub, A. N., & Kostyna, O. N. (2016). Group Behavior of UAVs in Obstacles Presence. *Proceedings IEEE 4th International Conference "Methods and Systems of Navigation and Motion Control" (MSNMC)*, 51 – 54. 10.1109/MSNMC.2016.7783104

Kulik, A. (2016). *Intelligent Transport System in Aerospace Engineering, in Intelligent transportation systems-problems and perspectives*. Springer International Publishing. Retrieved October 09, 2018, from http://www.springer.com/us/book/9783319191492

Li, W. (2018, March). The Constrained Rayleigh Quotient with a General Orthogonality Constraint and an Eigen-Balanced Laplacian Matrix: The Greatest Lower Bound and Applications in The Greatest Lower Bound and Applications in Cooperative Control Problems. *IEEE Transactions on Automatic Control, 12*. Retrieved from https://ieeexplore.ieee.org/abstract/document/8314760/

Liu, L., Liang, X., Zhu, C., & He, L. (2017). Distributed Cooperative Control for UAV Swarm Formation Reconfiguration Based on Consensus Theory. *2nd International Conference on Robotics and Automation Engineering*, 264 – 268. Retrieved October 09, 2018, from https://ieeexplore.ieee.org/document/8291392/

Ma, C.-Q., & Zhang, J.-F. (2010). Necessary and Sufficient Conditions for Consensusability of Linear Multi-Agent Systems. *IEEE Transactions on Automatic Control, 55*(5), 1263–1268. doi:10.1109/TAC.2010.2042764

Murray, R.M. (2007). Recent Research in Cooperative Control of Multi-Vehicle Systems. *ASME Journal of Dynamic Systems, Measurement, and Control*, 1 – 27.

Ni, W., & Cheng, D. (2010). Leader-following consensus of multi-agent systems under fixed and switching topologies. *Systems & Control Letters, 59*(3-4), 209–217. doi:10.1016/j.sysconle.2010.01.006

Ögren, P., Fiorelli, E., & Leonard, N. E. (2004). Cooperative Control of Mobile Sensor Networks: Adaptive Gradient Climbing in a Distributed Environment. *IEEE Transactions on Automatic Control, 49*(8), 1292–1302. doi:10.1109/TAC.2004.832203

Olfati-Saber, R. (2001). *Nonlinear control of underactuated mechanical systems with application to robotics and aerospace vehicles* (Ph.D. thesis). MIT.

Olfati-Saber, R., Dunbar, W. B., & Murray, R. M. (2003). Cooperative Control of Multi-Vehicle Systems using Cost Graphs and Optimization. *Proceedings of the 2003 American Control Conference*, 1 – 15. Retrieved October 09, 2018, from https://ieeexplore.ieee.org/document/1243403/

Olfati-Saber, R., Fax, J. A., & Murray, R. M. (2007). Consensus and Cooperation in Networked Multi-Agent Systems. *Proceedings of the IEEE, 95*(1), 215–233. doi:10.1109/JPROC.2006.887293

Pipattanasomporn, M., Feroze, H., & Rahman, S. (2009). Multi-Agent Systems in a Distributed Smart Grid: Design and Implementation. *Proceeding IEEE PES 2009 Power Systems Conference and Exposition (PSCE'09)*, 1 – 8. 10.1109/PSCE.2009.4840087

Ren, W., Beard, R. W., & Atkins, E. M. (2005). A Survey of Consensus Problems in Multi-Agent Coordination. *American Control Conference*, 1859 – 1864.

Rezaee, H., & Abdollah, F. (2016). Consensus Problem in General Linear Multi-Agent Systems Under Stochastic Topologies. *IFAC-PapersOnLine, 49*(13), 13–18. doi:10.1016/j.ifacol.2016.07.919

Shmelova, T., Sikirda, Y., Zemlyanskiy, A., Lazoprenko, V., & Danilenko, O. (2016). Artificial neural network for air traffic controller's pre-simulator training. *Proceedings of the National Aviation University*, (3), 13 – 23. 10.18372/2306-1472.68.10905

Xiao, F., Wang, L., Chen, J., & Gao, Y. (2009). Finite-time formation control for multi-agent systems. *Automatica, 45*(11), 2605–2611. doi:10.1016/j.automatica.2009.07.012

Xie, Y., Shi, Z., Sun, C., Deng, Y., & Wu, X. (2017). The Integration of Adaptive Cooperative Control for Position and Attitude of Distributed Spacecraft. *Proceedings of the 36th Chinese Control Conference*, 3409 – 3413. 10.23919/ChiCC.2017.8027886

Yu, X., Guo, Ge., & Gong, G. (2017). Cooperative control for a platoon of vehicles with leader-following communication strategy. *Chinese Automation Congress (CAC)*, 6721 – 6726. Retrieved October 09, 2018, from https://ieeexplore.ieee.org/document/8243988/

Yun, L., & Jian, Z. (2018). Design and Implementation of Cooperative Turning Control for the Towing System of Unpowered Facilities. *IEEE Access: Practical Innovations, Open Solutions*, *6*, 18713–18722. doi:10.1109/ACCESS.2018.2819692

Zhang, H., & Chen, J. (2014). Bipartite Consensus of Linear Multi-Agent Systems over Signed Digraphs: An Output Feedback Control Approach. *Proceedings of the 19th World Congress the International Federation of Automatic Control*, 4681-4686. 10.3182/20140824-6-ZA-1003.00608

Zuo, Z., Han, Q.-L., Ning, B., Ge, X., & Zhang, X.-M. (2018). An Overview of Recent Advances in Fixed-Time Cooperative Control of Multi-Agent Systems. *IEEE Transactions on Industrial Informatics*, *14*(6), 2322–2334. doi:10.1109/TII.2018.2817248

ADDITIONAL READING

Cooper, R. B. (1981). *Introduction to Queueing Theory*. Amsterdam: Elsevier North Holland.

Cormen, T. H., Leiserson, C. E., Rivest, R. L., & Stein, C. (2002). *Introduction to Algorithms*. McGraw-Hill Book Company.

Marghitu, D. B. (2001). *Mechanical Engineer's Handbook*. Orlando, FL: Academic Press.

Rodic, A. D. (2009). *Automation & Control – Theory and Practice*. Zagreb, Croatia: In-Tech.

Taha, A. H. (2003). *Operations Research: An Introduction*. London: Pearson Education Inc.

KEY TERMS AND DEFINITIONS

Agent: Some entity, a system that possesses properties, in addition to the properties of the object to exist and unite, necessary for interaction with the external environment.

Aggregate: System, the combination of something combined to fulfill a specific goal.

Attractive Field: This is a potential vector field around each body whose action opposite to repulsive field.

Control Law: The mathematical form of transformations of driving influences, disturbances, effects of feedbacks that determine the control actions.

Group: A set of related and interacting elements.

Movement: This called the change in position of a body in space relative to other bodies over time. In this case, the bodies interact according to the laws of mechanics.

Potential Field: A vector field that can be represented as the gradient of some scalar coordinate function.

Repulsive Field: This is a potential vector field around each body whose strength is inversely proportional to its distance.

UAV: Unmanned aerial vehicle or drone is an aircraft without crew on board.

Chapter 8
Forest Firefighting Monitoring System Based on UAV Team and Remote Sensing

Maryna Zharikova
Kherson National Technical University, Ukraine

Vladimir Sherstjuk
Kherson National Technical University, Ukraine

ABSTRACT

In this chapter, the authors propose an approach to using a heterogeneous team of unmanned aerial vehicles and remote sensing techniques to perform tactical forest firefighting operations. The authors present the three-level architecture of the multi-UAV-based forest firefighting monitoring system; features of patrolling, confirming, and monitoring missions; as well as functions of UAV in such missions. The authors consider an infrastructure for the UAV ground support and equipment used for the UAVs control. The method of the data integration into a fire-spreading model in a real-time DSS for the forest fire response is proposed. The proposed approach has been tested with the multi-UAV team that included three drones for the patrol missions, one helicopter for the confirmation mission, and one octocopter for the monitoring mission. The performance of such multi-UAV team has been studied in the laboratory conditions. The result of the experiment has shown that the proposed approach provides required credibility and efficiency of fire prediction and response.

BACKGROUND

The number and impact of disasters seem to be increasing in the last few decades as a result of the growth of population, urbanization, environmental degradation, global warming etc. Forest fires are natural disasters that cause deaths, injuries, and huge damages. A number and intensity of forest fires are significantly increasing year by year due to a growth of human activity and climate changes. Thus, they become bigger, more dangerous, and more costly to put out.

DOI: 10.4018/978-1-5225-7709-6.ch008

Forest fire response operations are becoming increasingly important but challenging because they are traditionally based on visual observations and decision-makers estimations. Due to exposure to smoke, flame, and high temperature, visual observation is usually ambiguous, imprecise, incomplete, and inconsistent, while the instrumental measurement is expensive, difficult, and dangerous. Besides that, decision-makers often operate under the high responsibility conditions in a lack of time. This stipulates the development of real-time decision-support systems (DSS) for the forest fire response.

However, developing such DSS is a complex and non-trivial task because the forest fire is a process, which arises unexpectedly, proceeds fleetingly, and evolves in space and time transiently, non-linearly. Consequently, the forest fire is poorly modeled and unpredictable process. This completely excludes the use of well-studied classical decision support approaches based on models, rules, etc. In such conditions, the efficiency of forest fire operations strongly depends on the availability of online monitoring and measurement tools. To build such tools, we can use a suite of the most modern methods and techniques, such as remote sensing, geolocation, geographical information systems, unmanned vehicles, etc. Their efficiency primarily depends on the synergy of their joint use.

An indispensable part of such DSSs is online monitoring aimed at the computation of the most critical parameters related to the forest fire spreading. A full range of data for the forest fire monitoring is generated by remote sensing techniques and have a form of data streams of great volumes that come from sensors on a continuous basis at a high rate and should be analyzed in a real-time.

The important task for decision-making is to diagnose the situation based on the online monitoring data, namely, to assign a given data to some category and then respond appropriately based on such classification (Zharikova & Sherstjuk, 2017). The problem of intelligent diagnosis in DSSs for disaster response has been dealt with in investigations of a number of authors. In (Balakrishnan & Honavar, 1998) the authors define a diagnosis system as one that is capable of identifying the problem by examining the observable symptoms (observations or assessments provided by monitoring). Such system is called intelligent if its diagnoses closely match those produced by a human.

It is necessary to give a short review of approaches to the development of intelligent diagnostic systems. One of the approaches is model-based used in the case when a domain model representing a structure of the system is available (Lucas, 2001; Kurien & R-Moreno, 2008).

In most practical tasks, precise models of the domain are unavailable. The most commonly used technique in such cases is the development of knowledge-based diagnosis systems [Bhagwat, 2015; Verbert, De Schutter & Babuska, 2015], which are based on heuristic knowledge about the domain and model the diagnostic reasoning of human experts. Knowledge-based systems have some limitations such as the inability to make creative responses as a human expert would in unusual circumstances and to adapt to changing environments.

Expert systems are a specified domain of such systems and are more specific than knowledge-based systems (Tan, Wahidin, Khalil, Tamaldin, Hu & Rauterberg, 2016; Amarosicz, Psiuk, Rogala & S. Rzydzik, 2016). They are typically built through a careful, tedious, and often expensive process of knowledge engineering (Martinez, 2017). There are some difficulties related to knowledge engineering because on the one hand the domain experts not always being able to explain their logic and reasoning sufficiently precisely to be encoded in a machine form, and on the other hand different experts often have a different understanding of the problem.

There is one approach that tries to work around the problem of extracting of domain expert knowledge. It is based on the development of a large repository of sample diagnosis or cases (Balakrishnan & Honavar, 1998; Qin, Wu, Jiang & Zhu, 2016). When facing a diagnosis problem, the case-based system attempts to identify scenarios with known diagnoses from the case repository. In comparison with expert systems where knowledge of the domain expert is modeled explicitly in the form of heuristic rules, in case-based systems, the knowledge is modeled implicitly in the form of cases.

Learning systems is the other attempt to overcome the difficulties associated with knowledge engineering (Rieck, Trinius, Willems & Holz, 2011). The learning system is a self-improving system capable of acquiring and integrating the knowledge automatically from experience. There exist some learning technologies used in learning systems such as neural networks (Goldberg, 2016), genetic learning (Gong, Li, Zhou, Li, Chung, Shi & Zhang, 2016), symbolic artificial intelligence (Hoehndorf & Queralt-Rosinach, 2017), statistical pattern recognition, syntactic pattern recognition, evolutionary computing. The fascinating technology for diagnosis systems is inductive learning systems that extract the domain knowledge transforming their inputs by means of generalizations (Symeonidisa, Chatzidimitriouc, Athanasiadisd & Mitkas, 2007). Inductive approaches are suited for domains, which are hard to formalize, or where experts are few (Donner-Banzhoff & Hertwig, 2014).

The domain of consideration is related to the processes with unpredictable behavior such as forest fires. Taking into account inaccurate or missing data describing them, incomplete scientific understanding of their behavior, they are poorly modeled and predicted. This completely excludes the use of such approaches to diagnosis as knowledge engineering, rule-based, or model-based approaches.

The performance of case-based systems significantly depends on a size of accumulated case base because the large case base incurs high computational cost during classification while the insufficient size of the case base does not provide a coverage of all possible situations for classification. Thus, it is difficult to provide the necessary representativeness of case bases in disaster response systems.

Learning systems are sufficiently non-linear systems, which are able to perform automatic classification of situations even in uncertainty conditions, but they do not provide the required credibility for decision-maker. Thus, automated identification of problems based on the observations obtained during online monitoring needs a quite complicated non-classical method, because the efficiency of forest fire response operations strongly depends not only on its performance but also on its credibility.

The next important issue of an operation of the forest fire response is a fighting planning based on the diagnosis. Decision-maker has to determine a necessary set of fire brigades, fire-means, and their optimal location, to provide their schedules, etc. For this, he should know, in what direction and at what speed the fire front will spread at the next time. Thus, he needs to estimate how some important parameters (e.g. actual shape of the fire front, its rate of spread, the height of the flame, etc.) evolve with time.

There are several approaches that are most often used for estimation of the forest fire parameters. A predictive fire modeling is the most used approach because chemistry and physics of fire are comparatively well studied.

One of the most used physical models of fire spread is the model, proposed by Rothermel (Rothermel, 1972), which is the primary fire spread model of many fire prediction systems. The Rothermel model uses numerical proportions of the combustion reaction intensity and allows obtaining exact values of some chemical and physical fire properties (e.g. spread rate) if the exact values of some initial data are known. Unfortunately, those data are not available for direct measurements in conditions of a real forest fire. Besides that, a spread rate is represented as a scalar, so it tells us little about spatial properties of fire propagation.

Vector models (Anderson, 1982) are more informative in sense of determining directions of the fire propagation. They use an ellipse as a standard geometrical shape for the fire front but assume that the fire spreads according to a well-defined growth law. Such models work only in narrow particular conditions, for which all the necessary coefficients of equations and initial data are well known. Fractal and wave extensions of vector models (Finney, 1999) have a little better accuracy, but they are much more computationally intensive, so they cannot be used in real-time applications.

Cellular automata models (Karafyllidis & Thanailakis, 1997) involve a discrete process of ignitions within the regular structure of the grid-based landscape. Such models are appropriate for the fire-spread modeling incorporated into spatial frameworks, but demand the constant conditions of weather, fuels, and topography. However, it is a very strong assumption, because usually, weather, fuels, and topography vary spatially, and weather varies temporally as well. Thus, the predictive fire modeling approach is not suitable sufficiently for decision-maker who solves the planning task for forest firefighting in real time.

Another known approach is online forest fire monitoring using remote sensing techniques (Yuan, Zhang & Liu, 2015). The forest fire monitoring is activity aimed at the real-time computation of the evolution of the most important parameters related to the fire propagation based on the online observations. Traditionally, satellite and airborne systems are used in order to have a broad overview of the fire evolution, but the monitoring activities are still carried out by people. A ground-based system for fire monitoring using static cameras has been presented in (Martínez de Dios, Arrue, Merino, Ollero & Gómez-Rodríguez, 2007). Given the fact that large-scale forest fires require a very large number of cameras, the price and complexity of such system are unacceptable.

Satellite-based systems (Olsson, Egberth, Engberg, Fransson & Pahlén, 2005) have been proposed for forest fire detection and monitoring, but the temporal and spatial resolutions of such systems are very low and fail to meet the requirements of forest-fire fighting. Manned aircrafts are large and expensive; their use highly depends on the weather conditions and requires the presence of aerodromes (Yuan, Zhang & Liu, 2015).

Recent years have seen a great progress in the field of using unmanned aerial vehicles (UAVs) for forest fire monitoring, detection, and even fighting (Ambrosia & Zajkowski, 2015). UAVs can perform long-time, monotonous and repeated missions beyond human capabilities. However, uncertainty and distortions of received image frames due to vibrations and turbulence, as well as the inability to measure directly the parameters required for decision-making are significant drawbacks of this approach (Merino, Caballero, Martinez-de-Dios, Maza & Ollero, 2012). It is also essential that the use of UAV-based monitoring approach makes it possible to assess the state of the forest fire at the observation time, but gives a little for the forest fire prediction due to significant uncertainties.

Thus, the integration of UAVs with remote sensing techniques can provide rapid, mobile, and low-cost powerful solutions for various forest fire tasks (Merino, Martínez de Dios & Ollero, 2015) Therefore, many researchers pay increased attention to using UAVs (Yuan, Liu & Zhang, 2017; Wardihani, Ramdhani, Suharjono, Setyawan, Hidayat, Widodo, Triyono & Saifullah, 2018).

The use of a single and complex UAV with sophisticated complex sensors has been investigated in FiRE project, while the cooperative use of a simpler UAVs' team was explored in European COMETS project and in a number of other projects surveyed in (Yuan, Zhang & Liu, 2015). Many works related to the detection, modeling and forecasting of forest fires based on UAVs were carried out at the Ukrainian research centers such as Kharkiv National University of Civil Protection, Ukrainian Research Institute of Civil Protection, and Kherson National Technical University (Sherstjuk, Zharikova & Sokol, 2018).

Despite the positive results, which have shown the possibility of using UAVs for the forest fire response (Sharifi, Mirzaei, Zhang & Gordon, 2015), many issues related to UAV-based forest fire monitoring systems, including their architecture, suitable platforms, sensors, remote sensing and image processing algorithms, still remain insufficiently investigated, so need further research.

Thus, the topic of our current interest is the research of ways of developing the online forest fire monitoring and diagnostic system for real-time response operations. The motivation and contribution of this chapter is to demonstrate a way to the solution of online monitoring and diagnosis problem in the forest fire-fighting system based on a team of unmanned air vehicles, remote sensing and image-processing techniques.

OUR APPROACH

The integration of UAVs with remote sensing techniques provides rapid, mobile, and low-cost powerful solutions for the forest fire response operations (Witayangkurn, Nagai, Honda, Dailey & Shibasaki, 2011).

Forest fire-fighting operation must start as early as possible and must be done as fast as possible because the smaller the fire is the easier it can be extinguished before it becomes too large. Early detection and rapid extinguishment of a forest fire can minimize damages caused by fire, so it needs to develop DSS for forest fires detection and monitoring. After the forest fire is detected, decision-maker has to estimate its features such as fire front location, fire site size and perimeter, flame length and height, hotspots and their location, and others. Obviously, the most significant parameter is the rate of spread, directly associated with fire intensity, because it characterizes a forest fire behavior. Such estimates projected onto a terrain map are essential for fire-fighting planning (Shahbazi, Théau & Ménard, 2014).

However, as we have seen from aforementioned literature review, decision-maker cannot obtain exact estimates in real time neither by automatic monitoring system due to their essential uncertainty, nor by predictive fire modeling due to their computational complexity and a lack of initial data. Thus, decision-maker cannot decisively predict fire behavior in good time despite its rapid propagation and long combustion cycle.

We note that in real conditions, decision-maker does not require precise estimates of parameters, but he needs a prediction, when and how the fire will spread, and the price of such prediction can be significant. Thus, we do not need prediction accuracy, but its credibility. Another difficulty is that we cannot directly measure some parameters required for the fire-spreading model, but can estimate them indirectly. Therefore, the use of exact numerical models does not make sense. The authors assume that the combination of automatic monitoring system based on UAVs and remote sensing techniques with an approximate model of forest fire spreading (Zharikova & Sherstjuk, 2015) can provide the required credibility of the prediction synergistically.

The detection of forest fire must be performed before fire break-out, while monitoring is related with an actual forest fire in progress. Accordongly, forest fire fighting system should solve two different types of tasks coherently, such as early detection of ignitions and online monitoring of fires. Forest fire detection requires using a heterogeneous team of UAVs with different kinds of sensors along with a central ground station. Monitoring provides to decision-maker the estimations of such important parameters as fire front location, flame length/height, and rate of spread.

However, in real conditions, decision-maker needs not only such estimations for fire-fighting planning but also understanding which valuable objects are exposed to the most dangerous situations, which of them need priority protection, and what the time limit is. Therefore, we need to develop an automated diagnosis method to identify the situation in UAV-based forest fire-fighting response system. Giving the fact that no above-mentioned approaches to intelligent diagnosis can be used independently, the authors propose in this chapter to develop a hybrid method, which combines the case-based approach with the rule-based approach to reduce a large case base.

Thus, the aim of this chapter is to develop an architecture of a multi-UAV-based online monitoring system for tactical forest fire-fighting operations and algorithms of its integration with the approximate fire-spreading model as well as a hybrid intelligent diagnosis method for such forest fire-fighting response system. The authors do not consider in the chapter some issues related to navigation, coordination, and cooperation of UAVs in their teamwork, leaving it for the future works. Instead of this, the authors consider a joint use of UAVs, remote sensing of different types and GIS-based common terrain model with the fire-spreading model. We define and describe functions and missions of the system and its architecture, the used types of UAVs and sensors, the image processing techniques, the common terrain model, the situation diagnosis method, and the algorithm of data integration with the fire-spreading model in the frame of the real-time DSS for the forest fire response.

THE SYSTEM ARCHITECTURE

The multi-UAV-based tactical forest fire monitoring system should perform the following functions:

- Fire detection (finding potential ignitions, detection of fire, triggering an event, and initializing further monitoring of the fire);
- Fire monitoring (determining the fire's location and extent, observing and tracking its evolution);
- Threat and risk analysis (determining the valuable objects at risks and evaluating the threats);
- Fire diagnosis (identifying and classifying the situation);
- Fire prognosis (predicting the future fire evolution in real-time).

The objectives are to use the heterogeneous team of UAV to detect and track forest fires, measure their features and predict their future evolution. In general, the forest fire monitoring system must provide real-time information to decision makers for response operation planning. In the future, the team of UAV can be immediately involved in firefighting.

System operation can be generally broken down into three successive stages (Merino, Caballero, Martinez-de-Dios, Maza & Ollero, 2012): fire search, fire confirmation, and fire observation. Accordingly, it is assumed that each UAV will perform a certain mission at each stage.

Three main type of missions are envisaged:

- Patrol mission (surveillance over the region and fire detection);
- Confirmation mission;
- Monitoring mission.

In the patrol mission, each involved UAV has its own flight plan that contains a pre-planned path as a sequence of waypoints. Flying along the pre-planned path, UAV observes terrain using onboard sensors and tries to identify fire automatically. It is clear that the flight plan generally is strictly bounded due to some limitations of UAVs' capability, such as duration, range, altitude, sensor resolution, etc.

Depending on the size and characteristics of the surveillance region, a various number of UAVs can be involved in patrol mission simultaneously along their own pre-planned paths. After the fire is detected, confirmation mission begins. In one of the options, the patrol UAV detected the fire can continue its patrol mission further, but it must trigger the event. Other UAVs with hovering capabilities should be sent to the detected fire location to hover at a safe distance and make confirmation. Another option is to change pre-planned paths of the patrolling UAVs to fly by a circle around the detected fire location in order to confirm it. If the detected fire is not confirmed, the patrolling UAVs resume their missions to surveillance of the region.

If the fire is confirmed, the fire monitoring mission starts. This mission usually requires multiple synchronous information obtained by UAVs from different points of view, therefore a couple of UAVs should be involved to execute the mission. It allows obtaining information about the fire continuously. Required estimations of the fire features should be delivered to decision-makers through the ground station and DSS to guide firefighting efforts better. Given the fact that performed missions differ in goals and requirements, we need the UAVs of different types equipped with the sensors of different types with a single ground command center.

The basic structure of the multi-UAV-based forest fire monitoring system is illustrated in Fig. 1.

The system includes the following components:

1. A multitude of UAVs (for the patrolling, confirming, monitoring missions) equipped with the onboard sensors;

Figure 1. Structure of the forest fire monitoring system

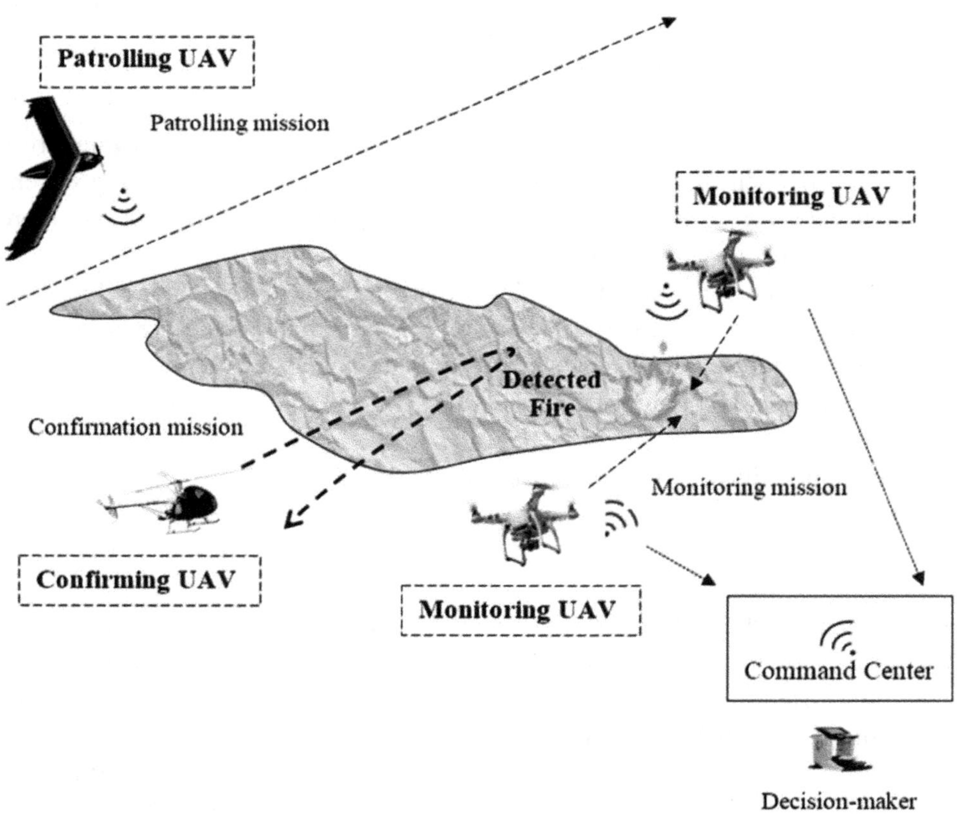

2. An infrastructure for the UAV ground support (launch and landing, maintenance) and equipment for the UAVs control;
3. Specific algorithms/techniques for remote sensing and image/signal processing;
4. A dedicated ground command center that includes communication/computation equipment, geographic information system (GIS), and decision support system;
5. Specific algorithms solving the fire detection, tracking, diagnosis, and prediction tasks.

Consider the used types of UAVs and sensors.

VEHICLES AND SENSORS

The common requirements on the used UAVs are the following:

- **All-Weather Suitability:** All UAVs should perform their functions in both night-time and daytime even in the most difficult weather conditions;
- **Self-Localization:** A common reference terrain model should be used by all involved UAVs for automatic geo-localization of their spatial positions;
- **Navigational Autonomy:** Sophisticated sensors (e.g. GPS receivers, inertial measurement units) should be used by all involved UAVs for automatic flying along paths given by ground command center;
- **Cooperation:** The UAVs should be able to coordinate their behavior and to cooperate with each other in order to solve their tasks optimally ;
- **Payload:** The UAVs should carry all required sensors for fire perception purposes.
- **Availability:** All UAVs should be equipped with onboard communication devices that guarantee receiving commands from a ground command center, sending information back to it, as well as exchanging information with the other UAVs.

The general structure of the used UAVs is illustrated in Fig.2.

In accordance with the different tasks to be performed during the missions, different types of UAV are used for those missions.

For the patrol mission in order to minimize the solution cost the authors propose to use fixed-wing micro-UAVs with electric propulsion system having flight ceiling up to 2000 m, cruise speed about 90 km/h, takeoff weight up to 5 kg. This type of UAV provides the flight duration about 2.5 – 3 hours, and flight range with an online connection up to 75 km. It is equipped with a low-cost non-thermal (5-13 μm band) infrared micro-camera and a simple 12-megapixel optical camera with electronically adjustable focus.

We can use the same type of UAV for the confirmation mission. However, the confirmation mission can be performed much easier, faster and more efficient if we send the UAV with hovering capabilities to the detected potential fire location. Equipped with more precise (and, thus, more expensive) infrared and optical sensors, such UAV can hover for some time at a safe distance from the potential fire site and monitor it for verification purposes. Therefore, the authors propose to use rotary-wing micro-UAVs with electric propulsion system having flight ceiling up to 180 m, cruise speed about 50 km/h, takeoff weight up to 7 kg. This type of UAV provides the flight duration about 1 hour, and flight range with the

Figure 2. Vehicle onboard control system

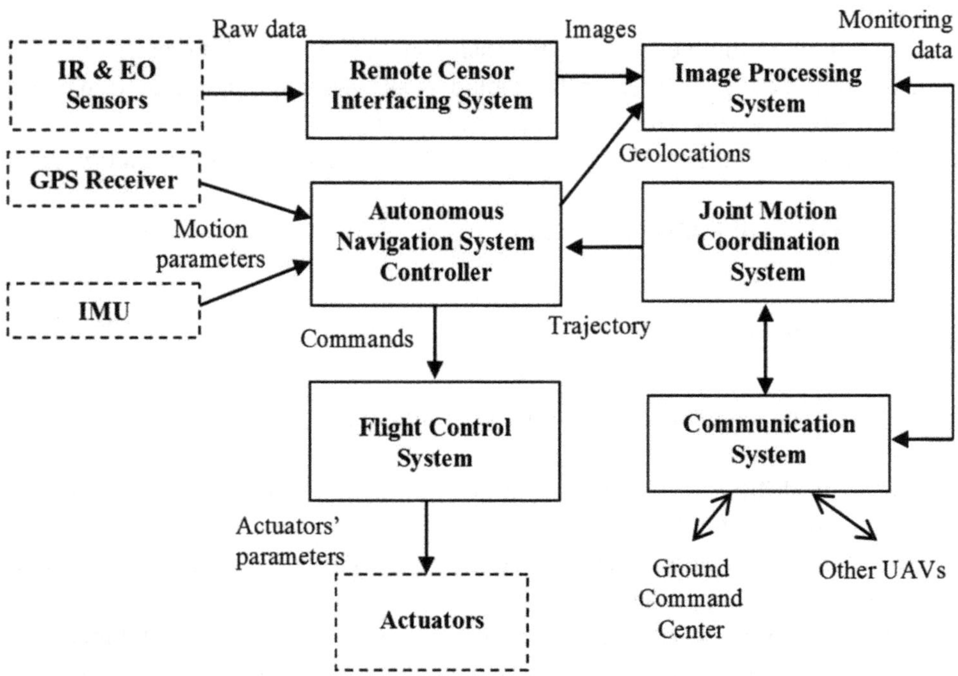

online connection up to 55 km. It is equipped with a precision gimbal that carries a wideband infrared camera and a 16-megapixel optical camera.

A relatively small, but safe enough hovering altitude provides the good applicability of this type of UAV. Due to the higher resolution of the sensors and the closest approach to the observed point, we can achieve a much greater observation accuracy. In addition, this type of UAV can change the point of view relatively quickly, providing a higher efficiency of fire detection and confirmation.

For the monitoring mission, it is essential to obtain clear images without distortions arising from turbulence and vibrations. Besides that, we need to estimate some parameters, which were not necessary for the previous missions. Obviously, monitoring UAVs should have hovering capabilities. Thus, the authors propose to use multi-rotary-wing UAVs (octocopters) having flight ceiling up to 120 m, speed about 20 km/h, and takeoff weight up to 15 kg. It provides the flight duration about 2 hours, and flight range with the online connection up to 25 km. It is equipped with a remotely-controlled precision damped gimbal that carries both a wideband non-thermal and a thermal (four-band, 5-13 μm) infrared cameras, as well as high-resolution optical camera with pan and tilt units. The thermal infrared camera has temperature measurement range up to 1250°C with the accuracy about 1°C at the top of the range.

This type of UAV also can use lidar sensor that provides advanced sensing capabilities, and wide-range weather sensors ensuring a very accurate measurement of temperature, humidity, wind speed, and direction, etc. It provides a sufficiently long hovering time but has a limited speed and flight range. Fortunately, fire-fighting brigades use UAVs of this type locally, immediately on the terrain, so the wide flight range is not necessary. The cost of the UAV of this type is much higher than UAV of the other types. However, they are used relatively little during a fire monitoring only, and cheaper solutions can be used to solve the patrol and confirmation tasks.

REMOTE SENSING AND IMAGE PROCESSING

The main goal of the forest fire monitoring is to continuously obtain information necessary for the approximate fire-spreading model proposed in (Zharikova & Sherstjuk, 2015; Zharikova, Sherstjuk & Baranovskij, 2015) in order to predict the future evolution of the fire. The model represents spatiotemporal relations among cells. It is known that fire spreading is affected by a range of factors, including the weather, the fuel load and topography (Karafyllidis and A.Thanailakis, 1997).

Only a few of critical variables for the fire-spreading model are directly measurable. For example, weather sensors provide us with the air temperature, humidity, wind speed, and direction. Data accumulated in GIS provide available information about a vegetation type, slope, aspect, and soil parameters. Using the remote sensing techniques, we can obtain necessary estimates of valuable parameters (Clarke, Brass & Riggan, 1994):

- The combustion reaction intensity can be quantitatively estimated through the energy release available by the thermal infrared sensor;
- A fuel load can be qualitatively estimated through the analysis of vegetation density distribution associated with the certain vegetation type;
- A fuel moisture can be qualitatively estimated through the saturation and brightness analysis for a specific color (e.g., green);
- The burnt areas can be qualitatively estimated through the contrast analysis (green-to-black).

During the monitoring mission, we need all above-mentioned estimations.

Common Terrain Model

The forest fires arise and spread through a certain area, represented by GIS as a certain terrain. The terrain model is based on the Digital Elevation Model (DEM) (Balasubramanian, 2017).

The obtained DEM is imposed onto the Euclidean space C with a metrical grid of coordinate lines sized δ using a norm ξ_l and a linear map f such that coordinate lines form a set D of cells of the size $\delta \times \delta$, $f: C \to D$. Thus, space C is discretized by a grid $D = \{d_{ij}\}$ of isometric square cells d_{ij}.

A grid of isometric square cells $D = \{d_{ij}\}$ constitutes a basic GIS layer. Therefore, the location of each valuable object as well as the location of the fire front are discrete and discretized by a certain set of cells.

Since the size $\delta \times \delta$ of the cell d_{ij} can vary, the terrain scale can also change.

Each cell $d_{ij} \in D$ is associated with a set of attribute values $A = \{a_1, \ldots a_n\}$, which is called the cell state, via the values function $\{f(d, a_i) | \forall a_i \in A\}$. Such attributes can be terrain-dependent or process-dependent.

During the monitoring, the system evaluates the ignition/burning possibility degree v_{ij} as well as the process-dependent attributes $\{a_1, \ldots a_m\} \in A$ for each cell $d_{ij} \in D$ using the remote sensing technique.

Thus, we obtain a fire front location and can evaluate a fire spreading process. Due to inaccuracy, incompleteness, and inconsistency of the observations, some estimations can be imprecise and some information is uncertain.

Therefore, the observed fire front F is blurred and can be determined by a plurality of soft sets: a set of burnt cells, a set of burning cells, a set of possibly burning cells, and a set of free of fire cells. Based on the location of the fire front F and a set of valuable objects $O = \{o_1, ...o_k\}$, the system performs an analysis evaluating the level of threat λ_i and risk r_i of fire front F with respect to each object $o_i \in O$.

On the next level, the terrain is divided into a finite set of disjoint spatial objects presented as geometric shapes, which outline boundaries of the certain areas. Such spatial object is named as geotaxon and represents a geo-referenced limited natural part of the terrain with the homogenious physical (or other) characteristics. For example, areas with the same features of the soil can be described as the geotaxons of a certain kind. Used GIS can contain an unlimited number of geotaxons' layers.

Image Processing During the Patrol Mission

During the patrol mission, the main sensor is the non-thermal infrared camera because obtained infrared images are not affected by smoke (the smoke is transparent for the used far infrared wavelengths). Besides that, the infrared camera is workable under either weak or no light conditions. It does not provide temperature measures but only estimations of the radiation intensity, represented by colors. If a certain cell does not have any radiation, its pixels in the image are black. The appearance and increase of radiation change the pixels' color from black to white. Thus, if there are no white pixel in the image, it is very likely that there is no fire in the corresponding cell. However, some pixels can have grey colors.

At first, the brightness data is averaged within each cell d_{ij}, so the whole cell takes a certain average brightness B_{ij}. Then the authors propose to use a partial order of grey colors (Fig. 3), which implies the ordinal color scale from black to white mapped on a numerical scale from 0 (black) to 1 (white).

Thus, based on the brightness B_{ij} of each cell d_{ij}, we obtain a value μ_{ij} expressing the degree of ignition/burning possibility at this cell.

Since various superheated or supercooled rocky areas, soils, and water surfaces can distort data obtained from the non-thermal infrared camera, we should simultaneously get data from the optical camera.

An obvious feature of the presence of a forest fire is the smoke, observed as a light gray figure like a cone elongated in the wind direction. Thus, the processing of color images obtained by the optical camera is aimed at searching smoke or flame within them. For this, firstly we define the discriminating interval in RGB space *LG* = [RGB(60%, 60%, 60%) - RGB(95%, 95%, 95%)], which excludes some distortions caused by lighting conditions.

The authors propose to use texture-based classification method with the color-diffusion evaluation that differentiates smoke and non-smoke based on the counting of the number of pixels with the color belonging to the interval *LG*, relative to the total number of pixels in the cell. These values could be averaged over a certain time interval (for example, 5 sec) for each cell d_{ij} and returned as a degree of ignition/burning possibility η_{ij} ranged from 0 to 1.

Figure 3. Partial order of grey colors

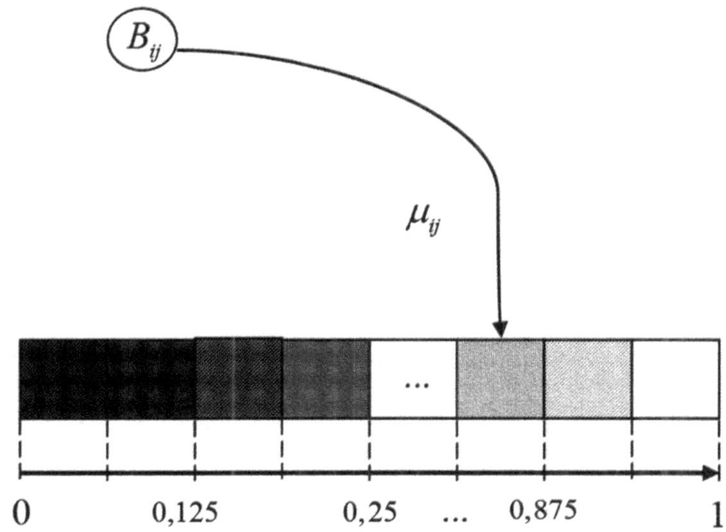

During the onboard automatic processing (Fig. 4), on the first step, the authors propose to perform stabilization of the images. On the next step, we process the infrared and visual images by the above-mentioned methods.

On the third step, the authors propose to perform geo-localization, then geo-rectification. The UAVs' locations can be obtained using the GPS. The positions and orientation of the infrared and optical cam-

Figure 4. Image processing during the patrol mission

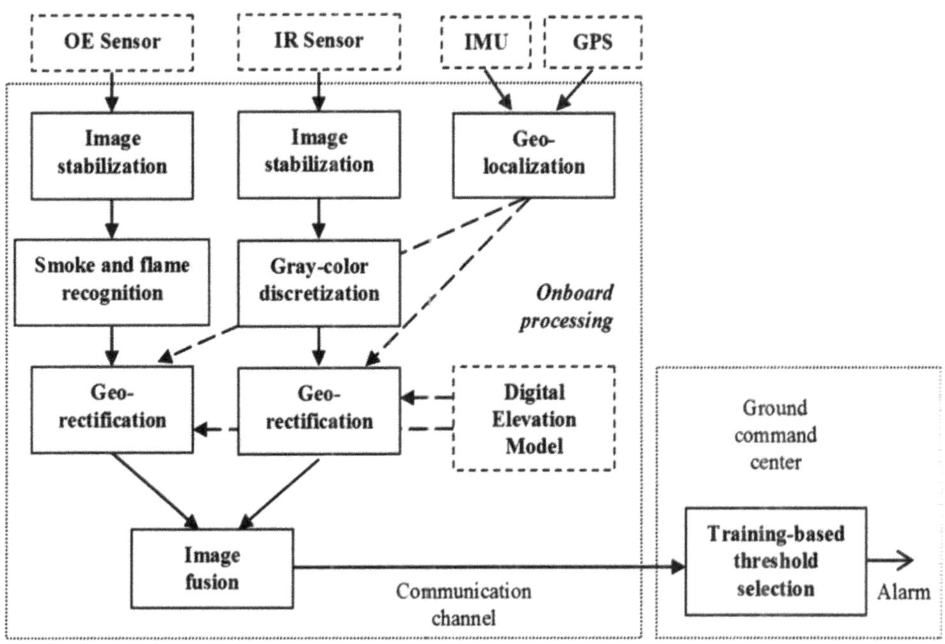

eras can be computed based on their orientation angles and IMUs' data. To determine the position of each image pixel we use the photogrammetric projective transformation that projects all points on the ground described by the DEM. DEM is also used to define the geospatial context (e.g., latitude/longitude/elevation) and timestamps for the geo-rectification. Thus, each image pixel becomes geo-referenced and can be mapped onto the grid *D*. If both cameras are calibrated and DEM is available, we obtain the geo-referenced images in the common terrain model discretized at the cell level.

On the final step, the authors propose to perform image fusion based on geo-referenced points. The values of the ignition/burning possibility μ_{ij} and η_{ij} that refer to the same cell d_{ij} for both images are combined by absorbing smaller values by larger: $\upsilon_{ij} = \max\left(\mu_{ij}, \eta_{ij}\right)$. Then the array of resulting values υ_{ij} should be transmitted to the command center.

On the receiving side, we use a training-based threshold selection method to reduce false alarms. Using the given threshold υ^*, we consider the fire detected only for the cells, which have greater or equal values of the ignition/burning possibility degree υ_{ij}. The training stage could be performed by an experienced decision-maker on a set of training images identifying the conditions, under which the uncertain assessments υ_{ij} can be considered as positive.

Image Processing During the Monitoring Mission

The monitoring mission is aimed at obtaining information about the fire features and behavior continuously. The monitoring UAVs should not be overloaded with substantial computing. UAV control system provides onboard only the image stabilization, geo-localization, and geo-rectification stages (Fig.5).

The results are transmitted to the ground command center for further image processing since most of the processing is related to the color-chromatic analysis of images. The weather sensors supply necessary information about air temperature and humidity, DEM provides sufficient information about the terrain, as well as the thermal infrared sensor supplies the model with the estimations of the fire intensity, which approximate the measured energy release of the combustion reaction. At the same time, the optical sensor provides us with information for the image processing, which starts from its transformation from RGB model to HSV model.

Using the HSV model, the authors propose to perform a slice of the pixels of the green spectrum.

Next, for each cell d_{ij} the authors suggest to evaluate the density of green pixels, using a partial order of green colors and implying the numerical green color density scale [$\lambda_{\min}$, $\lambda_{\max}$] by analogy with grey color as presented above. It gives us the estimation of the amount of the fuel within the cell.

Finally, it's proposed to estimate the saturation of all green pixels of the cell, which has a brightness more than 0,75. It gives us an estimate of the relative humidity on a scale of 0 to 1.

As a result of image processing, we obtain all estimations of fuel-related parameters needed to predict the fire propagation speed and direction. Since the speed of the prediction is significantly limited by the image processing speed, this stage was transferred to the ground station. Using the information from the infrared cameras, we can perform the fire front recognition. For example, the burnt cells can be discriminated by their color (close to black), as well as the burning cells can be discriminated by their level of radiation (temperature).

Based on soft rough sets, the model represents the blurred fire front. It determines by a plurality of soft sets: a set of burnt cells, a set of burning cells, a set of possibly burning cells, and a set of free of

Figure 5. Image processing during the monitoring mission

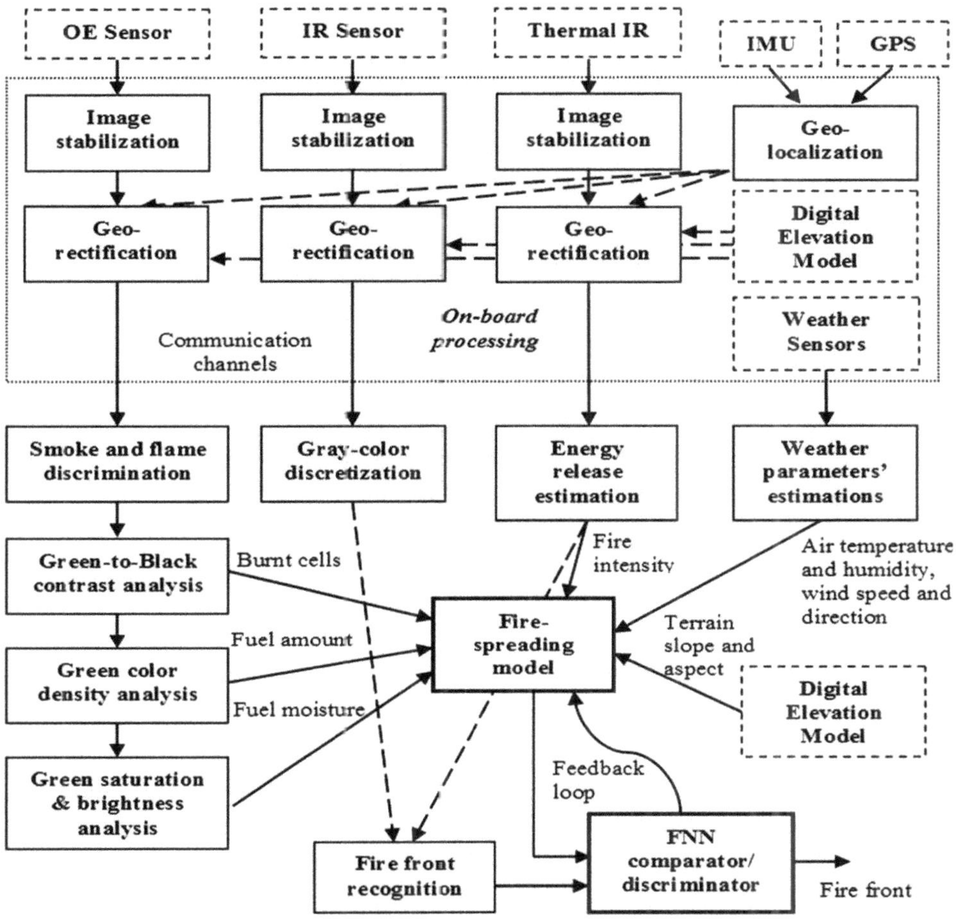

fire cells. Therefore, the fire front is represented by the boundary region of the rough set (Pawlak, Jerzy, Slowinski & Ziarko, 1995) of burning cells (Fig. 6).

Thus, at each moment we have two spatial descriptions of the fire front, the first is recognized as the result of infrared image processing, while the second is the result of the fire front prediction at the previous moment. Therefore, the authors suggest to compare periodically the recognized fire front with the predicted one. It allows us to adjust the approximate model in order to improve the accuracy of the prediction. Good results in improving the model performance were obtained by the feedback loop through the Fuzzy Neuron Network (FNN). The accuracy of the prediction and the quality of the model as a whole can be easily estimated by detecting deviations between predictions and subsequent observations. Obviously, using the FNN feedback can significantly reduce those deviations.

The ability of the developed system to work in near-real-time mode requires performing calculations based on the used fire-spreading model as quickly as possible because the time difference between the observation moment and the prediction moment should be small to neglect the influence of other factors.

Figure 6. Blurred fire front

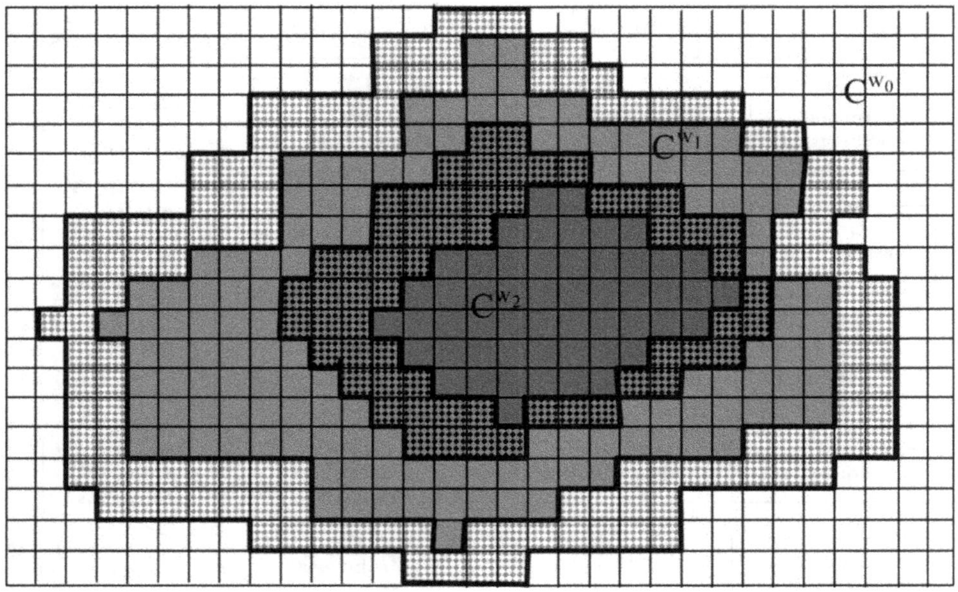

DIAGNOSIS OF THE SITUATION

The place and tasks of the intelligent diagnosis system are shown in Fig. 7.

Obviously, it performs the task of assigning a given input described as a set of observable symptoms to some output described as a category (class) of the given situation. Thus, a diagnosis system should be provided with a set of symptoms such as objects' and fire front locations, observed fire parameters (spread rate, flame length/height, etc.) and estimations obtained as a result of the threat and risk analysis (danger degree, levels of threat and risk). Besides that, such estimations as first and second derivatives of the fire front motion are also very important for the situation diagnosis.

Let $S = \{s_1, s_2, \ldots s_n\}$ be a set of classes of possible situations, p_i, $i = \overline{(1, n)}$, be a possibility of their appearance, and $X = \{x_1, x_2, \ldots x_m\}$ be a set of symptoms corresponding to the situations of the set S.

Let s^* be a current situation and x^* be a vector of symptoms for the current situation s^*. It is clear that due to the partial observability of environment some symptoms of x^* can be unknown or imprecise.

Suppose each symptom $x_j \in X$, $j = \overline{(1, m)}$ has a value (or interval of values) of a certain set $E_j \cup e_*$, $j = \overline{(1, m)}$, where e_* is a value of uncertainty. Thus, some symptoms of the vector x^* can be described as interval using the rough set approach (Pawlak, Jerzy, Slowinski & Ziarko, 1995), as an interval with a membership function using the fuzzy set approach (Zadeh, 1965), or even can have an empty value e_*.

Definitely, the diagnosis task is to identify the possible class $\hat{s} \in S$ that could explain the set of uncertain symptoms $x^* = \{x_j^*, \ldots x_l^*\}$, $x^* \subseteq X$, for the current situation s^*. This is a pattern classification task since the system has to determine the class or category that corresponds to the symptoms, where each class represents a diagnosis (Balakrishnan & Honavar, 1998).

Figure 7. Intelligent diagnosis in the forest fire fighting system

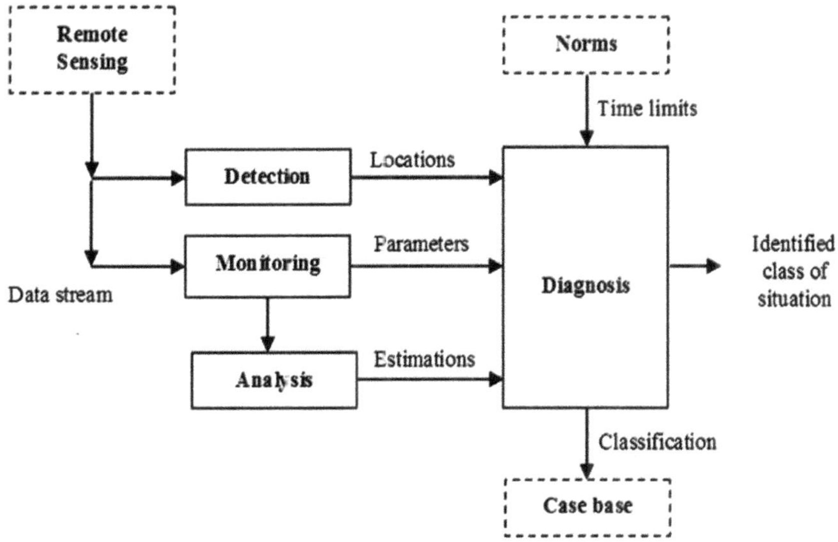

Thus, each situation $s \in S$ corresponds to a certain point (or neighborhood of points) within the Cartesian space of the symptoms $E_1 \times E_2 \times \ldots \times E_m$.

Obviously, each unrecognized situation s^* having symptoms $x^* = \left\{x_1^*, x_2^*, \ldots x_m^*\right\}$ can be matched to a set of possible situations S. Unfortunately, because of the uncertainty of the estimations of some symptoms, it is not always possible to find exact matches. Accordingly, it is necessary to search for coincidences of fuzzy or approximate (e.g., rough) values of symptoms.

First, we divide a set of possible classes of situations into disjoint subsets $\widehat{S}_1, \ldots, \widehat{S}_p \subset S$ such that $\widehat{S}_1 \cup \ldots \cup \widehat{S}_p = S$. Thus, we can use the method of measuring the proximity (also known as nearest neighbor method) of the unrecognized situation s^* to the reference set of classes $\left\{\widehat{S}_1, \ldots, \widehat{S}_p\right\}$ for developing a pattern classification algorithm. Each class of the reference set $\left\{\widehat{S}_1, \ldots, \widehat{S}_p\right\}$ should be considered as the basic case and must be described in the case base.

Next, the authors introduce the identification function $\tau : x^* \rightarrow \widehat{s}$, which allows us to recognize the class of the situation s^*. The algorithm of recognizing the class of the situation includes the following steps:

1. Calculation of distances $\left\{L_1, \ldots L_2, \ldots L_n\right\}$ between the current situation s^* and classes of well-defined situations $\left\{\widehat{S}_1, \ldots, \widehat{S}_p\right\}$:

$$L_i\left(s^*, \widehat{S}_i\right) = \sqrt{\sum_{j=1}^{m} l_j^2\left(x_j^*, x_j\right)},$$

where $l_j\left(x_j^*, x_j\right)$ is a proximity measure for symptom j with respect to the scale E_j.

2. Searching for a minimal distance α between the current situation s^* and situations $\left\{s_i\right\}_{i=1}^{q} \subseteq \left\{\widehat{S}_1,\ldots,\widehat{S}_p\right\}$ found in the previous step in the set of the classes of well-defined situations:

$$\alpha = \min_{i=(1,q)}\left(L\left(s^*,s_i\right)\right)$$

If $\alpha = 0$ this means the actual match of the current s ituation to one of several classes with respect to the corresponding symptoms, thus

$$s\left(t\right) = L_i\left(s^*,s_i\right) = \min_{i=(1,n)}\left(L\left(s^*,s_i\right)\right).$$

3. If $\alpha \neq 0$ this means incomplete observed data. Then the desired class coincides with a certain subset $\widehat{S}' \subset S$ of a set of well-defined situations, which are suitable for the incomplete description of s^*:

$$S' = \left\{s_i \mid L_i\left(s^*,s_i\right) = \alpha, i = \left(1,n\right)\right\}.$$

Thus, we range all situations $\left\{s_i\right\}_{i=1}^{q} \subseteq \widehat{S}'$ according to their total risk estimations $r = \sum_{i=1}^{k} r_i$ for all valuable objects and define the set of the most dangerous situation $\widehat{S}''$ within the selected subset $\widehat{S}'$:

$$\widehat{s}_r\left(t\right) : \left(s^*\left(t\right) \in \widehat{S}'\right) \cup \left[r\left(t\right) = \max_{i=(1,|\widehat{S}'|)}\left(r_i\right)\right].$$

4. Cutting three most dangerous situations in the range $\widehat{S}''$ obtained in the previous step, and further ranging them using the possibilities of their appearance p_i:

$$\widehat{s}_p\left(t\right) : \left(s^*\left(t\right) \in \widehat{S}''\right) \cup \left[p\left(t\right) = \max_{i=(1,|\widehat{S}''|)}\left(p_i\right)\right],$$

we obtain the approximated match of the current situation to the class of the most possible of the most dangerous situations with respect to the observed symptoms x^*.

IMPLEMENTATION OF THE DIAGNOSIS

The presented algorithm has been implemented in the tactical forest fire fighting system (Fig. 8).

The case base implicitly represents diagnostic knowledge. As shown in Fig. 8, the well-known classes of the possible situations are stored in the case base and used to obtain the diagnosis using the proposed

Figure 8. Implementation of the hybrid diagnosis method

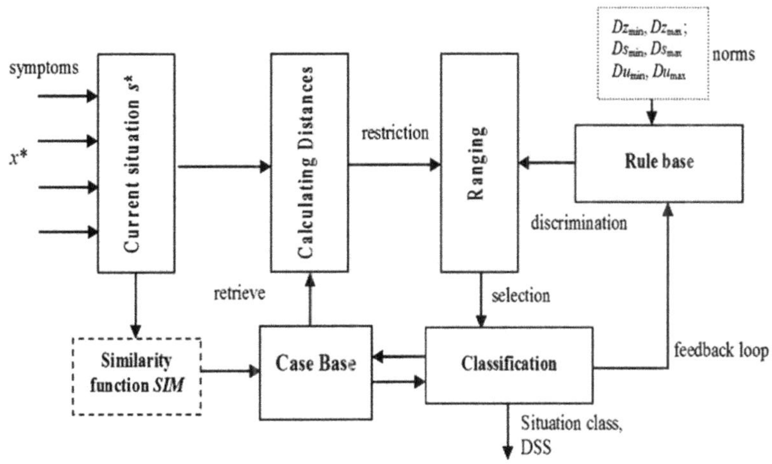

algorithm. The case structure includes a description of the current situation and corresponding vector of symptoms.

The rule base represents knowledge of the application domain and a time metric ξ_T on the time scale T such that $\forall t_i, t_j \in T \ \|t_i - t_j\|_T \rightarrow D_Z$, where $D_Z = \{D_{Z1}, ...D_{Zj}, ...D_{Zm}\}$ is the time limit set. Thus, the rule base contains a set of limit norms D_Z, which is necessary for discrimination of the subset of the coinciding situation $\widehat{S}'$. Using the set of limit norms, it is possible to classify the current situation with respect to its temporal features, e.g. normal, critical, danger, emergency, or catastrophic. When case base grows, general rules that cover wide subsets of the cases can be extracted and stored in the rule base. This reduces the size of the case base.

Using this approach, the authors obtain a hybrid method of knowledge representing, where the cases store the base points in the Cartesian space of symptoms while their neighborhoods are stored as a set of exceptions and constraints represented as rules in the rule base.

EXPERIMENT RESULTS

UAV onboard control system prototype was implemented using embedded microcontroller STM32F429 (180 MHz Cortex M4, 2Mb Flash/256Kb RAM internal, QSPI Flash N25Q512). Control center prototype used two servers HP ProLiant ML350 (Intel Xeon E5-2620, 8 cores up to 3 GHz). The system has been tested with the multi-UAV team (3 drones for the patrol missions, 1 helicopter for the confirmation mission, and 1 octocopter for the monitoring mission). It used the approximate fire-spreading model (Zharikova, Sherstjuk & Baranovskij, 2015) that allowed calculating the rate of fire spread and obtaining the other fire parameters necessary for decision-making as well as the fire front location on the terrain map. All UAVs' cameras were precisely calibrated.

The performance of the developed system was studied in the laboratory conditions. It depends mainly on the cell size. The simulation experiment has been conducted varying the cell size from 5 m to 25 m as well as varying a number of discretization level in image processing algorithms. Obtained results reflect that the bottleneck is the significant computational load of the UAV onboard control system. Further increase in processing power is limited by the available characteristics, and it requires to move some image processing to the ground command center.

However, the developed system has demonstrated the satisfactory probability of correct forest fire detection ($\approx$92%) in near-real-time conditions (processing time less than 2 min) with 10 m cell size and 16 levels of the color discretization. Using the developed system, the authors have achieved a good accuracy (up to 96%) of the fire spreading prediction for various terrains and weather conditions. Thus, the result of the experiment has shown that the developed system can provide required credibility and efficiency of fire prediction and response.

CONCLUSION

In the chapter, the multi-UAV-based monitoring system for tactical forest fire-fighting operations is presented. The functions and missions of the system and its architecture are considered, UAV types and sensors are described. Remote sensing algorithms and image processing techniques are presented. The data integration into the fire-spreading model in the real-time DSS for the forest fire response is proposed.

The intelligent diagnosis method for the tactical forest fire-fighting response system is also presented. The method is capable of identifying the problem by examining the observable symptoms and classifying the situation and makes it possible to respond appropriately based on the result of such classification. The proposed method is hybrid because it determines a solution for diagnosis task using the case-based approach but discriminates the solution using rule-based approach. The presented four-stage algorithm allows obtaining the class of the most possible of the most dangerous situations as diagnosing for the current situation with respect to the observed symptoms. Thus, the implementation of the proposed method in the multi-UAV-based automatic forest fire fighting system provides the required credibility and efficiency of the fire prediction and response.

It is shown that UAVs can be very helpful in fire-fighting response operations participating in fire monitoring and diagnosis tasks. The combination of the multi-UAV-based automatic monitoring system, hybrid diagnosis method, and remote sensing techniques with an approximate model of forest fire spreading can provide the required credibility and efficiency of fire prediction and response.

FUTURE RESEARCH DIRECTIONS

The direction of further research is related to the use of modern micro-ultraspectrometers on UAV board, which will allow to significantly improve the quality of fire recognition and the accuracy of its parameters estimation using more efficient processing of the observed information in certain narrow spectral bands. In addition, the effective use of fuzzy neural networks can significantly improve the efficiency of the tactical forest fire-fighting response system.

REFERENCES

Amarosicz, M., Psiuk, K., Rogala, T., & Rzydzik, S. (2016). Diagnostic Shell Expert Systems. *Diagnostica*, *17*(1), 33–40.

Ambrosia, V., & Zajkowski, T. (2015). Selection of Appropriate Class UAS/Sensors to Support Fire Monitoring: Experiences in the United States. In Handbook of Unmanned Aerial Vehicles (pp. 2723–2754). Springer Netherlands.

Anderson, H. (1982). *Predicting Wind Driven Wildland Fire Size and Shape*. Research Paper INT-305. USDA Forest Service, Intermountain Forest and Range Experiment Station.

Balakrishnan, K., & Honavar, V. (1998). Intelligent Diagnosis Systems. *Journal of Intelligent Systems*, *8*(3), 237–290.

Balasubramanian, A. (2017). *Digital elevation model (DEM) in GIS. Technical Report*. University of Mysore.

Bhagwat, A. (2015). Knowledge Based Service Diagnosis System. *International Journal of Computer Science and Technology*, *3*(5), 182–184.

Clarke, K., Brass, J., & Riggan, P. (1994). A Cellular Automaton Model of Wildfire Propagation and Extinction. *Photogrammetric Engineering and Remote Sensing*, *60*, 1355–1367.

Donner-Banzhoff, N., & Hertwig, R. (2014). Inductive foraging: Improving the diagnostic yield of primary care consultations. *The European Journal of General Practice*, *20*(1), 69–73. doi:10.3109/13814788.2013.805197 PMID:23815336

Finney, M. (1999). Mechanistic modeling of landscape fire patterns. In Spatial Modelling of Forest Landscape Change: Approaches and Applications (pp. 186–209). Cambridge University Press.

Goldberg, Y. (2016). A Primer on Neural Network Models for Natural Language Processing. *Journal of Artificial Intelligence Research*, *57*, 345–420. doi:10.1613/jair.4992

Gong, Y., Li, J., Zhou, Y., Li, Y., Chung, H. S., Shi, Y., & Zhang, J. (2016). Genetic Learning Particle Swarm Optimization. *IEEE Transactions on Cybernetics*, *46*(10), 2277–2290. doi:10.1109/TCYB.2015.2475174 PMID:26394440

Hoehndorf, R., & Queralt-Rosinach, N. (2017). Data Science and symbolic AI: Synergies, challenges and opportunities. *Data Science*, *1*(1-2), 27–38.

Karafyllidis, I., & Thanailakis, A. (1997). A model for predicting forest fire spreading using cellular automata. *Ecological Modelling*, *99*(1), 87–97. doi:10.1016/S0304-3800(96)01942-4

Kurien, J. & R-Moreno, M. (2008). Costs and Benefits of Model-based Diagnosis. *Proc. of the 2008 IEEE Aerospace Conference*, 1–14.

Lucas, P. (2001). Bayesian model-based diagnosis. *International Journal of Approximate Reasoning*, *27*(2), 99–119. doi:10.1016/S0888-613X(01)00036-6

Martinez, M. (2017). Knowledge Engineering for Intelligent Decision Support. *Proc. of the Twenty-Sixth International Joint Conference on Artificial Intelligence*, 5131–5135. 10.24963/ijcai.2017/736

Martínez de Dios, J., Arrue, B., Merino, L., Ollero, A., & Gómez-Rodríguez, F. (2007). Computer vision techniques for forest fire perception. *Image and Vision Computing*, *26*(4), 550–562. doi:10.1016/j.imavis.2007.07.002

Merino, L., Caballero, F., Martinez-de-Dios, J., Maza, I., & Ollero, A. (2012). An Unmanned Aircraft System for Automatic Forest Fire Monitoring and Measurement. *Journal of Intelligent & Robotic Systems*, *65*(1-4), 533–548. doi:10.100710846-011-9560-x

Merino, L., Martínez de Dios, J., & Ollero, A. (2015). Cooperative Unmanned Aerial Systems for Fire Detection, Monitoring, and Extinguishing. In *Handbook of Unmanned Aerial Vehicles* (pp. 2693–2722). Springer Netherlands. doi:10.1007/978-90-481-9707-1_74

Olsson, H., Egberth, M., Engberg, J., Fransson, J., & Pahlén, T. (2005). Current and Emerging Operational Uses of Remote Sensing in Swedish Forestry. In *Proc. of the 5th Annual Forest Inventory and Analysis Symposium* (pp. 39–46). US Forest Service.

Pawlak, Z., Jerzy, W., Slowinski, R., & Ziarko, W. (1995). Rough Sets. *Communications of the ACM*, *38*(11), 88–95. doi:10.1145/219717.219791

Qin, C., Wu, X., Jiang, J., & Zhu, A. (2016). Case-based knowledge formalization and reasoning method for digital terrain analysis – application to extracting drainage networks. *Hydrology and Earth System Sciences*, *20*(8), 3379–3392. doi:10.5194/hess-20-3379-2016

Rieck, K., Trinius, P., Willems, C., & Holz, T. (2011). Automatic analysis of malware behavior using machine learning. *Journal of Computer Security*, *19*(4), 639–668. doi:10.3233/JCS-2010-0410

Rothermel, R. (1972). *A Mathematical Model for Predicting Fire Spread in Wild Land Fuels*. Research Paper INT-115. USDA Forest Service, Intermountain Forest, and Range Experiment Station.

Shahbazi, M., Théau, J., & Ménard, P. (2014). Recent Applications of Unmanned Aerial Imagery in Natural Resource Management. *GIScience & Remote Sensing*, *51*(4), 339–365. doi:10.1080/15481603.2014.926650

Sharifi, F., Mirzaei, M., Zhang, Y., & Gordon, B. (2015). Cooperative Multi-vehicle Search and Coverage Problem in an Uncertain Environment. *Unmanned Systems*, *3*(1), 35–47. doi:10.1142/S230138501550003X

Sherstjuk, V., Zharikova, M., & Sokol, I. (2018). Forest Fire-Fighting Monitoring System Based on UAV team and Remote Sensing. *Proc. of the 38th Int. Conf. "Electronics and Nanotechnology"*, 99-106. 10.1109/ELNANO.2018.8477527

Symeonidisa, A., Chatzidimitriouc, K., Athanasiadisd, I., & Mitkas, P. (2007). Data mining for agent reasoning: A synergy for training intelligent agents. *Engineering Applications of Artificial Intelligence*, *20*(8), 1097–1111. doi:10.1016/j.engappai.2007.02.009

Tan, C. F., Wahidin, L. S., Khalil, S. N., Tamaldin, N., Hu, J., & Rauterberg, G. W. M. (2016). The Application of Expert System: A Review of Research and Applications. *Journal of Engineering and Applied Sciences (Asian Research Publishing Network)*, *11*(4), 2448–2453.

Verbert, K., De Schutter, B., & Babuska, R. (2015). Reasoning under uncertainty for knowledge-based fault diagnosis: A comparative study. *Proc. of the 9th IFAC Symposium on Fault Detection, Supervision and Safety of Technical Processes (SafeProcess 2015)*, 422–427. 10.1016/j.ifacol.2015.09.563

Wardihani, E., Ramdhani, M., Suharjono, A., Setyawan, T., Hidayat, S., Widodo, S., ... Saifullah, F. (2018). Real-time forest fire monitoring system using unmanned aerial vehicle. *Journal of Engineering Science and Technology*, *13*(6), 1587–1594.

Witayangkurn, A., Nagai, M., Honda, K., Dailey, M., & Shibasaki, R. (2011). Real-time monitoring system using unmanned aerial vehicle integrated with sensor observation service. *International Archives of the Photogrammetry, Remote Sensing and Spatial Information Sciences, XXXVIII-1/C22, ISPRS 2011 Workshop*.

Yuan, C., Liu, Z., & Zhang, Y. (2017). Aerial Images-Based Forest Fire Detection for Firefighting Using Optical Remote Sensing Techniques and Unmanned Aerial Vehicles. *Journal of Intelligent & Robotic Systems*, *88*(2-4), 635–654. doi:10.100710846-016-0464-7

Yuan, C., Zhang, Y., & Liu, Z. (2015). A Survey on Technologies for Automatic Forest Fire Monitoring, Detection and Fighting Using UAVs and Remote Sensing Techniques. *Canadian Journal of Forest Research*, *45*(7), 783–792. doi:10.1139/cjfr-2014-0347

Zadeh, L. (1965). Fuzzy sets. *Information and Control*, *8*(3), 338–353. doi:10.1016/S0019-9958(65)90241-X

Zharikova, M., & Sherstjuk, V. (2015). Development of the Model of Natural Emergencies in Decision Support System. *Eastern European Journal of Enterprise Technologies*, *4*(1), 62–69. doi:10.15587/1729-4061.2015.37801

Zharikova, M., & Sherstjuk, V. (2017). Threat assessment method for intelligent disaster decision support system. *Advances in Int. Systems and Computing*, *512*, 81–99.

Zharikova, M., Sherstjuk, V., & Baranovskij, N. (2015). The Plausible Wildfire Model in Geoinformation Decision Support System for Wildfire Response. Proc. of the 15th Int. Multidisciplinary Sc. Geoconference SGEM-2015, 2(3), 575–583.

Chapter 9
Preliminary Sizing and Performance Calculations of Unmanned Air Vehicles

Ali Dinc
https://orcid.org/0000-0002-3165-3421
American University of the Middle East, Kuwait

ABSTRACT

In this chapter, preliminary sizing and performance calculations of unmanned air vehicles (UAV), with relation to its propulsion system, are explained. Starting with the mission profile of the air vehicle, which is one of the important design drivers, the UAV is sized for the requirements. The requirements in design consist of a set of target values (e.g., payload amount to be carried, range, endurance time, cost, etc.). Additionally, the parameters within mission profile such as cruising altitude and speed of aircraft affect engine type, power level required, fuel quantity, and therefore general dimensions and the gross weight of the aircraft. It is often an iterative process to size the air vehicle and engine together. UAV is designed in a loop of calculations in which sizing, flight performance, and engine performance are done for each phase of the mission or flight profile to satisfy the overall design mission and requirements.

INTRODUCTION

UAVs are air vehicles, similar to regular aircraft, helicopters etc. except that they are unmanned or uninhabited. Likewise, UAVs fly and operate in an airplane environment, potentially in both segregated and non-segregated airspace. UAV design, design tools, and methodology are also similar to those of manned aircraft, however, there are some differences due to the different mission requirements. The difference in mission requirements such as flight altitude and speed during loiter, high maneuverability during combat (higher than a human pilot can withstand), payload type and characteristics, power and thrust needs, low observability, interoperability, swarming technologies etc. make UAV designs special and require new technologies.

DOI: 10.4018/978-1-5225-7709-6.ch009

BACKGROUND

In literature, there are numerous studies regarding the design of unmanned aerial vehicles. Austin (2010) presents a wide ranging knowledge of all aspects of UAVs. Gundlach (2014) covers a comprehensive range of all elements of unmanned aircraft systems, architectural options, and design drivers across diverse system classes. Fahlstrom and Gleason (2012) explains designers many aspects of the UAV systems including subsystem design choices, system-integration issues and lessons learned. In addition to many industry projects, some universities also perform projects on UAV design, manufacturing and fly to support undergraduate and graduate studies (senior design projects, master`s and Ph.D. dissertations etc.). Some university explained their studies and capabilities on UAVs as mentioned by Zbrutsky et al. (2013), Sanchez et al. (2018), Cinar et al. (2016). Some associations (e.g. AUVSI) organize student unmanned aerial systems competitions. The competitions typically require students to design, integrate, report on, and demonstrate a UAS capable of autonomous flight and navigation, remote sensing via onboard payload sensors, and execution of a specific set of tasks.

Similarly, some examples of parametric UAV conceptual design tools including engine performance models can be found in literature. Rapid Air System Concept Exploration (RASCE) is one example of design tools which was originally developed as an educational tool to support undergraduate student exploration (Chaput, 2010). It has generic propulsion models to calculate thrust and fuel flow as a function of power setting, speed and altitude for turboprop/turbofans (defined by bypass ratio, specific thrust, fuel-to-air ratio and thrust-to-weight) or internal combustion engine or electric motor with propeller installations.

Preliminary sizing and performance calculations of unmanned air vehicles are done based on many inputs and requirements. As a part of the technical requirements, we can start with the mission profile of UAV. The flight mission profile is one of the most important design inputs in aircraft design. The flight mission profile is a graphical representation of a scenario in which an aircraft needs to perform prescribed operations and includes parameters such as payload weight, range, speed, altitude etc. Mission requirements vary according to the type of aircraft. Flight mission profile varies according to the air vehicle category. When civil transport aircraft, military aircraft and unmanned air vehicles are classified into sub-categories, it is observed that each has a different flight mission profile. In other words, the flight mission profile is a schematic of the flight phases and the detailed descriptions of the activities performed on the flight. Therefore, it is very important in the conceptual design of the aircraft/UAV. While UAV/aircraft is sized per the technical specifications and requirements (payload, range, target cost, etc.), the parameters such as cruise altitude and speed within the flight mission profile affect the engine type/size, power level, the required fuel weight/storage volume and therefore the overall size and total weight of the aircraft.

Figure 1 shows a basic mission profile of a civil transport aircraft and Figure 2 depicts a typical reconnaissance UAV flight mission profile. Although both mission profiles look similar, at a quick glance, one can conclude that a reconnaissance UAV spends most flight time during loitering (reconnaissance/surveillance) whereas the transport aircraft in cruising.

Figure 1. Typical civil transport aircraft flight mission profile

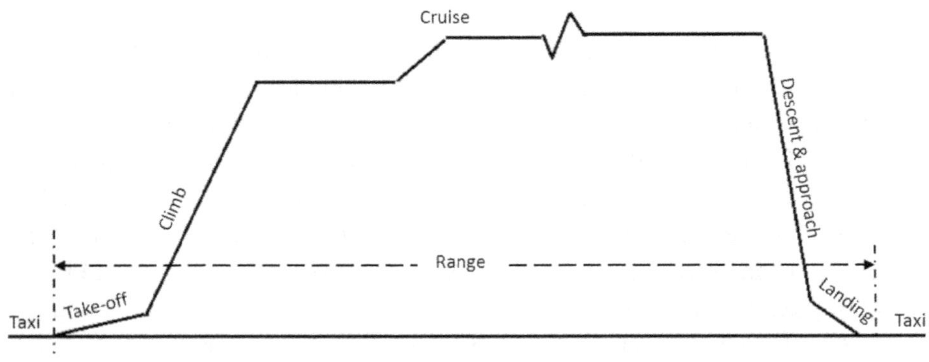

Figure 2. Typical reconnaissance UAV flight mission profile
(Adapted from [Federation of American Scientists, (1996)])

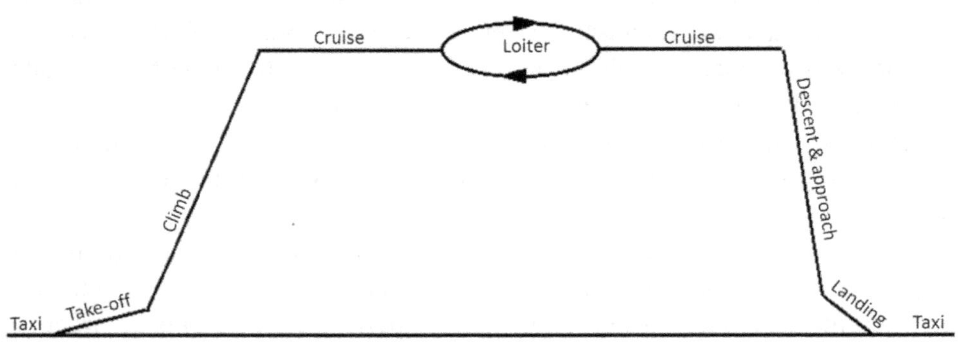

MAIN FOCUS OF THE CHAPTER

As a basis of this chapter, UAV and turbofan engine modeling as a genuine computer code was developed in a previous study conducted by Dinc (2010) according to the parametric equations for a simplified geometry UAV as depicted in Figure 5. Later, another code was developed for sizing of a turboprop engine powered UAV which has computational models for both airplane and turboprop engine (Dinc, 2015a). The aim of these two codes was to match the sizing and performance data of Global Hawk (a turbofan engine powered unmanned air vehicle) developed by Northrop Grumman (2010) and turboprop UAV `Predator-B` (General Atomics Aeronautical Systems, 2014) available in the literature. Therefore, the methodology and the codes developed were validated with literature data. As a practical example of the methodology explained in this chapter, calculation results for a sizing of a turboprop-powered UAV with its propulsion system, similar to Predator-B (General Atomics Aeronautical Systems, 2014) and TPE331−10 turboprop engine (Honeywell, 2014) are given in Appendix 1. Deviations from literature values on UAV weight & size calculations and engine performance results are about ±1.5%.

Based on the initial requirements of a UAV, to accomplish a target mission, designers do conceptual studies. Initial concepts or solutions are more than one to show as many alternatives as possible. For each concept UAV, overall geometry, wing, tail, propulsion type etc. can be different. Then the next step is to perform preliminary sizing and performance calculations for each different concept, in order to produce comparable data to make a viable selection at the end of the conceptual design phase. For example, Figure 3 shows two UAV concepts with different characteristics and propulsion system. In addition to geometric differences in wing, tail, fuselage UAV-(a) uses turboprop engine and UAV-(b) uses turbofan engine similar to Predator B and Global Hawk unmanned air vehicles. In order to make a selection between the two concepts, all the sizing and performance calculations should be done for each of them. Since the engine types are different, fuel consumptions will be different and to accomplish 24 hours loitering in a reconnaissance mission, total fuel weight will lead to a difference between two UAVs in size and weight. There will be other variances in flight performance parameters such as maximum cruise speed, maximum altitude etc. due to the difference in engine type.

After initial concepts are laid out (similar to Figure 3) illustratively, then it is time for starting calculations for sizing, weight and flight performance. Figure 4 is a flowchart of general steps for sizing and flight performance calculations of UAV. As it can be seen in the flowchart, calculations will be done iteratively, since only sizing will not be enough to check whether the UAV and its engine can satisfy take-off, climb, cruise, loiter etc. requirements at given speeds and altitudes.

The flowchart for UAV sizing methodology given in Figure 4 can be used in a computer code as a step by step approach to avoid lengthy and cumbersome hand calculations. In fact, a similar methodology has been used in many preliminary sizing and design tools. These design tools or sizing codes are generated to speed up the conceptual design phase using aircraft design methodology.

Inputs for Sizing

At the beginning of UAV sizing and performance calculations, a set of inputs derived from requirements is needed. Table 1 (in Appendix 1) shows a typical list of input parameters to be used in UAV sizing and performance. Those parameters are related with mission profile (altitude and speed etc.), overall geometry (wing, fuselage, tail dimensional ratios etc.), weight to carry (payload, fuel etc.), weight fractions (avionics, landing gear etc.), unit weights per area (wing, tail etc.), densities (fuel, avionics, engine etc.), aerodynamics (max lift, skin friction coefficients and stall speed ratios etc.) and engine or propulsion system parameters. In Table 2 (in Appendix 1), the typical values or range for some of the design

Figure 3. Different concept UAVs

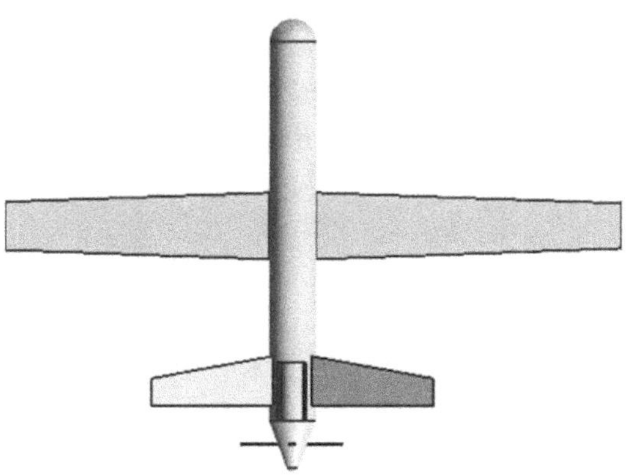

(a)

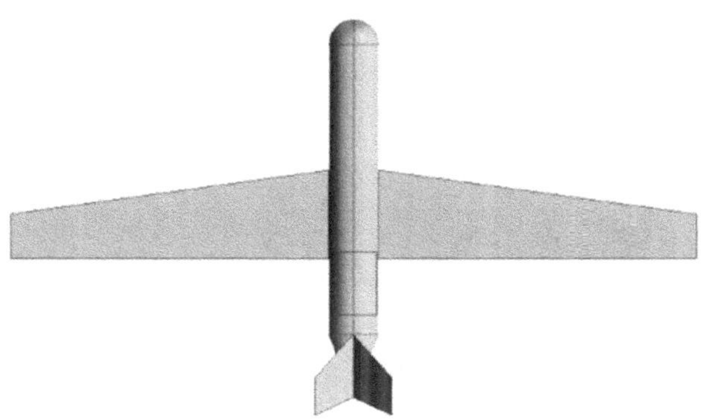

(b)

Figure 4. Flowchart of the method used in UAV sizing

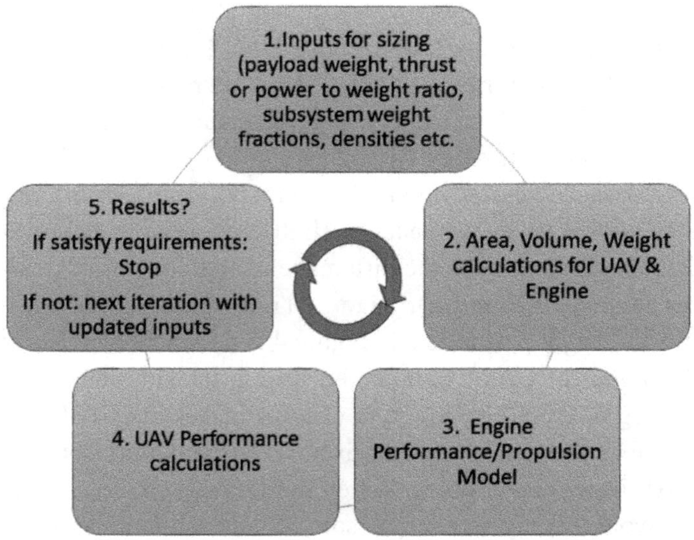

parameters are given. These values vary depending on the mission profile and therefore class/category of the UAV, type of engine and the technology level of the design company.

Those input parameters can be from many sources such as the customer requirement set (payload, altitude and speed etc.) and from the design choices or initial estimates (wing, fuselage, tail dimensional ratios etc.) and from empirical values from benchmarking or data from past experiences/designs (weight fractions, densities, aerodynamic coefficients etc.). Each input value is to be carefully evaluated and referenced to a source of information as a part of requirements management.

Area, Volume Weight Calculations

In sizing the UAVs, general aircraft weight equations can be used by removing parts related to crew and passengers (*W*crew = 0, Wpass = 0).

$$W_0 = W_e + W_{PL} + W_f \tag{1}$$

Therefore in Eq (1), W0, maximum aircraft gross weight can be calculated by summing up We aircraft empty weight, WPL total payload weight, Wf total fuel weight. Aircraft empty weight can be expressed as follows:

$$W_e = W_{af} + W_{lg} + W_{sys} + W_{eng} + W_{misc} \tag{2}$$

In Eq (2), Waf is airframe, Wlg is landing gear, Wsys is systems and Weng is engine, Wmisc miscellaneous other equipment weight respectively.

$$W_{af} = W_{fus} + W_{nac} + W_{wing} + W_{h.tail} + W_{v.tail} \tag{3}$$

$$W_{lg} = f_{lg} W_0 \tag{4}$$

$$W_{sys} = f_{sys} W_0 \tag{5}$$

In Eq (3), airframe weight Waf is the summation of fuselage, nacelle, wing and tail weights. For Weng, engine weight, real engine weight data or a statistical equation derived from literature, or a percentage of W0 can be used.

$$W_{fus} = u_{fus} S_{fus} \tag{6}$$

$$W_{nac} = u_{fus} S_{nac} \tag{7}$$

$$W_{wing} = u_{wing} S \tag{8}$$

$$W_{h.tail} = u_{htail} S \tag{9}$$

$$W_{v.tail} = u_{vtail} S \tag{10}$$

For Eqs (6-10), unit weights will be empirical inputs, whereas areas of wing, fuselage, nacelle and tails will come from geometrical calculations that will be discussed next. For unit weight values, benchmark data can be used from open literature if there is no past design experience.

In sizing and performance calculations conceptual geometries will be needed for evaluation. Figure 5 shows a sample simplified geometry UAV. An air vehicle fuselage can be defined by the most basic geometrical shapes such as cylinders, spheres, cones for simplification of geometry. In Figure 5, the UAV body consists of a half sphere (nose), cylinder (mid fuselage) and a beveled cone (rear fuselage), respectively. The total volume and area formulation of such a UAV can be done easily with simplified geometries.

In Figure 5, fuselage nose part consists of half a sphere. Therefore, the volume and area equations:

$$V_{f.nose} = \frac{1}{12}\pi \varnothing_{fus}^3 \tag{11}$$

$$S_{f.nose} = \frac{1}{2}\pi \varnothing_{fus}^2 \tag{12}$$

In the above two equations, $\varnothing_{fus}$ represents the fuselage diameter. The mid-fuselage consists of a cylinder. Respectively, the volume and area equations:

Figure 5. A simplified UAV conceptual geometry

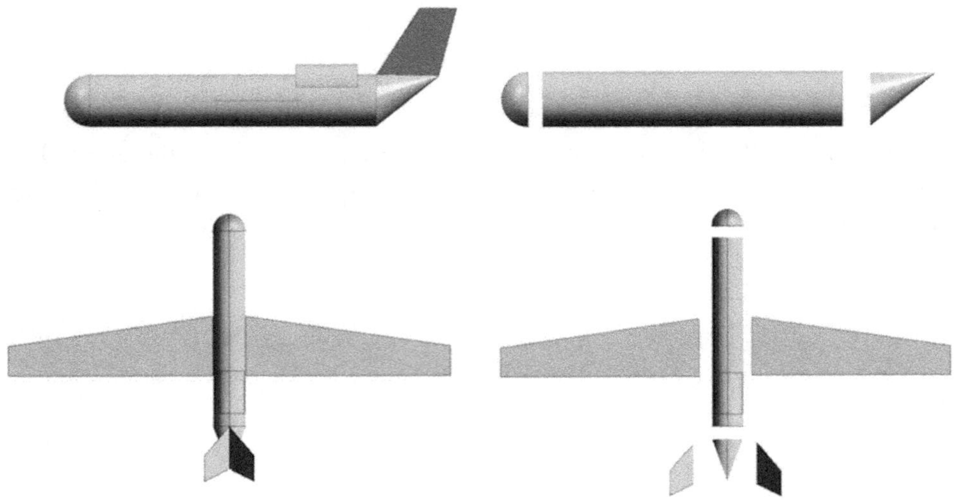

$$V_{f.mid} = \frac{1}{4}\pi \varnothing_{fus}^2 L_{fus.cyl} \tag{13}$$

$$S_{f.mid} = \pi \varnothing_{fus} L_{fus.cyl} \tag{14}$$

In Eqs (13-14), Lfus.cyl represents the length of the cylindrical part of the fuselage. In the area equation, only the side area of the cylinder is taken into account, the circular (top and bottom) sections are not included in the wet area of UAV.

The rear part of the body consists of a cone. The volume and area equations for a cone:

$$V_{f.rear} = \frac{1}{12}\pi \varnothing_{fus}^2 L_{fus.b} \tag{15}$$

$$S_{f.rear} = \frac{1}{2}\pi \varnothing_{fus} L_{fus.b}\sqrt{L_{fus.b}^2 + \varnothing_{fus}^2} \tag{16}$$

In Eqs (15-16), Lfus.b represents the length of the conical part of the fuselage. In the area equation, only the side surface is taken into account as a contributor to the wet area.

The total fuselage volume and wet area:

$$V_{fus} = \frac{1}{12}\pi \varnothing_{fus}^3 + \frac{1}{4}\pi \varnothing_{fus}^2 L_{fus.cyl} + \frac{1}{12}\pi \varnothing_{fus}^2 L_{fus.b} \tag{17}$$

$$S_{fus} = \frac{1}{2}\pi\varnothing_{fus}^{2} + \pi\varnothing_{fus}L_{fus.cyl} + \frac{1}{2}\pi\varnothing_{fus}L_{fus.b}\sqrt{L_{fus.b}^{2} + \varnothing_{fus}^{2}} \tag{18}$$

The engine is located inside the engine nacelle and can be assumed to be a cylinder. The volume and area formulas for the engine nacelle are as follows:

$$V_{nac} = \frac{1}{4}\pi\varnothing_{nac}^{2}L_{nac} \tag{19}$$

$$S_{nac} = \pi\varnothing_{nac}L_{nac} \tag{20}$$

In the above two equations, Lnac represents the length of the engine nacelle; $\varnothing_{nac}$ represents the diameter of the nacelle. If the engine nacelle is completely inside the fuselage, there will be no additional wet area (*Snac* = 0). However, if engine nacelle is located coaxially outside of the fuselage (e.g. Global Hawk) this additional area should be calculated and added to the total wet area of UAV (e.g. 50% of *Snac*).

If there are external fuel tanks or external hardware carried under the wing, the geometry of them can be assumed as a combination of half sphere, cylinder and cone (similar to fuselage) initially. The volume of fuel tanks can be calculated using weight and density of fuel used.

In order to size a typical wing (Figure 6), some dimensions like wingspan, chord length at root and tip need to be estimated using input parameters. In wing sizing, wing loading (WL) is an important design input. It is defined as the ratio of the total weight of the aircraft to the wing area as given in Eq(21). Wing loading value is normally a constant value input and it is used to calculate wing area by using Eq(22). Wing loading as constant input will automatically change the wing area if UAV weight changes in iterative calculations.

$$WL = \frac{W_0}{S} \tag{21}$$

$$S = \frac{W_0}{WL} \tag{22}$$

The aspect ratio (AR) is another parameter in the input set. It is equal to the square of wingspan (b) divided by wing area (S) as given in Eq(23). When aspect ratio (AR) and wing loading (WL) are known as input, wingspan (b) can be calculated with Eq(24) and using values of W0 in design iteration loops.

$$AR = \frac{b^2}{S} \tag{23}$$

$$b = \sqrt{\frac{W_0}{WL} AR} \tag{24}$$

Sizing the wing will require estimations on chord length at root and at tip (c_r, c_t) using taper ratio (λ) as an input:

$$c_r = \frac{2S}{b\left(1+\right)} \tag{25}$$

$$c_t = c_r \tag{26}$$

where,

$$\lambda = (c_t / c_r) \tag{27}$$

The next step is to calculate the wing volume with Eq(28), using a truncated pyramid volume formula. Another parameter introduced here is t2/t1 which is the thickness ratio of equal area and an equal chord length of an airfoil and a rectangle to simplify volume calculation (Figure 7). t2/t1 parameter can be calculated easily if the airfoil area and chord length are known. As an initial guess 0.6 is acceptable since the average wing thickness is approximately 60% of the maximum wing thickness for most common wing types. For a more accurate value of t2/t1 parameter, airfoil characteristics should be known. A portion of the wing volume is used for fuel storage, roughly 45-50% is acceptable for starters. Therefore, Vfuel /Vwing ratio can be taken as 0.45-0.50 in the input set (Table 1 in Appendix 1).

Figure 6. Truncated pyramid representation for wing volume calculation

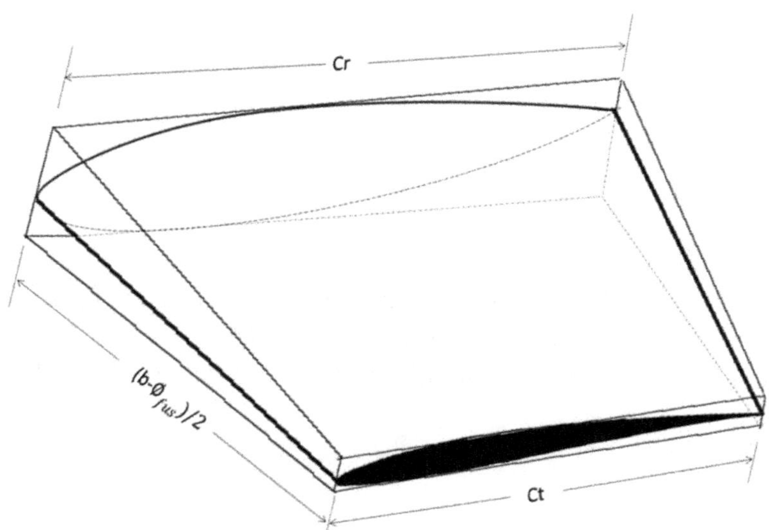

Figure 7. Equal area airfoil and rectangle with thickness ratio t_2/t_1

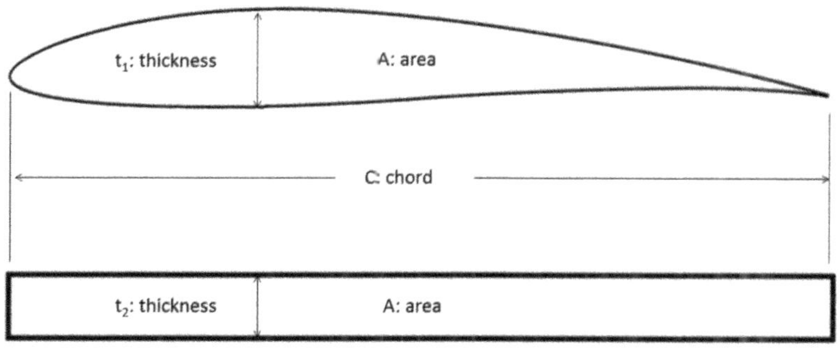

$$V_{wing} = \frac{t_2}{3t_1}\frac{c_r^3 - c_t^3}{c_r - c_t}\left(b - \varnothing_{fus}\right) \tag{28}$$

Until this point, area and volume formulations are given for main airframe parts (fuselage, wing, tail) of UAV. Area and volume values can be used to calculate the weights of these components using Eqs (6-10). There are also some empirical weight formulas available in literature such as Raymer (1999) which can be used alternatively.

Engine Performance Model

UAVs use many different types of propulsion systems. The UAV propulsion systems include electrical, internal combustion engine, gas turbine (turbojet, turbofan, turboprop, turboshaft etc.), rocket engines etc. Aircraft and UAV conceptual design tools need engine or propulsion performance models or data. The flight performance of UAV cannot be predicted accurately, without including detailed engine performance data or models. For this purpose, UAV designers use a data set of an existing engine or a detailed generic engine model for propulsion calculations (a rubber engine model).

A specific engine performance data set can be obtained from an engine company. Typically, this data includes thrust or power output and fuel consumption values of the engine at different altitude and speed conditions. For any given altitude and speed, engine thrust should be at least equal to the calculated aerodynamic drag force of UAV to be able to fly. If no detailed engine performance data is available from a manufacturer, then an aero-thermodynamic performance model can be established for a type of engine. As an example, Table 1 (in Appendix 1) has a section for engine parameters as input set. For propulsion system weight, statistical data can be used which are obtained from engine companies and literature. Weight data can be in the form of an equation (thrust or power versus engine weight) or a specific weight data about an engine. Additional weight data should be included if there are propeller and gearbox etc. in the propulsion system.

Figure 8 can be a general guide for selecting a propulsion system or engine type for a UAV. In general, faster speeds and higher altitude requirements in mission profile will result in the selection of higher fuel consumption engines. Higher fuel consumption will lead to bigger fuel tanks or storage and more fuel weight. This will change the size of the UAV, if there is a requirement for a certain time (12 or 24 hours

Figure 8. Propulsion systems fuel consumption comparison

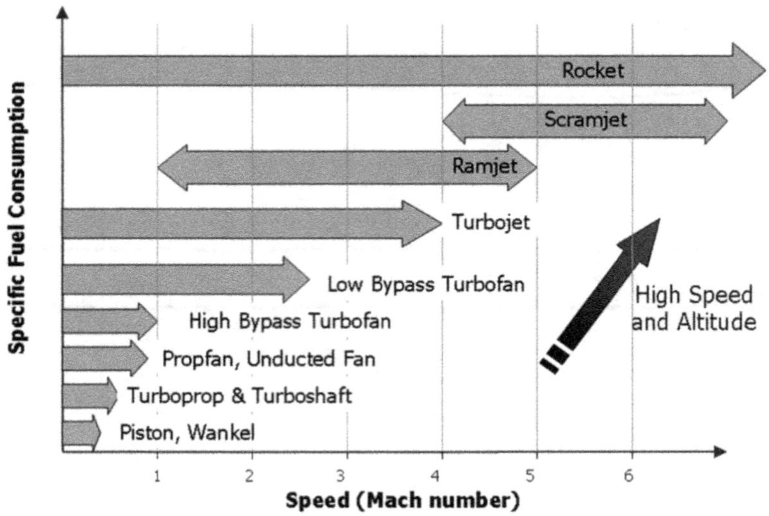

etc.) of endurance/loitering over a target area. Alternatively, higher fuel consumption will decrease the endurance/loitering time for a fixed amount of fuel. For electrical propulsion, instead of fuel amount, battery, fuel cell etc. capacity and weight should be considered.

Electric Propulsion

Today, aviation is responsible for 2% of global man-made CO2 emissions and about 12% of the CO2 emissions from all transportation sources (International Civil Aviation Organization, 2007). Additionally, there are other pollutants in exhaust gases such as nitrogen oxides (NOx), sulfur dioxide (SO2) and carbon monoxide. Therefore, there are a lot of efforts to reduce aviation related greenhouse gas emissions. One of the solutions is to use unconventional propulsion technologies such as hybrid electric and electric systems. Technology development efforts continue on reduction of emissions and towards all electric aircraft. In fact, numerous electric aircraft prototypes and demonstrators have been designed and flown in the last decades. Similarly, UAVs have been using these technologies. In small UAV applications, typically electric motors connected to lithium polymer batteries as energy source, are used. Alternatively, solar panels and fuel cells can be used to provide electricity. Eventually electric motors convert electrical energy to mechanical energy to drive a fan or propeller in order to generate thrust.

UAV Flight Performance

The UAV's performance is based on its design characteristics. The performance estimate of a UAV requires calculations of certain values such as lift and drag forces at different speeds and altitude values, climbing rate, maximum speed, the maximum range and endurance for a given amount of fuel etc. Although intensive test data and accumulated experience are needed for accurate estimates of performance in the later stages of design, the equations used in this study will be simple, basic and are sufficient for the preliminary stage.

According to the sample mission profile given in Figure 2, range, endurance (loiter time) and flight times are given below.

$$R_F = R_{CL} + R_{CR} + R_D \quad (29)$$

$$t_F = t_{CL} + t_{CR} + t_D \quad (30)$$

$$t_B = t_F + t_{GR} \quad (31)$$

$$R = \frac{V}{C}\frac{L}{D} ln \frac{W_{i-1}}{W_i} \quad (32)$$

$$R = \frac{550\eta_p}{C}\frac{L}{D} ln \frac{W_{i-1}}{W_i} \quad (33)$$

$$E = \frac{L/D}{C} ln \frac{W_{i-1}}{W_i} \quad (34)$$

Eq(32) is the modified Breguet range formula for constant velocity and altitude for a jet aircraft. Eq(33) is the range for propeller aircraft. In the above equations, definitions are as follows: R, range; RCL, climb distance (horizontal); RCR, cruising distance or range; RD, the descent distance (horizontal); tF, total flight time; tCL, climbing time; tCR, cruise time; tD, duration of descent; tB, block flight time; tGR, ground maneuvers time (taxi, takeoff and landing); V, the speed of the plane, C, specific fuel consumption; L, lift force; D, drag force; W, aircraft weight and ηp is propeller efficiency.

For the aerodynamic lift and drag force calculation, the below generic formulas are used:

$$L = \frac{1}{2}\rho V^2 c_l S \quad (35)$$

$$D = \frac{1}{2}\rho V^2 c_d S \quad (36)$$

Lift force formula consists of air density (ρ), flight speed (V), lift coefficient (c_l) and wing area (S). It is normally assumed that lift force equals to the instantaneous weight of UAV, during cruise and loiter. At a given speed and altitude of UAV, lift coefficient can be calculated with the below formula:

$$c_l = \frac{2W}{\rho V^2 S} \tag{37}$$

Drag force calculation is similar to lift formula, except that it has drag coefficient. Drag coefficient (c_d) has two parts, namely parasite (zero lift) drag (c_{d_0}) and lift-induced drag:

$$c_d = c_{d_0} + \frac{c_l^2}{\pi ARe} \tag{38}$$

$$c_{d_0} = c_{f_e} \frac{S_{wet}}{S} \tag{39}$$

In above equations, c_{d_0}, parasite (zero lift) drag coefficient; c_{f_e}, equivalent skin friction coefficient; Swet, total wet area of UAV; S, wing area; c_d, drag coefficient; AR, aspect ratio; e, Oswold efficiency. Typical equivalent skin friction coefficients are given in Table 4 (Appendix 1).

After calculating lift and drag forces, it is important to estimate the maximum lift-to-drag ratio (L/D) to achieve maximum endurance per Eq(34).

$$\left(\frac{L}{D}\right)_{max} = \frac{1}{2\sqrt{\frac{c_{d_0}}{\pi ARe}}} = \frac{1}{2\sqrt{c_{d_0} k}} \tag{40}$$

$$c_{l_{(L/D)max}} = \sqrt{\pi ARec_{d_0}} = \sqrt{\frac{c_{d_0}}{k}} \tag{41}$$

$$k = \frac{1}{\pi ARe} \tag{42}$$

$$q = \frac{1}{2}\rho V^2 \tag{43}$$

$$q = \frac{1}{2}\gamma PM^2 \tag{44}$$

Figure 9. Typical mission profile for a reconnaissance UAV

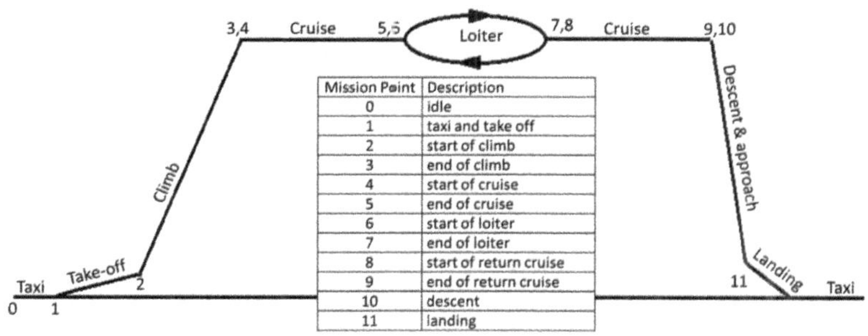

$$M = \sqrt{\frac{2q}{\gamma P}} \tag{45}$$

In above equations, q, dynamic pressure; γ, specific heat ratio; ρ, air density; V, airplane speed; P, static air pressure; M, Mach number.

Weight Calculation of UAV at Flight Mission Points

This section discusses the weight change of UAV due to fuel consumption throughout the flight. Figure 9 shows a typical mission profile for a reconnaissance UAV. Engines burn fuel from the start-up until the end. The amount of fuel which is consumed at different stages of flight determines the range, endurance of UAV as well as the weight change at each point of flight.

Figure 10 shows a representative weight change trend of a reconnaissance UAV (e.g. Predator-B) during a typical surveillance mission. UAV starts with 100% maximum take-off weight at point 0. As the taxi, take-off, climb, cruise etc. continue UAV weight decreases due to fuel consumed continuously.

Figure 10. Typical weight change of UAV during surveillance mission

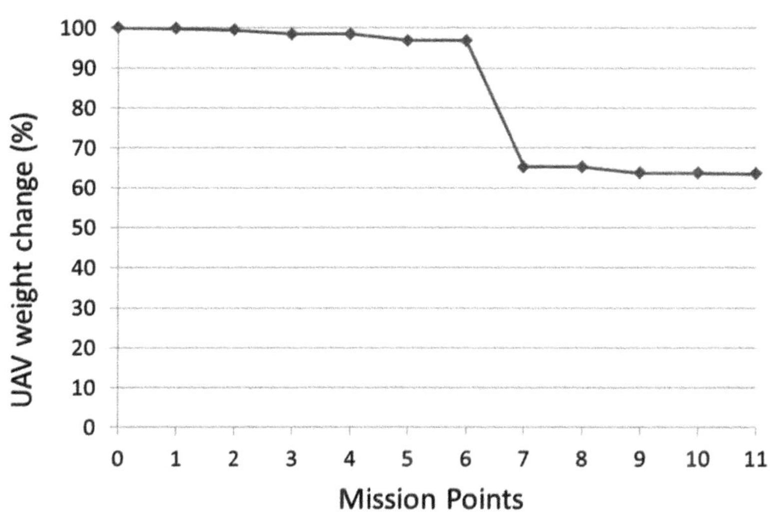

It is remarkable that the highest weight decrease occurs between mission points 6-7, due to 24+ hours loitering (flying at loiter speed and altitude for surveillance). Fuel weight (internal and external tanks) comprises about 37% of total weight of Predator-B UAV. Because of that, UAV weight decreases to 63% as depicted in Figure 10.

Weight change of UAV must be calculated in below steps to assess flight performance, lift-weight, drag-thrust etc. balance in different phases of flight.

Taxi (Point 0-1):

$$W_1 = W_0 - w_f t_1 \tag{46}$$

W_1, is the weight of UAV after startup and taxi; W_0, maximum weight of UAV; w_f, fuel flow; t_1, time spent for taxi.

Take-off (Point 1-2):

$$W_2 = W_1 - w_f t_2 \tag{47}$$

W_2, weight of UAV after take-off; w_f, fuel flow; t_2, take-off duration.

Climb (Point 2-3):

$$W_3 = W_2 - w_f t_3 \tag{48}$$

$$t_3 = \frac{h}{v} \tag{49}$$

W_3, weight of UAV after climb; w_f, fuel flow; t_3, climb duration; h, climb altitude; v, vertical speed or climb rate.

Cruise (Point 4-5):

$$W_5 = W_4 e^{\frac{-R C_{ave}}{V\left(\frac{L}{D}\right)_{ave}}} \tag{50}$$

W_5, weight of UAV at the end of cruise; R, cruise distance; V, speed of UAV; C_{ave}, average specific fuel consumption during cruise; $(L/D)_{ave}$, average lift to drag ratio during cruise; W_4, weight of UAV at the start of cruise.

Loiter (Point 6-7):

$$W_6 = W_7 e^{\frac{E C_{ave}}{\left(\frac{L}{D}\right)_{ave}}} \quad (51)$$

W_7, weight of UAV at the end of loiter; E, endurance time; C_{ave}, average specific fuel consumption during loiter; $(L/D)_{ave}$, average lift to drag ratio during loiter; W_6, weight of UAV at the start of loiter.

Return Cruise (Point 8-9):

$$W_8 = W_9 e^{\frac{R C_{ave}}{V\left(\frac{L}{D}\right)_{ave}}} \quad (52)$$

W_8, weight of UAV at the start of return cruise; R, return cruise distance; V, speed of UAV; C_{ave}, average specific fuel consumption during return cruise; $(L/D)_{ave}$, average lift to drag ratio during return cruise; W_9, weight of UAV at the end of return cruise.

Landing Loiter (Point 10):

$$W_{10} = W_{11} + w_f t_{10} \quad (53)$$

$$W_9 = W_{10} \quad (54)$$

W_{10}, UAV weight at the start of landing loiter; W_{11}, UAV weight at the end of landing loiter; t_{10}, landing loiter time.

Landing (Point 11):

UAV weight at landing can be calculated by subtracting the usable fuel weight from initial maximum weight at take-off:

$$W_{11} = W_0 - W_{fuel} + W_{fuel,reserve} \quad (55)$$

W_{11}, UAV weight at landing; W_0, UAV maximum gross weight; W_{fuel}, total fuel weight; $W_{fuel,reserve}$, reserve fuel weight.

At some mission points weight there is no weight change:

$$W_7 = W_8 \quad (56)$$

$$W_4 = W_3 \tag{57}$$

$$W_6 = W_5 \tag{58}$$

SOLUTIONS AND RECOMMENDATIONS

First solutions with this methodology can be done for a UAV with available data in literature. This way, the calculations or code will be validated for a typical UAV already designed and tested since there are a lot of empirical data in input set. Then, calculations can be done for another UAV which is new and different than the validation case UAV.

After all calculations are done at all mission points, it is necessary to check whether the results are satisfactory or not. For example, if the loiter time calculated is achievable with the given fuel amount or calculated drag is not more than thrust available etc. Additionally, UAV size and propulsion system output values should be reviewed. Is UAV too heavy or too big for the mission? Is engine power within the available range in the market (if rubber engine model used)? Depending on the acceptability of results, more iterations will be done until results satisfy the requirements.

FUTURE RESEARCH DIRECTIONS

As future research directions and subjects, the methodology described in this chapter can be improved in many ways. Additional and more complex mathematical modeling can be added to get more accurate results or results on uncovered areas. For example, a more complex aerodynamic model would improve the flight performance results, similarly, introducing models on structural calculations, emissions etc. would make the methodology more multidisciplinary. Any refinement with the real design data (weight estimation of fuselage, wing, tail etc.) by the manufacturing companies' knowledge and experience would further improve the accuracy of the results. Different propulsion models including electric propulsion can be added to the methodology to cover electrically powered UAVs. Adding optimization algorithms (genetic algorithm etc.) to the methodology such as maximization of loiter time or minimum fuel consumption will be useful and efficient for preliminary design efforts.

CONCLUSION

In this chapter, a methodology for preliminary sizing and performance calculations of unmanned aerial vehicles which have become more popular and useful systems in both military and civil arena has been explained. This methodology has been used in generating a computer code (Dinc, 2015a) to automate the calculations and create a fast design tool. As a case study, sizing of a turboprop unmanned air vehicle and its propulsion system is given in Appendix 1 (summary of input and output parameters). Therefore, students can use this chapter, appendices and references as a guide to create their own computer codes or design tools for educational purposes.

REFERENCES

Austin, R. (2010). *Unmanned Aircraft Systems*. West Sussex, UK: Wiley. doi:10.1002/9780470664797

Chaput, J. A. (2004). *Lecture Notes, Conceptual Design of UAV Systems*. University of Texas at Austin.

Chaput, J. A. (2010). Rapid Air System Concept Exploration - A Parametric Physics Based System Engineering Design Model. In *AIAA Aviation Technology*. Integration, and Operations Conference. doi:10.2514/6.2010-9301

Cinar, G., Emeneth, M., & Mavris, D. N. (2016). A Methodology for Sizing and Analysis of Electric Propulsion Subsystems for Unmanned Aerial Vehicles. *54th AIAA Aerospace Sciences Meeting*. 10.2514/6.2016-0216

Dinc, A. (2010). *Optimization with elitist genetic algorithm method in preliminary design of unmanned air vehicle & its propulsion system simultaneously* (PhD thesis). Graduate School of Sciences, Anadolu University. (in Turkish)

Dinc, A. (2015a). Sizing of a Turboprop Unmanned Air Vehicle and its Propulsion System. *Journal of Thermal Science and Technology*, *35*(2), 53–62.

Fahlstrom, P. G., & Gleason, T. J. (2012). *Introduction to UAV Systems* (4th ed.). Wiley.

Federation of American Scientists. (1996). *Air Combat Command Concept of Operations for Endurance Unmanned Aerial Vehicles*. Retrieved from http://www.fas.org/irp/doddir/usaf/conops_uav/part03.htm

General Atomics Aeronautical Systems. (2014). *Product Brochures*. Retrieved from http://www.ga-asi.com/products/aircraft/pdf/Predator_B.pdf

Gundlach, J. (2014). *Designing Unmanned Aircraft Systems: A Comprehensive Approach* (2nd ed.). AIAA. doi:10.2514/4.102615

Honeywell. (2014). *Product Brochures*. Retrieved from http://www51.honeywell.com/aero/common/documents/myaerospacecatalog-documents/BA_brochures-documents/TPE331-10_PredatorB_0292-000.pdf

International Civil Aviation Organization. (2007). ICAO. *Environment Reporter, 2007*. Retrieved from http://www.icao.int/environmental-protection/Documents/Env_Report_07.pdf

Kurzke, J. (2007). Design and Off-Design Performance of Gas Turbines. GasTurb 11 User Manual, Munich, Germany.

Northrop Grumman. (2010). *Global Hawk Product Brochure*. Retrieved from: http://www.northropgrumman.com/Capabilities/RQ4Block10GlobalHawk/Documents/GHMD-New-Brochure.pdf

Raymer, D. P. (1999). *Aircraft Design: A conceptual approach*. Washington, DC: AIAA.

Sanchez, G., Salazar, R., Hassanalian, M., & Abdelkefi, A. (2018). Sizing and performance analysis of albatross-inspired tilt-wing unmanned air vehicle. AIAA/ASCE/AHS/ASC Structures, Structural Dynamics, and Materials Conference, Kissimmee, FL.

Walsh, P. P., & Fletcher, P. (2004). *Gas Turbine Performance* (2nd ed.). Oxford, UK: Blackwell. doi:10.1002/9780470774533

Zbrutsky, O. V., Prokhorchuk, O. V., Kolesnichenko, V. B., & Marynoshenko, O. P. (2013). Unmanned Aerial Vehicles and Technologies by NTUU "KPI". *IEEE 2nd International Conference "Actual Problems of Unmanned Air Vehicles Developments" Proceedings.*

ADDITIONAL READING

Antoine, N.E. & Kroo, I.M. (2004). Aircraft optimization for minimal environmental impact. *AIAA Journal of Aircraft*, 41(4).

Boren, H. (1967). Development and Production Costs for Aircraft (DAPCA). *Memorandum RM-5221-PR*, RAND Corporation

Department of the Air Force Headquarters Air Force Civil Engineer Support Agency. (2009). *Airfield Planning and Design Criteria for Unmanned Aircraft Systems.* Available from https://www.wbdg.org/FFC/AF/AFETL/etl_09_1.pdf

Dinc, A. (2015b). Optimization of a Turboprop UAV for Maximum Loiter and Specific Power Using Genetic Algorithm. *International Journal of Turbo & Jet-Engines*, *33*(3), 265–273.

Federal Aviation Administration. (2015). *Unmanned Aircraft Systems (UAS) Regulations & Policies.* Available from https://www.faa.gov/uas/resources/uas_regulations_policy/

International Civil Aviation Organization. (1993). *Annex 16* (Vol. II). Montreal, Canada: Aircraft Engine Emissions.

NASA Aeronautics Research Mission Directorate. (2004). *Civil UAV capability assessment.* Available from www.nasa.gov/centers/dryden/pdf/111761main_UAV_Capabilities_Assessment.pdf

Nicolai, L., & Carichner, G. (2010). *Fundamentals of Aircraft and Airship Design* (Vol. 1). AIAA Education Series. doi:10.2514/4.867538

KEY TERMS AND DEFINITIONS

Endurance: The maximum length of time that an aircraft can spend in cruising flight.

Fuselage: The main body of an aircraft designed to accommodate the crew and the passengers or cargo.

Interoperability: The ability of military equipment or groups (e.g., UAVs) to operate in conjunction with each other.

Loiter: A phase of flight in which a UAV typically spends most of its time for reconnaissance over a target area.

Low Observability: A sub-discipline of military tactics and passive electronic countermeasures, which cover a range of techniques used with objects (e.g., UAVs) to make them less visible (ideally invisible) to radar, infrared, and other detection methods.

Mission Profile: A schematic of the flight phases (speed, altitude, etc.) and the detailed descriptions of the activities performed on the flight.

Nacelle: A housing that holds and covers an aircraft engine on the outside of an aircraft.

Sizing: Calculation of dimensions and weight of an aircraft and its subcomponents.

Swarming: A group of UAVs all moving together, communicating with each other while in flight, and can respond to changing conditions autonomously.

Turboprop Engine: A gas turbine engine used in aviation that produces shaft power and drives an aircraft propeller typically through a gearbox.

UAV: An aircraft that can autonomously fly or can be remotely controlled to perform a specific mission without flying crew inside.

APPENDIX 1

As a case study of methodology explained in this chapter, this appendix summarizes the inputs (Table 1) and calculation results (Table 3) for a sizing of a turboprop-powered UAV with its propulsion system (Dinc, 2015a), similar to Predator-B (General Atomics Aeronautical Systems, 2014) and TPE331−10 turboprop engine (Honeywell, 2014). Assumed mission profile for this UAV is given in Figure 9.

Table 1. Turboprop UAV sizing and performance input parameters

Mission Parameters	Parameter Description	Value
h_0	altitude for takeoff [km]	0
h_{11}	altitude for landing [km]	0
h_{34}	altitude at the end of climb or at the start of cruise [km]	6.1
h_{56}	altitude at the end of cruise or at the start of loiter [km]	9.14
h_{78}	altitude at the end of loiter or at the start of cruise [km]	9.14
h_9	altitude at the end of return cruise or at the start of descent [km]	6.1
R_{op}	operation radius of UAV (distance to travel until loiter area) [km]	472
V_{cr}	maximum cruise speed [km/h]	370
V_{loi}	maximum speed for loiter [Mach]	0.4
V_{cli}	maximum speed for climb [Mach]	0.4
V_{wind}	headwind [km/h]	0
PLA_{idle}	engine idle throttle percent	10%
t_{idle}	idle time before takeoff [h]	20
t_{pow}	time of maximum power for take-off [min]	5
t_{wait}	maximum waiting time spent on air before landing [min]	20
$f_{resfuel}$	landing reserve fuel ratio	0.03
Weight Parameters		**Value**
W_{Pi}	internal payload weight [kg]	363
W_{Pe}	external payload weight [kg]	454
W_{fi}	internal fuel weight [kg]	907
W_{fe}	external fuel weight [kg]	862
f_{lg}	landing gear weight ratio	0.043
f_{sys}	avionics and flight systems weight ratio	0.12
f_{misc}	other miscellaneous equipment weight ratio	0.02
f_{wm}	empty weight margin	0.03
u_{fus}	unit weight for aircraft fuselage and nacelle [kg/m^2]	19
u_{wing}	unit weight for wing [kg/m^2]	32
u_{htail}	unit weight for the horizontal tail [kg/m^2]	18

continued on following page

Table 1. Continued

Mission Parameters	Parameter Description	Value
u_{vtail}	unit weight for the vertical tail [kg/m²]	18
ρ_{fuel}	fuel density [kg/m³]	801
ρ_{eng}	uninstalled engine density [kg/m³]	449
ρ_{lg}	landing gear density [kg/m³]	256
ρ_{sys}	systems density [kg/m³]	256
ρ_{pl}	payload density [kg/m³]	272
Aerodynamic Parameters	**Parameter Description**	**Value**
C_{fe}	surface friction coefficient	0.0041
C_{Lmax}	maximum lift coefficient	1.8
$C_{L, TO}$	maximum lift coefficient for take-off	1.49
e	Oswald efficiency for wing	0.75
V_{climb}/V_{stall}	climb speed / stall speed	1.25
V_{loiter}/V_{stall}	loiter speed / stall speed	1.1
V_{ref}	reference speed for takeoff [km/h]	185.2
Airframe Parameters		**Value**
n_{fus}	number of fuselage of aircraft	1
$L_{fus}/Ø_{fus}$	total fuselage length [m]/ fuselage diameter [m]	10.75
$L_{fus.b}/L_{fus}$	fuselage back side cone length [m]/ total fuselage length [m]	0.1
$f_{fus.vol}$	fuselage volume utilization factor	0.7
$f_{fus.vol.m}$	fuselage volume margin	1.3
Wing Parameters		**Value**
AR	aspect ratio of wing	17.92
WL	wing loading [kg/m²]	210.8
λ	tip to hub chord length ratio of wing	0.444
t/c	thickness to chord ratio of wing	0.13
t_{ave}/t_{max}	average thickness of wing airfoil [m]/ maximum thickness of wing airfoil [m]	0.6
V_{fuel}/V_{wing}	fuel volume [m³] / wing volume [m³]	0.5
Tail Parameters		**Value**
S_h/S	horizontal tail area [m²] / total wing area [m²]	0.08
S_v/S	vertical tail area [m²] / total wing area [m²]	0.135
Nacelle Parameters		**Value**
$L_{nac}/Ø_{nac}$	nacelle length [m] / nacelle diameter [m]	2.7
$Ø_{nac}/Ø_{eng}$	nacelle diameter [m] / engine diameter [m]	1.25
$f_{nac.wet}$	nacelle wet area factor	0.5
L_{nac}/L_{fus}	nacelle length [m] / fuselage length [m]	0.5

continued on following page

Table 1. Continued

Mission Parameters	Parameter Description	Value
Engine parameters	Turboprop engine model	**Value**
n_{eng}	number of engines	1
$f_{th.ins}$	thrust installation loss factor	0.9
$f_{w.eng}$	engine installation weight factor	1.3
F_{net}/W_0	maximum net total thrust [kgf] / maximum aircraft gross weight [kg]	0.3359
T_4	total temperature at turbine entry [K]	1368.7
Π_C	compressor total pressure ratio	10.37
$\eta_{C,pol}$	compressor polytrophic efficiency	0.795
$\eta_{T,pol}$	turbine polytrophic efficiency	0.86
η_b	combustor efficiency	0.999
η_m	shaft mechanical efficiency	0.995
ΔP_{in}	intake total pressure loss	0
ΔP_b	total pressure loss in combustion chamber	0.03
$\Delta P_{t.duct}$	total pressure loss in the duct between high and low pressure turbines	0.025
$\Delta P_{j.pipe}$	total pressure loss of the low pressure turbine jet pipe	0.02
N_{cd}	nozzle discharge coefficient	1
N_{cx}	nozzle thrust coefficient	0.99
Π_N	exhaust nozzle pressure ratio	1.03
β_1	compressor middle stage bleed air ratio	0
ε_1	compressor middle stage air extraction ratio for low pressure turbine duct cooling	0
ε_{2a}	cooling air ratio for high pressure turbine	0.05
ε_{2b}	cooling air ratio for low pressure turbine duct	0
ε_3	cooling air ratio nozzle guide vanes	0.05
$\varnothing_{prop}$	propeller diameter [m]	2.8
Ω_{prop}	propeller speed [rpm]	1591
$\eta_{prop,d}$	propeller dynamic efficiency	0.8
$\eta_{prop,s}$	propeller static efficiency	0.7
W_{eng}	engine weight	Formula derived from statistical data

Source: (Dinc, 2015a)

Table 2. UAV sizing design parameters and typical values

Parameter	Typical values	Examples
Aspect ratio	15-25	Global Hawk: 25; Predator B: 17.5; Altair: 23.5
Fuel weight fraction (with piston engine)	0.1-0.4	Predator A: 0.28; Pioneer: 0.16
Fuel weight fraction (with gas turbine engine)	0.3-0.6	Global Hawk: 0.566; Altair: 0.47
Empty weight ratio (with piston engine)	0.3-0.8	Predator A: 0.51; Heron: 0.55
Empty weight ratio (with gas turbine engine)	0.3-0.5	Global Hawk: 0.36; Predator B: 0.44; Altair: 0.44
Payload weight fraction (with piston engine)	0.05-0.35	Pioneer: 0.16; Predator A: 0.2
Payload weight fraction (with gas turbine engine)	0.07-0.15	Global Hawk: 0.07; Predator B: 0.09
Landing gear weight fraction	0.033-0.057	
Electronic, mechanical and hydraulic systems weight fraction	0.05-0.07	

Source: adopted from (Chaput, 2004; Raymer, 1999)

Table 3. Sizing and performance calculations result summary for a turboprop UAV

Aircraft output parameters (sizing and performance calculation results)			
$Ø_{fus}$	fuselage diameter	1.02	m
L_{fus}	fuselage total length	10.97	m
S_{fus}	fuselage total wet area	34.04	m^2
V_{fus}	fuselage total volume	8.23	m^3
W_{fus}	fuselage weight	569.24	kg
b	wingspan	20.13	m
c_r	chord length at root of wing	1.56	m
c_t	chord length at tip of wing	0.69	m
S	wing area	22.61	m^2
V_{wing}	wing volume	1.97	m^3
W_{wing}	wing weight	610.16	kg
S_h	horizontal tail area	1.81	m^2
S_v	vertical tail area	3.05	m^2
$W_{h.tail}$	horizontal tail weight	27.13	kg
$W_{v.tail}$	vertical tail weight	45.79	kg
W_{af}	airframe weight	1252.31	kg
W_{lg}	landing gear weight	148.52	kg
W_{sys}	systems weight	414.48	kg

continued on following page

Table 3. Continued

$W_{eng,inst}$	engine weight, installed	237.66	kg
W_{misc}	miscellaneous weight	69.08	kg
$W_{e,mar}$	empty weight margin (result from f_{wm})	61.71	kg
W_e	aircraft empty weight	2183.6	kg
W_{0i}	gross weight of UAV (with only internal fuel & internal payload)	3454	kg
W_0	maximum gross weight of UAV (with internal+external fuel & payload)	4770	kg
t_{loi}	loiter time	28.96	hour
t_{CR}	cruise time	2.32	hour
t_{tot}	total flight time	32.20	hour
E	max endurance	31.28	hour

Mission points		**#1**	**#2**	**#3**	**#4**	**#5**	**#6**	**#7**	**#8**	**#9**	**#10**
h	km	0	0	6.096	6.096	9.144	9.144	9.144	9.144	6.096	0
M_0	mach	0.151	0.182	0.268	0.326	0.339	0.329	0.270	0.339	0.326	0.145
C_L		1.274	0.878	0.870	0.590	0.825	0.878	0.878	0.556	0.381	0.878
V_0	km/h	185.2	222.6	304.9	370.4	370.4	359.0	294.7	370.4	370.4	177.8
L/D		22.46	24.03	24.03	22.24	23.99	24.03	24.03	21.73	17.55	24.03
W	kg	4762	4743	4697	4697	4618	4618	3112	3112	3037	3037
D	kN	2078	1935	1917	2070	1887	1884	1269	1404	1696	1235
Engine output parameters (turboprop cycle analysis results)											
PW	kW	712	719	462	476	356	354	343	356	476	711
EPW	kW	739	752	485	504	375	372	358	375	504	737
$PSFC_{uninstalled}$	kg/(kW.h)	0.324	0.323	0.282	0.278	0.265	0.266	0.269	0.265	0.278	0.325
$ESFC_{uninstalled}$	kg/(kW.h)	0.312	0.309	0.269	0.263	0.252	0.253	0.258	0.252	0.263	0.313
T_{inst}	kN	11310	9503	4425	3730	2787	2862	3399	2787	3730	11757
$TSFC_{uninstalled}$	g/(kN.s)	5.673	6.782	8.184	9.864	9.408	9.133	7.562	9.408	9.864	5.453
w_i	kg/s	3.540	3.561	1.803	1.840	1.246	1.241	1.215	1.246	1.840	3.536
P_0	kPa	103.0	103.7	48.9	50.1	32.6	32.4	31.7	32.6	50.1	102.8
P_2	kPa	103.0	103.7	48.9	50.1	32.6	32.4	31.7	32.6	50.1	102.8
P_3	kPa	1067.7	1075.2	507.6	519.7	337.9	336.3	328.3	337.9	519.7	1066.3
P_{31}	kPa	1067.7	1075.2	507.6	519.7	337.9	336.3	328.3	337.9	519.7	1066.3
P_4	kPa	1035.6	1043.0	492.4	504.1	327.8	326.2	318.4	327.8	504.1	1034.3
P_{41}	kPa	1035.6	1043.0	492.4	504.1	327.8	326.2	318.4	327.8	504.1	1034.3
P_{416}	kPa	261.4	262.5	151.3	153.5	110.6	110.2	108.5	110.6	153.5	261.3
P_{44}	kPa	261.4	262.5	151.3	153.5	110.6	110.2	108.5	110.6	153.5	261.3
P_{46}	kPa	256.2	257.2	148.2	150.4	108.4	108.0	106.3	108.4	150.4	256.0
P_{48}	kPa	104.9	104.9	48.2	48.2	31.1	31.1	31.1	31.1	48.2	104.9
P_5	kPa	104.4	104.4	48.0	48.0	31.0	31.0	31.0	31.0	48.0	104.4
T_0	K	289.5	290.1	252.1	253.8	234.0	233.7	232.1	234.0	253.8	289.4

continued on following page

Table 3. Continued

T_2	K	289.5	290.1	252.1	253.8	234.0	233.7	232.1	234.0	253.8	289.4
T_3	K	660.0	661.3	579.3	583.0	539.4	538.7	535.1	539.4	583.0	659.8
T_{31}	K	660.0	661.3	579.3	583.0	539.4	538.7	535.1	539.4	583.0	659.8
T_4	K	1368.7	1368.7	1368.7	1368.7	1368.7	1368.7	1368.7	1368.7	1368.7	1368.7
T_{41}	K	1334.9	1335.0	1331.5	1331.7	1329.9	1329.9	1329.7	1329.9	1331.7	1334.9
T_{416}	K	1007.0	1006.3	1047.0	1045.1	1066.4	1066.7	1068.5	1066.4	1045.1	1007.1
T_{44}	K	990.9	990.4	1025.6	1024.1	1042.5	1042.8	1044.4	1042.5	1024.1	991.0
T_{46}	K	990.9	990.4	1025.6	1024.1	1042.5	1042.8	1044.4	1042.5	1024.1	991.0
T_{48}	K	818.5	817.3	807.4	803.5	800.0	800.8	804.8	800.0	803.5	818.7
T_5	K	818.5	817.3	807.4	803.5	800.0	800.8	804.8	800.0	803.5	818.7
T_7	K	818.5	817.3	807.4	803.5	800.0	800.8	804.8	800.0	803.5	818.7

Source: (Dinc, 2015a)

Table 4. Equivalent skin friction coefficients

Aircraft type	c_{f_e} - subsonic
Civil transport aircraft	0.0030
Military cargo	0.0035
Light aircraft-single engine	0.0055
Prop sea plane	0.0065

Source: adopted from (Raymer, 1999)

APPENDIX 2: NOMENCLATURE

a_i speed of sound at engine station "*i*" [m/s]
A_7 exhaust nozzle area [m^2]
$A_{7,c}$ nozzle exit area for choking condition [m^2]
AR aspect ratio of wing
b wingspan [m]
c chord length [m]
c_r chord length at root of wing [m]
c_t chord length at tip of wing [m]
C specific fuel consumption [kg/(kN.h)]
C_{corr} engine airflow correction factor
C_F thrust coefficient of propeller
C_{fe} surface friction coefficient
C_L lift coefficient
C_{Lmax} maximum lift coefficient
$C_{L, TO}$ maximum lift coefficient for take-off
C_{PC} specific heat for compressor [kJ/kg K]
C_{PT} specific heat for turbine [kJ/kg K]
C_{PW} power coefficient of propeller
d propeller diameter [m]
d_{cl} climb distance (in horizontal) [km]
D drag force [N]
e Oswald efficiency for wing
E Endurance [h]
EPW Equivalent shaft power [kW]
ESFC Equivalent power specific fuel consumption [kg/(kW.h)]
f fuel air ratio
$f_{fus.vol}$ fuselage volume utilization factor
$f_{fus.vol.m}$ fuselage volume margin
f_{lg} landing gear weight ratio
f_{misc} other miscellaneous equipment weight ratio
$f_{nac.wet}$ nacelle wet area factor
F_{net} net total thrust [N]
$f_{resfuel}$ landing reserve fuel ratio
f_{sys} avionics and flight systems weight ratio
$f_{th.ins}$ thrust installation loss factor
$f_{w.eng}$ engine installation weight factor
f_{wm} empty weight margin ratio
F_A nozzle thrust [N]
G_C Power required for compressor [kW]
G_{HPT} power generated by high pressure turbine [kW]
G_{LPT} power generated by low pressure turbine [kW]

h aircraft flight altitude [km]
h_0 altitude for takeoff (point 1) [km]
h_{10} altitude for landing (point 10) [km]
h_{34} altitude at the end of climb or at the start of cruise (points 3–4) [km]
h_{56} altitude at the end of cruise or at the start of loiter (points 5–6) [km]
h_{78} altitude at the end of loiter or at the start of cruise (points 7–8) [km]
h_9 altitude at the end of return cruise or at the start of descent (point 9) [km]
HPT high pressure turbine
J advance ratio
L lift [N]
L/D lift to drag ratio
L_{fus} total fuselage length [m]
$L_{fus.cyl}$ mid fuselage cylinder length [m]
$L_{fus.b}$ fuselage back side cone length [m]
L_{nac} nacelle length [m]
LPT low pressure turbine
M_0 flight Mach number
n propeller speed [rev/s]
n_{eng} number of engines
n_{fus} number of fuselage of aircraft
N_{cd} nozzle discharge coefficient
N_{cx} nozzle thrust coefficient
NGV turbine nozzle guide vanes
P_i total pressure at station "*i*" [kPa]
P_{amb} ambient static pressure [kPa]
PLA_{idle} engine idle throttle percent
PSFC power specific fuel consumption [kg/(kW.h)]
PW uninstalled engine shaft power [kW]
R range [km]
R_{CL} climb distance [km]
R_{CR} cruise range [km]
R_D descent distance [km]
R_F range [km]
R_{op} operation radius of UAV [km]
S total wing area [m^2]
S_{fus} fuselage total wet area [m^2]
$S_{f.nose}$ fuselage nose area [m^2]
$S_{f.mid}$ mid fuselage area [m^2]
$S_{f.rear}$ rear fuselage area [m^2]
S_h horizontal tail area [m^2]
S_{nac} nacelle wet area [m^2]
S_v vertical tail area [m^2]
SFC thrust specific fuel consumption [gr/(kN.s)]

t/c thickness to chord ratio of wing
t_{ave} average thickness of wing airfoil [m]
t_{max} maximum thickness of wing airfoil [m]
t_B block time [h]
t_{CL} climb time [h]
t_{CR} cruise time [h]
t_D descent time [h]
t_F flight time [h]
t_{GR} ground maneuver time [h]
t_{idle} idle time before takeoff [h]
T_{inst} installed thrust (propeller+jet)
t_{loi} loiter time [h]
t_{pow} time of maximum power for take-off [min]
t_{tot} total mission time [h]
t_{wait} maximum waiting time spent on air before landing [min]
T_{amb} ambient static temperature [K]
T_{bleed} Compressor middle stage total temperature for bleed [K]
T_i total temperature at station "*i*" [K]
T_{uninst} uninstalled thrust (propeller+jet) [N]
u_{fus} unit weight for aircraft fuselage and nacelle [kg/m^2]
u_{htail} unit weight for the horizontal tail [kg/m^2]
u_{vtail} unit weight for the vertical tail [kg/m^2]
u_{wing} unit weight for wing [kg/m^2]
UAS unmanned air system
UAV unmanned air vehicle
V_0 aircraft flight velocity [km/h]
V_{climb} climb speed [km/h]
V_{cr} maximum cruise speed [km/h]
V_{cli} maximum speed for climb [Mach]
V_{fuel} fuel volume [m^3]
V_{fus} fuselage total volume [m^3]
$V_{f.nose}$ fuselage nose volume [m^3]
$V_{f.mid}$ mid fuselage volume [m^3]
$V_{f.rear}$ rear fuselage volume [m^3]
V_{nac} nacelle volume [m^3]
V_{wing} wing volume [m^3]
V_{loi} maximum speed for loiter [Mach]
V_{stall} stall speed [km/h]
V_{ref} reference speed for takeoff [km/h]
V_{wind} headwind [km/h]
w_i engine airflow at station "*i*" [kg/s]
w_f fuel flow [kg/s]
WL wing loading [kg/m^2]

W aircraft weight during flight stages [kg]
W_0 maximum gross weight of UAV (with both internal and external fuel & payload) [kg]
W_{0i} gross weight of UAV (with with only internal fuel & internal payload) [kg]
W_{af} airframe weight [kg]
W_e aircraft empty weight [kg]
W_{eng} engine weight [kg]
W_f total fuel weight [kg]
W_{fe} external fuel weight [kg]
W_{fi} internal fuel weight [kg]
W_{fus} fuselage weight [kg]
$W_{h.tail}$ horizontal tail weight [kg]
W_{lg} landing gear weight [kg]
W_{misc} miscellaneous weight [kg]
W_{nac} nacelle weight [kg]
W_{Pe} external payload weight [kg]
W_{Pi} internal payload weight [kg]
W_{PL} total payload weight [kg]
W_{sys} systems weight [kg]
$W_{v.tail}$ vertical tail weight [kg]
W_{wing} wing weight [kg]
Greek Symbols
ΔP_b total pressure loss in combustion chamber
ΔP_{in} intake total pressure loss
$\Delta P_{j.pipe}$ total pressure loss of the low pressure turbine jet pipe
$\Delta P_{t.duct}$ total pressure loss in the duct between high and low pressure turbines
$Ø_{eng}$ engine diameter [m]
$Ø_{fus}$ fuselage diameter [m]
$Ø_{nac}$ nacelle diameter [m]
$Ø_{nac}$ nacelle diameter [m]
$Ø_{prop}$ propeller diameter [m]
β_1 compressor middle stage bleed air ratio
γ specific heat ratio
δ atmospheric pressure ratio
ε_1 compressor middle stage air extraction ratio for low pressure turbine duct cooling
ε_{2a} cooling air ratio for high pressure turbine
ε_{2b} cooling air ratio for low pressure turbine duct
ε_3 cooling air ratio nozzle guide vanes
η_2 compressor isentropic efficiency
η_{416} high pressure turbine isentropic efficiency
η_{46} low pressure turbine isentropic efficiency
η_b combustor efficiency
$\eta_{C,pol}$ compressor polytrophic efficiency
η_m mechanical efficiency

η_p propeller efficiency
$\eta_{prop,d}$ propeller dynamic efficiency
$\eta_{prop,s}$ propeller static efficiency
$\eta_{T,pol}$ turbine polytrophic efficiency
θ atmospheric temperature ratio
λ taper (tip to root chord length) ratio of wing
Π_C compressor total pressure ratio
Π_N exhaust nozzle pressure ratio
Π_T turbine total pressure ratio
ρ_{eng} uninstalled engine density [kg/m^3]
ρ_{fuel} fuel density [kg/m^3]
ρ_{lg} landing gear density [kg/m^3]
ρ_{pl} payload density [kg/m^3]
ρ_{sys} systems density [kg/m^3]
Ω_{prop} propeller speed [rpm]

Chapter 10
Unmanned Aerial Vehicle Applications for Military GIS Task Solutions

Azad Agalar Bayramov
Armed Forces War College of the Azerbaijan Republic, Azerbaijan

Elshan Giyas Hashimov
Armed Forces War College of the Azerbaijan Republic, Azerbaijan

Yashar Ali Nasibov
Geography Institute of Azerbaijan National Academy of Sciences, Azerbaijan

ABSTRACT

In the chapter, the authors present the results of unmanned aerial vehicle (UAV) applications for military geoinformation system (GIS) task solutions. The results of the visual modelling of the revealing process of invisible military objects from one of point of observation on the mountainous terrain of the Azerbaijan Republic by using UAV are presented. The observation conditions between two points of the selected mountain terrain during battle operation have been investigated using GIS technology. The quantitative method of the invisible area assessment and military objects in mountainous terrain are developed and offered by using UAVs. The numerical estimation method of a task support success of UAV reconnaissance flight in mountainous battle conditions has been offered and considered. Using UAVs for the purpose of orthophotomap making of the terrain and combat control the detailed 3D-model has been constructed.

INTRODUCTION

The development and application of small aviation automated systems is one of the most importance problem in the aviation and aerospace industries, in particular in military area. There are many examples of applications of the small modern aviation automated systems. In this chapter, the development and military application of small aviation automated systems (Unmanned Aerial Vehicles -UAV) is considered. The results of visual modelling of the revealing process of invisible military objects from one

DOI: 10.4018/978-1-5225-7709-6.ch010

of point of observation on the mountainous terrain of the Azerbaijan Republic by using UAV are presented. The observation conditions between two points of the selected mountain terrain during battle operation have been investigated using GIS technology. The traditional methods (e.g. ground mapping, space remote sensing) do not allow to carry out precision and efficient (operations) measurements. The quantitative method of the invisible areas and military objects assessment in mountainous terrain has been developed and offered by using UAVs. The numerical estimation method of a task support success of UAV reconnaissance flight in mountainous battle condition has been offered and considered. There have been made the comparison of obtained results by using various methods; it has demonstrated that by using small aviation automated systems helps efficiently to solve various civil and military tasks on the terrain. The future scope of presented investigations is improving and development of small aviation automated systems possibilities for more efficiently solve both civil and military tasks.

BACKGROUND

The observation condition is one of the tactic properties impacted on the application of weapons and battle technics, on the organization of war operations (Zaritski & Charkevich, 2007). This property is determined by the invisible level of terrain, visible distance and it is dependent on the relief condition. The right organized visual monitoring provide correct data. These data give to commandant the possibilities to make detailed assesstment of battle area and to adopt reasonable decision making. The more flank of hill, slope of ravine, valleys, heights, trees and bushes, various buildings the more disadvantageous observation condition on the terrain. During monitoring should to choose such observation point that mountain passes, patches, slopes of ravine and valleys are observed.

The topographic map is one of the main source giving information about terrain topographic elements: coordinates and mutual locations, sizes and other qualitative and quantitative peculiarities. Moreover, not reflected itself in the map changed elements have tactical importance, too. These elements impact much on the tactical battle properties. The main information sources about such objects are aerophotographs made during battle. Further, these aerophotographs can be applied in navigation systems of the high precise weapons, for instance, various kind of correlation-extreme navigation systems, to destroy invisible enemy targets. The large natural and artificial obstacles (forests, high hills, mountains, etc.) aggravate this problem, and create a necessity to applied UAVs (Butin, 2013; Varvarina, 2012). The UAVs are widely applied in reconnaissance searches for observation tasks (Petrov, 2013).

The modern GIS have been widely applied in Armed Forces for preparation and control of battle operations, for information providing of tactic activities, for improvement of topographic maps, for determination of location on the ground of the land forces etc. (Soкolov & Tixonov, 2008; . Olaf Conrad, 2010). In the real situations during battle operations the Geography information about situation on the terrain can be often changed. So, in this cause, the application of usual maps isn't effective. Only the modern automated control system can provide fast changing an information documenting. The modern GIS electron maps, video data editing systems, the especial GIS software in modern computer-driven control system (Windows 8, 10, 13, etc.) have been developed (ArcGIS, 2008; Karmanov, 2012). They provide a creation of vector, raster and matrix maps, and efficiently data about terrain updating.

In previously works (Bayramov, Hashimov & Khalilov, 2017; Bayramov and Hashimov, 2018) using GIS technology the relief digital model application and the terrain observation conditions were investigated for possibility of planning war operations. The observation condition is one of the terrain features for providing information about enemy troops position and facilities. This feature helps to determine the distance of sight between any observation points, the invisibility factors of terrain. It is dependent on nature of relief plant cover, human settlements and other objects, also meteorological conditions, influencing on the visibility. As a result of correct organization of observation, the obtained data help commander to estimate completely the military operation area and adopt a reasonable decision.

If there are many ravines, valleys, hills, trees and bushes, various buildings on the terrain then the visibility conditions are unfavourable. In mountain conditions during the supervision it should be selected such observation point from out of which it can be possible to see roads, precipices, paths and valleys. The observation conditions help a reconnaissance, organization of fire systems and control of military units. It helps to reveal surroundings terrain and targets. The observation conditions are characterized by optical visible distance in one of concrete sector (range) of elevations of the surroundings terrain and targets, also by the sizes of invisible areas and bounds. As a result of carried out and presented experiments, the estimation of observation situation in mountain terrain has been realized. The GIS analysis has been carried out at two stages. At the first stage, the visibility analysis on the line of sight between two oversight points has been carried out, then the visible and invisible areas under 0-180° angle in range 18000 m are determined, and the terrain profile has been constructed. At the second stage, the terrain elevation 3D model has been constructed. GIS and ArcGIS software have been used for our calculations.

UAV SMALL AVIATION AUTOMATED SYSTEM

The automated control system for UAV small aviation dynamic object can be based on the various principles. Since 1990s the possibilities and advantages of UAV ideology are very interesting and this ideology finds wide application to solution of various tasks, for example: conduct of policy, defense systems, transport, public health and other. The UAV's components are aviation automated systems too. In technical area the UAV definition can be divided on two groups: 1) UAV is consisted of components which solve one task and on the whole jointing own properties and possibilities; 2) UAV is such a system which during various operations combines and joints itself a synergy of commands, control functions, network and feedback information change. Presented part of chapter is devoted of the development of UAVs small aviation automated system making, the opportunity of military application and development of complex control algorithm. Developed 6-rotors UAV is taken as test UAV (Hashimov, Bayramov, Abdullayev & Mammadli, 2017).

The choice of battle tactic and its correction based on the input information is dependent on the professional level of staff commissioned and on the assigned decision making principles (Commander and Staff Officer Guide, 2001). If a radio communication between military units is broken, then commanders' correct decision making becomes very actual. In this case, with other units communication lost the commander must make an optimal decision based on the gotten information from last radio communication session, current situation and reconnaissance information. Therefore, it is very important to joint information received from UAVs.

Last 20 years the manage of mutually operated various groups was a reason to creation of UAVs small aviation ideology. The UAV theory is a mathematical modelling form to solve a task of support of the optimal mutually activities of independent units. At present, UAVs are applied widely for modelling of complex structural objects activities: engineer control systems, physical complexes, complex systems managment, etc.(Wenger, Antoniev & Gorod, 2013). Indicated features allow to simulate activities of military units as UAVs and it can be an usefull tool of optimal to solve tactic and reconnaissance tasks.

From the point of view of mathematics, moving in air, sea or space the various automated robots including UAVs can be considered an dynamic systems. Such systems can execute both peace and military tasks. The control of these systems can be based on the various principles. The effectiveness of task solution is dependent on number of UAV, on space location, on the environment conditions and events, etc. The task can be executed at own or enemy territory, on the mountainous or flat country. In these conditions, at each moment the rational decision making based on the UAV operation evaluation, UAV state and possibilities assessment are advantages of UAV ideology.

The making of UAV control system is conditioned by technical factors and possibilities: limitations of UAV flight dynamics, UAV task execution possibility (Hashimov, Bayramov, Abdullayev & Mammadli 2017). For example, let us consider developed 6-rotors UAV shown in figure 1 (Hashimov Bayramov, Abdullayev & Mammadli, 2017). The purpose of this UAV is to make terrain digital maps, video surveillance execution, transportation of goods and implementation of special assignment. This UAV has 6 rotors. Its main parts are made of carbon compocite material, it has a smart autopilot system. It has next technical characteristics:

- Takeoff weight -18 kg;
- Load-carrying ability -10 kq;
- Flight distance 20 km;
- Flight speed 80 km/hour
- Flight duration 1 hour.

The block chart of the UAV control system is shown in figure 2.

Microcontroller MC and its software operate all electronic parts and elements of UAV. GPS determines UAV space coordinates and during UAV flight the 3-coordinates gyroscope 3D-GY controls flight position. GPS determines UAV coordinates in space. Accelerometer block *X-Y-Z* ACC developes input date for generation of control algorithm. Using wireless radio communication system WRC and video surveillance system VSS it send obtained information to command post and receive orders. Using rotor control blocks RCB1, RCB2, RCB3, RCB4, RCB5 and RCB6 the microcontroller MC controls R1,

Figure 1. Developed 6-rotors UAV

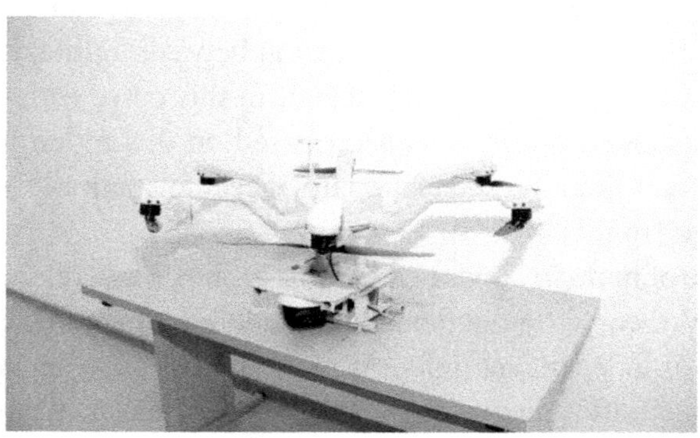

Figure 2. The block chart of the UAV control system: MC - microcontroller; 3D-GY –3-coordinates gyroscope; GPS – determines UAV coordinates in space; X-Y-Z ACC – accelerometer block; WRC - wireless radio communication system; VSS - video surveillance system; RCB1,...,RCB6 – rotor control blocks; R1,...,R6 – rotors; AC – accumulator

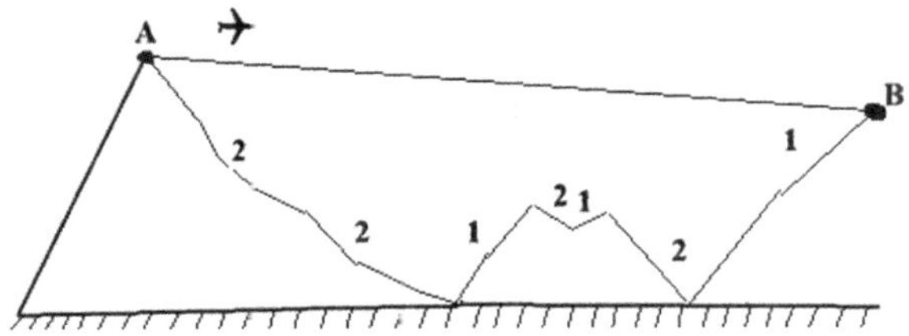

R2, R3, R4, R5 and R6 rotors flight condition. Accumulator AC provides electric energy all electronic systems and rotor engines.

In some cases, as contrasted to UAV usual control systems, in presented UAV small automated complex the wireless radio communication system block can be replaced infrared (IR) communication one. Using this system block the directly communication between UAVs and ground post can be provided. As a result, it troubles much enemy radio-electronic devices countermeasure to UAVs control system operation.

MODELING OF THE DETECTING PROCESS OF INVISIBLE OBJECTS BY USING UAV

In given part, by using of developed UAV (described in previous part of chapter), the results of visual modelling of the revealing process of invisible areas and military objects by using UAV are presented for the mountainous terrain of the Azerbaijan Republic. Autodesk-3ds-max program (Autodesk 3ds-max, 2016) has been used for modelling. The goal of the investigation of observation condition in operation field is determination the most advantageous observation point on the terrain or invisibility level. The observation condition is characterized by optical visible distance from one of the heights of targets sector (zone) and terrain, also by sizes of invisible area and boundary.

Invisible areas are determined by creation of terrain heights profile by next method (Hashimov, Bayramov & Xalilov, 2017). The observation point is shosen on the one of high place. From this point by along of observation direction in right and left sector borders two straight lines are drafted. The sector is divided by straight lines in several direction and is numbered. It is more preferably to draft these lines along on the tops of heights or watershed lines. Separate and enclosed profiles are drafted along of straight lines. In each profile the determined invisible area is shaded. The right and left borders of the shaded areas marked iterative on the profile line on the map. Considering terrain relief, marked points are jointed by smooth lines and invisible areas are shaded on the map. The wide sector has many profile's lines and if it is necessary prolong until ground. So, for given directions that to determine visible and invisible areas the terrain profiles are used.

Let us explain it by chart in the figure 3. In this figure a conditional mountainous terrain is shown. Here: ***A*** is a chosen observation point on the mountain (or hill), ***B*** is an enemy target point. Between ***A*** and ***B*** points along observation direction straight line by using of GIS technology the commander

Figure 3. The observation directions in mountainous terrain

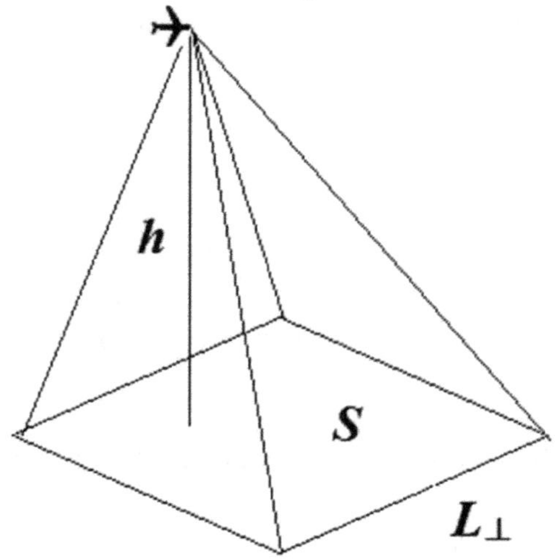

drafts terrain heights profile As a result, the visible (1) and invisible (2) areas are determined. Analyzing obtained information in staff observation point ***A*** the commanders can decision making that if enemy troops or combat technics are located in (2) areas then its will be invisible. However, if enemy forces are located in (1) areas, then they will be visible from point ***A***.

If C or D or E points are chosen as observation points then invisible areas will be determine the same methods: by along on CD, DE or EB lines. Such observation points in one region can be choosen on various peaks. That comandant makes right decision he must take such observation point from that the sum of the invisible areas is minimum. That invisibility level is minimum the UAV can used, that is, the UAV flight can be controled on the ***B***. But in such situation the risk of UAV destroy on ***B*** will be high. For safety UAV use it should fly up the ***A*** side on such safety height that the invisible areas on the ***B*** side are revealed and, at the same time, UAV cannot be destroy.

Let's, there are several peaks of mountain between observation point ***A*** and target point ***B*** (see figure 4). Let's, $\angle OC_1F=\alpha_1$, $\angle OC_2F = \alpha_2$,... $\angle OC_iF = \alpha_i$ are the mountain slopes angles, *H* is a height of observation point of mountain, x_0, x_1, x_2, x_i are the coordinates of the mountain peaks (beginning from observation point), *h* is a height of the UAV flight up the observation point, $\boldsymbol{OC_i}$ are the lines draft from UAV flight point ***O*** height until bottom of valley $\boldsymbol{C_i}$, *L* is a distance from UAV flight point ***E*** to observation point ***A***:

$$L \in \{h, x_i - x_o\} \tag{1}$$

$$L = \sqrt{\left(x_i - x_o\right)^2 - h_i^2} \tag{2}$$

$$h_i = \left(x_i - x_o\right)\left|tg\alpha_i\right| - H \tag{3}$$

Figure 4. Use of UAV for minimum making of invisibility level of the terrain

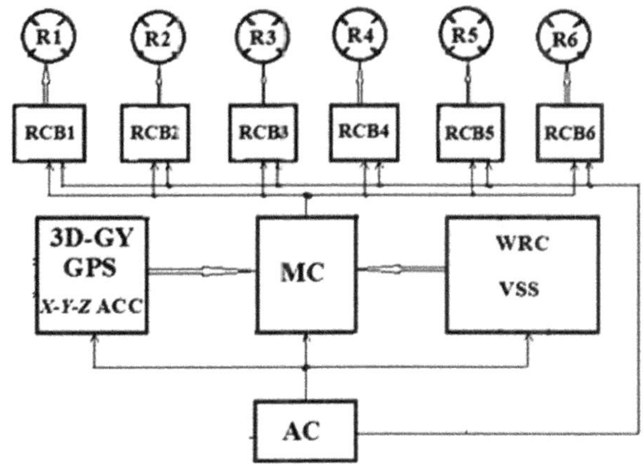

By using of (1), (2) and (3) equations, with aim of invisible from point ***A*** areas control the commanders can determine minimal safety flight height of the UAV and minimal safety flight distance:

$$\begin{cases} L \to \min \\ h \to \min \end{cases}$$

It should be note that, *H, h,* x_0, x_1,...x_i and tgα values are determined by GIS software for the each specific case. For vizual modeling of the revealing process of invisible spaces and military objects the authors have used Autodesk-3ds-max program 2016-Version. Autodesk-3ds-max program system has been developed by Autodesk company and has goal to edit 3D graphics and animation of various processes. By use Autodesk-3ds-max program the authors recieve an image which is very much alike real visibility. This program is much used in various areas and is the most effective among 3D programs. Some of the results of visual modelling from one of the choosen observation point in mountainous region of the Azerbaijan Republic are represented in figure 3. The angle of vision of a digital camera of the UAV is 145°.

The results of the UAV flight modelling on the height *h* above the starting point are presented in figure 5. The visible areas from observation point are marked by green color, and invisible areas are marked by dark color. The visible areas from observation point are marked by green color, and invisible areas are marked by dark color.

Figure 5. The UAV flight modelling on the height h above the start point

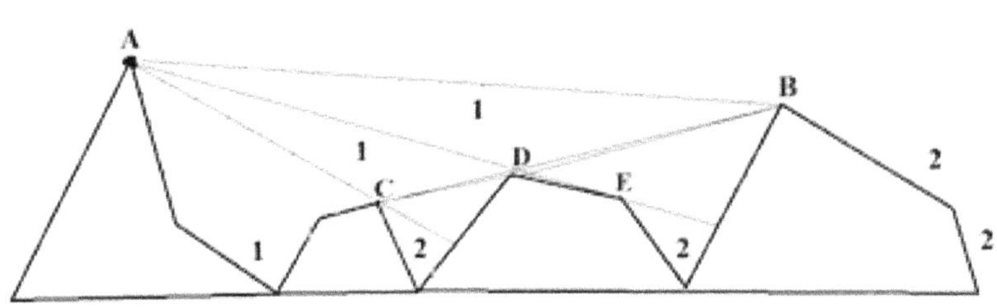

The authors have concluded, the higher and farther a UAV flight from observation point to enemy direction, the less invisible areas, and the less probability of hidden moving of the enemy combat trucks. The preliminary modelling of a visual monitoring during a UAV reconnaissance flight will help correct to determine the UAV flight traffic in real situation for specific terrain with aim of maximum survey of the terrain, minimization of invisible areas and reliable identification of the enemy hidden objects.

So, in given part, the results of visual modelling of the revealing process of invisible military objects by using a UAV are presented. Autodesk-3ds-max program has been used for modelling. It is shown that the higher and farther a UAV flight from observation point to enemy direction, the less invisible areas, and the less probability of hidden moving or concealment from view of the enemy battle trucks and troops. Obtained results will help correct to determine the UAV flight traffic in real situation for specific terrain with aim of maximum survey of the terrain, minimization of invisible areas and reliable identification of the enemy hidden objects.

For the purpose of investigation of the observation conditions by using GIS two viewpoints on the topographic map have been selected. The height of first point is 900.8 m, and the height of second point is 901.5 m. The distance between these viewpoints is 17941.16 m.The goal of this work is to study of the influence of relief, trees and bushes, human settlements, various buildings and the height of camera platform on the level of visibility (Hashimov, Bayramov, Nasibov & Amanov, 2015; Hashimov, Bayramov & Khalilov, 2015).

- **I-st experiment:** The result of analysis of the line of sight between two viewpoints is shown in figure 6 (Hashimov & Bayramov, 2017). The camera platform is located on the surface of ground (the height h = 1 m). The visible areas are marked by the green lines, the invisible areas are marked by the red lines. The visibility analysis of the line of sight between two viewpoints has been carried out with taken an account of Earth surface curvature.
- **II-d experiment:** The camera platform is lifted on the height h = 10 m above-ground level. Took 1-st viewpoint as basis, the visible and invisible areas under (0 ÷ 180°) angle in range of 18000 m to direction to second viewpoint are determined by GIS analysis. The result of analysis of the

Figure 6. The visibility analysis of the line of sight between two viewpoints

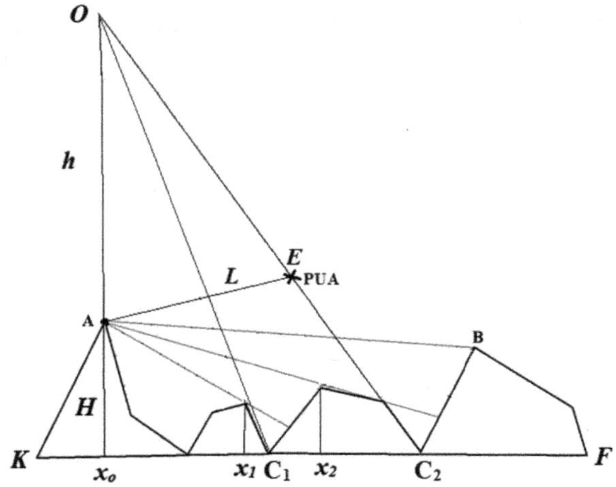

line of sight between two viewpoints is carried out and the 3D model of visible and invisible areas between two viewpoints (h = 10 m) is constructed (Hashimov & Bayramov, 2017).

- **III-rd experiment:** The camera platform is lifted on the height h = 20 m above-ground level. Took 1-st viewpoint as basis, the visible and invisible areas under 0-180° angle in range ≈ 18000 m to direction to 2-nd viewpoint are determined by GIS analysis.
- **IV-th experiment:** The camera platform is lifted on the height h = 30 m above-ground level. Took 1-st viewpoint as basis, the visible and invisible areas under (0 ÷ 180°) angle in range of ≈ 18000 m to direction to 2-nd viewpoint are determined by GIS analysis.

The investigated terrain were classified on heights and were divided on 13 zones (figure 7). Using GIS software the profile on the line of sight between two viewpoints has been constructed (figure 8). This profile is applied for determination of double-sided invisible areas between selected viewpoints, for investigation more precisely of terrain profile, for study of terrain slope, etc.

Figure 7. The digital heights model of terrain

Figure 8. Constructed profile between two viewpoints

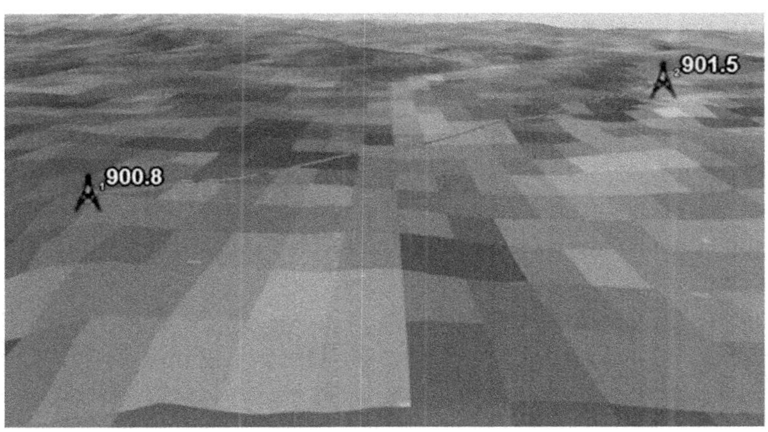

The investigation of observation condition on the war operation region has a goal to determine of visibility and invisibility levels (factors) on the most suitable viewpoint of terrain. By construction of the heights profile of terrain, the visible and invisible areas are determined below method. Let us explain it by using figure 3.

The profile of conditional mountain terrain is shown in this figure 3. The two viewpoints are selected on the high posts. Here: ***A*** is the first viewpoint selected on the top of mountain, ***B*** is the second viewpoint (enemy target) selected on the top of another mountain. Between ***A*** and ***B*** viewpoints on the line of sight the heights profile is constructed by using GIS software. As results, visible (1) and invisible (2) areas (see figure 3) are determined. From analysis of obtaining information the command staff located in the *A* post make at a decision that the enemy troops and armored machineries located in areas (2) will be unobserved. On the other hand, if enemy troops will be located in areas (1) then they will be observed from *A* viewpoint. If *C*, or *D*, or *E* points are selected as viewpoints then, by the same rule, the invisible areas can be determine on *CD*, *DE* or *EB* lines of sight.

Such viewpoints can be selected on the various peaks of the terrain. For the right and correct decision making the commander must choose such viewpoint that the sum of the invisible areas is minimal. For this, let us apply and use an invisibility factor *P* for various peaks:

$$P = \frac{1}{n-1}\sum_{i=1}^{n-1}\frac{1}{S_i^o}\sum_{j=1}^{n-2}S_{ij}^{in} \tag{4}$$

Here: S_{ij}^{in} is an invisible area between *i* and *j* peaks, S^o is a total area between *A* and *B* peaks; *n* is a number of peaks. After calculation P_1, P_2,... invisibility factors for various viewpoints to the lines of sights to the enemy post *B*, the minimum of them can be determined::

$$P_{min} = \min\{P_1, P_2, ...\}.$$

So, after calculation of the minimal invisibility factor P_{min} by this method, the commander can choose a right viewpoint. From this viewpoint (a peak) on the line of sight to ememy point *B* the invisible areas are minimal. Based on this method let us determine an invisibility factor *P* in our experiments.

- **I-st experiment:** the camera platform is located on the ground, then from analysis of data on the figure 1, authors have calculated the invisibility factor P_1 from *A* viewpoint on the line sight to *B* post:

$P_1 \approx 0{,}55$ or $P_1 \approx 55\%$

That is, 55% of total area between viewpoints *A* and *B* are invisible.

- **II-nd experiment:** the camera platform is lifted on the height of $h = 10$ m above-ground level. Then, from analysis of data authors have calculated the invisibility factor P_2 from *A* viewpoint on the line sight to *B* post ((h=10 m):

$P_2 \approx 0{,}42$ or $P_2 \approx 42\%$

That is, the camera platform is lifted on the height of $h = 10$ m above-ground level, then 42% of total area between viewpoints *A* and *B* are invisible.

- **III-nd experiment:** the camera platform is lifted on the height of $h = 20$ m above-ground level. Then, from analysis of data authors have calculated the invisibility factor P_2 from *A* viewpoint on the line sight to *B* post ((h=20 m):

$P_3 \approx 0{,}36$ or $P_3 \approx 36\%$

That is, the camera platform is lifted on the height of $h = 20$ m above-ground level, then 36% of total area between viewpoints *A* and *B* are invisible.

- **IV-th experiment:** the camera platform is lifted on the height of $h = 30$ m above-ground level. From analysis of data the authors have calculated the invisibility factor P_4 from *A* viewpont on the line sight to *B* post ((h=30 m): $P_4 \approx 0{,}22$ or $P_4 \approx 22\%$

That is, the camera platform is lifted on the height of $h = 20$ m above-ground level, then 22% of total area between viewpoints *A* and *B* are invisible. So, the authors have $P_4 < P_3 < P_2 < P_1$. The authors have concluded that if the height of camera platform is increased above-ground level then the invisible area is decreased:

- The experiments for estimation of observation situation in mountain terrain has been carried out;
- Using GIS technology and ArcGIS software the visibility analysis on the line of sight between two selected oversight points has been carried out;
- The visible and invisible areas under 0-180° angle in range 17941.16 m are determined;
- The terrain profile and the digital heights model of investigated terrain have been constructed;
- At the various heights of camera observation the 3D heights model of the visible and invisivle areas have been constructed;
- The invisibility factors between two delected viewpoints have been calculated;
- It is determined that if the height of camera platform is increased above-ground level then the invisible area is decreased.

ASSESSMENT OF INVISIBLE AREAS AND MILITARY OBJECTS IN MOUNTAINOUS TERRAIN

In the previous part of chapter, the terrain relief is constructed by using GIS technology. Now, a number of invisible areas and military objects will be estimate. During war activities, the evaluation of invisible areas and military armors in mountainous terrain is very important for assessment of operational-tactical situation in battlefield. This information can be used for prevention of an enemy sudden attack. In given part, the quantitative method of the invisible areas assessment and military objects in mountainous terrain are developed and offered by using UAV (Bayramov & Hashimov, 2018). The probabilities of invisible area or enemy object detection are calculated by using 1 and *n* UAVs. Obtained data by ArcGIS software (GIS technology) has been used for calculations and evaluations the number of invisible areas and military objects. The analysis of the dependence of revealing effectiveness, localization areas and military objects on the number of UAVs has been carried out. It is more efficiently to use one UAV for quantitative analysis of invisible areas, but it is more efficiently to use several UAVs for detection of armored vehicles and military objects. Our experiments in mountainous conditions had shown that at flight height 300 m and battle-front 2.5 km putting 3-5 UAVs is real. In this case, all of them perform operational mission and return to base. Finally, from the outcome of this study, the offered method can be applied for any mountainous conditions, for creation of a simulation and training UAVs war-game software.

Using the tactical features of the area gives the opportunity for the high effective application of technology, confidential of maneuver and enemy fire defense. The features of the area are defined with its relief, local and other geographical objects. The detection of invisible areas, military objects and armors in mountain terrain is very important for assessment of operational-tactical situation in battlefield. This information can be used for prevention of an enemy sneak attack. For this goal the various methods and technologies are used: GIS technology (Bayramov, Hashimov & Khalilov, 2017), photogrammetry method (Hashimov, Bayramov & Xalilov, 2017), seismic-location method etc. Large obstacles (large tracts of forest, high hills, mountains, etc.) exaggerate this problem and cause the importance of UAV application. For this reason, UAV is used widely for quantitative assessment of invisible areas and military objects in real battle situation (FMI 3-04.155, 2006). UAV operations support battlefield commanders as they plan, coordinate, and execute operations. Other UAV missions support the maneuver commander by contributing to the effective tactical operations of smaller units. UAV system has remote video terminals and operations video enhanced receiver, portable ground control stations, and helps set the conditions for the current and future forces success.

In this part, the calculation of number of the invisible (hidden) areas, military objects and armors on mountain terrain in the distance from observation post *A* to enemy post *B*, which can be detected by one or several UAVs, has been carried out for determination and comparison of the reconnaissance UAV possibilities. Flat country offered to the probabilistic analysis has been used as basis method for these calculations for mountain conditions. The calculations have been executed for one specifically taken mountain terrain where the analysis of hidden areas was carried out previously by using GIS technology.

UAV reconnaissance missions obtain information about tactical activities and enemy resources. The number of invisible areas or enemy military objects, the locations of which can be detected, depends on some parameters: the kind of battle activities; the number of UAV reconnaissance; the distribution density of invisible areas in battlefield; the size of battlefield; the size of surveillance area by optical-electronic device; the probability of invisible areas or enemy objects detection. During surveillance of mountain terrain, it should be taken into account next factors, too: the height of terrain from observation post; the height of terrain from enemy location; the level of terrain invisibility, presence of forests, bushes, etc. In addition, it should be taken into account camouflaging of military objects by natural mountain shelters. The number of invisible areas or enemy objects detected in the battle theatre by UAV with single flight can be calculated by below equation (FMI 3-04.155, 2006; Robert, 2008):

$$N = P \cdot \rho \cdot S_o, \tag{5}$$

Here:

N is the number of detected invisible areas or enemy objects;
P is the probability of invisible area or enemy object detection;
ρ is the distribution density of invisible areas or enemy objects on battlefield per km^{-1};
S_o is the size of reconnaissance area per km^2;

The probability of invisible area or enemy object detection P depends on the probability P_I of object highlighted (detected) on the background of mountain landscape during single UAV flight. In turn, P_I depends on the information content about sought-for area or object. If the information content displayed on a screen of the UAV control post is sufficient for confirmation this object, then $P_I = 1$. Therefore, the probability of object detection by using optical-electronic device of UAV can be determined by below equation (FMI 3-04.155, 2006; Robert, 2008):

$$P = P_c \frac{1 - K}{1 - \xi \cdot \eta}, \qquad (6)$$

Here:

P_c is the probability of object camouflaged on the background of mountain landscape;
K is the factor of camouflage level of an object;
ξ is the likelihood degree of false positions;
η is the number of false hidden areas (objects) per one real area (object).

For the probability of object camouflaged on the background of mountain landscape during UAV flight can be adopted that $P_c = 1$. In our case, it can be adopted that the covering area of forests, bushes or big stones on investigated mountain terrain, which really camouflage the invisibility level of terrain, on the height above sea level of 2500-2900 meters is equal 5%, then $K = 0.05$; it can be adopted that $\xi = 0.9$ and $\eta = 0.01$.

The application of digital photogrammetry helps us to carry out our experiments and calculations for real conditions, to determine the invisible and visible areas of some mountain terrain of the Azerbaijan Republic, to estimate coefficients K, ξ and η, to plane and to determine UAVs flights. In conditions of other countries, the coefficients can be diverse. After calculation by using (6) the authors have obtained $P = 0.96$. Depending on the conditions, a reconnaissance flight of UAV in surveillance area can be carried out by several UAV of the same type, so the (6) changes to:

$$P_n = 1 - (1 - P_1)^n, \qquad (7)$$

Here:

P_n is a probability of the invisible area or the enemy object detected by using n UAV;
n is a number of the reconnaissance UAV;
P_I is a probability of the area or the object detected (highlighted) by using single UAV.

The distribution density of invisible areas or enemy objects in battlefield can be determined as ratio of common number of potentially invisible areas or objects to total area of reconnaissance area, that is,

$$\rho = \frac{m}{S_o}, \qquad (8)$$

Here: m is a number of the potentially invisible areas or objects on reconnaissance area.

The size of reconnaissance area, where UAV carries out a flight, depends on technical features of UAV optical system and flight parameters: a view angle of optical system, a flight height, a flight endurance, a UAV speed and a distance from take-off point to front of reconnaissance area. The total size of reconnaissance area S_o is calculated by below equation:

$$S_o = L_{\parallel} \cdot L_{\perp}, \qquad (9)$$

Here:

$L_{\parallel}$ is a length of the reconnaissance area in km;
$L_{\perp}$ is a width of optical system view area, that is, a width of the reconnaissance area in km.

The length of the reconnaissance area is calculated by using equation

$$L_{\parallel} = t \cdot V, \qquad (10)$$

$$t = T - 2 \cdot t_1, \qquad (11)$$

Here:

t is UAV flight duration in reconnaissance area in hour;
V is UAV flight speed in km/hour;
T is the maximum UAV flight duration in hour.

Where: T is the maximum UAV flight duration.

$$t_1 = \frac{SS}{V}, \qquad (12)$$

Here:

t_1 is UAV flight duration from take-off point to front of reconnaissance area in hour;
SS is a distance from UAV take-off point to front of reconnaissance area in km.

The height relief outline of invisible and visible areas of some mountain terrain in the Azerbaijan Republic has been obtained and investigated experimentally by ArcGIS software (GIS technology), see

Figure 9. Let us consider the UAV flight chart (see Figure. 10) in mountain terrain between points *A* and *B* obtained from Figure 10: 1 – visible areas, 2 – invisible areas (from point *A*).

By using above indicated method, let us estimate a number of invisible areas, armored vehicles and military objects located between points *A* and *B*, which can be detected by one or several (for instance, 10) UAVs. “Trimble UX5 HP” UAV has been taken for consideration (Trimble UX5 HP Unmanned Aircraft System). This UAV “allows to quickly create a high quality orthomosaics and 3D models for applications such as survey grade mapping, field leveling, site and route planning, progress monitoring and asset mapping”.

For 15 mm lens at flight height 300 m the area/flight equals $S = 6{,}5\ km^2$ (see Figure 11). and $L_{\perp} = 2{,}55$ km. Let us take $SS = 10$ km and $V = 85$ km/h, then by using equations (9)÷(12) the authors have obtained:

Figure 9. The height relief of visible (solid line) and invisible (dash line) areas between observation point A and target point B; bold line is observation one

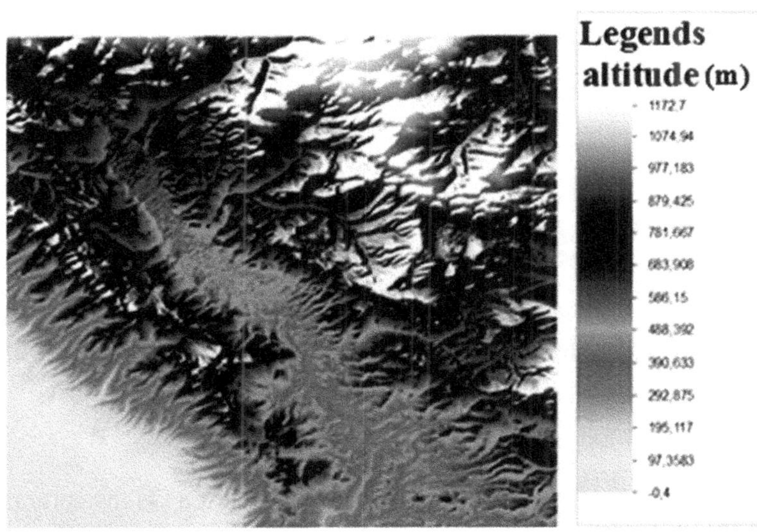

Figure 10. UAV flight chart between points A and B

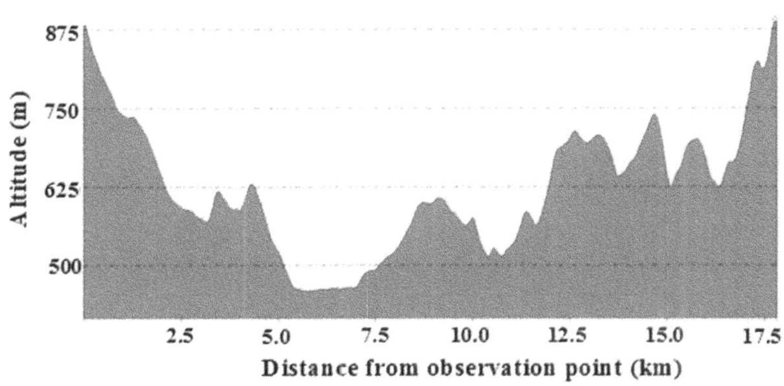

Figure 11. The chart of the area/flight at observation UAV flight

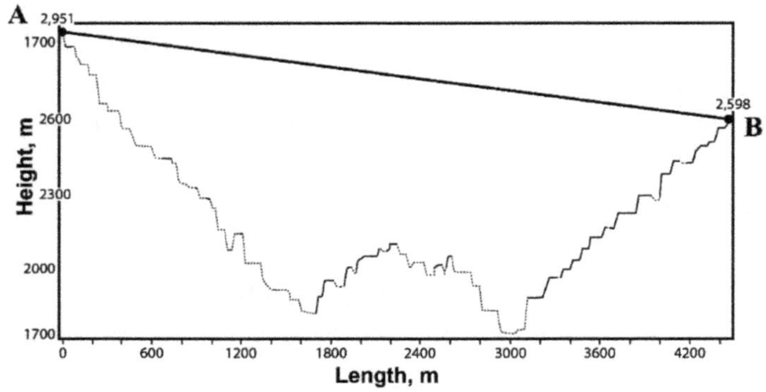

$$t_1 = \frac{SS}{V} = 0.12 \text{ h}$$

$$t = T - 2 \cdot t_1 = 0.67 - 0.24 = 0.43 \text{ h}$$

$$L_{||} = t \cdot V = 0.43 \cdot 85 = 37 \text{ km}$$

$$S_o = L_{||} \cdot L_{\perp} = 94 \text{ km}^2$$

$$S_o = 94 \text{ km}^2$$

From analysis of information given in Figure 9 authors have assessed that $\rho = 0.98$ km^{-1}. Then, authors have calculated from equation (5) that by using one UAV the commander can detect

$N_1 = P \cdot \rho \cdot S_o = 88$ invisible areas.

If the commanders use 10 identical UAVs then, based on equation (7) the authors have obtained $N_{10} = 92$. So, the detection efficiency of invisible areas is not too increased, and it is reasonable to use one UAV. Now, let us admit that there are 10 units of armored vehicles per 1 km^2 of invisible area. Let us consider that a factor of camouflage level of armored vehicles K=0.9, a likelihood degree of false positions $\xi = 0.9$, a number of false hidden η =0.01, then the detection probability of armored vehicles by one UAV in accordance with equation (6) $P = 0.1$. The number of hidden armored vehicles is detected during one UAV flight in accordance with equation (5) $N_1 = 94$ units. If the commanders take 10 identical UAVs then, in accordance with equation (8) $N_{10} = 611$ units. That is, it is more efficiency to use 10 UAVs for detection invisible armored vehicles.

The obtained results could be explained if the commanders use initial GIS information about terrain, then the probability of one UAV detection of the invisible areas increase. Therefore, the application of 10 UAVs doesn't improve much the probability of invisible areas detection. The authors take into account that the probability of revealing of one armored vehicle is less than the probability of revealing of one invisible area. Therefore, if commanders solve a problem of the camouflaged armored vehicles detection on the invisible areas, it is natural, that commanders use several UAVs, then the commanders can detect more camouflaged armored vehicles or military objects.

It should be noted, in the Azerbaijan Republic the real war activities had shown that it is not a hypothetical occurrence of multiple objects. The high cruise speed, the high flight height, unexpectedness, the flight noiseless, the sky blue color of UAV, fully automated flight, the geometrical shape and the special structural material of UAV provide the upper level of UAV defense. Our experiments in war conditions (unpublished) had shown that at flight height 300 m and battle-front 2.5 km putting 3-5 UAVs (in mountainous conditions) is real. In this case, all of them perform operational mission and return to base.

The offered method of application of the reconnaissance UAV possibilities allows to take into account correctly their significance and place in the informational support system, to reveal the nature and quantitative measures affecting on the efficiency of enemy targets detection system, and also to give the proof of priority and the importance of the fight against improvement such targets. Based on this method the simulation and training UAVs war-game software for any mountainous conditions was developed.

So, the quantitative method of the assessment of invisible areas and military objects in mountain terrain by using UAF is offered. This method has been applied for quantitative assessment in some mountain terrain of the Azerbaijan Republic. For calculations there have been used data early obtained by GIS technology. It was established that for quantitative analysis of invisible areas it is more efficiently to use one UAV, but for detection of armored vehicles and military objects it is more efficiently to use several UAV. The probabilities of detection of the invisible areas and objects depend generally on the terrain relief and UAV speed.

THE NUMERICAL ESTIMATION METHOD OF A TASK SUCCESS OF UAV RECONNAISSANCE FLIGHT

In the previous part of chapter, by using GIS technology, a number of invisible areas and military objects has been estimated. Now, by using the numerical estimation method a number of required PUA will be assess for success reconnaissance flight. The numerical estimation of performance of battle task in mountainous region by using UAV reconnaissance flight is very important for theatre-of-war planning of battle operations. In this part, the numerical estimation method of a task support success of UAV reconnaissance flight in mountainous battle condition has been offered and considered. The minimum of permissible safety height of UAV flight by using GIS technology has been determined. Only the UAV tactic reconnaissance flight is considered in this part of chapter. For solution of this task, the various input parameters have been taking into account: security events set of the troops, security events set of UAV reconnaissance flights, set of the UAV performance characteristics, set of UAV operating conditions, set of mountainous landscape characteristics. The calculations have been carried out for the specific conditions of one of the mountainous broken ground region of the Azerbaijan Republic.

The numerical estimation of performance of battle task in mountainous region by using UAV reconnaissance flight is very important for theatre-of-war planning of battle operations. The numerical estimation is mean a level of the battle task performance, which is varied from 0 to 1. For solution of this task it is necessary to know next input parameters:

- Security events set of the troops;
- Security events set of UAV reconnaissance flights;

- Set of the UAV performance characteristics;
- Set of UAV operating conditions;
- Set of mountainous landscape characteristics.

Below, the method of numerical estimation method of a success level of the battle support based on the application of UAV reconnaissance flight in mountainous battle conditions has been offered and considered. The battle (strategically) support is activities complex realized for the purpose of creation of the favorable conditions for successful battle activities and enemy impact degradation for any conditions. The battle support (security) includes reconnaissance, electronic warfare, tactical camouflage, engineer support, radiation-chemical and biological protection, aiming control, hydro-meteorological support and protection (Encyclopaedia of missile strategic army, 2009). Only UAV tactical reconnaissance will be considered below.

Taking into an account that the considered process is stochastic, then the numerical estimation of events realization can be carried out by using probabilistic factors: a degree of success of the event of battle support, a guaranteed probability of realization of the event of battle support, a probability of realization of the *M* events among of all *Z* events etc. For the purpose of finally solution (choice reasonable option of battle support) it is necessary to use a validity criterion (Podinovski, 2007). The investigation task is: by using UAV to find reasonable option of a battle support *U** belong among *U* set when a mathematical expectation of the number of successfully performed events of battle support will be the largest that is

$$U^{*} \in U : M_{suc}(t) = \sup M_{suc}^{*}\left(t\right), \qquad (13)$$

Here: $\sup M_{suc}^{*}\left(t\right)$ is determined by using method of the examination of obtained calculated values of the mathematical expectation in sample of its general set. Calculated values of the mathematical expectation of the number of successfully performed events are calculated by using next formula (Kotov & Dmitriev, 2014; Robert, 2008)

$$M_{suc}\left(t\right) = \sum_{i=1}^{Z} P_{suc_i}\left(t\right), \qquad (14)$$

Here: $P_{suc_i}\left(t\right)$ is a probability of successfully performed *i* event of UAV battle support; *Z* is a common number of performing events.

The providing of accident-free UAV application is the main task during UAV flight in mountainous difficult broken relief. The mathematical model of UAV flight in regime of the terrain following was offered in (Gulevich, Melnikov & Suvorov, 2016). The minimal permissible safety height of flight is the main operation factor determined UAV flight safety in regime of the terrain following. The minimal permissible safety height of UAV flight is determined from the conditions of UAV non-collision with ground. It is dependent from UAV mobility and maximum overloads, the rate of overloads forming, the range of permissible flight speed:

$$H_{min} = f(n, k_{max}, t, \Delta V),$$

Here:

n is a value of the normal component of overload determined by the limiting value of lift force which can be formed at instantaneous values of flight height and speed;
k_{max} is a maximum overload (UAV ultimate strength limit) determined by the maximum permissible value of lift force;
t is a duration of normal overload formation;
ΔV is a range of flight speed.

The duration of normal overload formation t defines possibilities of violent maneuvers making demanded an agility of flight trajectory curvature in UAV plane of symmetry. It is very important during flyby around mountainous difficult broken relief of the terrain. The duration of normal overload formation t and type of dependence $\Delta n = f(t)$ is dependent from UAV moment of inertia relatively transverse axis; damping moment; static steadiness moments, efficiency of elevation rudder, sluggishness of UAV moving control.

Let's determine the minimum of permissible safety height of UAV flight by using GIS technology. The authors use experimental data obtaining in our previous work. The task of determination of the height relief for specific region of the Azerbaijan Republic by using GIS technology had been considered in this work. Let's consider the features of UAV reconnaissance flight between observation point *A* and point *B* (see figure 9). Using analysis of these data commanders made next conclusions:

- At UAV flight on 4400 m distance the mountain with 250 m height is the highest obstacle 1600 m away from point *A*;
- The hill with 50 m height is the highest among hills 1100 m away from point *A*;
- The cruising speed of used wing-shaped "Trimble UX5 HP" UAV is 24 m/sec;
- The widths of the largest depression in point of 1600-1700 m from away *A* and in point of 2900-3100 m from away *A* are from 100 m until 200 m;
- The estimated flight time these depressions are from 5 until 10 sec.

Using these date commanders can estimate that the minimum permissible safety height of UAV flight in regime of the terrain following taking into account UAV speed change is ≈ 50 m. Now, let's consider and estimate the probability (a success level) of UAV reconnaissance flight realization. Let us write the expression of an objective function of successful realization of one battle task based on one UAV application. Then, for the battle conditions, the authors get the next expression:

$$P_{suc}(t) = K_{per}(t)\cdot K_{env}(t)\cdot P_{unf}(t)\cdot P_{im}(t)\cdot P_{hid}(t)\cdot P_{inf}(t)\cdot P_{unaf}(t) \quad (15)$$

The authors don't take into account correlations between factors in (15).

Here:

$K_{per}(t)$ is a factor of the personnel combat readiness for the battle tasks realization;
$K_{env}(t)$ is a factor of the nature features (mountainous landscape, wind etc.);
$P_{unf}(t)$ is a probability of UAV fault-free operation;
$P_{im}(t)$ is a probability of objects and events identification;

$P_{hid}(t)$ is a probability of hidden UAV operation;
$P_{inf}(t)$ is a probability of information transfer to central post;
$P_{undes}(t)$ is a probability of UAV un-destruction.

For getting the specific expressions in (15) authors need to choose the theoretical laws of random values distribution. In considered case authors can choose exponential and normal distribution laws (Podinovski, 2007; Kotov, 2014). At the same time, for determination of the values of coefficients and probabilities in objective function (15) the authors need to use empirical data obtained, for example, in tactical exercises. In case of realization of the M events of battle support by using one UAV, the probabilities of successful realization of the indicated events are determined by formula

$$P_{suc_M}\left(t\right) = \prod_{i=1}^{M} P_{suc_i}\left(t\right), \tag{16}$$

In case of realization of the K events of battle support from Z events the authors can use Bernoulli's formula:

$$P_{sucKZ}(t) = C_{KZ} \cdot P_{sucK}(t) \cdot [1 - P_{suc}(t)]^{Z-K}, \tag{17}$$

Here: C_{KZ} is the number of combination of executed battle support events

$$C_{KZ} = \frac{Z!}{K!\left(Z-K\right)!} \tag{18}$$

The guaranteed probability $P_G(Z,t)$ of successful realization Z events of UAV battle support in combat task conditions can be determined by using Poisson's formula (Gulevich, Melnikov & Suvorov, 2016):

$$P_G\left(Z,t\right) = \exp\left(-t\right) \sum_{K=0}^{Z} \frac{\left(t\right)^K}{K!} \tag{19}$$

Here: $\lambda = 1/\text{T}$, T is average time of battle task realization.

Let us carry out specific calculations for practice event. Let us assume that personal is ready at all to realize combat task, that is, $K_{per}(t) = 1$. The reconnaissance event is realized in conditions of mountainous difficult broken relief of the terrain. Given region is characterized by leap direction and speed of air flow till 15-20 m height, and by unstable state of atmosphere till 100 m height. The UAV has obstacle's sensors. The UAV's frame is made on based of very firm and light composite fiber carbon material. This UAV's frame can be stand high overloads and can execute spin round. Based on the expert judgements and height relief the authors can assume that $K_{env}(t) = 0{,}90$.

The UAV used by us has high reliability (Trimble UX5 HP Unmanned Aircraft System), therefore $P_{unf}(t)$=0,92. UAV has high-precision optics, and it provides a high probability of events and objects identification that is $P_{im}(t) = 1$. Using the results of investigations authors determine a probability of UAV hidden operation $P_{hid}(t) = 0{,}80$.

Taking into account that the distance of information transfer is less 5000 m, then the authors can assume that the probability of information transfer to central post is $P_{inf}(t) = 1$. Taking into account that UAV executes flight not above enemy region then the probability non-destruction of UAV is $P_{naf}(t) = 1$.

Thereby, taking into account above obtained parameters and coefficients in (15) the authors can determine the probability of successful realization (level realization) of reconnaissance task in conditions of mountainous difficult broken relief of the terrain:

$$P_{suc}(t) = 1{\cdot}0{,}90{\cdot}0{,}92{\cdot}1{\cdot}0{,}80{\cdot}1{\cdot}1 = 0{,}66$$

or

$$P_{suc}(t) = 66\%$$

Mountainous very difficult broken relief of the terrain and atmospheric meteorological condition causes such not high probability. When N identical UAVs execute one battle task (Kotov, 2016) then the authors have

$$P_{suc}\left(t\right) = 1 - \prod_{i=1}^{N}\left[1 - P_{suci}\left(t\right)\right] \qquad (20)$$

When $N = 2$ UAVs, then $P_{suc}(t) = 88\%$.
When $N = 3$ UAVs, then $P_{suc}(t) = 96\%$.

Thereby, 3 UAVs are sufficient for successful, with high probability level, execution of combat reconnaissance task. Thus, the numerical estimation method of a combat task success of UAV reconnaissance flight in mountainous battle condition has been developed. For specific conditions of the mountainous difficult broken relief of one of the terrain of the Azerbaijan Republic, it is determined that the probability of successful execution of combat reconnaissance task by using one UAV is 66%. 3 UAVs are sufficient for successful, with high probability level (96%), execution of combat reconnaissance task. The calculations have shown that the minimum permissible safety height of UAV flight in regime of the terrain following taking into account given parameters is $\approx$ 50 m. Obtained results can be useful when determination of the application limits of UAV complexes in the system of execution of the combat support tasks of troops.

At the conclusion, authors note that for terrain battle reconnaissance the military application of geographic information system and photogrammetry technologies can be implemented if, first of all, photogrammetry specified photographing carry out, shot photographies in real-time mode are sent by the radio communication to center, an operator implements efficiently the terrain orthophotomap making (determination of coordinates of the sought-for stationary object, efficiently decision making, etc.) or constructs the terrain detailed 3D model for orginization in future the battle operations and combat control.

CONCLUSION

The development and application of small aviation automated systems is one of the most importance problem in the aviation and aerospace industries, in particular in military area. In chapter, the development and military application of small aviation automated systems (UAV) is considered.

The revealing of invisible military objects on the terrain is one of the most important tasks for successful realization of battle operation. For solution of this task, various methods are used, including UAV. The results of visual modelling of the revealing process of invisible military objects from one of point of observation on the mountainous terrain of the Azerbaijan Republic by using UAV are presented. Autodesk-3ds-max program has been used for modelling. It is shown that the higher and farther UAV flight from observation point to enemy direction, the less invisible areas, and the less probability of hidden moving or concealment from view of the enemy battle trucks and troops. Obtained results will help correct to determine UAV flight traffic in real situation for specific terrain with aim of maximum survey of the terrain, minimization of invisible areas and reliable identification of the enemy hidden objects.

The observation conditions between two points of mountain terrain during battle operation have been investigated by using GIS technology. The experiments for estimation of observation situation in mountain terrain has been carried out for one of the chosen region of the Azerbaijan Republic. The visible and invisible areas have been revealed on the line of sight between two selected oversight points. The invisibility factors have been calculated and studied in depended on UAV's camera platform height. The visible and invisivle areas are determined under 0-180° angle in range 17941.16 m. The height profile between observation points and the digital heights model of investigated terrain have been constructed and analysed. At the various heights of camera observation platform of UAV the 3D heights models of the visible and invisible areas have been constructed. The invisibility factors between two detected viewpoints have been calculated. It is determined that if the height of camera platform is increased above-ground level then the invisible area is decreased. ArcGIS software has been used for development and calculation measured results. The experiments for estimation of observation situation in mountain terrain has been carried out.

The quantitative method of the invisible areas assessment and military objects in mountainous terrain are developed and offered by using UAVs. The probabilities of invisible area or enemy object detection are calculated by using single and *n* UAVs. The analysis of the dependence of revealing effectiveness, localization areas and military objects on the number of UAVs has been carried out. It is more efficiently to use one UAV for quantitative analysis of invisible areas, but it is more efficiently to use several UAVs for detection of armored vehicles and military objects. Our experiments in mountainous conditions had shown that at flight height 300 m and battle-front 2.5 km putting 3-5 UAVs is real. In this case, all of them perform operational mission and return to base. The numerical estimation method of a task support success of UAV reconnaissance flight in mountainous battle condition has been offered and considered. This method is based on the probabilistic approach. The minimum of permissible safety height of UAV flight by using GIS technology has been determined. The calculations have been carried out for the specific conditions of one of the mountainous broken ground region of the Azerbaijan Republic. For specific conditions of the mountainous difficult broken relief of one of the terrain of the Azerbaijan Republic, it is determined that the probability of successful execution of combat reconnaissance task by using one UAV is 66%. 3 UAVs are sufficient for successful, with high probability level (96%), execution of combat reconnaissance task. The calculations have shown that the minimum permissible safety height of UAV flight in regime of the terrain following taking into account given parameters is ≈ 50 m. Obtained results can be useful when determination of the application limits of UAVs complexes in the system of execution of the combat support tasks of troops.

Making a comparison between obtained results there have been concluded that by using small aviation automated systems and GIS software the various tasks of studying of the terrain profile and mapping for civil and military scopes can be efficiently solved.

REFERENCES

ArcGIS 3D analiz, eğitim dökümanı. (2008). İşlem şirketler qrubu, Ankara.

Autodesk 3ds-max visualization, simulation program. (2016). Retrieved from https://www.autodesk.com/products/3ds-max/overview

Bayramov, A. A., & Hashimov, E. G. (2018). Assessment of invisible areas and military objects in mountainous terrain. *Defence Science Journal*, *68*(4), 343–346. doi:10.14429/dsj.68.11623

Bayramov, A. A., Hashimov, E. G., & Khalilov, B. M. (2017). GIS technology and terrain orthophotomap making for military application. *Journal of Defense Resources Management.*, *8*(2), 81–90.

Butin, V. V. (2013). Remote manned flying objects as Remote Sensing data source. *Geomamics*, *3*, 24–27.

Commander and Staff Officer Guide. Army Tactics, Techniques, and Procedures ATTP 5-0.1. (2011). Headquarters Department of the Army.

FMI 3-04.155. (2006). *Army unmanned aircraft system operations headquarters*. FMI.

Gulevich, S. P., Melnikov, D. N., & Suvorov, A. P. (2016). *Mathematical modeling of UAV flight in the regime of the terrain following. In Perspectives of UAVs development and application. Pros. of Scientific Practical conference* (pp. 70–76). Kolomna: Ministry of Defense of the Russian Federation.

Hashimov, E. G., & Bayramov, A. A. (2017). Investigation of the observation conditions on the terrain of war operation using GIS technology. *Advanced Information Systems*, *1*(1), 65–69. doi:10.20998/2522-9052.2017.1.12

Hashimov, E.G., Bayramov, A.A., Abdullayev, F. & Mammadli A. (2017). Development of the multirotor Unmanned Aerial Vehicle. *National Security and Military Sciences, 3*(4), 21-31.

Hashimov, E. G., Nasibov, Ya. A., Bayramov, A. A., & Amanov, R. R. (2015). Application of the relief digital model for battle operations plannig. *Herbi Bilik.*, *4*, 63–69.

Karmanov, A. G. (2012). *GIS tutorial. Sankt-Peterburg Naional Reseach Universityof Information Technologies, mechanics and optics.*

Kotov, A. A. (2016). *Methodical estimation of task success of the combat support when robot system application. In Perspectives of UAVs development and application. Pros. of Scientific Practical conference* (pp. 135–140). Kolomna: Ministry of Defense of the Russian Federation.

Kotov, A. A., & Dmitriev, S. Yu. (2014). Main statements of the method of successful forming of data base taking into account aging and update. *Quality. Innovation. Education*, *5*, 51–55.

Nazarov, A. S. (2010). *Photogrammetry*. Minsk: TetraSystem.

Olaf, C. (2010). *SAGA: System for Automated Geoscientific Analyses. Physical Geography, University Hamburg*. GEOSTAT.

Petrov, M. V. (2013). UAV practical application experience Swinglet Cam. *Geoprofi*, *2*, 42–44.

Podinovski, V. V. (2007). *Introduction in the theory of importance measure*. Moscow: Fizmatgiz.

Robert, B. A. (2008). Basic probability theory. Dover Publications, Inc.

Soкolov, A. V., & Tixonov, M. L. (2008). GIS military application. *OBSERVEF*, *5*, 37–45.

Solovtsova, N. E. (Ed.). (2009). *Encyclopaedia of missile strategic army*. Moscow: RVSN.

Trimble UX5 HP Unmanned Aircraft System. (2015). Retrieved September 6, 2015, from http://www.kmcgeo.com/Datasheets/UX5HP.pdf

Varvarina, E. A. (2012). Remote manned flying objects using for construction of orthophotoplanes of the linear objetcs. *Earth Science*, *2*, 66–70.

Wenger, N., Antoniev, A., & Gorod, A. (2013). The International Space Station: Applying system of systems methodology. *Proceedings of 2013 8th International Conference on System of Systems Engineering*, 235 - 242. 10.1109/SYSoSE.2013.6575273

Zagorski, M.Y., Bogdanov, V.L., Garmanov, V.V., Koroleva, V.P., & Ryabov, Y.V. (2015). Mathematical model of stereomate films and the renewal algorithm of terrestial relief on the basis on their. *Modern Problems of Space Earth Remote Sensing, 3,* 56.

Zaritski, V. N., & Charkevich, L. A. (2007). *General Tactics*. Tambov: TGTU.

Chapter 11
Automated System of Stabilization and Position Control of Aviation Equipment

Olha Sushchenko
National Aviation University, Ukraine

ABSTRACT

In this chapter, the author presents the problems of design of the robust automated system for stabilization and control of platforms with aircraft observation equipment. The mathematical model of the triaxial stabilized platform is developed. The procedure of synthesis of robust stabilization system based on robust structural synthesis is represented. The above-mentioned procedure uses loop-shaping approach and method of the mixed sensitivity. The matrix weighting transfer functions are obtained. The optimization programs in MatLab are developed. The developed procedures are approved based on the results of simulation by means of the appropriate Simulink model. The obtained results can be useful for unmanned aerial vehicles and aircraft of special aviation, which are used for monitoring technical objects and aerial photography. The technical contributions are procedures of the robust controller design represented as the flowchart. The proposed approach is validated by application of the theoretical suppositions to the concrete example and appropriate simulation results.

INTRODUCTION

The motivation of the proposed work is to fill gaps in solving conflict requirements such as accuracy and robustness to the aviation automated systems of stabilization of systems with observation equipment.

Nowadays the robust control is widely used in the design of automated aviation systems. Such an approach is well studied and widespread in the design of aircraft control systems. A novelty of the presented research is the application of the known methodology of design of the robust automated control systems in a new area such as the development of inertially stabilized gimballed platforms operated on aircraft. Respectively, the new type of moving vehicle (platform) and the new area of the application (inertial stabilization) require new special researches. The new mathematical models and design procedures

DOI: 10.4018/978-1-5225-7709-6.ch011

based on the H_∞-method including loop-shaping and method of mixed sensitivity, which is given in this chapter, present the theoretical contributions. The technical contribution is an improvement of accuracy of observation processes, which take place on aircraft during mapping, monitoring of terrain and so on.

The necessity to provide stabilization of airborne observation equipment is one of the important problems of modern aviation. The typical observation equipment operated on modern unmanned aerial vehicles (UAVs) and special aircraft includes television camera, digital camera, and laser scanner. Above listed types of observation equipment allow solving problems cartography and aerial survey by means of the digital photogrammetry and aerial photography using television camera of the high resolution. The cartographic survey can be implemented by means of the laser scanner. Using of these devices provides development of the relief model by means of sending laser impulses and analysis of reflected signals. The high quality of an image can be achieved by means of the system for stabilization and control of orientation by line-of-sights of observation devices. These devices are mounted on UAVs and aircraft of special aviation to provide a solution of observation problems.

Functioning of observation equipment operated on UAVs and aircraft of special aviation is accompanied with parametric and coordinate disturbances. First of all, these disturbances are caused by object aerodynamics and the influence of wind. The modern trend of the design of stabilization systems operated in difficult conditions of real operation is using of robust control. Such an approach provides stabilization of observation equipment in conditions of aircraft angular motion caused by various disturbances.

It should be noted that approaches to robust control of aircraft motion are well developed and widely used in modern aviation. Robust control of systems, which carry out functions of observation equipment stabilization, has some features defined by a type of the vehicle, on which the observation equipment is mounted, and operating conditions (Sushchenko & and Tunik, 2013). So, development of procedures for the design of an automated system for stabilization of observation equipment is a very important and urgent problem of the modern aviation.

The main goal of the design of robust automated systems lies in a search of the control law, which can provide accuracy of stabilization in some given range under conditions of uncertainty. This uncertainty can be caused by many factors such as external disturbances, inaccuracy of the mathematical model and unmodeled dynamics.

The objective of the proposed work is to develop the design procedure of the discrete robust automated stabilization system, which implements an increase of quality of observation processes on the board of an aircraft.

REVIEW OF PREVIOUS LITERATURE

The main directions of the design of inertially stabilized platforms are given in (Hilkert, 2008). The problems of robust control are not mentioned in this publication although robustness is very important for aviation inertially stabilized platforms, which function in difficult conditions of real operation. The features of the design of aviation robust automated stabilization system depend significantly on the type of the researched vehicle. For example, papers (N. Wu & J. Yu. 2018) and (S. Kim, S. Cho & H. Him at all, 2018) are dealt with such different moving vehicles as the hypersonic vehicle and multirotor respectively.

It should be noticed that principles of the design of aviation automated control systems differ significantly for such objects as aircraft and moving platforms with a payload of the different type. For example, procedures of the design of the marine robust inertial platforms with navigation measuring instruments are presented in the paper (Sushchenko, 2017). The given chapter continues researches (Sushchenko&Goncharenko, 2016), which are dealt with the design of aviation robust platforms with observation equipment.

The difference between present research and previous publications is the development of the detailed mathematical model for the design of the triaxial inertially stabilized platform. Using stabilization by all three axes improves significantly the accuracy of observation processes. Furthermore, the algorithm of procedure of the robust structural synthesis is represented that can be useful for the design of the robust stabilization systems of the wide class.

BACKGROUND

Design of robust automated systems for stabilization of inertial platforms with observation equipment requires solving some problems such as the development of a mathematical model of the inertial platform with payload and a design procedure for the design of robust controller of the stabilization system.

Taking into consideration the paper (Hilkert, 2008), it can be formulated that the main stages of the development of the mathematical model of the inertial platform with the payload are as follows.

1. Creation of the model of the platform dynamics.
2. Creation of the model of the platform kinematics.
3. Development of the model of the platform with payload in the state space taking into consideration Euler equations of dynamics and kinematic.
4. Development of the full model of the inertial platform with payload taking into consideration nonlinearities inherent to real systems.

According to (Skogestad & Postlethwaite, 2001) design of the robust controller for stabilization system includes the following stages.

1. Choice of the method of robust controller synthesis based on features of the aircraft type and operating conditions.
2. Development of the augmented mathematical model using transfer weighting functions.
3. Forming of the optimization functional based on a combination of quality indices including performance, robustness, and control losses.
4. Synthesis of a robust controller.
5. Simulation of synthesized stabilization system and checking of obtained results.

One of the most widespread methods of design of robust systems is H_∞-synthesis (Skogestad & Postlethwaite, 2001). This approach represents the robust structural synthesis and can be used for the design of the system for stabilization of observation equipment (Sushchenko & Goncharenko, 2016).

MATHEMATICAL MODEL OF PLATFORM WITH PAYLOAD

The mathematical model of the inertially stabilized platform can be developed based on the equation of motion in the inertial space (Hilkert, 2008). For precision applications, such a model must take into consideration all stabilization channels (yaw, pitch, and roll).

To provide the high accuracy of observation processes it is necessary to implement stabilization of observation equipment by three aircraft axes. For this, the stabilized platform with observation equipment and inertial sensors is mounted in the triaxial gimbals. Such an approach provides the possibility of platform rotation in the wide range such as 360° by the yaw and ±70° by the pitch and roll. In the general case, the stabilization system can operate in the mode of the coarse leveling and the fine leveling. Implementation of these modes requires using inertial sensors such as rate gyroscopes and accelerometers.

So, consider the stabilization system, which consists of three channels. For such a system the stabilization plant of the yaw channel is the external frame of the gimbals. This frame implements rotation by an angle of the yaw. To represent the stabilization plant mathematically, it is necessary to take into consideration a mass and moments of inertia of pitch and roll frames, and platform with payload (observation equipment). The stabilization plant of the pitch channel is the internal frame of the gimbals. This frame implements rotation by an angle of the pitch. In this case, it is necessary to take into consideration a mass and moments of inertia of the roll frame and platform with the payload. Respectively, the plant stabilization of the roll channel includes the roll frame and platform with the payload. It should be noted that three rate gyroscopes and two accelerometers are mounted on the platform too. Gyroscope's measuring axes are orthogonal. The initial position of gyroscope axes coincides with the yaw, pitch, and roll axes. Measuring axes of accelerometers coincide with pitch and roll axes respectively. Such a set of inertial sensors (gyroscopes and accelerometers) provides the high accuracy of stabilization of observation equipment.

The mathematical model of the stabilization plant must include a description of its dynamics and kinematics.

Dynamic of the platform can be described by the Euler equations (Sapio, 2017)

$$\begin{aligned}
&\dot{\omega}_x J_x + \omega_y \omega_z (J_z - J_y) - (\omega_y^2 - \omega_z^2) J_{yz} - (\omega_x \omega_y + \dot{\omega}_z) J_{xz} + (\omega_x \omega_z - \dot{\omega}_y) J_{xy} = M_x;\\
&\dot{\omega}_y J_y + \omega_x \omega_z (J_x - J_z) - (\omega_z^2 - \omega_x^2) J_{xz} - (\omega_z \omega_y + \dot{\omega}_x) J_{xy} + (\omega_x \omega_y - \dot{\omega}_z) J_{yz} = M_y;\\
&\dot{\omega}_z J_z + \omega_x \omega_y (J_y - J_x) - (\omega_x^2 - \omega_y^2) J_{xy} - (\omega_x \omega_z + \dot{\omega}_y) J_{yz} + (\omega_x \omega_z - \dot{\omega}_x) J_{xz} = M_z,
\end{aligned} \tag{1}$$

where ω_x, ω_y, ω_z are projections of the platform angular rates on its own axes; J_x, J_y, J_z are moments of inertia of the platform with payload and inertial sensors; J_{yz}, J_{xz}, J_{xy} are the centrifugal moments of inertia relative to the gimbals axes; $\dot{\omega}_x$, $\dot{\omega}_y$, $\dot{\omega}_z$ are projections of the angular acceleration of the platform on its own axes; M_x, M_y, M_z are moments acting along the gimbals axes. It should be noted that the moments of inertia J_x, J_y, J_z in the expression (1) are equivalent moments of inertia, which take into consideration moments of inertia of the platform with payload and inertial sensors and moments of inertia of stabilizing motors J_m. The equivalent moments of inertia are defined by the expressions (Sushchenko & Goncharenko, 2016)

$$J_i = J_{pi} + n_r^2 J_m,\ i = x, y, z\ , \tag{2}$$

where J_{pi}, J_m are moments of inertia of the platform with payload and stabilizing moments respectively; n_r is the transfer ratio of the reducer. Relationships for the centrifugal moments of the platform can be determined in a similar way.

Moments acting on the platform include such components as moments of the dry friction in bearings of the gimbals, moments created by the stabilizing motors and disturbances moments. To create the mathematical model of the stabilization system in the mode of the fine leveling it is necessary to take into consideration that every control contour includes stabilizing motor, the pulse-width modulator (PWM), and rate gyroscope. Components of moments acting along the gimbals axes can be determined in the following way

$$M_{1i} = M_{fr}\text{sign}\omega_i\ ;$$

$$M_{2i} = \tilde{n}_m U_{ai}\ /\ R_a\ ;$$

$$M_{3i} = M_{dist\ i}\ ;$$

$$i = x_2, y_0, z_1\ , \tag{3}$$

where M_{fr} is the nominal friction moment in the bearings of the gimbals; c_m is a coefficient of loading at the motor shaft; U_{ai} are voltages of control windings of the armature motor; R_a is the resistance in the circuit of the motor armature; $M_{dist\ i}$ are the disturbance moments.

Voltages in the control winding of the motor armature are described by the expressions

$$T_a \dot{U}_{ai} + U_{ai} = k_{PWM} U_{PWMi} - n_r c_e \omega_i,\ i = x_2,\ y_0,\ z_1\ , \tag{4}$$

where T_a is the time constant in the armature circuit; c_e is a coefficient of proportionality between an angular rate of the stabilizing motor and electromotive force; k_{PWM} is a transfer constant of the linearized PWM; U_{PWMi} are voltages at the output of PWM.

Control voltages at the outputs of rate gyroscope $U_{\omega i}$ can be described in the following way

$$T_g^2 \ddot{U}_{\omega i} + 2\xi T_g \dot{U}_{\omega i} + U_{\omega i} = k_g \omega_i\ ,\ i = x, y, z\ , \tag{5}$$

where T_g is the time constant of the rate gyroscope; ξ is a coefficient of damping; k_g is a transfer coefficient of the rate gyroscope.

To describe kinematics of the stabilization system it is necessary to introduce reference frames, which describe a position of the stabilized platform in the inertial space. The position of the platform-axis reference frame $X_pY_pZ_p$ in the inertial space relative to the aircraft-axis reference frame $X_AY_AZ_A$ is defined by the sequence of three rotation angles ψ, ϑ, γ as it is shown in Fig. 1.

It is known that the mutual position of the moving body in the space is described by the matrices of directional cosines. For the above-mentioned reference frames, these matrices become

$$\mathbf{A}_1 = \begin{array}{c|ccc} & X_1 & Y_1 & Z_1 \\ \hline X_A & \cos\psi & 0 & \sin\psi \\ Y_A & 0 & 1 & 0 \\ Z_A & -\sin\psi & 0 & \cos\psi \end{array};$$

$$\mathbf{A}_2 = \begin{array}{c|ccc} & X_2 & Y_2 & Z_2 \\ \hline X_1 & 1 & 0 & 0 \\ Y_1 & 0 & \cos\vartheta & -\sin\vartheta \\ Z_1 & 0 & \sin\vartheta & \cos\vartheta \end{array};$$

$$\mathbf{A}_3 = \begin{array}{c|ccc} & X_p & Y_p & Z_p \\ \hline X_2 & \cos\gamma & -\sin\gamma & 0 \\ Y_2 & \sin\gamma & \cos\gamma & 0 \\ Z_2 & 0 & 0 & 1 \end{array} \tag{6}$$

Figure 1. Mutual position of the reference frames of the triaxial platform

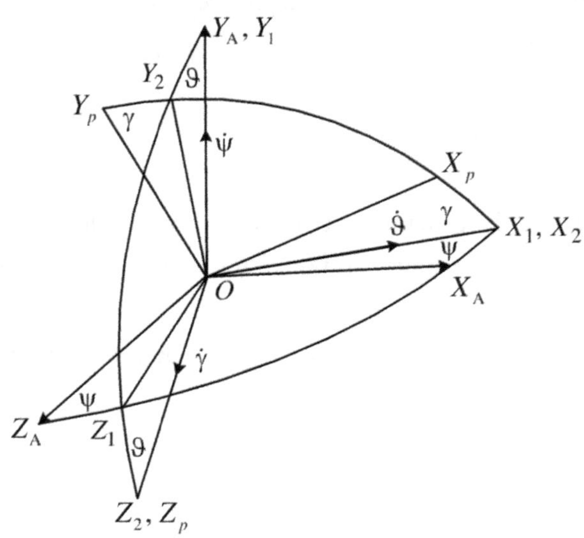

Based on relationships (6) matrices of directional cosines between reference frames $X_pY_pZ_p$, $X_AY_AZ_A$ and reference frames $X_AY_AZ_A$, $X_pY_pZ_p$ take the form

$$\mathbf{A} = \begin{array}{c} \\ X_A \\ Y_A \\ Z_A \end{array} \begin{array}{ccc} X_p & Y_p & Z_p \\ \cos\psi\cos\gamma + \sin\psi\sin\vartheta\sin\gamma & -\cos\psi\sin\gamma + \sin\psi\sin\vartheta\cos\gamma & \sin\psi\cos\vartheta \\ \cos\vartheta\sin\gamma & \cos\vartheta\cos\gamma & -\sin\vartheta \\ -\sin\psi\cos\gamma + \cos\psi\sin\vartheta\sin\gamma & \sin\psi\sin\gamma + \cos\psi\sin\vartheta\sin\gamma & \cos\psi\cos\vartheta \end{array} \quad (7)$$

$$\mathbf{A}^{\mathrm{T}} = \begin{array}{c} \\ X_p \\ Y_p \\ Z_p \end{array} \begin{array}{ccc} X_A & Y_A & Z_A \\ \cos\psi\cos\gamma + \sin\psi\sin\vartheta\sin\gamma & \cos\vartheta\sin\gamma & -\sin\psi\cos\gamma + \cos\psi\sin\vartheta\sin\gamma \\ -\cos\psi\sin\gamma + \sin\psi\sin\vartheta\cos\gamma & \cos\vartheta\cos\gamma & \sin\psi\sin\gamma + \cos\psi\sin\vartheta\sin\gamma \\ \sin\psi\cos\vartheta & -\sin\vartheta & \cos\psi\cos\vartheta \end{array} \quad (8)$$

The platform stabilizing angular rates can be determined in the following way. Let us suppose that the angular rate of the gimballed platform, on which the observation equipment and inertial sensors are mounted, is an angular rate of the platform caused by the control moments defined by the controller of the stabilization system. Taking this assumption into consideration the condition of the accurate stabilization takes the form

$$\Omega + \omega = 0 \quad (9)$$

If an angular rate of the aircraft, on which the stabilization system is operated, is determined by the expression $\omega_o = \omega_x \mathbf{X}_A + \omega_y \mathbf{Y}_A + \omega_z \mathbf{Z}_A$, and transition to a new position is implemented by rotation angles ψ, ϑ, γ, angular rates of all frames of the gimbals can be determined in the following way

$$\begin{bmatrix} \omega_{\psi x} \\ \omega_{\psi y} \\ \omega_{\psi z} \end{bmatrix} = \mathbf{A}_1^{\mathrm{T}} \begin{bmatrix} \omega_x \\ \omega_y \\ \omega_z \end{bmatrix} + \mathbf{A}_1^{\mathrm{T}} \begin{bmatrix} 0 \\ \dot{\psi} \\ 0 \end{bmatrix}; \begin{bmatrix} \omega_{\vartheta x} \\ \omega_{\vartheta y} \\ \omega_{\vartheta z} \end{bmatrix} = \mathbf{A}_2^{\mathrm{T}}\mathbf{A}_1^{\mathrm{T}} \begin{bmatrix} \omega_x \\ \omega_y \\ \omega_z \end{bmatrix} + \mathbf{A}_2^{\mathrm{T}} \left(\mathbf{A}_1^{\mathrm{T}} \begin{bmatrix} 0 \\ \dot{\psi} \\ 0 \end{bmatrix} + \begin{bmatrix} \dot{\vartheta} \\ 0 \\ 0 \end{bmatrix} \right);$$

$$\begin{bmatrix} \omega_{x1} \\ \omega_{y1} \\ \omega_{z1} \end{bmatrix} = \begin{bmatrix} \omega_{\gamma x} \\ \omega_{\gamma y} \\ \omega_{\gamma z} \end{bmatrix} = \mathbf{A}_3^{\mathrm{T}} \begin{bmatrix} \omega_{\vartheta x} \\ \omega_{\vartheta y} \\ \omega_{\vartheta z} \end{bmatrix} + \mathbf{A}_3^{\mathrm{T}} \begin{bmatrix} 0 \\ 0 \\ \dot{\gamma} \end{bmatrix} = \mathbf{A}_3^{\mathrm{T}}\mathbf{A}_2^{\mathrm{T}}\mathbf{A}_1^{\mathrm{T}} \begin{bmatrix} \omega_x \\ \omega_y \\ \omega_z \end{bmatrix} + \mathbf{A}_3^{\mathrm{T}} \left(\mathbf{A}_2^{\mathrm{T}}\mathbf{A}_1^{\mathrm{T}} \begin{bmatrix} 0 \\ \dot{\psi} \\ 0 \end{bmatrix} + \mathbf{A}_2^{\mathrm{T}} \begin{bmatrix} \dot{\vartheta} \\ 0 \\ 0 \end{bmatrix} + \begin{bmatrix} 0 \\ 0 \\ \dot{\gamma} \end{bmatrix} \right) =$$

$$\mathbf{A}^{\mathrm{T}} \begin{bmatrix} \omega_x \\ \omega_y \\ \omega_z \end{bmatrix} + \begin{bmatrix} \dot{\vartheta}\cos\gamma + \dot{\psi}\cos\vartheta\sin\gamma \\ -\dot{\vartheta}\sin\gamma + \dot{\psi}\cos\vartheta\cos\gamma \\ \dot{\gamma} - \dot{\psi}\sin\vartheta \end{bmatrix} \quad (10)$$

The first members in the formulae (10) are projections of the angular rate of the aircraft, on which the platform with observation equipment is operated. The second members of the formulae (10) are projections of the platform relative angular rate. In the considered situation, we will believe that aircraft's axes coincide with the axes of the gimballed platform. Taking this assumption into consideration the sequence of the platform rotations during stabilization will correspond to Fig. 2.

In accordance with Fig. 2, the vector of the stabilizing angular rate can be determined by the expression

$$\Omega = \dot{\alpha}\mathbf{Y}_p + \dot{\beta}\mathbf{X}_1 + \dot{\phi}\mathbf{Z}_2, \qquad (11)$$

where $\dot{\alpha}$, $\dot{\beta}$, $\dot{\phi}$ are angular rates of rotations of the external frame, internal frame and platform respectively.

Based on (9) the conditions of the precision stabilization can be represented in the following form

$$\Omega_{x_p} + \omega_{x_p} = 0;$$

$$\Omega_{y_p} + \omega_{y_p} = 0;$$

$$\Omega_{z_p} + \omega_{z_p} = 0 \qquad (12)$$

Projections of the stabilizing angular rate of the platform on its own axes taking into consideration Figures 1, 2 and expressions (10) are determined by the relations

Figure 2. The sequence of rotations of the gimballed platform during stabilization

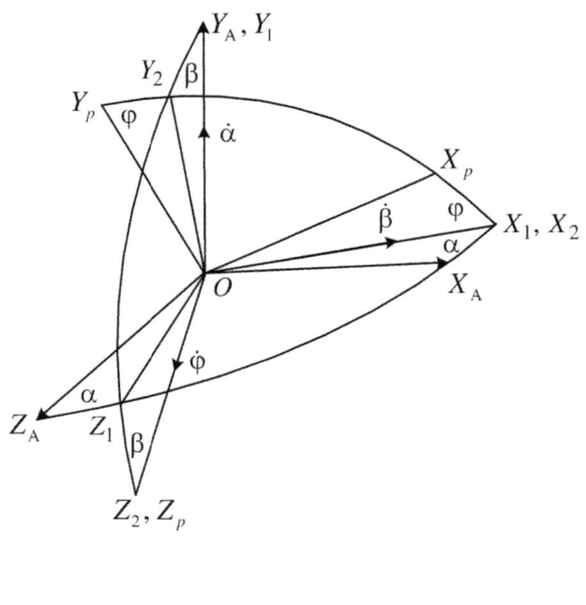

$$\Omega_{x_p} = \dot{\beta}\cos\gamma + \dot{\alpha}\cos\vartheta\sin\gamma\,;$$

$$\Omega_{y_p} = -\dot{\beta}\sin\gamma + \dot{\alpha}\cos\vartheta\cos\gamma\,;$$

$$\Omega_{z_p} = \dot{\phi} - \dot{\alpha}\sin\vartheta \tag{13}$$

Based on equations (11), (12) and relationships (13) it is possible to obtain expressions

$$-\omega_{x_p} = \dot{\beta}\cos\gamma + \dot{\alpha}\cos\vartheta\sin\gamma\,;$$

$$-\omega_{y_p} = -\dot{\beta}\sin\gamma + \dot{\alpha}\cos\vartheta\cos\gamma\,;$$

$$-\omega_{z_p} = \dot{\phi} - \dot{\alpha}\sin\vartheta \tag{14}$$

Using expressions (14) after conversions, it is possible to obtain the differential equations, which characterize the process of changing angles of the gimbals frames position during stabilization

$$\dot{\alpha} = -\frac{1}{\cos\vartheta}(\omega_{xp}\sin\gamma + \omega_{yp}\cos\gamma)\,;$$

$$\dot{\beta} = -\omega_{xp}\cos\gamma + \omega_{yp}\sin\gamma\,; \tag{15}$$

$$\dot{\phi} = \frac{\sin\vartheta}{\cos\vartheta}(\omega_{xp}\sin\gamma + \omega_{yp}\cos\gamma) - \omega_{zp}$$

On the one hand, stabilizing control moments are formed relative to the gimbals axes. On another hand, the moments of the platform in accordance with (1) are determined in projections onto the platform axes. Therefore it is necessary to perform agreement of angular rate projections on different axes. The expressions for determination of the stabilizing control moments can be determined after the following transformations

$$\begin{bmatrix} M_{xp} \\ M_{yp} \\ M_{zp} \end{bmatrix} = \mathbf{A}_3^{\mathrm{T}} \begin{bmatrix} 0 \\ 0 \\ M_{z\gamma} \end{bmatrix} = \begin{bmatrix} 0 \\ 0 \\ M_{z\gamma} \end{bmatrix};\ \begin{bmatrix} M_{xp} \\ M_{yp} \\ M_{zp} \end{bmatrix} = \mathbf{A}_3^{\mathrm{T}}\mathbf{A}_2^{\mathrm{T}} \begin{bmatrix} M_{x\vartheta} \\ 0 \\ 0 \end{bmatrix} = \begin{bmatrix} \cos\gamma M_{x\vartheta} \\ -\sin\gamma M_{x\vartheta} \\ 0 \end{bmatrix};$$

$$\begin{bmatrix} M_{xp} \\ M_{yp} \\ M_{zp} \end{bmatrix} = \mathbf{A}_3^{\mathrm{T}}\mathbf{A}_2^{\mathrm{T}}\mathbf{A}_1^{\mathrm{T}} \begin{bmatrix} 0 \\ M_{y\psi} \\ 0 \end{bmatrix} = \begin{bmatrix} \cos\vartheta\sin\gamma M_{y\psi} \\ \cos\vartheta\cos\gamma M_{y\psi} \\ -\sin\vartheta M_{y\psi} \end{bmatrix} \tag{16}$$

Based on expressions (16), relationships for determination of moments for control of platform motion are represented in the following way

$$M_{xp} = \cos\gamma M_{x\vartheta} + \cos\vartheta\sin\gamma M_{y\psi};$$

$$M_{yp} = -\sin\gamma M_{x\vartheta} + \cos\vartheta\cos\gamma M_{y\psi};$$

$$M_{zp} = M_{z\gamma} - \sin\vartheta M_{y\psi}. \tag{17}$$

Relationships (1) – (8), and (15), (16) present the mathematical model of the system for stabilization of observation equipment operated on the aircraft.

The modern control theory foresees design of control robust systems in two stages (Skogestad & Postlethwaite, 2001). The first stage includes robust structural synthesis based on the linearized model in the state space. The second stage lies in checking the synthesized system by means of simulation using the full mathematical model taking into consideration all the non-linearities inherent to real systems. Depending on simulation results the first stage can be repeated after changing initial data or weighting coefficients or weighting transfer functions of the optimization criterion.

So, implementation of the robust structural synthesis requires using the linearized model in the state space. Such a model can be obtained based on relationships (1) – (8), and (15), (16) using the following assumptions:

1. Neglect of the centrifugal moments of the platform and difference of the axial moments that allows significant simplification of the expression (1);
2. Accounting only small angles of the platform rotation that allows simplifying the expressions (7), (8), (15), (16);
3. Changing non-linear dry friction moments by the linearized moments (Sushchenko & Goncharenko, 2016);
4. Using the linearized PWM model;
5. Neglect of disturbance moments acting on the platform in the expression (1).

The expressions (1) – (8), (15) – (17) and above listed suppositions allow describing the platform motion by the set of the linearized differential equations

$$\dot{\alpha} = \omega_x;$$

$$\dot{\beta} = \omega_y;$$

$$\dot{\gamma} = \omega_z;$$

$$\dot{U}_{\omega x} = U_{\omega dx};$$

$$\dot{U}_{\omega y} = U_{\omega dy};$$

$$\dot{U}_{\omega z} = U_{\omega dz};$$

$$\dot{U}_{ax2} = [-U_{ax2} + k_{PWM} U_{PWMx2} - n_r c_e \omega_{x2}] / T_a;$$

$$\dot{U}_{ay0} = [-U_{ay0} + k_{PWM} U_{PWMy0} - n_r c_e \omega_{y0}] / T_a;$$

$$\dot{U}_{az1} = [-U_{az1} + k_{PWM} U_{PWMz1} - n_r c_e \omega_{z1}] / T_a;$$

$$\dot{\omega}_{x2} = [-f_x \omega_x - c_m U_{ax2} / R_a] / J_x;$$

$$\dot{\omega}_y = [-f_y \omega_y + c_m U_{ay0} / R_a] / J_y;$$

$$\dot{\omega}_z = [-f_z \omega_z + c_m U_{az1} / R_a] / J_z;$$

$$\dot{U}_{\omega dx} = [-2\xi T_g U_{\omega dx} - U_{\omega x} + k_g \omega_x] / T_g^2;$$

$$\dot{U}_{\omega dx} = [-2\xi T_g U_{\omega dy} - U_{\omega y} + k_g \omega_y] / T_g^2;$$

$$\dot{U}_{\omega dx} = [-2\xi T_g U_{\omega dz} - U_{\omega z} + k_g \omega_z] / T_g^2, \qquad (18)$$

where $U_{\omega dx,}\ U_{\omega dy}, U_{\omega dz}$ are derivatives of voltages $U_{\omega x,}\ U_{\omega y}, U_{\omega z}$; f_x, f_y, f_z are coefficients of the linearized friction moments.

Based on the expressions (18) and above listed assumptions the linearized model of the stabilization plant can be described by the vector of states

$$x^{\mathrm{T}} = [\alpha \quad \beta \quad \gamma \quad U_{\omega x} \quad U_{\omega y} \quad U_{\omega z} \quad U_{ax2} \quad U_{ay0} \quad U_{az1} \quad \omega_x \quad \omega_y \quad \omega_z \quad U_{\omega dx} \quad U_{\omega dy} \quad U_{\omega dz}], \qquad (19)$$

where $U_{\omega d_x}, U_{\omega d_y}, U_{\omega d_z}$ are derivatives of voltages $U_{\omega_x}, U_{\omega y}, U_{\omega_z}$, and the quadruple of matrices of the state space A, B, C, D

$$A=\begin{bmatrix}
0&0&0&0&0&0&0&0&0&1&0&0&0&0&0\\
0&0&0&0&0&0&0&0&0&0&1&0&0&0&0\\
0&0&0&0&0&0&0&0&0&0&0&1&0&0&0\\
0&0&0&0&0&0&0&0&0&0&0&0&1&0&0\\
0&0&0&0&0&0&0&0&0&0&0&0&0&1&0\\
0&0&0&0&0&0&0&0&0&0&0&0&0&0&1\\
0&0&0&0&0&0&-1/T_a&0&0&-n_r c_e/T_a&0&0&0&0&0\\
0&0&0&0&0&0&0&-1/T_a&0&0&-n_r c_e/T_a&0&0&0&0\\
0&0&0&0&0&0&0&0&-1/T_a&0&0&-n_r c_e/T_a&0&0&0\\
0&0&0&0&0&0&c_m/(R_a J_x)&0&0&-f_x/J_x&0&0&0&0&0\\
0&0&0&0&0&0&0&c_m/(R_a J_y)&0&0&-f_y/J_y&0&0&0&0\\
0&0&0&0&0&0&0&0&c_m/(R_a J_z)&0&0&-f_z/J_z&0&0&0\\
0&0&0&-1/T_g^2&0&0&0&0&0&k_g/T_g^2&0&0&-2\xi/T_g&0&0\\
0&0&0&0&-1/T_g^2&0&0&0&0&0&k_g/T_g^2&0&0&-2\xi/T_g&0\\
0&0&0&0&0&-1/T_g^2&0&0&0&0&0&k_g/T_g^2&0&0&-2\xi/T_g
\end{bmatrix},$$

where f_x, f_y, f_z are coefficients of the linearized dry friction moments;

$$B^{\mathrm{T}}=\begin{bmatrix}
0&0&0&0&0&0&k_{PWM}&0&0&0&0&0&0&0&0\\
0&0&0&0&0&0&0&k_{PWM}&0&0&0&0&0&0&0\\
0&0&0&0&0&0&0&0&k_{PWM}&0&0&0&0&0&0
\end{bmatrix};$$

$$C=\begin{bmatrix}
0&0&0&1&0&0&0&0&0&0&0&0&0&0&0\\
0&0&0&0&1&0&0&0&0&0&0&0&0&0&0\\
0&0&0&0&0&1&0&0&0&0&0&0&0&0&0
\end{bmatrix};$$

$$D=\begin{bmatrix}0&0&0\\0&0&0\\0&0&0\end{bmatrix} \quad (20)$$

The mathematical model in the state space (19), (20) can be used in the procedure of the robust structural synthesis.

ROBUST PARAMETRIC OPTIMIZATION

The various and conflicting requirements are given to the inertially stabilized platforms operated on aircraft in difficult conditions of real operation. The most important conflict is the necessity to ensure the accuracy and robustness at the same time. To solve this conflict it is necessary to find the compromise solution. It should be noted that requirements to the system accuracy during the influence of the deterministic and stochastic disturbances are conflicting too.

One of the approaches to solving the multi-criterion problem is vector optimization. For the systems of the considered type in some applications, it is convenient to use robust parametric optimization because it allows successful modification of existing systems via changing their adjustable parameters.

For the system parametric optimization, it is necessary to choose the optimization criterion, which takes into consideration the various aspects of the system functioning. For calculation of the local performance indices of the robust stabilization systems the H_2-norms are used. It should be noted, that the H_2-norms represent the square roots from the integral quadratic performance criteria. In the general form, these criteria look like

- For the deterministic dynamic systems and signals (Kwakernaak, 1993):

$$J_d = \int_0^\infty (x^{\mathrm{T}} Q x + u^{\mathrm{T}} R u) dt \,, \tag{21}$$

where Q, R are the weighting matrices, which take into consideration the weights of the state variables and the external disturbances;

- For the stochastic dynamic systems and signals (Kwakernaak, 1993)

$$J_s = \int_0^\infty (\mathrm{M}[x^{\mathrm{T}} Q x + u^{\mathrm{T}} R u]) dt \,, \tag{22}$$

where M is the symbol of the mathematical expectation.

For calculation of robustness criteria of the stabilization system, the H_∞-norms of the closed loop systems complementary sensitivity matrix function $\Phi_T(j\omega)$ (Skogestad & Postlethwaite, 2001] are used

$$H_\infty = \sup_{\omega \in [-\infty,\infty]} \bar{\sigma}(j\omega), \tag{23}$$

where $\bar{\sigma}$ is the maximum singular value of the closed-loop system matrix transfer function $\Phi_T(j\omega)$ at the circular frequency ω.

The function $\bar{\sigma}(j\omega)$ in (23) is called the singular characteristic of the multi-dimensional system

$$\bar{\sigma}(j\omega) = \max_{i \in 1,\ldots,n \;\; \forall \omega \in [-\infty,\infty]} \sqrt{eig_i \Phi^*(j\omega)\Phi(j\omega)} \,, \tag{24}$$

where *eig* means the operation of the eigenvalues determination. The H_∞-norm (24) characterizes the upper boundary of the system singular frequency characteristic maximum value.

Combination of the H_2 and H_∞-norms in the optimization criterion allows solving a problem of design of the stabilization system with the optimal performance level based on the quadratic criteria for deterministic and stochastic cases under the condition of preserving this performance level during the action of parametric disturbances (Tunik, Rye & Lee, 2001). That is why the combined optimization criterion for the studied system must include the performance indices of the nominal and parametrically disturbed systems for the deterministic and stochastic cases. The optimization criterion must also include the robustness indices of the nominal and disturbed systems.

Then the combined criterion becomes (Tunik, Rye & Lee, 2001)

$$\begin{aligned} J_{H_2/H_\infty} &= \lambda_2^{nom\,d} \,||\, \Phi_{S1}(\mathbf{K}, \mathbf{x}, \mathbf{u}, j\omega)^{nom\,d} \,||_2 + \lambda_2^{nom\,s} \,||\, \Phi_{S2}(\mathbf{K}, \mathbf{x}, \mathbf{u}, j\omega)^{nom\,s} \,||_2 + \\ & \lambda_\infty^{nom} \,||\, \Phi_T(\mathbf{K}, \mathbf{x}, \mathbf{u}, j\omega)^{nom} \,||_\infty + \sum_{i=1}^{n} \lambda_{2_i}^{par\,d} \,||\, \Phi_{S1}(\mathbf{K}, \mathbf{x}, \mathbf{u}, j\omega)_i^{par\,d} \,||_2 + \\ & \sum_{i=1}^{n} \lambda_{2_i}^{par\,s} \,||\, \Phi_{S2}(\mathbf{K}, \mathbf{x}, \mathbf{u}, j\omega)_i^{par\,s} \,||_2 + + \sum_{i=1}^{n} \lambda_{\infty_i}^{par} \,||\, \Phi_T(\mathbf{K}, \mathbf{x}, \mathbf{u}, j\omega)_i^{par} \,||_\infty + PF, \end{aligned} \qquad (25)$$

where $\lambda_2^{nom\,d}$, $\lambda_2^{nom\,s}$, λ_∞^{nom}, $\lambda_{2_i}^{par\,d}$, $\lambda_{2_i}^{par\,s}$, $\lambda_{\infty_i}^{par}$ are the weighting coefficients of the appropriate norms;

$||\, \Phi_{s1}^{nom\,d} \,||_2$, $||\, \Phi_{s2}^{nom\,s} \,||_2$, $||\, \Phi_{s1}^{par\,d} \,||_{2_i}$, $||\, \Phi_{s2}^{par\,s} \,||_{2_i}$ are the H_2-norms of the matrix sensitivity functions for the nominal and disturbed by the parametric structured disturbances closed-loop systems including the deterministic and stochastic cases; K is the vector of the controller parameters to be optimized; $||\, \Phi_T^{nom} \,||_\infty$, $||\, \Phi_T^{par} \,||_{\infty_i}$ are the H_∞-norms of the matrix complementary sensitivity functions for the nominal and disturbed by the parametric structured disturbances closed-loop systems; n is an amount of models of the system disturbed by the parametric structured disturbances; PF is the penalty function, that guarantees system stability during the optimization process.

The H_∞-norms taken into consideration in the combined optimization criterion (25) provide a certain insensitivity of the synthesized system to the parameter changes in the range of the accepted domain. It is known that the requirements to the control performance and robustness are mutually conflicting. Therefore the H_2 / H_∞ optimization of the stabilization system lies in the search of the compromise between the accuracy and robustness of the system.

Usage of the combined criterion for parametric optimization execution allows finding a solution able to ensure a compromise between the requirements to the system accuracy and robustness. Such an approach to synthesis problem solving is called multi-purpose (Tunik, Rye & Lee, 2001). The goal of the robust parametric optimization is the minimization of the criterion (25) for the various combinations of the system parameter values. It is important to provide the presence of the system state variables in the range given by the specification of requirements and to obtain the aforementioned compromise between robustness and performance.

The criterion (25) represents a conversion of the set of scalar local criteria in the single global criterion. In this problem the separate scalar criteria are conflicting. It is known that minimization of the performance criterion (the H_2-norm of the nominal system sensitivity function S) leads to maximization of the robustness criterion (the H_∞-norm of the nominal system complementary sensitivity function T) (Kwakernaak, 1993). Moreover, the performance criteria based on the H_2-norms of the sensitivity function S for the deterministic and stochastic cases are conflicting too. For the deterministic case, the minimization of the closed-loop system sensitivity function H_2-norm for improvement of its performance means increasing of its bandwidth. At the same time for the stochastic case, it means increasing of the variance of the closed-loop system error during the white noise action at its input, thus degrading its accuracy.

Taking into consideration above stated factors the criterion (25) represents the scalar conversion of the vector criterion by means of the weighting coefficients λ_i, which in the multiobjective optimization theory are called the coefficients of the local criteria importance. Therefore it is convenient to consider the H_2 / H_∞-criterion more in detail from the multi-objective optimization viewpoint.

It is possible to represent the local criteria (25) in the vector form (Tunik & Sushchenko, 2013)

$$J_{H_2}^d = \begin{bmatrix} \| \Phi_{S1}(K, x, u, j\omega)^{nom\ d} \|_2 \\ \| \Phi_{S1}(K, x, u, j\omega)_1^{par\ d} \|_2 \\ \dots \\ \| \Phi_{S1}(K, x, u, j\omega)_n^{par\ d} \|_2 \end{bmatrix}; \tag{26}$$

$$J_{H_2}^s = \begin{bmatrix} \| \Phi_{S2}(K, x, u, j\omega)^{nom\ s} \|_2 \\ \| \Phi_{S2}(K, x, u, j\omega)_1^{par\ s} \|_2 \\ \dots \\ \| \Phi_{S2}(K, x, u, j\omega)_n^{par\ s} \|_2 \end{bmatrix}; \tag{27}$$

$$J_\infty = \begin{bmatrix} \| \Phi_T(K, x, u, j\omega)^{nom} \|_\infty \\ \| \Phi_T(K, x, u, j\omega)_1^{par} \|_\infty \\ \dots \\ \| \Phi_T(K, x, u, j\omega)_n^{par} \|_\infty \end{bmatrix}. \tag{28}$$

The vectors of the weighting coefficients for the local criteria (26) – (28) look like

$$\lambda_d = [\lambda_2^{nom\ d} \quad \lambda_{21}^{par\ d} \quad \dots \quad \lambda_{2n}^{par\ d}]; \tag{29}$$

$$\lambda_s = [\lambda_2^{nom\ s} \quad \lambda_{21}^{par\ s} \quad \dots \quad \lambda_{2n}^{par\ s}]; \tag{30}$$

$$\lambda_\infty = [\lambda_\infty^{nom\ s} \quad \lambda_{1\infty}^{par\ s} \quad \dots \quad \lambda_{n\infty}^{par\ s}]. \tag{31}$$

Using the expressions (26) – (31), it is possible to rewrite the criterion (25) in the form

$$J_{H_2/H_\infty} = \lambda_d J_{H_2}^d + \lambda_s J_{H_2}^s + \lambda_\infty J_\infty + PF. \tag{32}$$

The first three local criteria in the global criteria (32) are conflicting as it was mentioned above.

Denote K_p the Pareto-optimum solution of the design problem for the controller with the vector of the adjustable parameters K. In this case, it is reasonable to find the problem solution acceptable from the point of view of the engineering requirements to the designed controller. This solution must be a compromise between performance and robustness on the one hand and between performance in deterministic and stochastic cases on the other hand. The compromise is grounded on the basis of the system designer's preferences. Such a solution can be considered as some engineering analog of the Pareto-optimal solution. A search of this solution is implemented in the space of coefficients defined by the vectors of weighting coefficients λ_d, λ_s and λ_∞ in criterion (32) and elements of the weighting matrices Q and R of the H_2-norms, calculated for the local criteria (21) and (22).

Denote weighting coefficients $\Lambda = \begin{bmatrix} \lambda_d & \lambda_s & \lambda_\infty \end{bmatrix}$. Then the problem for the design of the H_2 / H_∞-controller for the platform stabilized in the inertial space taking into consideration the expression (32) may be formulated in the following way (Tunik & Sushchenko, 2013):

$$K_p = \operatorname{argmin}_p J_{H_2/H_\infty}(K, Q, R, \Lambda, x, u, j\omega); \tag{33}$$

$$K \in D,$$

$$D : \operatorname{Re}[\operatorname{eig}_i(I + L(s))] < 0,$$

$$i \in 1, \ldots, i_0, \tag{34}$$

where D is the stability region in the space of parameters; $L(s) = P(s)W(s)$, $P(s)$ is the controller; $W(s)$ is the transfer function of the plant taking into consideration the actuator and the stabilized plant; i_0 is the order of the set of the differential equations in the Cauchy form;

$$x_i < x_{i0},\ i = 1, \ldots, n_0;$$

$$u_j < u_{j0},\ j = 1, \ldots, m_0; \tag{35}$$

where $n_0 \times 1$ is a dimension of the vector of states; $m_0 \times 1$ is a dimension of the control vector.

In the optimization problem (33) the constraints (34) are defined by the conditions of the closed-loop system stability and the constraints (35) are defined by the specification of requirements to the designed system. The symbol $\min_p$ denotes a minimum, which is determined by the concrete requirements to the designed system.

The process of the solution search represents a procedure of the multiple minimizations of the criterion (25) by one of the known methods. The Nelder-Mead method or the genetic algorithm can be used as such methods. The advantage of the genetic algorithm is the possibility to search the global minimum in every concrete case.

There are two trends of the optimization by the vector criteria (Jahn, 2011). The first trend uses the interactive procedures; the second foresees the development of the specialized criteria for the system of the certain class. For the studied problem connected with the compromise solution search and use of the specific criteria for the robust systems, it is expedient to use the optimization based on a combination of both trends.

The interactive heuristic procedure of the vector optimization for the system of the studied type is divided into two steps. At the first step, the vector optimization of the system acceptable from the point of view of the specification requirements is realized by means of a search of the appropriate weighting matrices Q and R. This process is accompanied by a check of the constraints (35). If these constraints are not satisfied, the elements of the weighting matrices q_i (for $x_i > x_{i0}$) and r_j (for $u_j > u_{j0}$) are changed. The procedure is repeated while the non-equalities (35) will be true. Satisfaction of the expressions (35) for all x_i and u_j means determination of the solution acceptable from the point of view of the specification requirements given to the system. At the second step, the vector optimization with a variation of the criteria weighting coefficients defined by the vector Λ is implemented. This provides the search of solution ensuring compromise between the system robustness and accuracy. The operation is carried out for the weighting matrices Q and R obtained at the first step.

A compromise between the local criteria by means of the weighting coefficients Λ is defined by a designer's preferences, which are directly connected with the operation conditions of the system for stabilization of the observation devices and its design features. For example, in most cases, the preference is shown to the criterion $J^s_{H_2}$ if the operation is carried out in conditions of the intensive random disturbances. In other cases, it is desirable to decrease the criterion J_∞ for the sufficient parametric disturbances. To achieve the desired compromise, it is sufficient to carry out some cycles of the minimization procedure execution in conditions of variation of the criteria weighting coefficients (components of the vector Λ).

It should be noted, that the given procedure has the heuristic nature. It is impossible to prove its convergence but it is checked for the sufficient quantity of the practical situations that have been proved its efficiency. As it follows from the constraints (34), during the parametric optimization procedure and the vector K^* optimal values search, it is necessary to ensure the closed-loop system stability during variation of the plant characteristics and the controller parameters. To achieve this purpose the penalty function is added to the optimization criterion (25). This function provides a location of the closed-loop system poles in the left half-plane of the complex variable.

METHODS OF ROBUST STRUCTURAL SYNTHESIS

In many practical design situations, it is more preferable to use the vector structural synthesis. Such an approach is especially effective for systems, which are developed in the first time. Methods of the robust structural synthesis can be classified by a number of degree-freedoms, methods of loop-shaping and way to introduce the weighting transfer functions. In accordance with the book (Bosgra, Kwakernaak

& Meinsma, 2005), the typical structures of control systems with different quantity of degree-freedoms are presented in Figures 3 – 6.

Notations r, u, y, e in Figures 3 – 6 describe the reference signal, control signal, measured input signal, error signal respectively, K_1 is a pre-filter or the feedforward controller, K_2 is the feedback controller. The controller K includes components C_1, C_2; K_0 is the pre-filter.

The structure of the one-degree-of-freedom controller can be described by relationships $w = r$ and $z = \begin{bmatrix} e & u & y \end{bmatrix}^{\mathrm{T}}$. In this case, the matrix transfer function of the plant, optimization criterion and problem statement of the robust structural synthesis can be determined in the following way (Gu, Petkov & Konstantinov, 2005)

$$P = \begin{bmatrix} I & -G \\ 0 & I \\ 0 & G \\ I & -G \end{bmatrix},$$

Figure 3. Control system with two degrees of freedom

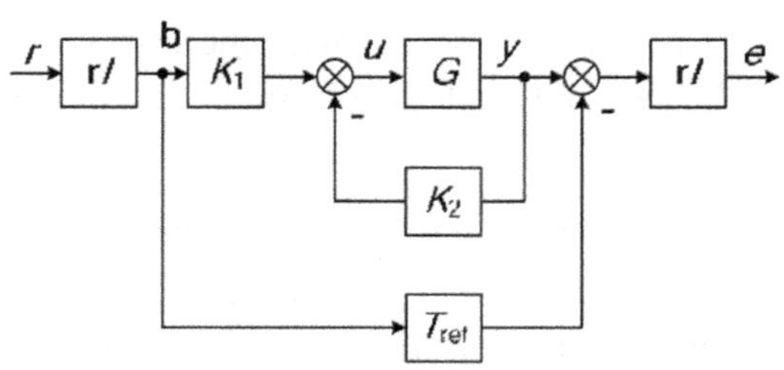

Figure 4. Control system with one degree of freedom

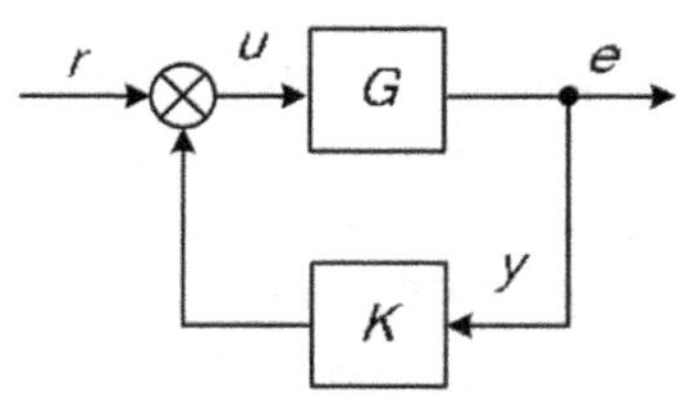

Figure 5. Control system with 1.5 degrees of freedom

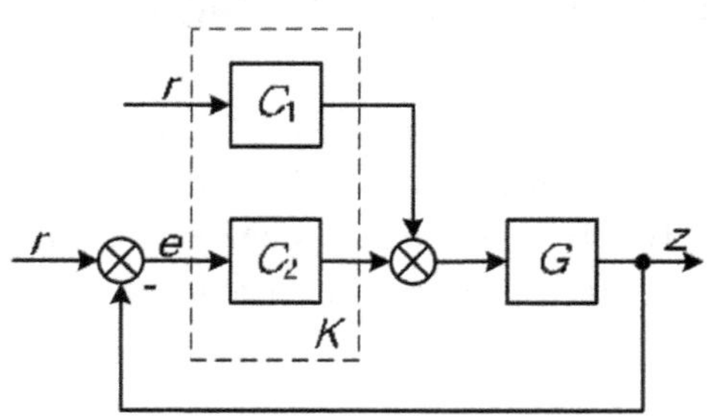

Figure 6. Control system with 2.5 degrees of freedom

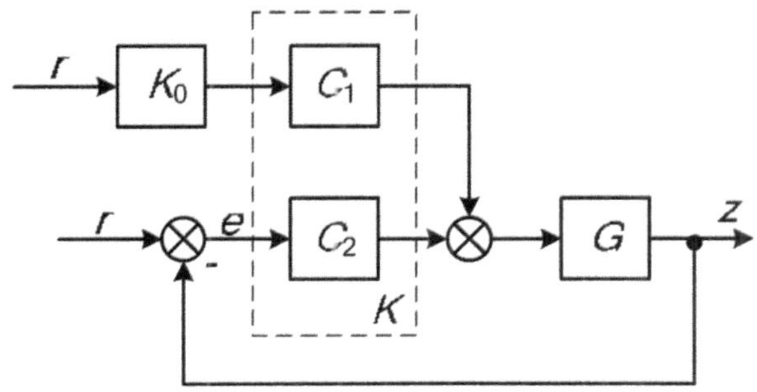

$$J(G,K) = \left\| \begin{bmatrix} (I+GK)^{-1} \\ K(I+GK)^{-1} \\ GK(I+GK)^{-1} \end{bmatrix} \right\|_{\infty},$$

$$K_{\mathrm{opt}} = \arg \inf_{K_{\mathrm{opt}} \in K_{\mathrm{per}}} J(G,K) \tag{36}$$

For the two-degree-of-freedom controller with the vector of signals $[\beta \quad y]^T$, the plant, optimization criterion and problem statement can be described by the expressions (Gu, Petkov & Konstantinov. 2005)

$$P = \begin{bmatrix} -T_{ref} & G \\ 0 & I \\ 0 & G \\ I & 0 \\ 0 & 0 \end{bmatrix},$$

$$J = \left\| \begin{bmatrix} -T_{ref} + GK_1(I+GK_2)^{-1} \\ K_1(I+GK_2)^{-1} \\ GK_1(I+GK_2)^{-1} \end{bmatrix} \right\|_{\infty},$$

$$K_{opt} = \arg \inf_{K_{opt} \in K_{per}} J(G,K) \tag{37}$$

In expressions (36), (37) the first components are sensitivity functions, the second ones – sensitivity functions of control, and the third ones – functions of complementary sensitivity.

Methods of robust structural synthesis by the way of obtaining desired characteristics can be divided into methods of forming frequency characteristics based on analysis of their transfer functions and characteristics of system's signals. There is a method based on forming transfer function of the plant that is the open-loop system. The finite goal of this approach is an improvement of frequency characteristics of the closed-loop system transfer function. Such a procedure is based on the robust stabilization based on the definition of parametrical disturbances by means of the co-prime factorization (Skogestad & Postlethwaite, 2001). The method based on analysis of system signals is convenient for solving multi-objective problems. But this method requires using the complex mathematical model of the plant, models of uncertainties, and determination of a class of the system input signals and norms of signal errors.

Methods of robust structural synthesis can use different ways to introduce weighting transfer functions such as using of transfer weighting functions to form the desired transfer functions of the open-loop system, using of pre- and post-compensators to form the desired transfer function of the open-loop system and also using of weighting functions of the system's signals. It should be noted that systems after introducing weighting transfer functions become augmented (Skogestad & Postlethwaite, 2001).

Structural schemes of augmented systems determined by means of all the above-listed methods are presented in Figure 7 – 9.

Figure 7. Introducing weighting transfer functions by augmentation of the closed-loop transfer functions

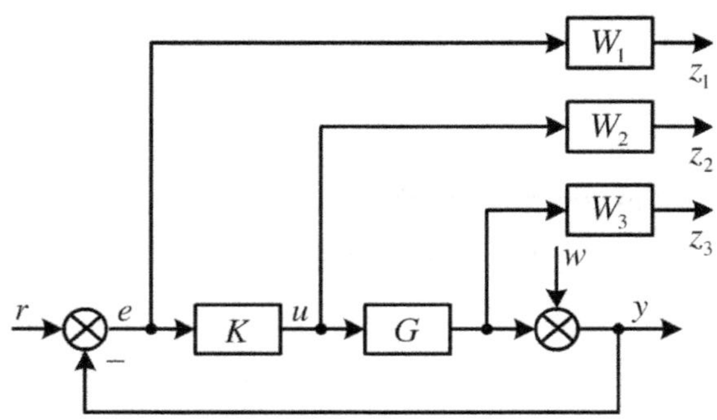

Figure 8. Introducing weighting transfer functions by using pre- and post-compensators

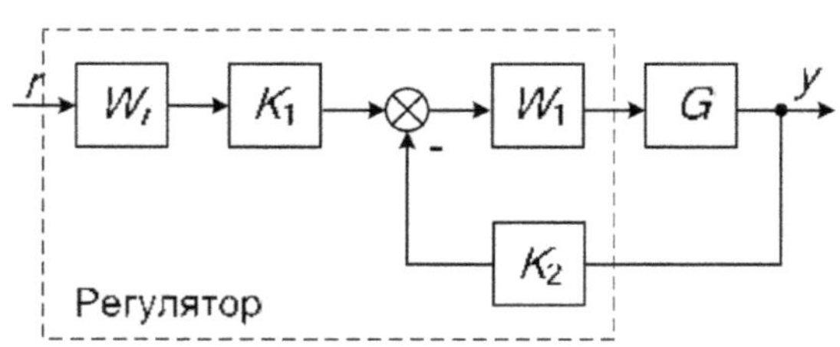

Figure 9. Introducing weighting transfer functions by using of weighting functions of signals

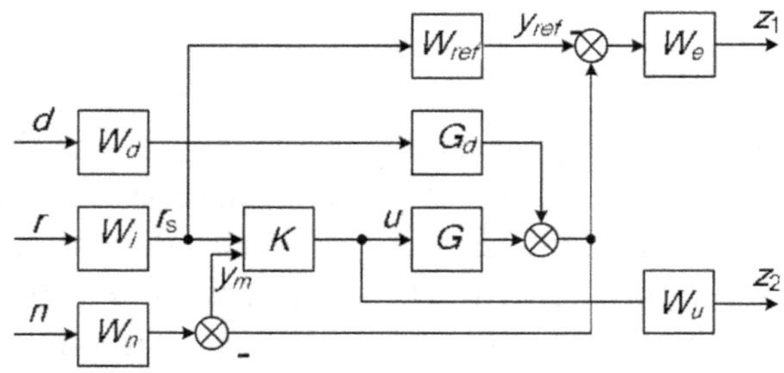

Introducing weighting transfer functions in accordance with Fig. 7 is typical for the method of the mixed sensitivity. The signals z_1, z_2, z_3 are additional outputs. The weighting transfer functions W_1, W_2, W_3 are used to decrease stabilization error, to bound the signal power, and to provide robustness of the system.

For the method of the mixed sensitivity and scheme presented in Fig. 7, the transfer function of the augmented generalized plant and optimization criterion take the form

$$P = \begin{bmatrix} W_1 & -W_1G \\ 0 & W_2 \\ 0 & W_3G \\ I & -G \end{bmatrix},$$

$$J = \left\| \begin{bmatrix} W_1(I+GK)^{-1} \\ W_2K(I+GK)^{-1} \\ W_3GK(I+GK)^{-1} \end{bmatrix} \right\|_\infty$$

or

$$J = \left\| \begin{bmatrix} W_1S \\ W_2R \\ W_3T \end{bmatrix} \right\|_\infty \quad (38)$$

As follows from (38), the optimization criterion includes indices of accuracy, control's losses, and robustness. This means that considered method of the robust structural synthesis satisfies conflicting requirements, which are given to the system of the considered type.

Choice of the weighting transfer functions requires using of the heuristic methods, for example, trial and error, which takes into consideration the experience of the system designer. Introducing of weighting transfer functions in accordance with Fig. 8 characterizes augmentation of the open-loop transfer function. If W_1, W_2 are pre- and post-compensators respectively, the augmented plant with the bounded frequency characteristics G_s can be determined in the following way (Skogestad & Postlethwaite, 2001)

$$G_s = W_2GW_1. \quad (39)$$

Usage of transfer functions for the method based on the analysis of a system's signals is shown in Fig. 9. Here G and G_d are the nominal models of the plant and the disturbance model, and K is the designed controller. Weighting characteristics of such external signals as disturbance, the reference signal and measuring noise W_d, W_i, W_n can be defined both by constant values and by transfer functions. The weighting characteristic W_{ref} is the desired closed-loop transfer function from the weighting reference signal r_s to the output signal y. The weighting characteristics W_e and W_d describe the desired frequency spectrum of the error $z_1 = y - y_{ref}$ and control signal $z_2 = u$ respectively.

For the system presented in Fig. 9, it is desired to minimize the H_∞-norm of the transfer function from the input signals r, d, n to the output signals z_1, z_2 (Skogestad & Postlethwaite, 2001)

$$\begin{bmatrix} z_1 \\ z_2 \end{bmatrix} = \begin{bmatrix} W_e S & -W_e S W_d & W_e T W_n \\ W_b KS & -W_b KS W_d & -W_b KS W_n \end{bmatrix} \begin{bmatrix} r \\ d \\ n \end{bmatrix}. \tag{40}$$

Finally, this problem can be formulated as H_∞-synthesis problem, where signals w, z, y, u can be determined in the following way

$$w = \begin{bmatrix} r \\ d \\ n \end{bmatrix};$$

$$z = \begin{bmatrix} z_1 \\ z_2 \end{bmatrix};$$

$$y = \begin{bmatrix} r_s \\ y_s \end{bmatrix} \tag{41}$$

The expressions (38) – (41) describe ways of using the weighting transfer functions for loop-shaping in the process of the robust structural synthesis.

STRUCTURAL ROBUST SYNTHESIS OF SYSTEM FOR STABILIZATION OF UAV OBSERVATION EQUIPMENT

One of the modern approaches to the structural synthesis of the robust systems is the H_∞-synthesis. Basic principles of this method are presented in many textbooks, for example, (Zhou & Doyle, 1999), (Gu, Petkov & Konstantinov, 2005). The standard configuration of the robust system designed by means of the H_∞-synthesis is shown in Fig. 10. The system consists of the plant G and controller K. It can be characterized by the vector of the output parameters z to be optimized, the vector of the external input signals w, the vector of control signals u, and the vector of the measured signals y entering to the input of the controller. The problem statement of the H_∞-synthesis is based on the concept of the so-called generalized or interconnected system.

The synthesized system consists of the plant and controller described by the matrix transfer functions $G(s)$, $K(s)$ respectively. These transfer functions must be fractionally-rational and proper. The generalized plant represents a system with two inputs and two outputs. The vector w represents the external output, which, in the general case, consists of disturbances, measurement noise, and reference signals. The input vector u represents the control signals. The output vector z determines the quality of the control processes. For example, it can be characterized by the tracking error, which must be equal to zero in the ideal case. The output vector y represents the vector of the observed signals, which are used for feedback organization. The transfer function from input w to output z, is denoted W_w^z. Respectively, the main task of the H_∞-synthesis is the choice of such a controller $K(s)$, which can minimize the $\| W_w^z \|_\infty$-norm.

The control system, shown in Fig. 10, can be described in the following way

$$\begin{bmatrix} z \\ y \end{bmatrix} = P(s)\begin{bmatrix} w \\ u \end{bmatrix} = \begin{bmatrix} P_{11}(s) & P_{12}(s) \\ P_{21}(s) & P_{22}(s) \end{bmatrix}\begin{bmatrix} w \\ u \end{bmatrix}; \quad (42)$$

$$u = K(s)y.$$

Representation (42) in the state space becomes

$$\dot{x}(t) = Ax(t) + B_1 w(t) + B_2 u(t);$$

$$z(t) = C_1 x(t) + D_{11} w(t) + D_{12} u(t);$$

$$y(t) = C_2 x(t) + D_{21} w(t) + D_{22} u(t);$$

$$u(t) = Ky(t) \quad (43)$$

The mathematical model (43) in the matrix form can be represented in the following form

Figure 10. Standard H_∞-configuration

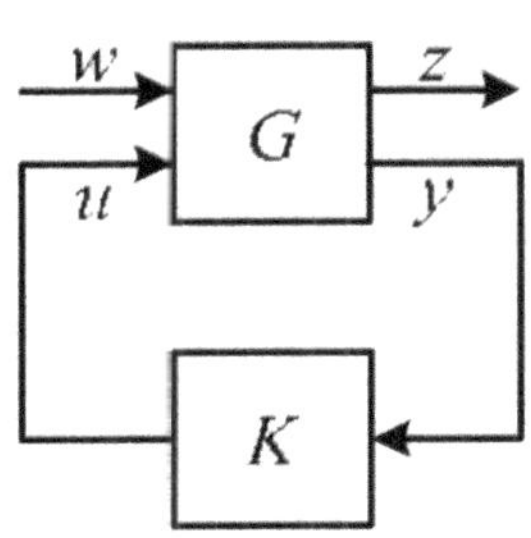

$$\begin{bmatrix} \dot{x}(t) \\ z(t) \\ y(t) \end{bmatrix} = \begin{bmatrix} A & B_1 & B_2 \\ C_1 & D_{11} & D_{12} \\ C_2 & D_{21} & D_{22} \end{bmatrix} \begin{bmatrix} x(t) \\ w(t) \\ u(t) \end{bmatrix} \tag{44}$$

The mathematical model (44) is used for the robust structural synthesis. The solution of the H_∞-synthesis problem is based on the solutions of the Riccati equations, at that, the following conditions must be satisfied (Skogestad & Postlethwaite, 2001).

1. The pair of matrices A, B_1 must be stabilizable and the pair of matrices A, C_1 must be detectable.
2. The pair of matrices A, B_2 must be stabilizable and the pair of matrix A, C_2 must be detectable.
3. $D_{12}^{\mathrm{T}} \begin{bmatrix} C_1 & D_{12} \end{bmatrix} = \begin{bmatrix} 0 & I \end{bmatrix}$.
4. $\begin{bmatrix} B_1 \\ D_{21} \end{bmatrix} D_{21}^{\mathrm{T}} = \begin{bmatrix} 0 \\ I \end{bmatrix}$.

Conditions 1 and 2 guarantee the absence of imaginary eigenvalues of the Hamilton matrices, which correspond to the Riccati equations. Condition 3 defines the orthogonality of the signals $C_1x(t)$ and $D_{12}u(t)$. According to the H_2-problem, this condition means that the weighting control matrix in the norm of the vector $z(t) = C_1x(t) + D_{12}u(t)$ is unitary and the components $z(t)$ of the state vector $x(t)$ and control vector $u(t)$ do not influence this norm. Condition 4 shows the orthogonality of the signals $B_1w(t)$ and $D_{21}w(t)$. So, conditions 3, 4 are usual for the H_2-problem and are spread on the case of the H_∞-synthesis (Skogestad & Postlethwaite, 2001).

It is worth notice, that in such statements of problems, the plant is believed to be a set of devices and units. This set of devices and units consists of the control object, actuator, measuring system, and some additional units.

The system presented in Fig. 10 can be transformed to the form shown in Fig. 11.

Figure 11. Structural scheme of the designed system

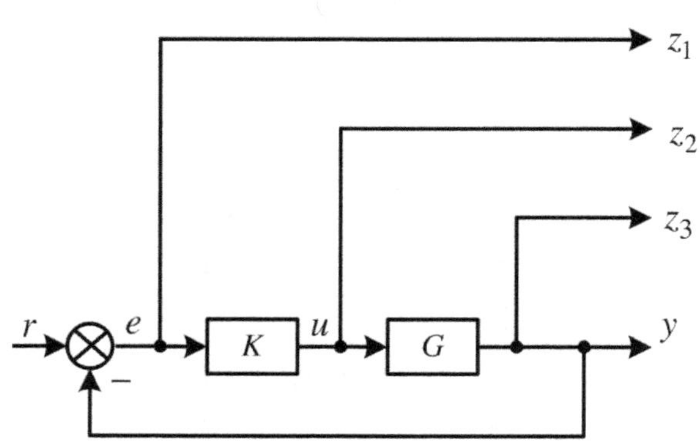

Interconnections between signals of the system represented in Fig. 11 can be described by the following expressions

$$w = r;$$

$$z = \begin{bmatrix} e \\ u \\ y \end{bmatrix};$$

$$P = \begin{bmatrix} P_{11} & P_{12} \\ P_{21} & P_{22} \end{bmatrix} = \begin{bmatrix} I & -G \\ 0 & I \\ 0 & G \\ I & -G \end{bmatrix}, \quad (45)$$

where P is the transfer function of the interconnected system.

Based on (45) the transfer function from the input w to the output z becomes

$$T_w^z = P_{11} + P_{12}K(I - P_{22}K)^{-1}P_{21}.$$

The optimization problem of the robust structural synthesis can be stated in the following form

$$K_{opt} = \arg \inf_{K_{opt} \in K_{con}} J(G, K), \quad (46)$$

Here

$$J(G, K) = \left\| \begin{bmatrix} (I + GK)^{-1} \\ K(I + GK)^{-1} \\ GK(I + GK)^{-1} \end{bmatrix} \right\|_\infty. \quad (47)$$

The optimization problem (46) can be solved by the method of mixed sensitivity (Skogestad & Postlethwaite, 2001). One of the modern approaches to solving the problem of the robust structural synthesis is based on loop-shaping using weighting transfer functions as it is shown in Fig. 12.

In this problem the optimization criterion of the method of mixed sensitivity (47) becomes

$$J(G, K) = \left\| \begin{bmatrix} W_1(I + GK)^{-1} \\ W_2K(I + GK)^{-1} \\ W_3GK(I + GK)^{-1} \end{bmatrix} \right\|_\infty = \left\| \begin{bmatrix} W_1S \\ W_2R \\ W_3T \end{bmatrix} \right\|_\infty, \quad (48)$$

Figure 12. Functional scheme of the designed system augmented by means of the weighting transfer functions

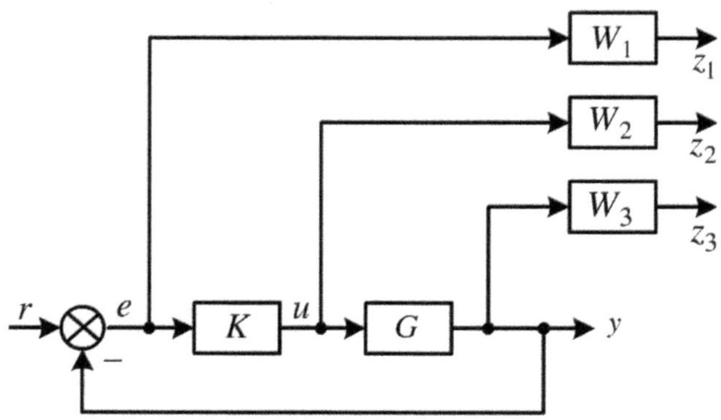

where W_1, W_2, W_3 are weighting transfer functions; S, R are functions of sensitivity by the given signal and control and T is the complementary sensitivity function.

The expression (48) presents the H_∞-norm of the function sensitivity of the augmented system.

The implementation of the H_∞-synthesis includes such stages as the development of the mathematical model (both linearized and taken into consideration non-linearities inherent to real systems), choice of weighting transfer functions, augmentation of the plant and structural synthesis of the robust controller. All the stages require complex transformations and calculations. To decrease the time of design works it is necessary to use the computer-aided designed procedures based on the Robust Control Toolbox, which is a part of MatLab software (Sushchenko, 2017).

One of the most difficult stages of the procedure of the robust structural synthesis is a choice of the weighting transfer functions based on the heuristic methods. These functions provide the forming of the augmented plant and loop-shaping approach. There are following general recommendations for determination of weighting transfer functions for augmented plant forming (Skogestad & Postlethwaite, 2001)

$$W_1 = \frac{s / M + \omega_0}{s + \omega_0 A};$$

$$\mathrm{W}_2 = \mathrm{const};$$

$$W_3 = \frac{s + \omega_0 / M}{As + \omega_0}, \tag{49}$$

here A is the desired error in the stable mode; ω_0 is the desired bandwidth; M is sensitivity peak.

Design experience of the stabilization systems of the considered type has been shown the necessity of using integrator in the expression of the weighting transfer function W_1. As a rule, in the practical applications, the weighting transfer functions are determined by methods of trials and errors. In accordance with Fig. 12, the weighting transfer functions impose a penalty on the value of the error, control and output signals. Changing the parameters of the weighting transfer function A, ω_0, M, it is possible to achieve the desired frequency characteristics of the system. As a whole, usage of the weighting transfer functions provides loop-shaping of the designed system. Such an approach allows forming the designed system with the desired frequency characteristics. During synthesis of the researched system, the following matrices of the weighting transfer functions (49) were obtained

$$W_1 = \begin{bmatrix} \frac{1}{s}\frac{0,1s+20}{s+0,005} & 0 & 0 \\ 0 & \frac{1}{s}\frac{0,02s+10}{s+0,002} & 0 \\ 0 & 0 & \frac{1}{s}\frac{0,01s+10}{s+0,001} \end{bmatrix};$$

$$W_2 = \begin{bmatrix} 0,04 & 0 & 0 \\ 0 & 0,04 & 0 \\ 0 & 0 & 0,04 \end{bmatrix};$$

$$W_3 = \begin{bmatrix} \frac{s}{0,005s+50} & 0 & 0 \\ 0 & \frac{s}{0,002s+20} & 0 \\ 0 & 0 & \frac{s}{0,001s+10} \end{bmatrix}.$$

Nowadays the usage of the discrete controllers is the most important for the practical applications. There are two known approaches to the creation of the discrete controllers (Balas, Chiang, Packard & Safonov, 2008).

The first approach is based on the discretization of the model of the plant augmented by means of the weighting transfer functions. Further, the w-transformation, structural synthesis and z-representation of the controller by means of the inverse w-transformation are implemented. All above-listed operations can be automated by means of the embedded function *dhinfopt* of MatLab Robust Control Toolbox. The second approach lies in the automated design of the continuous robust controller. Further discretization of the controller is carried out by means of the embedded function, for example, *c2d*. Such a discretization must satisfy some requirements. In the first place, the Tustin bilinear transformation as a discretization method must be used. Such a choice ensures discretization accuracy. In the second place, the discretization sampling must be in several times higher than the bandwidth frequency of the system.

The discrete controller of the researched system was designed based on the second approach. The flow chart of this process is represented in Fig. 13.

As stated above this procedure design can be implemented by software of Robust Control Toolbox.

Figure 13. Flowchart f the discrete controller design procedure

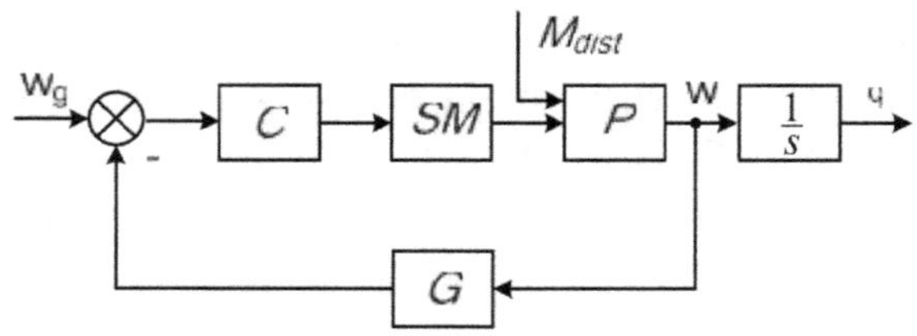

CASE STUDY

The stabilization system of UAV observation equipment is considered as the servo system. The structural scheme of this system is shown in Fig. 14.

Synthesis of the robust controller was implemented by means of MatLab function *hinfopt*. At first, an order of the synthesized controller was equal to 8. A decrease of controller order was implemented by means of function *schmr.* This allowed decreasing the controller order to 5. Discretization of the controller was implemented by means of function *c2d.*

Figure 14. Structural scheme of the rate servo system: C is the controller; SM is the servomotor; P is the plant; M_{dist} is the disturbance moment; ω is the angular rate of the line of sight; θ is the angular position of the line of sight

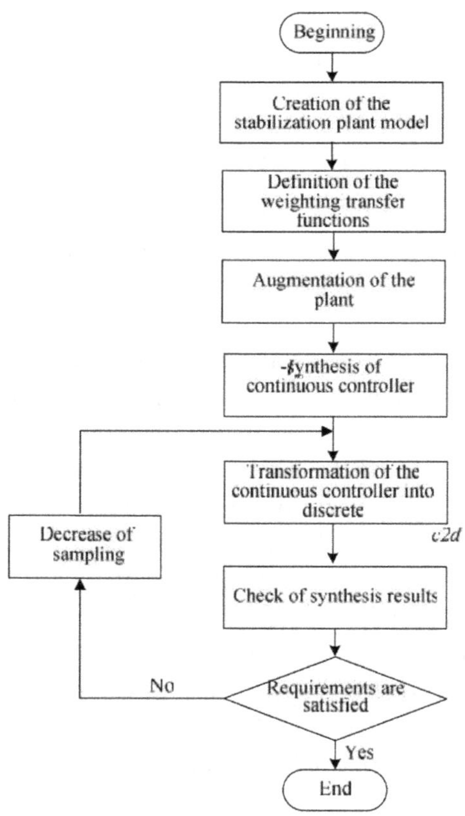

Structure of the synthesized robust controller of the stabilization system can be represented in the form of the quadruple of matrices in the state space

$$A_c = \begin{bmatrix} 186,3 & 352,6 & 45,83 & 11,33 & 43,7 \\ 711,8 & -966,3 & -179,7 & -12,17 & -11,6 \\ -117,75 & 126,7 & -18,41 & -15,84 & -15,62 \\ -343,7 & 517,3 & 136,8 & 19,39 & -23,43 \\ -218,3 & 268,8 & 20,53 & 18,75 & -25,89 \end{bmatrix};;$$

$$B_c^T = \begin{bmatrix} 20,98 & -37,1 & -8,3 & -5,36 & -7,25 \\ 17,6 & -32,1 & -7,3 & -4,2 & 6,23 \\ 15,6 & -30,5 & 5,4 & -3,1 & -5,21 \end{bmatrix};$$

$$C_c = \begin{bmatrix} 8,31 & -9,8 & -4,78 & -1,43 & 2,35 \\ 9,6 & -6,4 & 12,2 & -1,2 & 16,3 \\ 11,2 & -7,5 & 9,2 & -11,1 & -15,21 \end{bmatrix};$$

$$D_c = \begin{bmatrix} -0,9 & -0,543 & -0,217 \\ -0,6 & -0,312 & -0,123 \\ -0,4 & -0,213 & -0,223 \end{bmatrix}.$$

The Simulink-model for system simulation is represented in Fig. 15. Results of the synthesized system simulation are presented in Figures 16 – 19.

Results of simulation presented in Figures 16 – 19 correspond to the operation of the observation equipment during UAV flight. The external disturbance is an angular rate of UAV, on which the observation equipment is mounted. Presented results prove the possibility to achieve acceptable indices of stabilization accuracy and speed of operation in real operating conditions. H_2 and H_∞-norms of the synthesized robust automated system are equal to values $H_2 = 0.526$ and $H_\infty = 0.258$.

These results prove the efficiency of the researched designing procedure of the robust controller based on the H_∞-synthesis including loop-shaping and the method of mixed sensitivity.

The proposed procedures provide definite advantages during automated stabilization system design. The obtained results show that the designed system satisfied requirements for accuracy in conditions of significant parametrical disturbances. It is known that the automated control systems designed by traditional methods can be unstable in conditions of the influence of parametrical disturbances. In the latter case, the system can not carry out its functions. The robust systems have not such a disadvantage. So the proposed approach to the design of the automated stabilization system provides its operability even in conditions of the real operation.

Figure 15. Simulink-model of the stabilization system

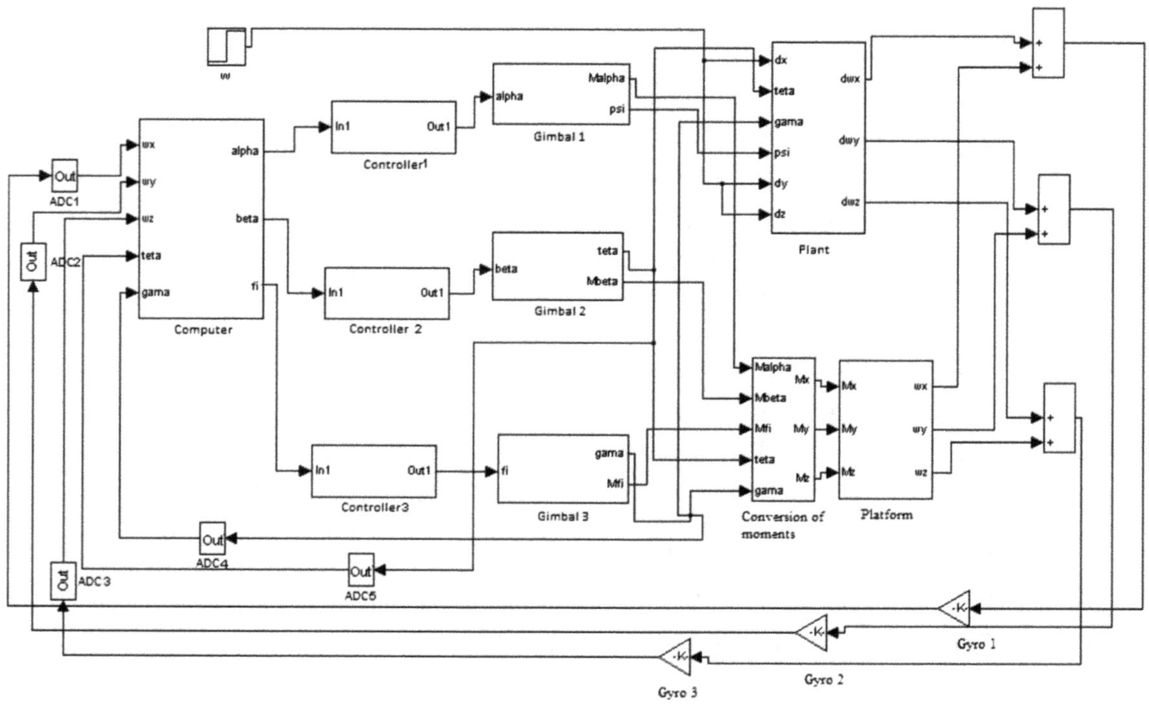

Figure 16. Results of simulation of platform motion of the coarse leveling: a is the transient process of the reference angular rate

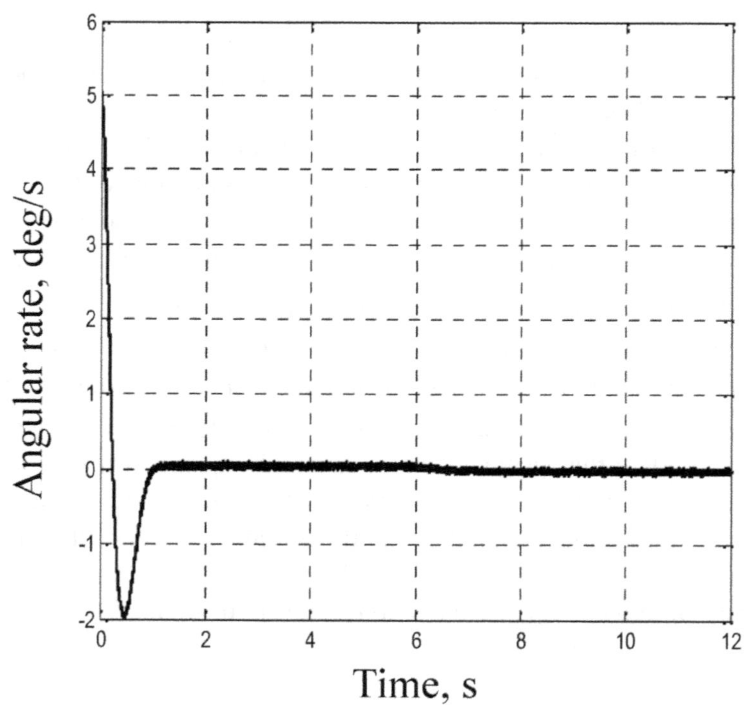

Figure 17. Results of simulation of platform motion of the coarse leveling: the transient process of platform angular position determination

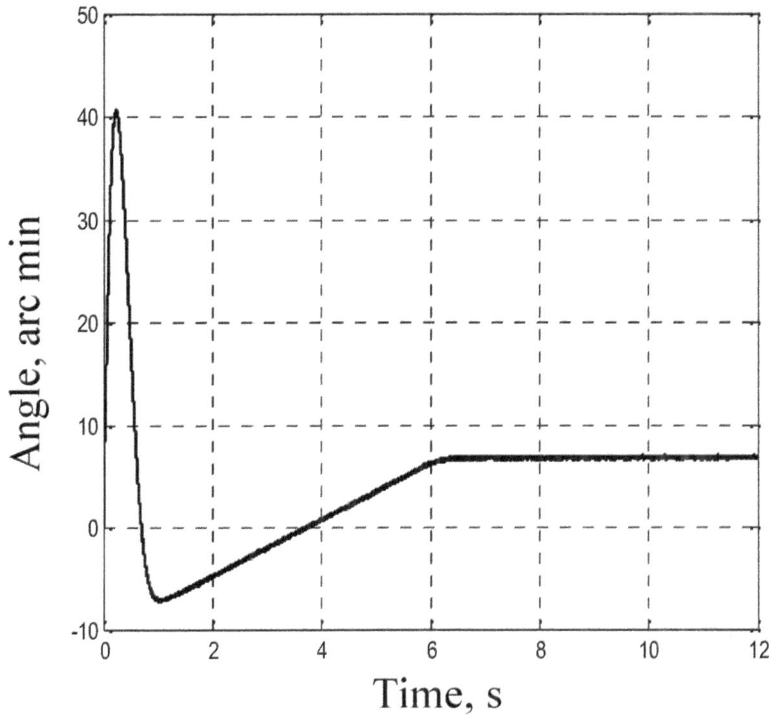

Figure 18. Results of simulation of platform motion of the fine leveling: the transient process of the reference angular rate

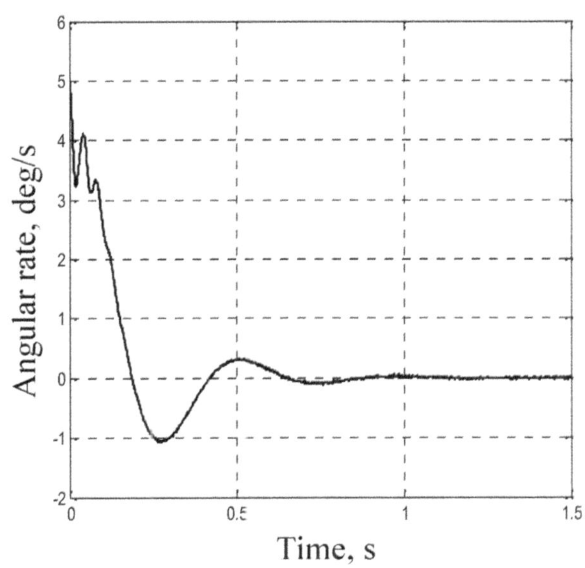

Figure 19. Results of simulation of platform motion of the fine leveling: the transient process of platform angular position determination

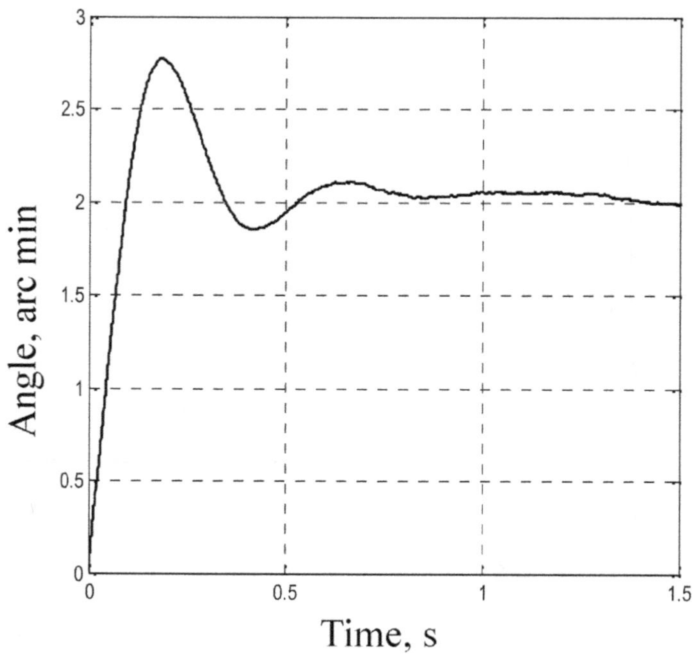

FUTURE RESEARCH DIRECTIONS

Future research directions will be associated with the design of robust systems of the considered type using redundant nonorthogonal inertial measuring unit based on inertial sensors designed on micro-electromechanical systems (MEMS) technology. Such an approach will be especially effective for the operation of UAV observation equipment. Using the redundant nonorthogonal measuring units is a low-cost way to improve the accuracy of observation processes.

CONCLUSION

In this chapter, the author presents the way to solve problems of design of robust automated systems stabilization of observation equipment operated on aircraft. The approach to the robust structural synthesis of the stabilization system of the observation devices is represented. The mathematical description of the system providing stabilization and control by the orientation of the observation devices lines of sight is obtained. The basic features of the design procedure are described and the appropriate algorithm is presented. The robust discrete controller is developed. The matrix weighting transfer functions for the robust structural synthesis are chosen. The simulation results are represented.

Adequacy of scientific results represented in the chapter is grounded on the detailed analysis of the references, the mathematical correctness of the obtained results and correspondence of simulation results to scientific assessments.

REFERENCES

Balas, G., Chiang, R., Packard, A. & Safonov, M. (2005-2008). *Toolbox User's Guide*. The Math Works Inc.

Bosgra, O. H., Kwakernaak, H., & Meinsma, G. (2005). Notes for a course of the Dutch Institute of Systems and Control. Academic Press.

Gu, D., Petkov, P., & Konstantinov, M. (2005). *Robust Control Design with MATLAB*. London: Springer-Verlag.

Hilkert, J. M. (2008). Inertially stabilized platform technology. *IEEE Control Systems Magazine*, *28*(1), 26–46. doi:10.1109/MCS.2007.910256

Jahn, J. (2011). *Vector Optimization. Theory, Application and Extensions*. Berlin: Springer-Verlag.

Kim, S., Cho, S., Kim, H., & Shim, J. (2018). Robust Control of an Equipment-Added Multirctor Using Disturbance Observer. *IEEE Transactions on Control Systems Technology*, *26*(4), 1524–1531.

Kwakernaak, H. (1993). Robust control and H_∞. *Automatica*, *29*(2), 255–273. doi:10.1016/0005-1098(93)90122-A

Sapio, V. (2017). *Advanced Analytical Dynamics. Theory and Application*. Cambridge, UK: Cambridge University Press. doi:10.1017/9781316832301

Skogestad, S., & Postlethwaite, I. (2001). *Multivariable Feedback Control*. New York: Jonh Wiley and Sons.

Sushchenko, O. (2017). Approach to computer-aided design of UAV robust inertial platforms. In *Proceedings of IEEE 4th International Conference on Actual Problems of Unmanned Air Vehicles Developments (APUAVD 2017)* (pp. 53 – 57). Kyiv, Ukraine: IEEE. 10.1109/APUAVD.2017.8308775

Sushchenko, O. A. (2017). Design of navigation and stabilization loops of precision attitude and heading reference system. *Proceedings of the National Aviation University*, 48–56. 10.18372/2306-1472.72.11981

Sushchenko, O. A., & Goncharenko, A. V. (2016) Design of robust systems for stabilization of unmanned aerial vehicle equipment. International Journal of Aerospace Engineering.

Sushchenko, O. A., & Tunik, A. A. (2013). Robust stabilization of UAV operation equipment. In *Proceedings of IEEE 2nd International Conference on Actual Problems of Unmanned Air Vehicles Developments (APUAVD 2013)* (pp. 176 – 180). Kyiv, Ukraine: IEEE. 10.1109/APUAVD.2013.6705318

Tunik, A. A., & Sushchenko, O. A. (2013) Usage of vector parametric optimization for robust stabilization of ground vehicles information-measuring devices. *Proceedings of the National Aviation University*, 23 – 32. 10.18372/2306-1472.57.5530

Wu, N., & Yu, J. (2018) Robust Controller Design of Hypersonic Vehicles in Uncertainty Models. In *Proceedings of 3rd International Conference on Electromechanical Control Technology and Transportation (ICECT-2018)* (p. 288-293). Huazhong, China: Academic Press. 10.5220/0006969302880293

Zhou, K., & Doyle, J. (1999). *Essentials of Robust Control*. Prentice Hall.

ADDITIONAL READING

Anderson, B. D., & Moore, J. B. (1989). *Optimal Control Linear Quadratic Methods.* Prentice Hall.

Burns, R. S. (2001). *Advanced Control Engineering*. Oxford, UK: Butterworth-Heinemann.

Merhav, S. (1996). *Aerospace Sensors Systems and Applications*. New York: Springer-Verlag. doi:10.1007/978-1-4612-3996-3

Yang, W. Y., Cao, T. S., Cao, I., & Chung, J. (2005). *Applied Numerical Methods Using MATLAB*. New York: John Wiley and Sons. doi:10.1002/0471705195

KEY TERMS AND DEFINITIONS

Gimbals: A set of frames suspended on bearings that allows rotation and control of a plant.

Gyroscope: An instrument or device that measures rotational motion in the inertial space is called a gyroscope.

Robust System: An automated system is called robust if they keep stability and performance indices in some permissible range in conditions of disturbances influence without using adaptation methods.

Stabilization System: A system that keeps the constant value of a controlled quantity under influence of changed disturbances, is called stabilization system.

Unmanned Aerial Vehicle: Aircraft without crew on the board is called an unmanned aerial vehicle.

Chapter 12
Information Technologies for Learning Principles of Fault-Tolerant Systems

Juan Pablo Martínez Bastida
National Aerospace University – Kharkiv Aviation Institute, Ukraine

Olena Havrylenko
National Aerospace University – Kharkiv Aviation Institute, Ukraine

Andrey Chukhray
National Aerospace University – Kharkiv Aviation Institute, Ukraine

ABSTRACT

In this chapter, the authors present a methodology for developing a model-tracing cognitive tutor. The methodology is based on Bayesian probabilistic networks for generating pedagogical interventions. The presented probabilistic model increases fidelity assessment due to its ability of independently diagnosing the degree of mastery for every knowledge component involved in students' actions; fidelity assessment in education is the ability to represent students' cognitive states as close as possible for analysis and evaluation. The cognitive tutor was developed to promote a self-regulated learning approach with an open learner model. The open learner model let students change the learning flow by changing the assigned tasks. The authors explain in detail the structural construction and employed algorithms for developing a model-tracing cognitive tutor in the domain of fault-tolerant systems. Preliminary results and future work are also discussed to assess effectiveness of the proposed approach and its implication in actual educational programs.

INTRODUCTION

In this chapter, we discuss several concepts and their implementation for developing a Model-tracing cognitive tutor (MTCT). An Open learner model (OLM) is featured to support student self-assessment for promoting development of "help-seeking" meta-cognitive skills. This is based on the hypothesis about

DOI: 10.4018/978-1-5225-7709-6.ch012

students that do not look for help when they need it, are less probable to learn, i.e. those students do not get closer to a teacher to obtain assistance or other king of communication, information or learning supports. Because they lack of meta-cognitive skills for "help-seeking", besides a help-seeking student is known to become a better learner (Aleven, McLaren, Roll, & Koedinger, 2016).

Different features are presented to foment a Self-regulation learning (SRL) process that leads to improve domain-level learning outcomes. Considerations for fidelity prototyping are also discussed, and their importance for assessment stages. The aim of this chapter is to discuss and implement a methodology as a fundamental framework for developing MTCTs.

The proposal is compounded by several principles, approaches and paradigms that were implemented to build a cognitive tutor for learning the foundations of fault-tolerance systems, a very specific domain but they can be used for other domains as well. TITUS is a developed tutor by means of the proposed approach, it supports the idea of learning by doing, self-assessment and help-seeking meta-cognitive skills (Aleven, McLaren, Roll, & Koedinger, 2016; Arroyo et al., 2014). A discussion about how pedagogical interventions are generated in a rule-based cognitive model is also carried out. Finally, preliminary results are assessed to evaluate effectiveness of the proposed approach.

BACKGROUND

Since the beginning, humans have developed technology to solve sundry problems. Nevertheless, the creativity has led us onto a raise to develop more complex and bigger technological solutions in order to facilitate our lives. Nowadays, every person in this planet and out in the space directly or indirectly is attached to such amount of technical processes and systems that their lives depend on them. Thus, it is critical to integrate fault-tolerance support (Kulik, 1991; Martinez Bastida, & Chukhray, 2016) in every design, because of its absence increments the possibility of failure and jeopardize the reliability of such systems. Thus, proficiency about fault-tolerant systems should be faster increased with higher effectiveness in students and current professionals, in order to reduce failure risks in systems and technical processes that have a deep impact on our lives.

Intelligent Tutoring Systems (ITS) have shown to improve and accelerate learning processes. There are a wide amount of documented work about their application. They have been successfully applied in diverse domains such as physics (Aleven, Nkambou, Bourdeau, & Mizoguchi, 2010; Aleven et al., 2016), mathematics (Long, & Aleven, 2015; Arroyo et al., 2014), programming (Russel, & Norving. 2010; Wang, & Heffernan, 2012), and engineering subjects (Dzikovska, Steinhauser, Farrow, Moore, & Campbell, 2014) for instance. However, prototyping effective ITSs has shown to be a difficult complex task due to the interdisciplinary process that must be done. Moreover, the necessary time to unravel each module in the ITS structure and implement them all. There are an "endless" quantity of information and documented works about how a learning program is structured (Paquette, Lebeau, Beaulieu, & Mayers, 2015; Wieber, Lukas, & Gollwitzer, 2015; Long & Aleven, 2015; Pelánek & Jarusek, 2015; Conati, Gertner, & VanLehn, 2002; VanLehn, 2008; VanLehn et al., 2011), the way assessment is carried out (VanLehn, 2008), how graphic interfaces should be designed (Aleven et al., 2010), and other important factors that should be taken into account during the prototyping and implementation of an ITS. This cloud of information helps to improve and supports ITS development, but it creates gaps between researchers, developers and the necessary models and algorithms that might be implemented for efficiently designing an ITS.

At its early stages, advanced computer tutoring (ACT) (Bessing, Gilbert, Ourada, & Ritter, 2009) was concerned about developing cognitive skills. This kind of tutors were called Cognitive tutors. According to it, a cognitive skill manly consists on goal-related knowledge that is acquired by actions and consequences while solving relating tasks. Other kind of ITSs have performed a model-tracing functionality that have shown its effectiveness (Bessing et al., 2009), they are able to interact with a *cognitive model* (CM) representation of the student. MTCT analyses and interprets uncertain information and data mining obtained while students use the tutor, and it is also called as CM. (Paquette et al., 2015; Aleven et al., 2010; Aleven, et al., 2016; Bessing et al., 2009; Anderson, Corbett, Koedinger, & Pelletier, 1995).

Cognitive models are a core component in developing MTCTs (Paquette et al., 2015; Aleven et al., 2010; Bessing et al., 2009) they require a proper understanding of the knowledge involved in student actions of a given domain. It also might frame student's strategies, problem-solving principles (Aleven et al., 2010), and cognitive and meta-cognitive skills. This representation contains uncertainty information and it is very important in order to increase fidelity assessment of mastery. *Fidelity assessment* in education is a parameter that helps to know how close the CM reflects the student's mental state. It has a strong relationship between the degree of mastery and actions in the domain (Anderson, Corbett, Koedinger, & Pelletier, 1995). Interpretation of actions is a feature that differentiates ITSs from other educational systems, because it helps to maintain individual student models, that leads to a personalized learning program. Personalized programs have demonstrated an increment of performance in students up to 98% (Bervery, 2009).

Scientific community dedicated to ITS, is focused on maintaining or enhancing effectiveness of tutors at domain-level learning (Wieber, Lukas, & Gollwitzer, 2015; Long, & Aleven, 2015). In the literature, it is possible to find terms like OLMs and Self-Regulated Learning (SRL). OLM refers to the ability of a certain learning model to be accessed or modified by the student. This is, a student can choose or change the way she learns. Meanwhile, SRL is related to student's metacognition ability to manage learning environments for self-development. According to (Wieber, Lukas, & Gollwitzer, 2015), generation a varied set of pedagogical interventions and metacognitive support in a MTCT, enforces self-assessment that might lead to a better performance, but it does not assure learning effectiveness. Therefore, there is a vast space of study in developing effective solutions that promote the concepts discussed above. OLMs aggregates characteristics for supporting SRL by giving students possibilities to manage the way they learn and monitor their progress.

Estimation of knowledge mastery requires uncertainty reasoning since cognitive processes are being assessed and such processes cannot always be observed. Bayesian networks (BNs) have widely been used in ITS due to their ability to manage uncertainty (Aleven et al., 2010; Aleven, McLaren, Roll, & Koedinger, 2016; Chi, VanLehn, & Litman, 2010; Conati, Gertner, & VanLehn, 2002). BN is a probabilistic model that helps to estimate degree of mastery by means of conditional probabilities. A *conditional probability* is the probability that something might happen in base of observable facts that already occurred. In education, it is estimated the probability that certain student masters a knowledge in base of her actions when that knowledge is required in a problem-solving environment. Hence, a proper representation of the knowledge involved (fidelity) in a student action, is very important for assessing knowledge mastery. In this work, it is discussed a BN topology that increases fidelity assessment. This topology can assess every knowledge involved in student actions by diagnosing them separately (Chukhray, 2013). Once mastery is estimated, it is used to infer decisions for generating pedagogical interventions according to an OLM.

STRUCTURAL APPROACH

Cognitive tutors commonly employ a constrain-based step-by-step couching program. Trainees master knowledge and develop skills in basis of a problem-solving scheme. However, conceptualization of the knowledge involved in students' cognitive processes has always been a challenge for developers. A high comprehension of the knowledge involved in a cognitive process (cognitive fidelity), permits a better prototyping fidelity for promoting SRL (Long, & Aleven, 2015). This means, better self-assessment, pedagogical interventions, task development, self-development, etc. Higher cognitive fidelity helps a better task analysis (Aleven et al., 2010) for a more effective pedagogical modeling. OLM is a component that supports students' self-assessment and study choice (Long, & Aleven, 2015). It provides students graphical or other kind of tools for monitoring their performance. General structure for developing a MTCT named as Technical Intellectual Tutoring System (TITUS) is depicted in Figure

Information in the student model is used to perform assessments, and it is used to generate the proper pedagogical actions and decisions to maintain a continuous and productive learning process. For instance, it needs to know when the student's knowledge mastery has been obtained so it let student advance to the next learning module or difficult level. Nevertheless, it also maintains an OLM that gives students the possibility to take part and decide the learning path by letting them change the task-problem to solve. *Prototyping fidelity* has become such an important term in modeling cognitive processes, that without it, a proper representation of the knowledge components involved in a problem-solving environment cannot be marked-up. This might decrease cognitive fidelity and block cognitive task analysis, thus *assessment fidelity* becomes unreliable, and compromises the other functionalities. In the following sections, design considerations and development of such features are discussed. Cognitive tutors that implement SRL principles commonly include all or a few of the following features (Aleven et al., 2010; Long, & Aleven, 2015; VanLehn et al., 2014):

1. Rich set of tasks.
2. Give student control over selecting what to learn.
3. Personalized learning program.
4. Knowledge mastery assessment.
5. Tools for promoting self-assessment.
6. Tools for self-monitoring progress and performance.
7. Prompt hints and feedback about misused knowledge.
8. Promote problem-solving strategies.
9. Promote meta-cognitive skills development as help-seeking.

- **User Interface** is a bidirectional graphic portal where the student will execute his actions and attempts to solve different kinds of tasks-problems and obtain hints, feedbacks and other pedagogical information.
- **Control Module** is on charge to process and administrate the flow of information from and to other blocks in TITUS. Necessary pedagogical processes are involved with the *learning methods* and the *help and support* modules, where pedagogical interventions are generated. Task analyzer (step analyzer), a part of the *Bayesian inference* block, constantly monitors student's actions and returns the proper support, (i.e. hints, feedbacks, by means of the *help and support* module).

Figure 1. General architecture of TITUS

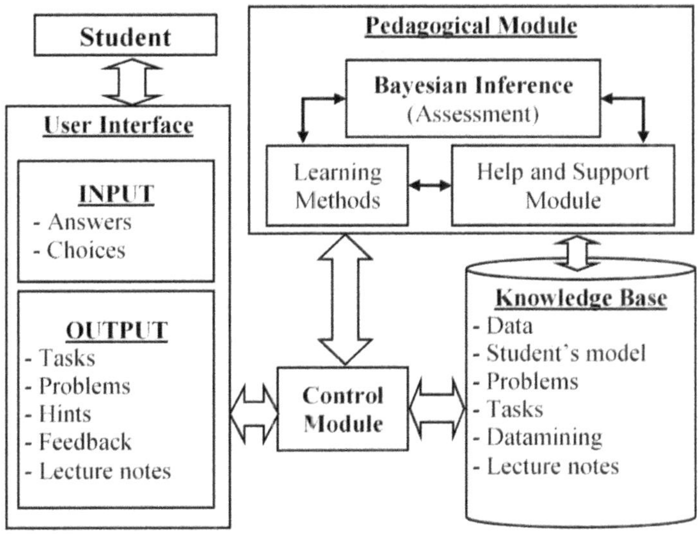

- **Knowledge Base** is constituted by the expert knowledge in the domain and information that is saved while students are using TITUS. This information is required, updated and reused as well. It also contains the cognitive model, task-problems model, lessons notes, etc. *Knowledge base* records the complete system state and student model. Student model is a database that contains information about the student: the correct answers, wrong answers, attempts by task-problem, time for answering, and other important information for a representation of the student's cognitive state.

TASK-PROBLEM MODELING

Cognitive tutors need a vast task-problem set for covering the entire domain. The greater the set is the higher possibilities of choice exist. Task-problem modeling requires expertise domain analysis with deep cognitive fidelity in order to increase assessment reliability. Task domain in TITUS covers a curriculum for learning the foundations of fault tolerance systems under the signal-parametric approach (Kulik, 1991; Martinez Bastida, & Chukhray, 2016). It provides a learning program of fine-grained task-problems properly classified in sequential modules and complexity levels. Each complexity level has as minimum two task-problems, five complexity levels have been defined: very easy, easy, average, difficult, and very difficult. This feature satisfies the minimal requirements for having a choice (Pelánek, & Jarusek, 2015). Moreover, every task-problem in each module must train the complete set of relevant knowledge components (KCs) that are included in it. Task-problem model is an interwoven network between KCs and tasks. *Task Model* (*MT*) is represented in (1) and its boundaries in (2)-(4), where T is a task, KW defines a knowledge component, i is the task identifier, $j \in [1, 5]$ represents the levels of complexity (1: very easy, 5:very difficult), k is the module for the task T, and l is the identification number for the knowledge component. An example of the MT above explained is depicted on Figure 2. MT is a concentric model, at the center is the knowledge domain, next level includes the atomic representations of every relevant knowledge component for each module; outer level contains task-problems that train those knowledge components, and the outside ring represents the classification modules. For simplification, an example of the *MT* with only 11 KCs and 16 tasks is depicted on Figure 2.

$$MT:\left\{T_{ijk}\right\}\rightarrow\left\{KW_{kl}\right\} \tag{1}$$

$$\forall k,\forall j\left\{T_{ijk}\right\}\neq\varnothing,T_{ijk}\geq 2 \tag{2}$$

$$\forall k,\forall j,\forall l\left\{T_{ijk}\right\}=MT^{-1}\left(KW_{kl}\right)\neq\varnothing,T_{ijk}\geq 2 \tag{3}$$

$$\forall k\cup MT\left(T_{ijk}\right)=\left\{KW_{kl}\right\} \tag{4}$$

Student model (*MS*) is dynamically built by the student's actions. Thus, *MS* is a dynamic representation of the degree of mastery in the student. However, if the *MS* is robust enough could represent a student's psychological profile that might be used to prompt a better pedagogical support. *MS* can be represented by (5) and (6), where *S* represents the student, *q* is her identifier, $P \subset \Re$ in the interval [0, 1] that represents the probability of mastery, *N* is the attempts (steps) realized.

$$MS1:\left\{S_{q}\right\}\times\left\{T_{ijk}\right\}\rightarrow N \tag{5}$$

$$MS2:\left\{S_{q}\right\}\times\left\{KW_{kl}\right\}\rightarrow P \tag{6}$$

In case, a student S_q uses TITUS for the first time, every value is initialized, S_q: $\forall i,\forall j,\forall k$ $MS1\ (S_q, T_{ijk}) = \{0\}$, $\forall l,\forall k$ $MS2\ (S_q, KW_{kl}) = \{0.5\}$. Once, initialization is set, first module is selected and complexity level is set to the intermediate one. Thus, next task (*NT*) is chosen by means of (7), a task that has the lowest average probability of mastery in the current module (*MT*).

$$NT=MT^{-1}\left(KW_{kl}\middle|\, MS2\left(S_{q},KW_{kl}\right)\rightarrow min\right) \tag{7}$$

If $\left\|NT\right\|>1$, search of the next task will be based on attempts (*NT'*), represented by (8). In case $\left\|NT'\right\|>1$, selection will be implemented by (9) and a task will randomly be selected (*NT**). Certainly last case is possible at the first time a student starts using TITUS.

$$NT'=\left(NT\middle|\, MS1\left(S_{q},T_{ijk}\right)\rightarrow min\right) \tag{8}$$

Figure 2. Task Model structure (example)

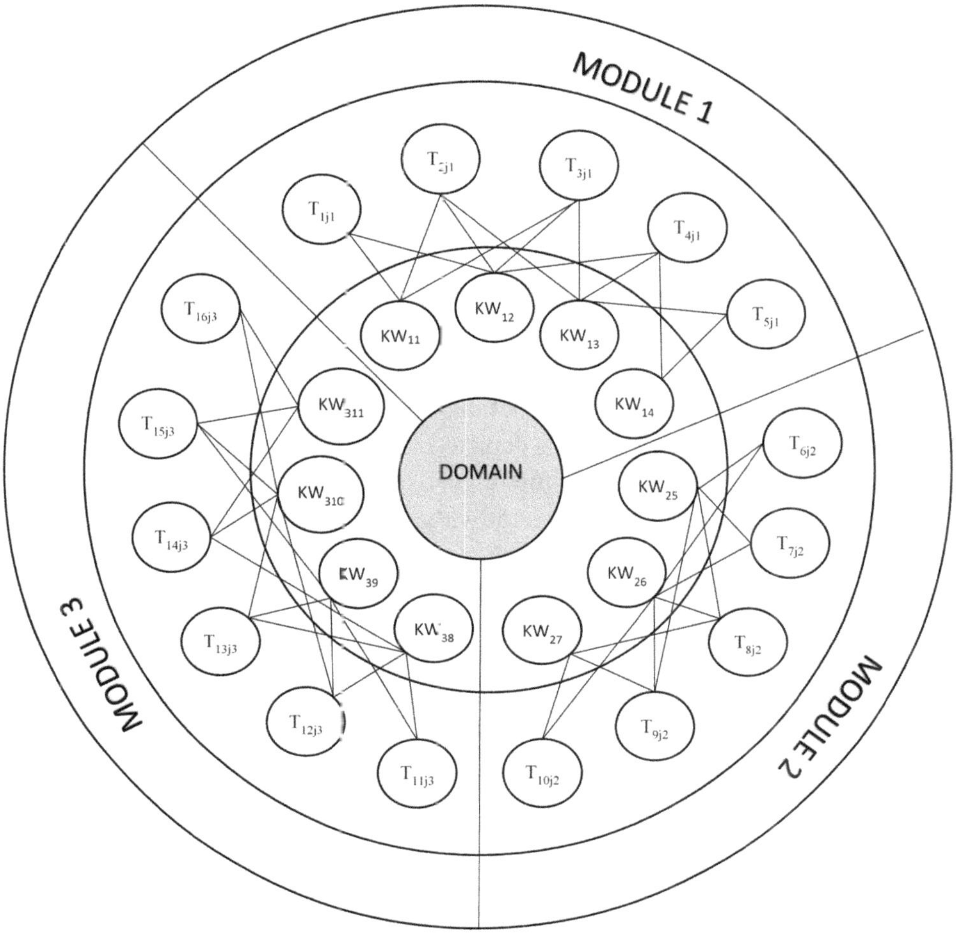

$$NT^{*} = RAND\left(NT\right) \tag{9}$$

Meanwhile the student has not mastered KCs in the current module yet, tasks will keep being choosing by (7)-(9). A module is taken as completed when the KCs that conform it are all mastered, a threshold value (pKW = 0.85) helps estimating that (VanLehn, Burleson, Chavez Echeagaray, Christopherson, Gonzalez Sanchez, Hastings, Hidalgo Pontet, & Zhang, 2011). Expressions (10) and (11) are used for determining probability of mastery. Once threshold is overcome, the tutor starts selecting tasks of the closest following module: $k+1$ until $k < max(k)$.

$$MIN\left[MS2\left(S_{q}, KW_{kl}\right)\right] > pKW \tag{10}$$

$$AVG\left[MS2\left(S_{q}, KW_{kl}\right)\right] > pKW \tag{11}$$

ASSESSMENT MODELING

Knowledge assessment intrinsically implicates uncertainty, because it must reflect effectiveness of cognitive processes involved in a problem-solving environment. Cognitive analysis is directly related to strategies for a possible solution path in base of a given context. Once strategies are applied, they deliver evidence of the degree of knowledge mastery in a specific situation. Hence, the challenge is to generate pedagogical inferences according to these outcomes. Bayesian networks (BNs) are probabilistic models that have widely been employed in ITSs (Long, & Aleven, 2015; Pelánek, & Jarusek, 2015). BNs based on the Knowledge Tracing approach equally affect prior probabilities of mastery in KCs. Thus, when multiple KCs are involved in a task-problem solution and one of them is misused, all probabilities of mastery will be equally decreased in every involved KC, without taking into account if they were or were not misused.

This presents a low-fidelity assessment model due to an improper representation of the cognitive process in the task solution. Basic BN model is depicted on Figure 3. It implements a Diagnostic Model (*DM*) that improves fidelity assessment. This topology assumes that each cognitive task solving process depends on individual KCs. Thus, each KC is an individual cognitive process when a student attempts to complete a task, so KCs can be independently applied one from another, then their posterior conditional probability of mastery should be assessed separately.

BN depicted on Figure 3, consists of four nodes: K_t, S_{t+1}, DM and K_{t+1}; where K_t is the probability of mastery of certain KC or skill at t time; S_{t+1} is a step (student's action) at moment $t+1$; DM is a diagnostic model that is directly linked to the step and influences the assessment of mastery; and K_{t+1} is the probability of mastery at $t+1$ moment. $\neg K_t$, $\neg S_{t+1}$, $\neg DM$ and $\neg K_{t+1}$ are the respective complementary probabilities of mastery. Conditional Probability Tables (CPT) for Figure 3 are presented in Tables 1-4.

Figure 3. BN with Diagnostic Model for Knowledge Tracing

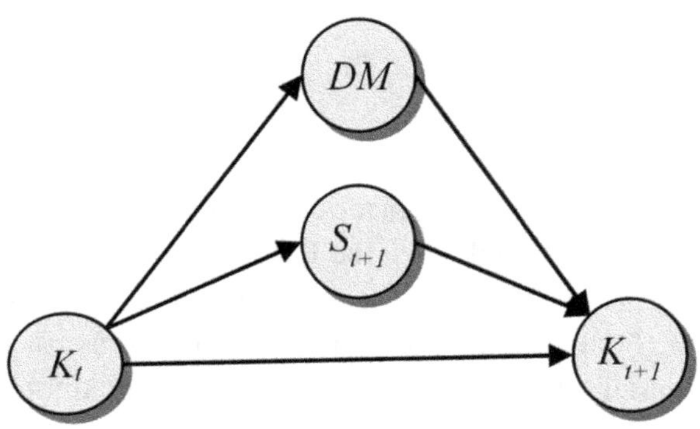

Table 1. CPT for node K_t

Mastered	$P(K_t)$
Unmastered	$P(K_t)$

Table 2. CPT for node S_{t+1}

K_t	Mastered	Unmastered
Mastered	$P(S_{t+1} \mid K_t)$	$P(S_{t+1} \mid K_t)$
Unmastered	$P(S_{t+1} \mid K_t)$	$P(S_{t+1} \mid K_t)$

Table 3. CPT for node DM

K_t	Mastered	Unmastered
Activated	$P(DM \mid K_t)$	$P(DM \mid K_t)$
Deactivated	$P(DM \mid K_t)$	$P(DM \mid K_t)$

Table 4. CPT for node K_{t+1}

K_t	Mastered			
S_{t+1}	Correct		Incorrect	
DM	**Activated**	**Deactivated**	**Activated**	**Deactivated**
Mastered	$P(K_{t+1} \mid K_t, S_{t+1}, DM)$	$P(K_{t+1} \mid K_t, S_{t+1}, DM)$	$P(K_{t+1} \mid K_t, S_{t+1}, DM)$	$P(K_{t+1} \mid K_t, S_{t+1}, DM)$
Unmastered	$P(K_{t+1} \mid K_t, S_{t+1}, DM)$	$P(K_{t+1} \mid K_t, S_{t+1}, DM)$	$P(K_{t+1} \mid K_t, S_{t+1}, DM)$	$P(K_{t+1} \mid K_t, S_{t+1}, DM)$
K_t	**Unmastered**			
S_{t+1}	**Correct**		**Incorrect**	
DM	**Activated**	**Deactivated**	**Activated**	**Deactivated**
Mastered	$P(K_{t+1} \mid K_t, S_{t+1}, DM)$	$P(K_{t+1} \mid K_t, S_{t+1}, DM)$	$P(K_{t+1} \mid K_t, S_{t+1}, DM)$	$P(K_{t+1} \mid K_t, S_{t+1}, DM)$
Unmastered	$P(K_{t+1} \mid K_t, S_{t+1}, DM)$	$P(K_{t+1} \mid K_t, S_{t+1}, DM)$	$P(K_{t+1} \mid K_t, S_{t+1}, DM)$	$P(K_{t+1} \mid K_t, S_{t+1}, DM)$

Probability $P(K_{t+1})$ of mastery certain knowledge component at a moment *t* after a student's correct step is applied is presented in (14). Conditional probabilities $P(K_t|S_{t+1}, \neg DM)$ and $P(\neg K_t|S_{t+1}, \neg DM)$ in (14) are obtained with (12) and (13) respectively, where α is a normalization coefficient. The evidences in a student's actions are denoted by $P(S_{t+1}) = 1$ (correct step) and $P(DM) = 0$ (deactivated).

$$P\left(K_t|S_{t+1},\neg DM\right) = \alpha \sum_{K_{t+1}} P\left(K_t,S_{t+1},\neg DM,K_{t+1}\right) = \alpha \sum_{K_{t+1}} P\left(K_t\right)P\left(S_{t+1}|K_t\right)P\left(\neg DM|K_t\right)P\left(K_{t+1}|K_t,S_{t+1},\neg DM\right) \quad (12)$$

$$P\left(\neg K_t|S_{t+1},\neg DM\right) = \alpha \sum_{K_{t+1}} P\left(\neg K_t,S_{t+1},\neg DM,K_{t+1}\right) = \alpha \sum_{K_{t+1}} P\left(\neg K_t\right)P\left(S_{t+1}|\neg K_t\right)P\left(\neg DM|\neg K_t\right)P\left(K_{t+1}|\neg K_t,S_{t+1},\neg DM\right) \quad (13)$$

$$\begin{aligned}P\left(K_{t+1}\right) = {} & P\left(K_t|S_{t+1},\neg DM\right) P\left(K_{t+1}|K_t,S_{t+1},DM\right) P\left(S_{t+1}\right) P\left(DM\right) + \\ & +P\left(\neg K_t|S_{t+1},\neg DM\right)P\left(K_{t+1}|\neg K_t,S_{t+1},DM\right)P\left(S_{t+1}\right)P\left(DM\right)+ \\ & +P\left(K_t|S_{t+1},\neg DM\right)P\left(K_{t+1}|K_t,S_{t+1},\neg DM\right)P\left(S_{t+1}\right)P\left(\neg DM\right)+ \\ & +P\left(\neg K_t|S_{t+1},\neg DM\right)P\left(K_{t+1}|\neg K_t,S_{t+1},\neg DM\right)P\left(S_{t+1}\right)P\left(\neg DM\right)+ \\ & +P\left(K_t|S_{t+1},\neg DM\right)P\left(K_{t+1}|K_t,\neg S_{t+1},DM\right)P\left(\neg S_{t+1}\right)P\left(DM\right)+ \\ & +P\left(\neg K_t|S_{t+1},\neg DM\right)P\left(K_{t+1}|\neg K_t,\neg S_{t+1},DM\right)P\left(\neg S_{t+1}\right)P\left(DM\right)+ \\ & +P\left(K_t|S_{t+1},\neg DM\right)P\left(K_{t+1}|K_t,\neg S_{t+1},\neg DM\right)P\left(\neg S_{t+1}\right)P\left(\neg DM\right)+ \\ & +P\left(\neg K_t|S_{t+1},\neg DM\right)P\left(K_{t+1}|\neg K_t,\neg S_{t+1},\neg DM\right)P\left(\neg S_{t+1}\right)P\left(\neg DM\right)\end{aligned} \tag{14}$$

Therefore, a *step analyzer* assesses each relevant KC in the current step in order to determine the corresponding pedagogical actions. Values for Conditional Probability Tables for nodes K_t, K_t, S_{t+1}, DM and K_{t+1} are the following shown in Tables 5-8. According to Tables 1-4, it is possible to obtain the posterior probability of mastery by employing equations (12)-(14). Probabilities $P(K_t|S_{t+1},DM)$ and $P(K_t|S_{t+1},DM)$ are computed in (15) and (16) respectively.

$$P\left(K_t \mid S_{t+1},\neg DM\right) = \alpha\left(0.5 * 0.9 * 0.9 * 1 + 0.5 * 0.9 * 0.9 * 0\right) = \alpha * 0.405 \tag{15}$$

$$P\left(\neg K_t \mid S_{t+1},\neg DM\right) = \alpha\left(0.5 * 0.1 * 0.8 * 0.6 + 0.5 * 0.1 * 0.8 * 0.4\right) = \alpha * 0.04 \tag{16}$$

From (15) and (16), we can obtain: $\alpha(0.405+0.04) = 1$; $\alpha = 2.24719$; $P(K_t|S_{t+1},DM) = 0.910112$;

Table 5. CPT for node K_t

Mastered	0.5
Unmastered	0.5

Table 6. CPT for node S_{t+1}

K_t	Mastered	Unmastered
Mastered	0.9	0.1
Unmastered	0.1	0.9

Table 7. CPT for node DM

K_t	Mastered	Unmastered
Activated	0.1	0.2
Deactivated	0.9	0.8

Table 8. CPT for node K_{t+1}

K_t	Mastered			
S_{t+1}	Correct		Incorrect	
DM	Activated	Deactivated	Activated	Deactivated
Mastered	1	1	0.6	0.7
Unmastered	0	0	0.4	0.3
K_t	Unmastered			
S_{t+1}	Correct		Incorrect	
DM	Activated	Deactivated	Activated	Deactivated
Mastered	0.6	0.6	0	0
Unmastered	0.4	0.4	1	1

$P(K_t|S_{t+1},DM) = 0.089888$. Finally, $P(K_{t+1})$ and $P(\neg K_{t+1})$ are obtained by (17) and (18). They denote the posterior probabilities of mastery and non-mastery relevant KCs.

$$P\left(K_{t+1}\right) = 0.910112 * 1 * 1 * 1 + 0.089888 * 0.6 * 1 * 1 + 0.910112 * 1 * 1 * 0 + \\ +0.089888 * 0.6 * 1 * 0 + 0.910112 * 0.6 * 0 * 1 + 0.089888 * 0 * 0 * 1 + \\ 0.910112 * 0.7 * 0 * 0 + 0.089888 * 0 * 0 * 0 = 0.964045 \tag{17}$$

$$P\left(\neg K_{t+1}\right) = 1 - P\left(K_{t+1}\right) = 0.035955 \tag{18}$$

GeNIe© and SMILE *(Decision Systems Laboratory GeNIe 2.0)* were used to model and test the probabilistic assessment. Figure 4 shows an example of the assessment model. A case when a student has misused all KCs in a task-problem is shown on Figure 4(a), thus both DMs have been activated, and conditional probabilities of mastery for each KC are equally decreased. A case when only one relevant KC was misused is presented on Figure 4(b). It can be observed that student's step was "Other" because the *step analyzer* determined that knowledge component K23_0 was misused and activated the corresponding DM23, resulting in a less posterior probability of mastery in K23_1.

COMPLEXITY LEVEL ADJUSTMENT AND PEDAGOGICAL FEEDBACK

Once a task has been chosen, TITUS expects the student's action. After a student has committed it, the *step analyzer* is triggered and assesses probability of mastery of the relevant KCs in the task: $Sol_i(NT) \in \{0,1\}$, $NT \in \{ NT, NT', NT^* \}$, and updates the attempt counter as well. Complexity level is adjusted according to the following algorithm:

Step 1: If the module *k* is starting, then complexity level *j* is set to middle one

Figure 4. Experimental test of the Probabilistic Assessment with GeNIe©, (Decision Systems Laboratory GeNIe 2.0)

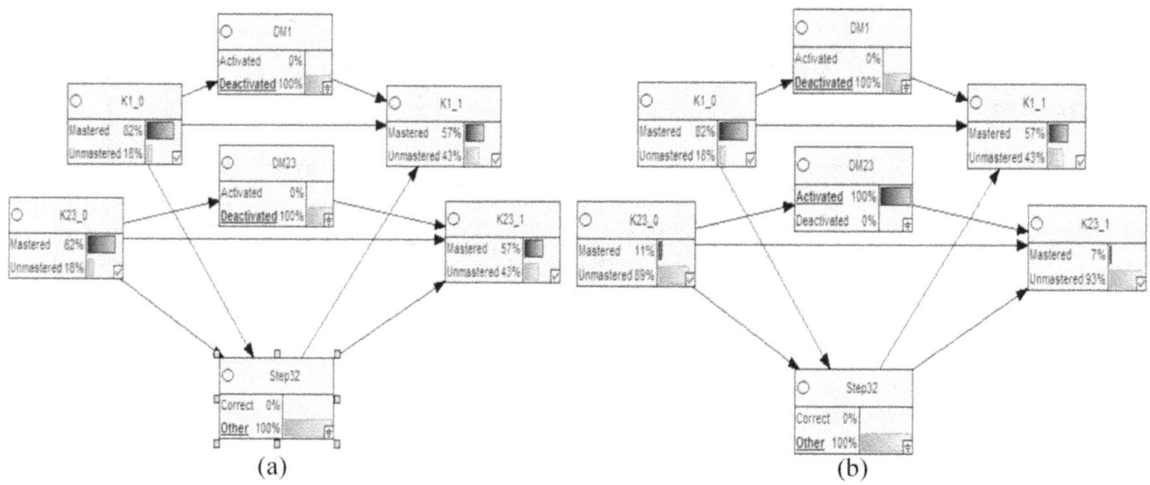

Step 2: If step Sol_i *(NT)* is right then complexity level j is increased $j = j+1$ meanwhile $j < max(j)$, if $j = max(j)$ then $j = j$, and continues with task selection model (5), if no any condition applies, then continue to the following point

Step 3: If Sol_i *(NT)* is wrong, then complexity level is decreased $j = j-1$ meanwhile $j > min(j)$, if $j = min(j)$ then $j = j$, and continues with task selection model (5)

Pedagogical feedback is a "service" that may be offered at the moment a student makes steps. Although, a hint could be supplied before, during or after committing a step to support or assist students. Hints are intended to avoid frustration or remarking misconceptions or error patterns. However, in this work, it is only proposed a general method for supplying pedagogical feedback after a student has submitted a step.

Nevertheless, it can be used as a base for developing other supporting pedagogical methodologies, but this would increase software complexity to make it capable of tracking every minimal student's action and interpreting it to "translate" it into a pedagogical intervention. The algorithm for the *pedagogical feedback support* (see below) is executed when the student's step is submitted and the *step analyzer* has already assessed the relevant KCs involved in the current task. The algorithm applies for $k\ Sol_i\ (NT)$, $NT \in \{ NT, NT', NT^* \}$.

Step 1: After every $\{Sol_i\ (NT)\} \rightarrow [T_{ijk}]$, every $1\{KW_{kl}\}$ are analyzed

Step 2: If step {Soli (NT)} $\leftrightarrow$ 1 is right, then total and right attempts counters are incremented $N = N + 1$, MS1: $\{S_q\}x\{T_{ijk}\} \rightarrow (\{N_{ikr}\}+1)$, and minimal feedback is given $\{min(FB_l)\}$: $\{Sol_i\ (NT)\} \rightarrow 1$, meanwhile $N < 1$, otherwise continue to the next point

Step 3: If step {Soli (NT)} $\leftrightarrow$ 0 is wrong, then total and wrong attempts counters are incremented $N = N + 1$, MS1: $\{S_q\}x\{T_{ijk}\} \rightarrow (\{N_{ikw}\}+1)$, and second level feedback is given $1\{FB_l\} = 2$, for every misused KC $1\{KW_{kl}\}$ in the current task T_{ijk}, meanwhile $\{N_{ikw}\} \in [2, 3]$, otherwise continue to the next point

Step 4: If step $\{Sol_i (NT)\} \leftrightarrow 0$ is wrong, then total and wrong attempts counters are incremented $N = N + 1$, MS1: $\{S_q\}x\{T_{ijk}\}\rightarrow(\{N_{ikw}\}+1)$, and third level feedback is given $1\{FB_l\} = 3$, for every misused KC $1\{KW_{kl}\}$ in the current task T_{ijk}, meanwhile $\{N_{ikw}\} > 3$

In addition, pedagogical feedback algorithm computes times a student has properly employed a specific KC (N_{ikr}); how many time she has misused it (N_{ikw}), and accordingly a feedback is returned $(FB_l) \in \{1$: minimal feedback, 2: hint about error, 3: specific error feedback$\}$. For the first time a KC is misused, a minimal feedback ($FB_l \rightarrow 1$) is returned, such as "correct" or "incorrect". For the second and third time, it will return an error-specific hint ($FB_l \rightarrow 2$), e.g. "You should pay more attention on the value of the transfer coefficient", "The class of fault you have chosen is not correct", or "Static characteristics for this class of fault are depicted on the figure, identify them", etc. It has been determined second level feedback should be given twice as a very simple mechanism to minimize feedback abuse. Nevertheless, other more advanced mechanisms may be implemented. If a KC is misused more than 3 times, TITUS will return an error specific feedback that will guide the student to review and study the corresponding material, or the related information to overcome the deficiencies on the corresponding KCs. This is a mechanism to prevent it occurring again and supporting SRL, and development of help-seeking skills. TITUS gives only delayed pedagogical feedback support in accordance to the policies explained above, and it will only give them right after the student has committed a step.

SELF-REGULATION LEARNING APPROACH

A fully functional Intellectual Technical Tutoring System (TITUS) for learning fault-tolerant principles under the signal parametric approach was developed. Self- assessment and adaptive study choice based on probabilistic models were implemented.

TITUS is conformed by two programming modules. Module 1 gives students a comfortable platform to study 32 kinds of faults reported in a Gyroscopic sensors unit (GSU). The GSU is made of three gyroscopic sensors, two angular velocity sensors and one angle sensor. Faults can be individually emulated for each gyroscopic sensor while the student is observing their affectations in the system in real time. Module 1 also permits students to perform a diagnosis of the GSU. This procedure helps students to understand algorithms for diagnosing the GSU. Students can dynamically observe and analyze in real time the gyroscopic sensor's signals, how they behave under different kinds of disturbances as well as the necessary processed signals to perform the diagnosis of the GSU. Although, students can apply any kind of the reported faults to any of the three gyroscopic sensors at any time.

Module 2 is the part of TITUS that gives students a learning and study support. This module helps them to understand and learn all concepts and necessary knowledge about fault-tolerant theory, diagnosis algorithms and methodologies applied to the GSU. Module 2 is a good tool that guides students in developing their knowledge with interactive ways, due to its graphic interface where different tasks to solve are presented. Tasks challenge and help students in a continuous constructive way for constantly increasing of their knowledge and develop skills. Module 2 adapts and controls the flow of tasks to be solved by the students based on their answers; this feature creates an individual task-learning program for each student in accordance to their own performance, but also students can take control of the study flow.

Diagnosis and Faults Study Platform (Module 1)

The Diagnosis and Faults Study Platform (DFSP) is able to emulate 32 kinds of faults for each gyroscopic sensor; for instance, the student will be able to emulate a voltage drift on the range of -15 V to +15 V or change the value of the transfer coefficient for any of the three gyroscopic sensors independently and/or simultaneously. The DFSP has been designed to work in connection with the GSU as well as without it. When the DFSP is working without the GSU, it uses real data from gyroscopic sensors that has already been saved in advance. This feature permits students to make use of the DFSP without the necessity to be connected to the GSU. Therefore, students can use it anywhere and whenever they want for studying and practicing. Thus, the DFSP is a viable tool that supports students in their leaning process about fault tolerant principles applied to a GSU under the signal-parametric approach. The graphic interface of the DFSP has different controls to emulate faults in the GSU, obtain the diagnosis and apply the algorithms for recovering the system performance. The different sections on the graphic interface of the DFSP are shown on Figure 5, and Table 9 describes each one of them.

Cognitive Tutor (Module 2)

Training program in the MTCT was structured in three sequential modules with 29 relevant knowledge components. KCs trained by the 43 tasks are shown in Table 10. They are grouped by complexity levels. Moreover, some tasks have more than one variant; this feature increases task choice up to 212 different options that MTCT may present students. Figure 6 depicts MTCT main window.

Figure 5. Sections on the Graphic interface of DFSP

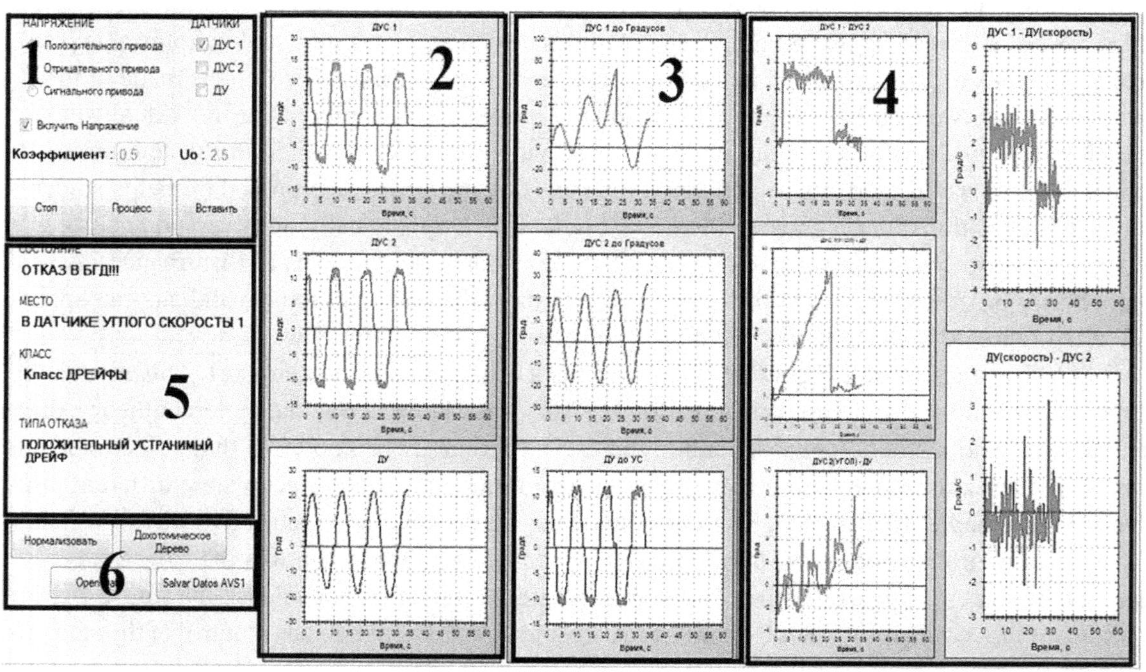

Table 9. Description for sections on the graphic interface of DFSS

Section No.	Description
1	Controls for sensor selection and parameters for emulation of faults.
2	Gyroscopic sensors' signals, from up to down, AVS_2, AVS_1 and AS.
3	Necessary conversion signals for each gyroscopic sensor.
4	Parametric signals of the system.
5	Panel for diagnosing the state of the system.
6	Data controls.

Figure 6. MTCT main window

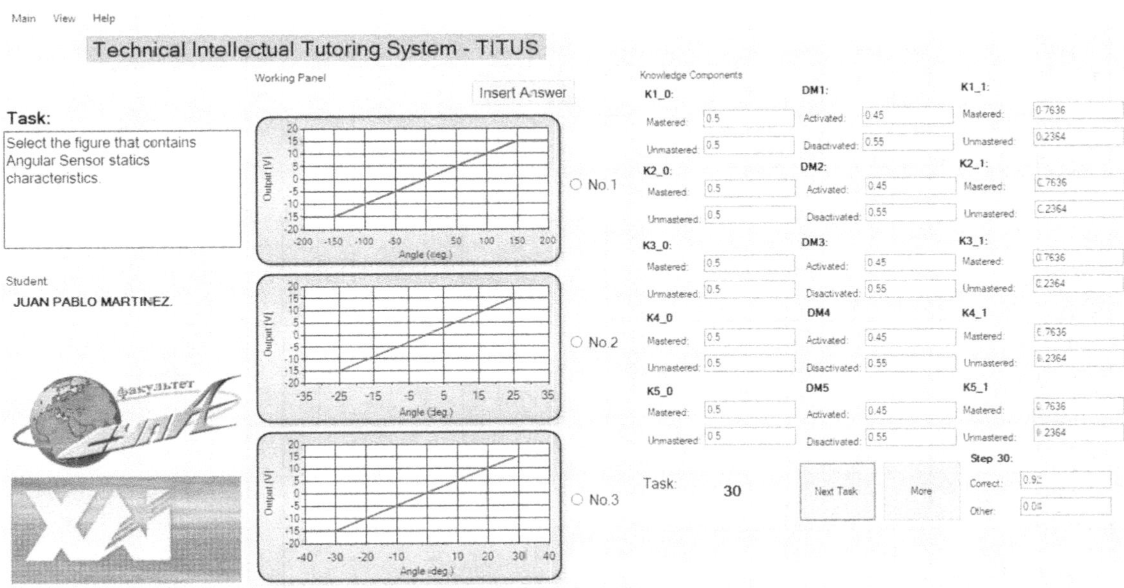

Figure 6 presents tasks to students on the half-left side of it. Required task on Figure 6, trains five KCs as we can see on the right panel. Students can monitor their degree of mastery on the right panel. They also may change the displayed task by clicking the "Next Task" button. These features promotes OLM and SRL principles. However, MTCT provides an engine that let students the freedom of choosing a next task to solve, but also it can select by itself a proper task in accordance to student's performance. Therefore, study choice either is taken by the student or by the tutor, and it promotes a SRL. We believe that flexibility in tools and support supplied by the tutor, increments students interest and reduces frustration.

For instance, a student could be on the situation a task like that one depicted on Figure 7. KCs required for the task on Figure 7 are the same as on Figure 6, but they have a "friendlier" presentation. It is possible to appreciate the big differences in presentation of the task-problem, and the way KCs are required. This might generate stress in learners and higher levels of frustration in case a student might be stuck.

Feedback is given according to student's performance and the policies above discussed. When a student has repeatedly misused a KC, Cognitive tutor in TITUS will suggest a guidance window. It

Table 10. Knowledge components trained by each task

Task	K1	K2	K3	K4	K5	K6	K7	K8	K9	K10	K11	K12	K13	K14	K15	K16	K17	K18	K19	K20	K21	K22	K23	K24	K25	K26	K27	K28	K29	Op.
1	X	X	X	X	X																									1
2						X	X	X	X	X																				1
3											X	X																		1
4													X	X																6
5											X				X															1
6											X					X														1
7											X				X															1
8																	X	X												9
9																	X		X											6
10																	X			X										6
11																	X				X									9
12											X	X																		1
13											X					X														1
14											X				X															1
15													X	X																6
16																	X	X												9
17																	X		X											6
18																	X			X										6
19																	X				X									9
20													X	X																6
21																	X	X												9
22																	X		X											6
23																	X			X										6
24																	X				X									9
25											X	X																		1
26											X					X														1
27													X									X								1

continued on following page

Table 10. Continued

Task	K1	K2	K3	K4	K5	K6	K7	K8	K9	K10	K11	K12	K13	K14	K15	K16	K17	K18	K19	K20	K21	K22	K23	K24	K25	K26	K27	K28	K29	Op.
28													X									X								1
29													X									X								1
30		X	X	X	X																									1
31							X	X	X	X																				1
32	X																						X							16
33						X																		X						20
34	X																						X							22
35						X																		X						22
36																									X	X				1
37																									X		X			1
38																									X			X		1
39																									X				X	1
40																									X	X				1
41																									X		X			1
42																									X			X		1
43																									X				X	1
Total number of tasks																														212

Figure 7. MTCT main window (example 2)

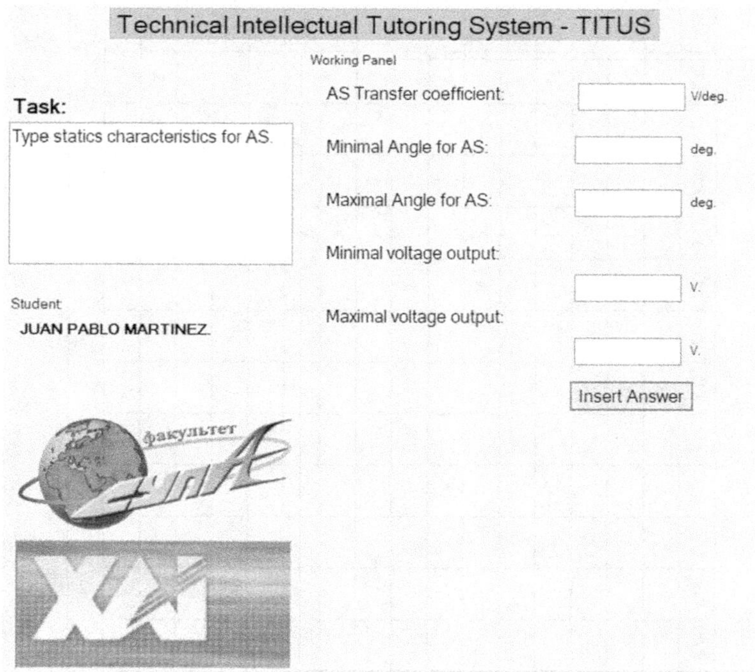

Figure 8. Help window

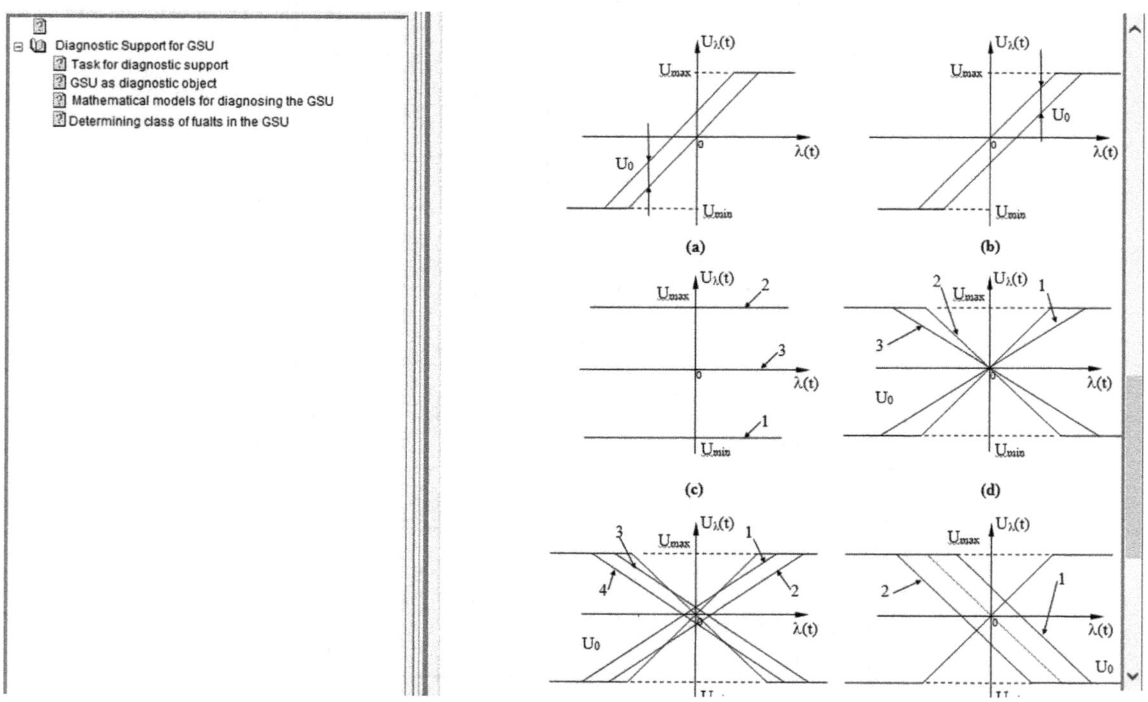

will specify where to look for the information to overcome the misconceptions. Help window has all the necessary theory and information in the domain, and it is depicted on Figure 8. This information is summarized and cataloged in modules according to the learning domain in TITUS. Students can read and work with help-seeking skills. Such skills do not only serve to find information but for developing other metacognitive skills like problem-solving patterns.

MTCT also provides a feedback service to reduce the possibility of being stuck and frustrated. Help-abuse management is beyond this work. Nevertheless, we are looking forward to implement it in the near future, by means of datamining techniques, expression recognition, and probabilistic model implementations in order to increase assessment fidelity.

It also provides a set of tools for a more detailed assessment that may be used by students or human tutors. On Figure 9 is depicted the window that shows information about student's performance. It let students monitor different kind of information about their performance:

- Degree of mastery for each KC.
- Times Diagnostic Models were activated.
- Attempts by task-problem.
- Correct attempts.
- Other attempts.
- Spent time.

We can observe on Figure 10 by means of a plot that shows how many errors were committed, which DMs were activated due to a misuse of the corresponding KCs during a task-problem solving. This might

Figure 9 Self-assessment window

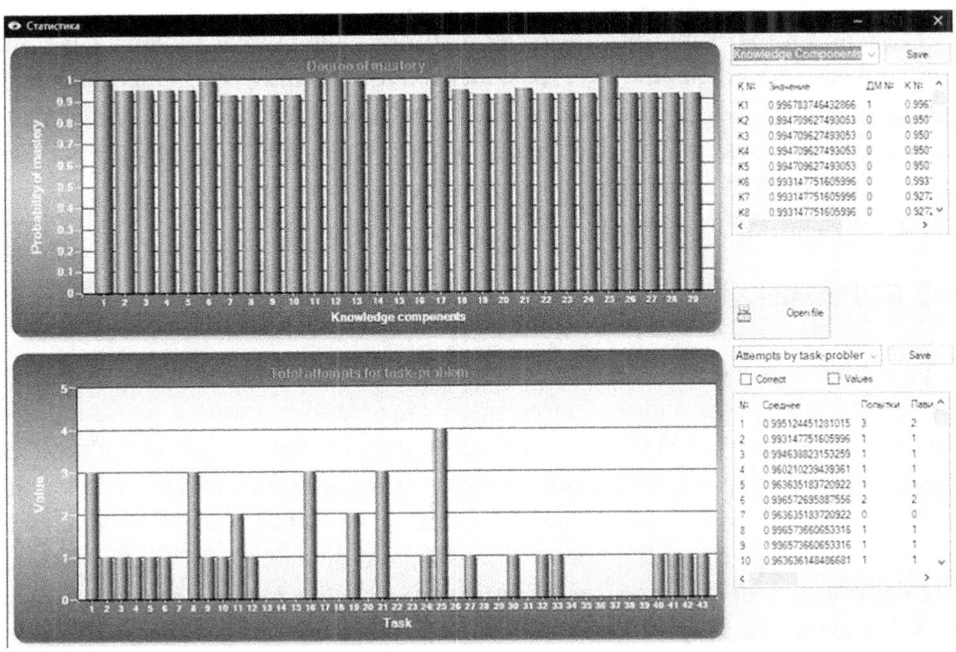

Figure 10. Diagnostic Model activations and solution attempts

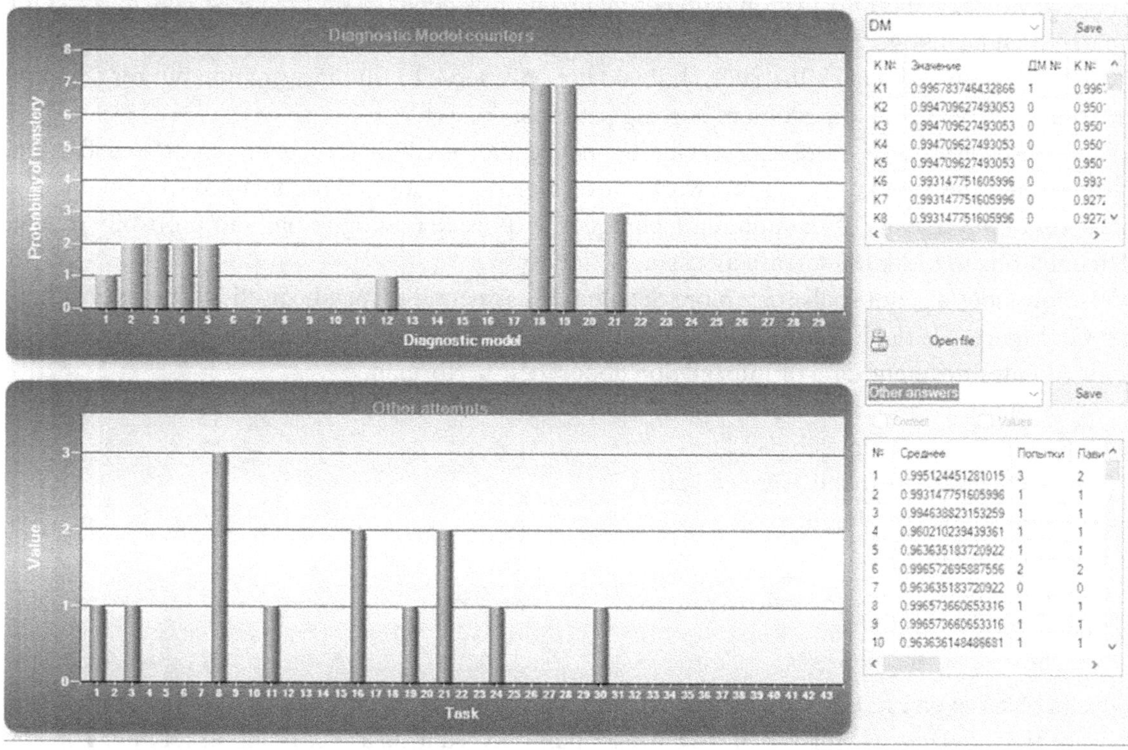

help to understand the relationship between conflicted KCs and tasks that train them in order to perform future adjustments or implementations for delivering pedagogical interventions.

Assessment window also permits directly to compare total attempts that a student has committed and how many of them are correct. In other words, students may use this or be guided to detect their own wrong actions, in conjunction with the others features above discussed to perform the corrective actions for overcoming those deficiencies. This feature can be seen on Figure 11 on the downer plot. Besides, information that is more detailed can be read on tables beside plots.

LEARNING SCHEME

A practical application in a real tutoring environment with TITUS is shown on Figure 12. The "Learning" connection between teacher and student represents the classical tutoring process that is carrying out in classrooms. Student use the MTCT to learn and mastery the foundations of fault-tolerant systems under the signal-parametric approach. Learning process is supported by DFSP what is plugged to the GSU. Thus, the student can realize experiments described on a practice plan or supplied by the teacher. Students can analyze, study and understand qualitatively and quantitatively the fault-tolerant process and diagnosis algorithms under the signal-parametric approach. In addition, the teacher can employ the MTCT like a feasible tool for supporting lectures.

Figure 11. Comparison of correct/incorrect attempts graph

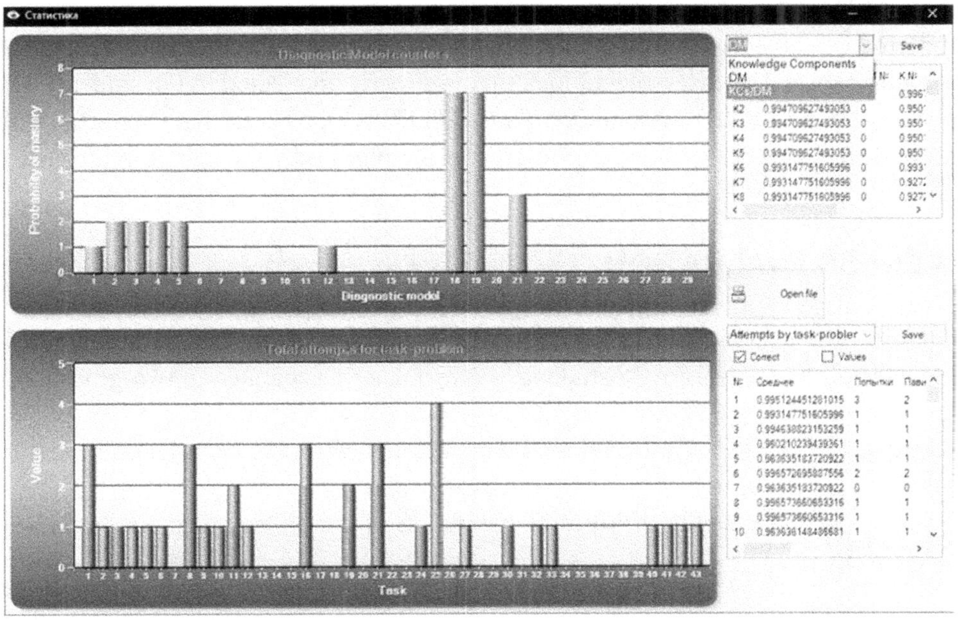

Figure 12. Learning scheme

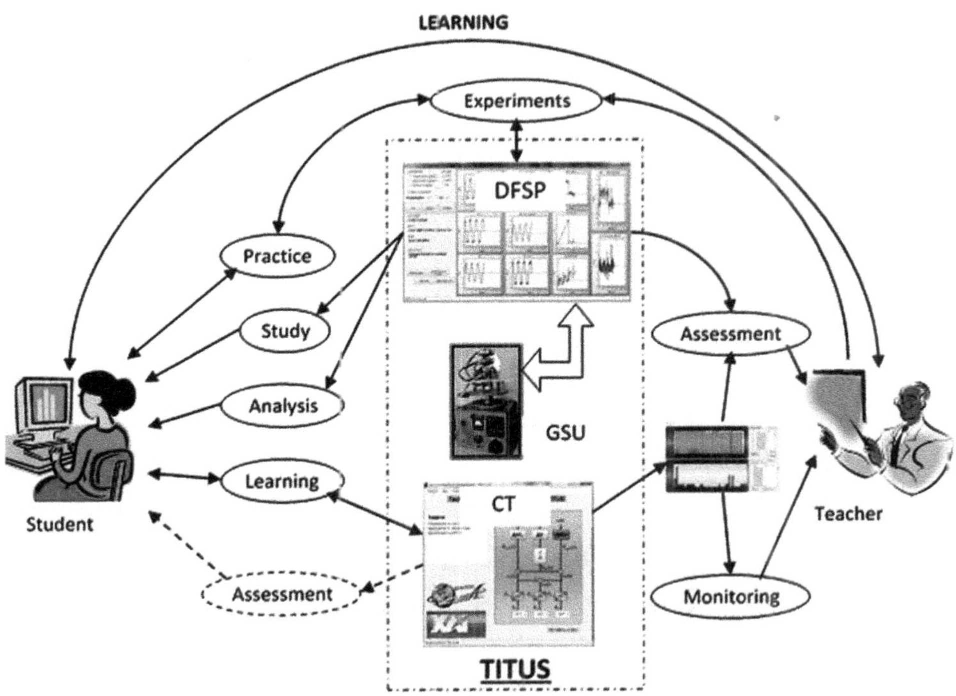

Moreover, teacher can use TITUS to assess and monitor students' performance. It can help human tutors to detect problematic knowledge components or misconceptions and reinforce them on regular lectures. Problematic students can be detected, that need more specific attention and apply the proper pedagogical adjustments. The assessment block that is directly connected to the student, represents the capabilities of TITUS on a SRL environment. Thus, TITUS is a complex that supplies feasible tools to both students and teachers for learning fault-tolerant principles under the signal-parametric approach.

EFFECTIVENESS EVALUATION

Experimental results for evaluating effectiveness of pedagogical interventions provided by TITUS, were obtained by means of the analysis of 38 students' performances, separated in two groups as follows:

1. 19 students used TITUS with a minimal feedback support in a closed learner model during the learning process (Group A).
2. 19 students used TITUS with a full implementation of the pedagogical support and OLM with SRL (Group B).

Experimental results from Group A are depicted on Figure 13. Average degree of mastery for each of the 29 knowledge components in the domain is clearly below the threshold pKW (0.85). It states that mastery of the knowledge components in the domain is less probable. On the other hand, when Group B used TITUS, the probability of mastery for every KC considerably increased, and this result is shown on Figure 14. Times when student misused a knowledge component are shown on Figure 15.

Total times each DM was activated because of a misuse of the corresponding KC are shown on Figure 15. These outcomes reflect conflicted KCs but also could be due to an improper representation for the task that might lead to confusion or misunderstanding.

MTCT counts the times that every task is presented to students, this statistics are show on Figure 16. Incorrect attempts are also depicted, defining tasks where students committed more misused knowledge components and how many times MTCT presented certain tasks. Adaptability of the proposed approach

Figure 13. Probability of mastery from Group A

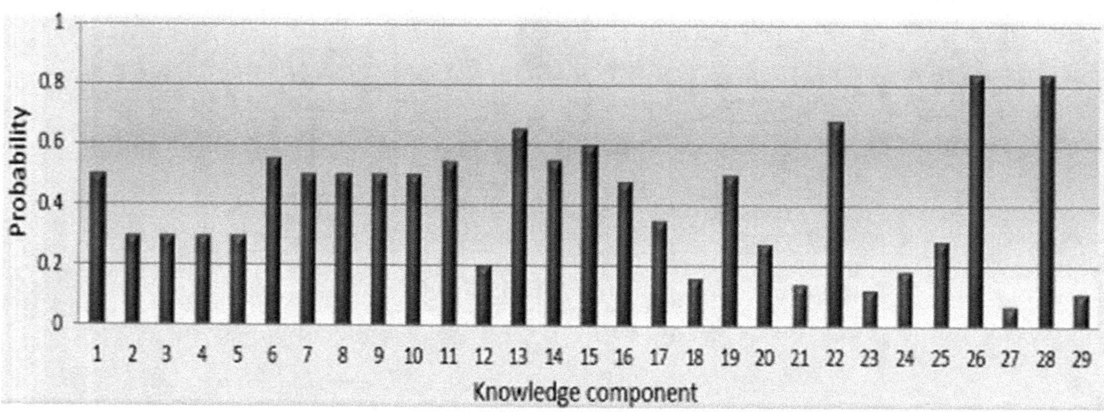

Figure 14. Probability of mastery from Group B

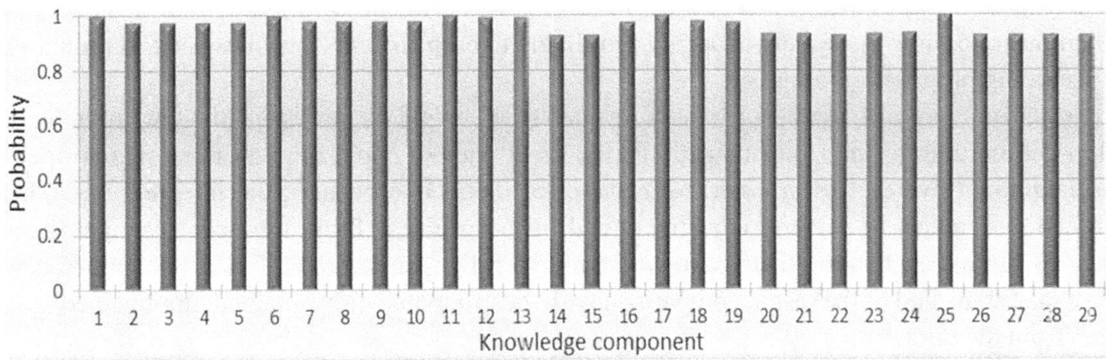

Figure 15. Times DMs were activated for each Knowledge Component

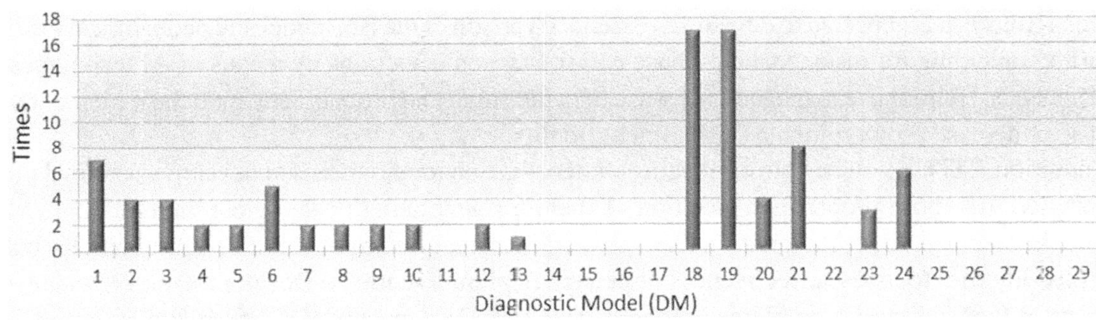

Figure 16. Total and incorrect attempts for each task

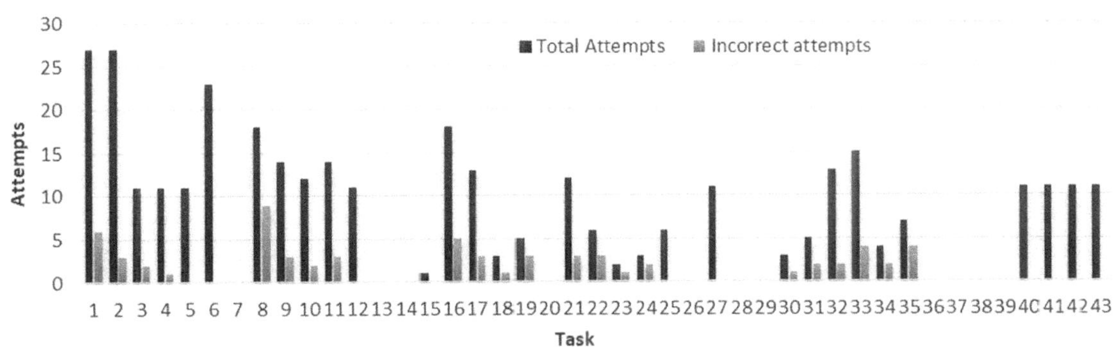

Table 11. Average probability of mastery the KCs in tested groups

Group	Average probability
A	0.4068
B	0.9662

can be seen, due to the differences of times that MTCT presented each task, for instance, task 6 was presented about 23 times meanwhile task 15 was presented only once or tasks 28, 37 were never shown. This demonstrates how the system adapted accordingly to each students performance and built an individual learning program for each one.

Attempts on Figure 16 show which tasks resulted problematic for students, and how study plan was developed according to the student's actions. Incorrect attempt could be indicators of improper task representations. Nevertheless, it could be a matter of ineffective pedagogical interventions. Thus, it could lead researchers to perform adjustment and further studies. Table 11 shows average degree of mastery by groups, and shows that group which used a fully featured MTCT clearly obtained a better performance at the end of the test in comparison with the group that only obtained minimal pedagogical support from the tutor.

FUTURE RESEARCH DIRECTIONS

In the near future, it is expected to develop an extended version of the BN model and pedagogical feedback support by including for instance, help-abuse and frustration detections by means of AI techniques and visual analysis. Help-abuse management by use of datamining and probabilistic models implementations in order to increase prototyping and assessment fidelity.

In addition, TITUS will be able of tracking students' behavioral patterns to develop a personal profile that may led to a more proper interpretation of them for generating pedagogical interventions. Moreover, further studies about the relationship between incorrect attempts and task presentations will be performed in order to assess effectiveness of pedagogical interventions. Development of a wider set of pedagogical tools and tasks-problems will be carried out. Furthermore, TITUS should be tested under different environments for a deeper evaluation.

CONCLUSION

This chapter proposes an approach for developing a learning platform that provides adaptive generation of pedagogical intervention support. It follows an Open Learner Model that is based on Self-Regulation Learning principles. It supports students to master the concepts and basic theory about fault-tolerant systems under the signal-parametric approach, by means of several pedagogical techniques. It provides a module for studying and practicing all principles in the domain, giving means for a knowledge reinforcement.

It was presented the design of a task-problem model and considerations that were taken into account to its construction and implementation. It was defined in three modules; each module has a set of task-problems with different levels of difficulty. Task-problem model is a set to task-problem with an interwoven network of Knowledge Components over the task domain. Structural considerations and characteristics were also discussed.

An assessment model based on probabilistic models presented as a Bayesian network for providing pedagogical interventions was presented as well. It helps to provide learners a cognitive pedagogical support, like hints and feedbacks. It has the ability to build a cognitive model from each student and provide individual pedagogical interventions based on it, in order to actively adapt the learning process according to the student's performance. Selection of study path is provided on-demand by the student that let her to decide to "skip" the current task-problem requirement in the current module. At the same time, the cognitive tutor also provide a selection task-problem engine that permits the MTCT to automatically choose a task among the 212 possible ones.

This automatic process is directly related to the probabilistic assessment and the analysis of other datamining features that conform a more fine-grained assessment. Different kind of support tools were developed and discussed as well, in order to promote self-assessment. Learners may use them during a self-learning program or in a classic classroom program.

Furthermore, a help support window and its implications were discussed. How directions and guidelines are employed to promote students to develop metacognitive skills for "help-seeking", because active learners are better learners.

Results demonstrate effectiveness of the approach based on the increment of mastery in learners. This effectiveness was obtained by developing a learning platform that includes a MTCT and a practicing module into a complex called TITUS. It was altogether employed with regular students in a master degree program of the task domain. Students that received a complete pedagogical support obtained a 42% better performance in comparison to those ones that received minimal pedagogical assistance, and it proves the positive educational impact in students when the proposed approach is implemented in a complex as TITUS.

REFERENCES

Aleven, V. (2010). Rule-Based Cognitive Modeling for Intelligent Systems. Advances in Intelligent Tutoring Systems, 33-62. doi:10.1007/978-3-642-14363-2_3

Aleven, V., McLaren, B., Roll, I., & Koedinger, K. (2016). Toward Meta-cognitive Tutoring: A model of Help-Seeking with a Cognitive Tutor. *International Journal of Artificial Intelligence in Education, 16*, 101–130.

Aleven, V., Nkambou, R., Bourdeau, J. & Mizoguchi, R. (2010). Rule-Based Cognitive Modeling for Intelligent Systems. *Advances in Intelligent Tutoring Systems*, 33-62.

Anderson, J. R., Corbett, A. T., Koedinger, K. R., & Pelletier, R. (1995). Cognitive Tutors: Lessons Learned. *Journal of the Learning Sciences, 4*(2), 167–207. doi:10.120715327809jls0402_2

Arroyo, I., Woolf, B. P., Burelson, W., Muldner, K., Rai, D., & Tai, M. (2014). A Multimedia Adaptive Tutoring System for Mathematics that Addresses Cognition, Metacognition and Affect. *International Journal of Artificial Intelligence in Education, 24*(4), 387–426. doi:10.100740593-014-0023-y

Bervery, P. W. (2009). *Building Intelligent Interactive Tutors*. Burlington, MA: Morgan Kaufmann.

Bessing, S. B., Gilbert, S. B., Ourada, S., & Ritter, S. (2009). Authoring Model-Tracing Cognitive Tutors. *International Journal of Artificial Intelligence in Education, IOS Press, 19*, 189–210.

Biswas, G., Segedy, J. R., & Bunchongchit, K. (2016). From Design to Implementation to Practice a Learning by Teaching System: Betty's Brain. *Journal of Artificial Intelligence in Education, 26*(1), 350–364. doi:10.100740593-015-0057-9

Chi, M., VanLehn, K., & Litman, D. (2010). Do micro-level tutorial decisions matter: Applying reinforcement learning to induce pedagogical tutorial tactics. In V. Aleven, J. Kay & J. Mostow (Ed.), *Intelligent Tutoring Systems: 10th International Conference, ITS 2010* (pp. 224-233). Heidelberg, Germany: Springer. 10.1007/978-3-642-13388-6_27

Chukhray, A. G. (2013). *Metodology for learning algorithms*. Kharkiv, Ukraine: National Aerospace University, KhAI.

Conati, C., Gertner, A., & VanLehn, K. (2002). Using Bayesian Networks to Manage Uncertainty in Student Modeling. *User Modeling and User-Adapted Interaction, Kluwer Academic Publishers, 12*(4), 371–417. doi:10.1023/A:1021258506583

Decision Systems Laboratory GeNIe 2.0. (n.d.). Decision Systems Laboratory, University of Pittsburg. Retrieved from http://genie.sis.pitt.edu

Dzikovska, M., Steinhauser, N., Farrow, E., Moore, J., & Campbell, G. (2014). BEETLE II: Deep Natural Language Understanding and Automatic Feedback Generation for Intelligent Tutoring in Basic Electricity and Electronics. *International Journal of Artificial Intelligence in Education, 24*(3), 284–332. doi:10.100740593-014-0017-9

Frank, J., & Thürmer, J., Lukas, & Gollwitzer Peter, M. (2015). Promoting the translation of intentions into action by implementation intentions: Behavioral effects and physiological correlates. *Frontiers in Human Neuroscience, 9*, 395. doi:10.3389/fnhum.2015.00395 PMID:26236214

Kulik, A. S. (1991). Fault diagnosis in dynamic Systems via signal-parametric approach. *IFAC/IMACS Symposium of fault detection, supervision and a technical process, SAFE PROCESS 91, 1*, 157-162.

Long, Y., & Aleven, V. (2015). Active Learners: Redesigning and Intelligent Tutoring Systems to Support Self-Regulated Learning. *Proceedings of the 8th European Conference on Technology Enhanced Learning*, 490-495.

Martinez Bastida, J. P., & Chukhray, A. G. (2016). An active diagnostic algorithm for a gyroscopic sensors unit. *2016 II International Young Scientists Forum on Applied Physics and Engineering (YSF)*, 29-32. doi: 10.1109/YSF.2016.7753793

Paquette, L., Lebeau, J. F., Beaulieu, G., & Mayers, A. (2015). Designing a Knowledge Representation Approach for the Generation of Pedagogical Interventions by MTTs. *International Journal of Artificial Intelligence in Education, 25*(1), 118–156. doi:10.100740593-014-0030-z

Pelánek, R., & Jarusek, P. (2015). Student Modeling Based on Problem Solving Times. *Journal of Artificial Intelligence in Education, 25*(4), 493–519. doi:10.100740593-015-0048-x

Russel, S., & Norving, P. (2010). *Artificial Intelligence: A Modern Approach*. Prentice Hall.

Shanabrook, D. H., Arroyo, I., Woolf, B. P., & Burleson, W. (2012). Visualization of Student Activity Patterns within Intelligent Tutoring Systems. *Proceeding of the 11th International Conference on Intelligent Tutoring Systems (ITS'2012)*, 46-51. 10.1007/978-3-642-30950-2_6

VanLehn, K. (2008). Intelligent Tutoring Systems for Continuous, Embedded Assessment. In The future of assessment: Shaping teaching and learning (pp. 113–138). Lawrence Erlbaum Associates.

VanLehn, K., Burleson, W., Chavez Echeagaray, M. E., Christopherson, R., Gonzalez Sanchez, J., Hastings, J., ... Zhang, L. (2011). The Affective Meta-Tutoring Project: How to motivate students to use effective meta-cognitive strategies. In *Proceedings of the 19th International Conference on Computers in Education, ICCE 2011* (pp. 128-130). Chian Mai, Thailand: ICCE.

VanLehn, K., Burleson, W., Girard, S., Chavez-Echeagaray, M. E., Hidalgo-Pontet, Y., & Zhang, L. (2014). The Affective Meta-Tutoring project: Lessons learned. Intelligent Tutoring Systems *12th International Conference, ITS 2014*, 94-103.

Walker, E., Rummel, N., & Koedinger, K. R. (2014). Adaptive Intelligent Support to Improve Peer Tutoring in Algebra. *International Journal of Artificial Intelligence in Education, 24*(1), 33–61. doi:10.100740593-013-0001-9

Wang, Y., & Heffernan, N. (2012). Leveraging first response time into the knowledge tracing model. In *Proceedings of the International Conference on Educational Data Mining* (pp. 176-179). Academic Press.

Wieber, F., Lukas, J., & Gollwitzer, P. M. (2015). Promoting the translation of intentions into action by implementation of intentions: Behavioral effects and physiological correlates. *Frontiers in Human Neuroscience*, 9. PMID:26236214

KEY TERMS AND DEFINITIONS

BN: Bayesian network, a probabilistic model that helps to estimate degree of mastery by means of conditional probabilities.

Cognitive Fidelity: A high comprehension of the knowledge involved in a cognitive process.

Cognitive Model: A representation of the cognitive state of a given learner, see also student model.

Conditional Probability: Probability that something might happen in base of observable facts that already occurred.

CPT: Conditional probability table.

DFSP: Diagnosis and faults study platform.

DM: Diagnostic model.

Fidelity Assessment: Parameter that helps to know how close the cognitive model reflects the student's mental state.

GSU: Gyroscopic sensors unit.

ITS: Intelligent tutoring system.

KC: Knowledge components.

MTCT: Model-tracing cognitive tutor.

OLM: The ability of a certain learning model to be accessed or modified by a student.

SRL: Student's metacognition ability to manage learning environments for self-development.

Student Model: Information about a student that represents a cognitive state about a given task domain, see also cognitive model.

TITUS: A MTCT for training the foundations of fault-tolerant systems under the signal-parametric approach.

Chapter 13
Model of Calculating the Weight and Centering of the Airship With Measurement Errors

Ann Tymoshenko
Kirovograd Flight Academy of the National Aviation University, Ukraine

Larisa Saganovskaia
Kirovograd Flight Academy of the National Aviation University, Ukraine

Oksana Danylko
Kirovograd Flight Academy of the National Aviation University, Ukraine

Sergei Osadchy
Central Ukrainian National Technical University, Ukraine

ABSTRACT

In this chapter, the authors present an imitation model for calculating the weight and centering of the airship, taking into account the errors that arise in the process of determining these parameters. Also in this chapter there is the analysis of existing means, methods, and methods for determining the weight and alignment of the airship in domestic airlines and air companies. The authors made the research of the sources of errors and identified their types. This chapter provides a new model of the takeoff weight and balance of the aircraft determining process, as well as the weight distribution of passengers law. These materials allow simulating one of the main operations in the preparation for departure process.

INTRODUCTION

Flight safety is a complex character that reflects both air transport and aviation operations safety; it determines the ability to operate flights without a threat to health and life of people (Flight Manual).

The important task of providing safety of air transportation is the correct calculation weight and balance of the aircraft. While loading the airplane it is necessary to take into account two factors:

DOI: 10.4018/978-1-5225-7709-6.ch013

- Operational weight of aircraft;
- The position of its center-of-mass.

Structure of some lightweight aircraft provides centering in possible limits placing the load at different locations, however, the majority of aircraft have the strict load charts.

The incorrect distribution of load can cause:

1. Decrease flight performance when overloaded;
2. Decrease aircraft controllability at the displacement of the center of gravity;
3. Overloading leads to increasing the sluggishness and decreasing rate of climb and also to the stalling speed increase.

Performance of climb would get worse if an increase in the rate of climb is ignored. Ignoring other characteristics would even cause the catastrophic results.

BACKGROUND

For in order to determine what characteristics must satisfy the model we are developing, we have conducted an investigation of existing automated systems used in aviation. In works Shmelova (2005), Artemenko (2010) is presented the model of the choice of a spare aerodrome, as well as the importance of the question of the correct calculation of all parameters of the flight for safe flight performance. The issue of decision-making by the operator of the air navigation system is considered in the works Harchenko et al. (2011, 2012), Sikirda (2004), the authors of these works built the model of decision making by the operator of complex systems. The intellectual decision-making systems are considered in the works Gerasimov (2007), Gluhih (2000), Moskvichov (1998), Nedelko (2002). The main task of these works is in order to increase the safety of flights, the intelligent modules - decision support systems, in which the information support of the aviation operator is realized, should be included in the composition of the automated air traffic control systems (ATC).

Analysis of the Causes of air accidents for forty-three years, given in Kasak (2010) shows that the culprit of disasters in the 50% of cases is the aircraft crew. Overload of the aircraft occurs in the 0,8% of such cases.

Analyzing aviation events according to the site Aviation Safety Network Database that is related to the wrong loading and centering of aircraft (AC) over, we set next reasons that led to it, we illustrated it in Figure 1.

In spite of different repetition of reasons for passenger and freight transportations, they have an approximately identical influence on the safety of flights. Unsupported loads are more often observed at freight transportations. The conditions of origin of these events are that all standard sizes (weight of passengers, the weight of members of a crew, weight, and density of fuel and oil and others like that) undertake for a calculation. If there are the errors during a calculation related to an inexact determination of these sizes.

Figure 1. Reasons of incidents what connected with loading and centering

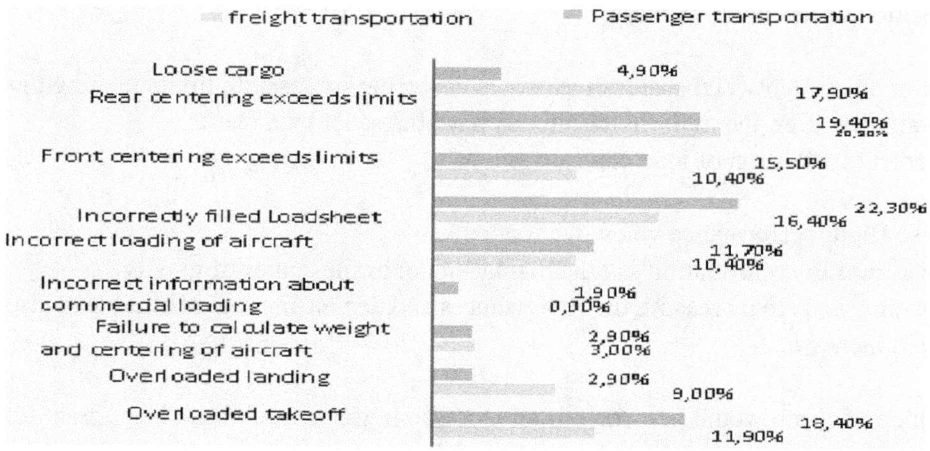

Therefore there is a problem of correct calculation of preset parameter taking into account these errors. To that end, it was suggested to work out the model of the system of loading and centering of the airship, what would take into account all these errors, in independence from the method of determining the weight of centering of the airship. The use of this model will allow working out CAS of support of decision-making on loading and centering of the airship, which can be used.

MAIN STAGES OF MODEL DEVELOPMENT

Calculating the weight and centering of passenger aircraft is usually executed by specialized airport services or airlines, although on occasion these calculations are executed by the members of flight crews, or employees on providing of flights. In the small aviation of calculating the weight and centering is the direct duty of pilot or employee that provides preparation to flight.

Most modern passenger aircraft is constructed so that if all passenger seats will be occupied, cargo compartment is fully loaded, and fuel tanks are fully filled by a fuel, then weight aircraft will exceed maximally possible for taxi and flight. Such construction aircraft allows operating combination the "commercial loading — a fuel", giving the advantage to the factor more important for the aim of this flight. If the maximal distance of flight is a more valuable factor, then it is necessary to limit loading, and if it is necessary to transport a maximum of loading, then to tuck in less than fuel and that is why will diminish flight distance.

During our research, we'd study all components of the weight of aircraft, presented in Table 1.

Calculating the landing weight at the planning of flights comes true after the special graphic arts, that is present in a presence in the instruction of "Flight Planning and Performance Manual", that is given out for every concrete type of airplane.

The next constituent of safe implementation of flight is a calculation of centering aircraft. There are a few methods of calculating the center of gravity of AC, such as:

Table 1. Components of the weight of the aircraft

Dry Operating Weight (DOW)	Balance empty weight (BEW)
	The weight of crew, luggage, catering, press
Zero Fuel Weight (ZFW)	Dry Operating Weight(DOW)
	Commercial load
Takeoff Weight (TOW)	Zero Fuel Weight (ZFW)
	Fuel
Taxi Weight (TW)	Takeoff Weight (TOW)
	Fuel for run engine and taxing

1. There is weighing of the airplane and the center of gravity settles accounts in length units along him longitudinal axis and conditional plane (Datum).
2. Graphical method of centering chart.
3. Method of moments.
4. The method of the use of the side systems of determining the weight and center of gravity AC.

Studying the question of wrong determination of loading and centering aircraft we set basic dependences in Table 2.

In the aviation calculating of mass and centering of the airplane is the direct duty of employee from providing of flights (FOO/FD). For providing of safe and effective implementation of flight of FOO/FD during the planning of flights must entertain all factors that influence the process of implementation of flight.

To increase the efficiency and safety of the airline introduced automated process control systems. The automated system of commercial software Flight consists of a number of subsystems:

- Reservation of seats
- Registration of passengers and baggage
- Calculating commercial aircraft loading
- Collection Development enplaned

Automated system for calculating enplaned is a logical continuation of computer reservation systems locations, passenger check-in, and baggage handling. They provide a high level of service for passengers

Table 2. Dependence of performance characteristics of AC from the position of the center

Dependence aircraft performance on the position of the center of gravity	
"heavy" nose of aircraft	"heavy" tail of aircraft
• AC has better durability • AC leaves corkscrew better • Difficult to control aircraft, especially on takeoff or landing	• Less stall speed • Better cruise speed and fuel consumption • AC has less durability • AC leaves corkscrew much worse

comfort, the speed of commercial transactions, maximum employment and passenger loading baggage and cargo compartments, taking into account safety.

The automated system consists of:

- Complex hardware
- Mathematical and software
- Information Assurance
- Organizing software
- **Complex Hardware:** Related computational and organizational machinery, terminal and communications equipment for the collection, storage, processing and transmission of data.
- **Mathematical and Software:** Mathematical methods, models, algorithms and programs for solving the problems of information processing with the use of computer technology.
- **Dataware:** Input, output, and regulatory information and documentation used in the system.
- **Organizational Support:** Officials and operating instructions used in the process of information processing. The automated system calculates the load in two stages - preliminary and final. This provides:
- Calculation of the limit of enplaned
- Calculation of alignment and weight of the aircraft by flight stage
- Control of keeping restrictions limiting payload, valid alignment and weight of the aircraft by flight stage
- Issue a summary of the load manifest

The main purpose of the automated system is to reduce the time to ensure commercial flight and increase flight safety by eliminating human error manual calculation Doc. 9806 (2010).

To date, there are systems for automating the process of loading the aircraft:

1. SMMC (system for measuring mass and centerline of aircraft)
2. “CENTRING” system
3. SZV
4. WB – Guarantee.

SMMC (system for measuring mass and centerline of aircraft). Purpose. Used to determine the value of hand weight and center of mass (alignment) before takeoff of aircraft, calculates the change the center of gravity in flight and provides the necessary information to the crew and interacting systems avionics.

Features. Remote MSWBA provides job modes and display information calculator, and performing manual input standby mode. Such type of scale is really popular in the aviation sphere and is used in different airports around the world.

Further provides:

- Conducting elements embedded control systems with the issuance of the information about failures in interacting systems side;
- Conduct tests and ground modeling MSWBA - AIRCRAFT ;
- Data processing and measurement of fuel and inertial systems.

Automated payments system alignment and loading of aircraft "CENTRING" in response to the order of the branch target specific organizations, with the participation of civil aviation design bureau of the Ministry of Aviation Industry.

In accordance with the decisions of FAS (FATA) Russian safety, as well as on the basis of the ruling of the Ministry of the system "CENTRING" is included in the list of mandatory equipment for the certification of international and federal airports.

1. Appointment

System "Centering" is designed to automate the functions and balance calculations download for all types and modifications of the passenger, cargo and aircraft convertible domestic and foreign production. The system serves all types of domestic aircraft design bureau in Russia and the CIS (Tupolev, Ilyushin, Yakovlev, Antonov, Sukhoi), as well as aircraft company Boeing, Airbus, Embraer, Bombardier, Falcon, Fokker, Hawker, Dach, SAAB, etc. Currently, in the database system about three thousand types of options/ modifications/layouts aircraft virtually all Russian and foreign airlines operating in the airports of the Russian Federation. System "open" for the inclusion of any additional types and modifications of aircraft supplied by the documentation specified type for input into the database of the automated systems (ANM560/565).

2. Key features:

- Creating and maintaining a normative reference database (by type, modifications, and layouts AC, as well as the aircraft fleet).
- Calculation and control characteristics of commercial flight.
- Calculation and control the weight of the airplane at all stages of the flight.
- Calculation of alignment on all types of aircraft and modifications included in the base system.
- Making recommendations on a changed flight, including calculation of the necessary ballast and guidelines for its placement.
- Formation and the issuance of final documentation requirements within the industry of the Russian Federation (centering and pivot - boot statements) standards and IATA (type "Loadsheet", "LDM", etc.).

The system is built in the shape workstation Manager centering aircraft based on the personal computer, operating independently or as part of a corporate network. Actually, this system is used in main airports in Ukraine.

WB-Guarantee

Program complex calculation parameters centering AC

Key features. Powerful system of calculation of alignment, aircraft parameters, optimize the distribution of loading on board the aircraft, preparation of supporting documents, telegrams formation, a database of the sun and archive flights. Ergonomic design helps you quickly learn to work with the system, the interface provides a pleasant work. Integrated control system eliminates many errors at data entry. Integrated help system provides information on the current results of the calculations, including in the form of a balanced schedule and helps to quickly make the right decision.

- Compliance with international and Russian standards
- Integrated editor database on aircraft (ANM050, ANM051)
- client-server architecture, which gives advantages access speed and reliability of data storage
- Integrated support for SQL- database server InterBase, Firebird, technology-based IBX
- Integration with SITA Network based Liasion OCX
- Cross- Wednesday Linux / Windows-based TCP / IP
- Modern Windows interface that simplifies data processing.

Features:

- The system of calculation
- Multilingual interface
- Summary statement boot, loading pattern (Russian, English)
- Telegrams about downloading LDM, CPM
- Pre-loading, the final load distribution
- Optimization of loading
- Archive flights
- Database/editor functions
- Document Description
- Details of the document holder
- Details of the carrier
- Automatically generated documents
- Addresses mailing telegrams
- The weight of passengers and carry-on baggage
- Standard/actual weight of luggage
- A weight of the crew and crew baggage
- Basic formulas
- Position galleries/accommodation kitchens
- Alignment classes/zones / series
- Free class code
- Passenger zone
- Layouts
- Plan landing/equipment locations
- Definitions decks/compartments/freight items
- Summary restrictions on downloading compartments, including passenger area
- Summary restrictions on downloading items
- Overlapping positions/mandatory loading
- Accessories
- Distribution of flight and cabin crew/crew baggage
- Install the stabilizer from entering
- Procedures for the fuel/water injection
- Density of fuel
- Fuel tanks
- Ballast tanks fuel/water

- Restrictions on the center of gravity (wells)
- Additional area restrictions ("wells")
- The operating weight/index
- Onboard facilities aircraft
- Variable load manifest format summary

For determination of requirements, that must satisfy the developed model of the system of determination of loading and centering aircraft, we learned the basic systems and facilities of determination of these parameters and their defects. Table 3 and Figure 2 summaries it results.

One of the objectives of our study is to determine by the expert the requirements that the model we are developing must meet.

Accuracy and reliability ranking procedure is largely dependent on the number of objects. Such objects smaller than the better its "distinctiveness" from the standpoint of the expert, and hence, the more reliably the rank of the object can be set. The number of objects being ranked n should not be more than 20, and most reliable this procedure when n < 10.

For the survey were selected by 10 experts, which are relevant to the calculation of loading and centering AC; they were asked to complete a questionnaire sample.

Matrix components group preferences for the expert group (m = 10) and define the opinion of the group. (Table 4)

The coefficient of variation of v<33% means that the opinions of experts on assessment of priority requirements N° N° 1, 4, 5, 6, 7, 8 coincide. Requirements for N° N° 2, 3 expert opinions do not coincide. Since the six requirements for the coefficient of variation v <33%, then we can consider the opinion of experts (Rgr) agreed on the importance of procedures and normally distributed.

Table 3. Home facilities of determining the weight and centering

Name	Description	Use	The methods used to determine the weight and balance
"Centring" system	It is used to automate functions and balance load calculation	IL-62,76,86,96,TU-134,154,204, 214, JAK-40,42, AN-12,24,26,30,32, AN-72,74,124, B -737,757,767, 777, A-310, 319, 320, 321, DC-10, CRj-100/200, Rj-85 etc.	Graphics and analyses method Method of moments
SMMC – system for measuring mass and centerline of aircraft	Determines the stopping mass and the center of mass of the aircraft before departure and the common changes of the above-mentioned values during the flight. The information is transmitted to the Flight Instruments	IL-62,76,86,96,TU-134,154,204, 214, JAK-40,42, AN-12,24,26,30,32, AN-72,74,124, B-737,757,767, 777, A-310, 319, 320, 321, DC-10, CRj-100/200, RJ-85 etc.	Method of moments
SZV	Data providing and Calculation of SZV, Printing (with preview).	For the Airports and the Air companies.	Methods of moments
WB- Guarantee	Powerful system to calculate the main parameters of the aircraft, optimize load distribution on the aircraft, preparation of supporting documentation, the formation of cables, keeping the database and archive for the aircraft flights	IL-62,76,86,96,TU-134,154, TU-204, 214, JAK-40,42, AN-12,24,26,30,32, AN -72,74,124, B-737,757,767, 777, A-310, 319, 320, 321, DC-10, CRj-100/200, Rj-85 etc.	Graphics and analyses method Method of moments

Figure 2. Cause and effect of scheduling

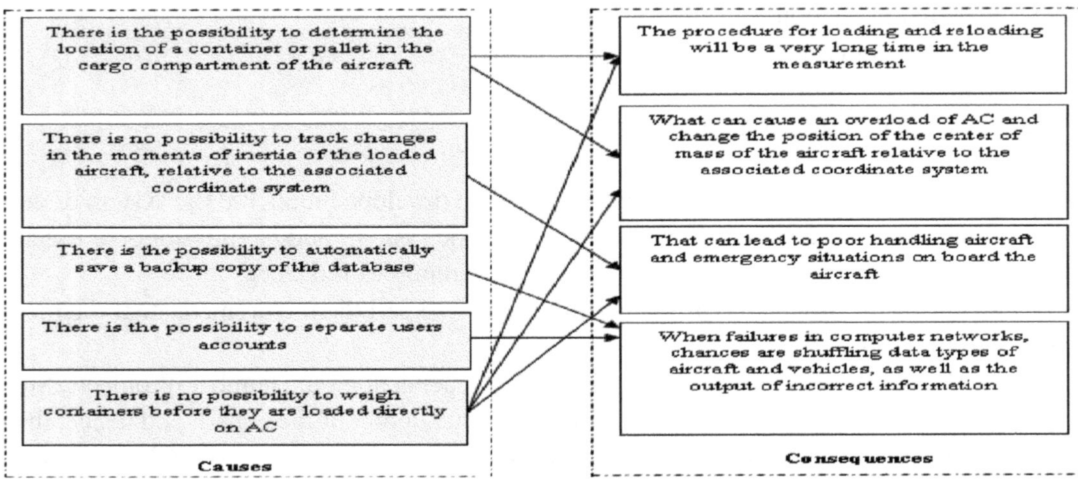

For the evaluation of consistency across all the requirements necessary to use the Kendall coefficient of concordance or repeat survey of experts.

$$W = \frac{12S}{m^2(n^2 - n)} \tag{1}$$

Table 4. Group preference matrix

	Requirements to the system download and centering AC							
experts	**w1**	**w2**	**w3**	**w4**	**w5**	**w6**	**w7**	**w8**
1	2	3	1	4	6	5	7	8
2	1	2,5	2,5	5	8	4	6	7
3	3	1	2	6	5	4	8	7
4	2	2	2	5	6	4	8	7
5	2	3	1	4	6	5	7	8
6	2	1	3	4	6	5	7	8
7	2	3	1	4	6	5	8	7
8	2	2	4	5	8	2	6	7
9	1	3	3	5	6	3	7	8
10	2	4	2	6	5	2	8	7
Rgr i	1,9	2,45	2,15	4,8	6,2	3,9	7,2	7,4
R'gr i	1	3	2	5	6	4	7	8
Di	0,32222	0,91388	1,00277	0,62222	1,06666	1,43333	0,62222	0,26666
Gi	0,56764	0,95595	1,00138	0,78881	1,03279	1,19721	0,7888	0,51639
Vi,%	29,8761	39,014	46,5761	16,4335	16,6579	30,6979	10,955	6,97834

$$S = \sum_{i=1}^{n} \left\{ \sum_{j=1}^{m} x_{ij} - a_{ij} \right\}^2$$ - the sum of squared differences (deviations) (2)

$$a_{ij} = \frac{1}{2} m(n+1)$$ - the average value for the total number of ranks (3)

In order to identify what requirements must meet the AC load scheduling system, the expert survey conducted by specialists of AC load. Also, to identify the impact of AC load and proper stowage on it to change the aircraft moments of inertia with respect to the related coordinate system considered three different aircraft weight and load capacity (AN - 26, An - 124, Boeing 767- 300t). Using the guidelines for calculating the gyroscopic devices Babaev (1967) and guidelines "Gyro on a rocket" Nazarov (1964) we calculated the moments of inertia for these aircraft. In the study, the planes were replaced by equivalent cylinders and rated them for their moments of inertia of the empty and the full cylinder. Using the table of coefficients of equations oriented lateral movement of heavy, medium and light aircraft for two regimes of flight determined that increasing the moment of inertia of the aircraft, all coefficients are reduced in inverse proportion, leading to a deterioration of handling and aircraft emergency on board the aircraft.

The criteria for assessment:

w1 - Full-featured tasks intended purpose, computing centering AC for the effect of all components of the service and commercial load;
w2 - The presence in the system of (functions) effective control of all the parameters and characteristics of the load;
w3 - Compliance requirements and constraints for specific types and modifications of aircraft designated a Manual on loading and centering the aircraft;
w4 - Versatility constructing algorithms and databases on the types of AC;
w5 - Incorporating the system into the existing traditional technology of the industry;
w6 - Calculation requirements and recommendations of international standards IATA/ICAO in this area, including in terms of technology and supporting documentation;
w7 - Mandatory filing of appropriate aircraft manufacturing corporations (EDO, ASTC), a joint trial;
w8 - Having developed a set of measures to ensure the industrial mode of operation of the system for mass use.

With the help of the expert, assessment identified the criteria that must satisfy the planning system AC load. The results of calculations of consistency of the expert group on requirements planning system AC load are presented in Table 5.

The significance of the resulting coefficient - 0.863305. This ratio was significantly (and a significance level of 0.01 and a significance level of 0.05), it is possible to argue that the opinions of experts agreed.

The next stages of our research are the determination of sources of the appearance of errors and verification of submission of these errors to the law of normal distribution of casual sizes.

As we talked before there are a few methods of calculating the weight and centering of AC. The first in a frequency of the user is a graphic method of centering chart. The essence of the method consists in that, to inflict on the corresponding scales of the chart presented in Figure 3, sizes of the weight of

Table 5. Calculation of the consistency of the expert group on requirements planning system AC load

Nº	Criteria	Parameters					
		System priorities Experts Rgr	Dispersion Di	Standard deviation σi	Coeff. variation υi,%	Coeff. concordance W	Weighting coefficient. α,%
1	w_1	1,9	0,3222	0,56765	29,88	W = 0,8633	20
2	w_2	2,45	0,91389	0,95598	39,02		18
3	w_3	2,15	1,00278	1,00139	46,58		19
4	w_4	4,8	0,6222	0,7881	16,43		12
5	w_5	6,2	1,06667	1,03279	16,66		8
6	w_6	3,9	1,43333	1,19722	30,7		14
7	w_7	7,2	0,6222	0,7881	10,96		5
8	w_8	7,4	0,2667	0,5164	6,98		4

equipment, luggage, passengers, load, fuel, members of the crew. The main feature of this technology consists in an acceptance for the calculation of actual not weight, but her the standardized value. For example, the weight of the passenger is considered even 75 kg, member of crew 80 kg, a density of fuel 0,75 kg/L etc.

1. The weight of the equipped aircraft

$$P_{CII}^{C} = P_{0}^{II} + \phi_{P0} + P_{CH} + \phi_{C1} \quad (4)$$

$P^{П}_{0}$ – a weight of empty aircraft;
φ_{p0} – error filling the form;
P_{CH} – a weight of the equipped aircraft;

φ_{C1} – measurement error.

2. The weight of luggage

$$P_{b} = P_{b0} + \varphi_{b} \quad (5)$$

P_{b0} – the true weight of luggage; φ_{b} – measurement error.

3. The weight of passengers

$$P_{fi} = P_{mi} + \phi_{mi} \quad (6)$$

P_{mi} – the standard weight of passengers; φ_{mi} – measurement error.

Figure 3. Loadsheet and Balance Chart

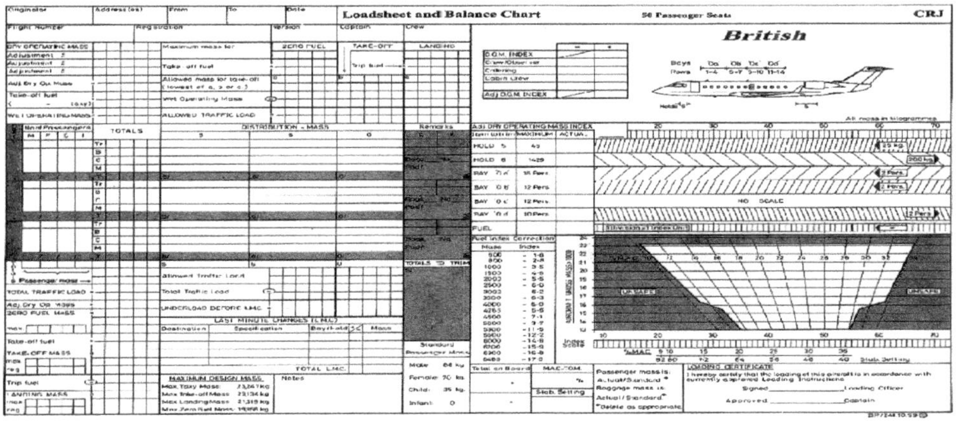

4. The weight of the crew

$$P_E = \sum_{i=1}^{n} P_{\Pi i} + \sum_{i=1}^{n} \varphi_{\Pi i} \tag{7}$$

$\Sigma P_{\Pi i}$ – commonly accepted the weight of crew members; $\varphi_{\Pi i}$ – calculation error.

5. The weight of fuel

$$P_T = P_{T0} + \phi_{T0} \tag{8}$$

P_{T0} – fuel weight at standard atmospheric conditions; φ_{T0} – measurement error.

6. The take-off weight

$$P_b^C = P_{C\Pi}^C + P_b + \sum_{i=1}^{N} P_{fi} + P_E + P_T \tag{9}$$

Using Figure 4, consider the second technology uses the method of moments and consists of the sequential performance of eight actions:

1. Determine the weight and center of an empty aircraft

$$x_T = \frac{\overline{x}_{T0} MAC}{100 \cos \alpha} + X_B - (y_B - y_T) \sin \alpha \tag{10}$$

2. Enter in table 6 the weight and center of the empty aircraft, all types of cargo

Table 6. The consolidated table of static moments and weights of the individual elements of the aircraft

Item name	PX, Nm	X, m	P, N	Y, m	PY, Nm
1	2	3	4	5	6

3. Put in the table against each weight the coordinates of its center relative to the axis *xOy*
4. Calculate the static moments *PX* та *PY*
5. 5. Find the sum of static moments and weights

$$S_y = \sum P_i Y_i \tag{11}$$

$$S_x = \sum P_i X_i \tag{12}$$

$$P_\Sigma = \sum P_i \tag{13}$$

6. Determine the coordinates of the center of gravity of the loaded plane relative to the axis *xOy*

$$X_T = \frac{S_x}{P_\Sigma} \tag{14}$$

$$Y_T = \frac{S_y}{P_\Sigma} \tag{15}$$

7. Determine the coordinates of the center of gravity relative to the coordinate system connected to the wing x_0By_0

$$X_{T0} = X_T - X_B + (Y_B - Y_T)\sin\alpha \tag{16}$$

8. Find the center of the gravity of the aircraft.

B - a wing sock;
O – a nose of the aircraft;
C - the aircraft center of mass;
x – a center line;
y – a vertical;
x_{T0} – the aircraft balanced data.

Figure 4. Associated with the aircraft coordinate systems

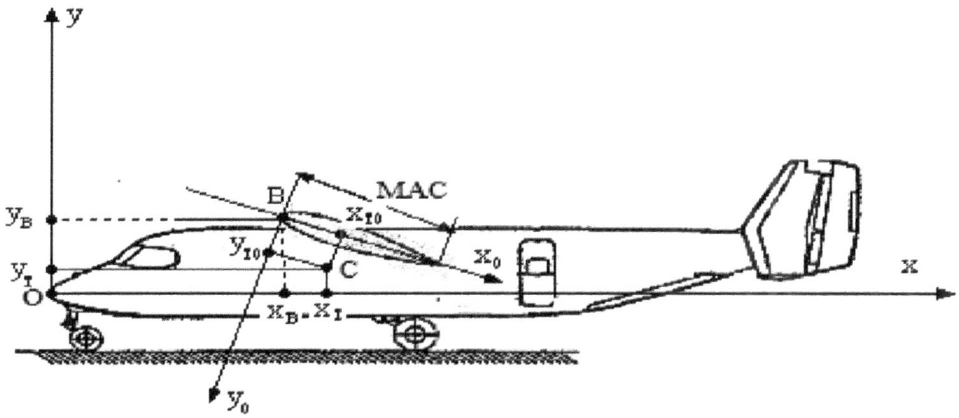

In the course of our research, we studied the existing methods, ways, and means of loading and centering the aircraft and identified the following sources of the error presented in Table 5.

The study of features initial of data for implementation of these actions allowed defining, that introductions of all data for calculations accompany errors.

In an order to define or it is possible to use the value of the got errors in at creation of the model, will consider and practically will check these errors of a law of normal distribution of casual sizes submit, on the example of the weight of passengers. In our research took parts 200 passengers weight of that was varied from 43-125 kg. In an order to define answer or not, the law of normal distribution the got is given to take part in our research, we used the next methods:

1. The next types of averages Register:

We deal with multidimensional totality, that is why calculated AV arithmetic, for determination of central tendency of the totality of data, $\overline{E}$= 63,3 kg;

a. Calculated the value of Moda (Mo) and median (Me), after, for the greater informing about the central tendency of totality, there is 60 kg in our case of Mo and Me =63,8 kg.

Table 7. The sources of errors

The sources of errors	The types of errors
The graph-analytical method of centering graphics	
The actual weight is not taken into account, but its standard value (the weight of the passenger is 75 kg, the velocity of the fuel is 0.75 kg / l)	The methodological errors
A person makes a transition from a numerical value to an analog	The instrumental errors
The method of moments	
Caused by noise measurement	The methodological errors
Deviation of the actual weight of this or that element on board from the accepted to the calculated value	The instrumental errors

Value, of $\overline{E}$, Mo and Me considerably does not differ one from another that is why it is possible to draw conclusion about normality of distribution of the obtained data (Krutova & Popova, 1989).

2. Calculated indexes of variety of the obtained data:
    a. The scope of values of R= 30,9 kg;
    b. The standard deviation σ = 5,95 kg, specifies on that variation of data in relation to a mean value is not large, that is why it is possible to talk about normality of distribution of the obtained data.
3. The calculated coefficient of variation of the obtained data υ = of 9,4% to these values can be said, that got data distributed on a normal law and variants are closely grouped near a mean value Lvovskiy (1989).
4. The built histogram in Figure 5 of experimental data, it is also possible to say after her form, that the obtained data submit to the normal law of distribution of casual sizes.

Rapid hypothesis of normality of distribution for the wide class of selections of 3<n<1000 it is also possible to execute by means of method that is offered in Zaks (1976), using the scope of varying of R. Count up the relation of R/σ and compare to the critical highest and lower bounds of this relation, that is driven to Zaks (1976). If R/σ less low bound and more overhead, then normal distribution does not have, it is especially important, that a condition was executed after π=0,10(10% of the level of meaningfulness). If n = 200 and π=0,10 low and overhead bounds in a corresponding table equal: 4,9 and 6,5, id est 4,9<5,19<6,5. That is why the hypothesis of normality of distribution is confirmed on this criterion.

For the decision of task of our research will take advantage of the method of statistical tests on the methodology expounded in Evlanov (1979). For its application the structure in Figer 6 of conceptual design in terms of Tomashevskiy (2005) of the process of measuring of flight weight and centering of the airplane is made. In accordance with this chart of design functions as follows. On the known requirements to the mass of empty airplane, the vector of G, position of center-of-mass, vector of B

$$B = \begin{bmatrix} MAC^n \\ x_{T0}^n \end{bmatrix} \tag{17}$$

Figure 5. The weight distribution diagram

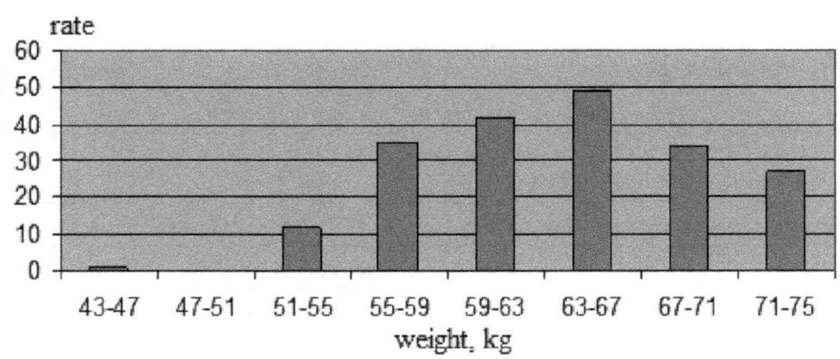

where MAC^n is a set value of the length of the aerodynamic chord of the wing;

x^n_{T0} - the horizontal centering of the empty airplane of admittances is set on the rejection of the marked parameters at the production of airplane α_i, set by the producer of the airplane, the weight and centering of empty airplane are determined in the subsystem of calculation of gravimetric descriptions of empty airplane (GDEA).

After known weighing P_0 and centering of $\bar{x}_0$ empty airplane the vector of gravimetric descriptions of the empty airplane of X_0 is formed:

$$X_0 = \begin{bmatrix} P_0 \\ \bar{x}_0 \end{bmatrix} \tag{18}$$

Obviously, that the algorithm of forming the vector (14) does not depend on the method of determining the flight weight and centering of the airplane.

The vector of X_0 is given to the entrance of model of forming the actual values of measuring parameters of GDEA. Except for the marked vector the vectors of actual weight of equipment, passengers and load operations on the entrances of GDEA, that is formed in the generator of the system of random numbers with normal distribution of SRN; actual coordinates of equipment of passengers and load, that is formed in the generator of coordinates of KSRN. At the output of the model for the formation of actual values of the measurement parameters, the vector of actual weight characteristics of the Xt is formed:

$$X_t = \begin{bmatrix} P^C_{b0} \\ x^C_{b0} \end{bmatrix} \tag{19}$$

After known weighing P_0 and centering of $\bar{x}_0$ empty airplane the vector of gravimetric descriptions of the empty airplane of X_0 is formed:

$$X_0 = \begin{bmatrix} P_0 \\ \bar{x}_0 \end{bmatrix} \tag{20}$$

The vectors of the weight of equipment, passengers and load, actual gravimetric descriptions of airplane and coordinates are given on the entrance of model of the checking of MCH system. Structure of MCH and its parameters determined by the method of being of flight weight and centering of the airplane. GCE is the block of generating of control errors. Its structure also depends on the method of being of flight mass. The exit of the model is a vector of estimations of flight gravimetric descriptions of AC of X_i

$$X_i = \begin{bmatrix} P^C_{bi} \\ x^C_{bi} \end{bmatrix} \tag{21}$$

i – the number of method of calculating weight and centering of aircraft.

Figure 6. The structure of the model for determining the take-off weight and centering of the aircraft

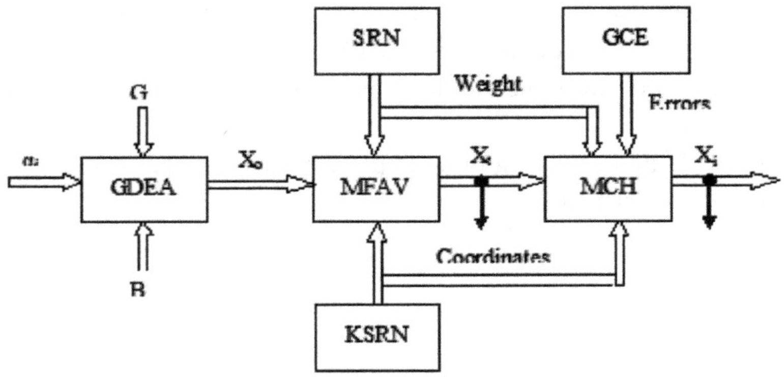

FUTURE RESEARCH DIRECTIONS

Further researches are sent to being of corresponding probabilities, thus, it must take place as a result of implementation of next sequence of executions for each of methods of determining the weight and centration of airplane:

- To work out the algorithms of work of blocks of SRV, KSRV, MFAV, GCE, MCH;
- To work out the structures of simulation models of blocks of SRV, KSRV, MFAV, GCE, MCH
- To work out programmatic realization of simulation model of determining the flight weight and centration of airplane;
- To carry out planning of realization, collection and treatment of results of calculable experiments;
- To conduct the comparative analysis of the got statistical estimations of numerical descriptions of vectors of X_t, X_i
- To define to the closeness of probabilities of $h(P_{b0}^C)$ and $h(x_{b0}^C)$, and also conditional possibilities $f(P_{bi}^C|P_{b0}^C)$, $f(x_{bi}^C|x_{b0}^C)$

CONCLUSION

An important task of the safety of air transportation is the proper loading of the aircraft. When placing the goods must consider two factors:

- The weight of the aircraft flight;
- The position of its center of mass.

In the study, we analyzed the existing ways, means and methods of calculating weight and balance of the aircraft, as well as existing systems automate the process of loading and centering plane, we found that one of the causes of aviation accidents related to improper loading and alignment, is that the calculation includes all standard size (weight of passengers, crew weight, the weight and density of fuel oil, etc.). That is already occurring in the calculation errors related to imprecise definitions of these variables. Therefore there is the problem of calculating the correct set parameters to reflect these errors. For this purpose, it was proposed to develop a model of load and balance of the aircraft, which takes into account all these errors, regardless of the manner of determining the weight of the aircraft centering. In this paper, we present a model structure determination of load and balance the aircraft in view of errors.

Using this model will develop an automated system for decision support for loading and balance of the aircraft, which can be used in domestic airlines.

REFERENCES

Artemenko, O. (2010). *Modeling of the automated pre-flight information preparation system* (Doctoral dissertation). Kiev Institute of Automatic, Ukraine.

Artemenko, O., & Shmeleva, T. (2005). Modeling the decision-making process for departure in the automated pre-flight information preparation system. *Artificial Intelligence*, *4*, 441–446.

Aviation Accident Statistics. (2018). *National Transportation Safety Board*. Retrieved from www.ntsb.gov/aviation/aviation.htm

Babaev, N. (1967). Calculation and design of elements of gyroscopic devices. *Engineering*, 480.

Basic principles on human factors in the user manual Audit Aviation Safety. (2010). Doc. ICAO 9806.

Evlanov, L. (1979). Control of dynamic systems: monograph. Nauka, 132.

Flight Manual. (n.d.). Retrieved from http://www.avialogs.com/index.php/en/aircraft/europe-and-consortiums/airbus/a319-320-321/

Gerasimov, B. (2007). Intelligent decision support systems. European University.

Gluhih, I. (2000). Integrated automated systems of intelligent decision support in air traffic control (theory, models, algorithms, decision-making). Samara, 34.

Harchenko, V., Shmelova, T., & Sikirda, U. (2011). Graph-analytical models of decision-making by a person of an air navigation system operator. *Journal of the National Aviation University*, *1*, 5–17.

Harchenko, V., Shmelova, T. & Sikirda, U. (2012). *Decision making by the operator of the air navigation system: the monograph.* Kirovograd Flight Academy of National Aviation University.

Kasak, V. (2010). *The system of methods for the integration of living entities in special situations during the flight: the monograph.* The National Aviation University.

Krutova, V. & Popova, V. (1989). Fundamentals of Scientific Research. *High school,* 400.

Lvovskiy, E. (1989). Statistical methods for constructing empirical formulas. *High school,* 239.

Moskvichov, V. (1998). Information support for controller decisions for controlling the forced landing of aircraft outside the aerodrome. Kyiv: Academic Press.

Nazarov, B. (1964). Gyro on a rocket. *Voenizdat,* 104.

Nedelko, V. (2002). Ensuring the effectiveness of information support of decision-making in automated systems of air traffic services with elements of artificial intelligence. Kyiv: Academic Press.

Osadchiy, S., & Tymoshenko, G. (2017). Methods for Determine the weight and the Center of gravity of UAV. *Materials of the fourth International conference "Actual problems of Unmanned Aerial Vehicles Developments"*, 139-142.

Shmeleva T., Artemenko O., & Tymoshenko G. (2012). Construction of a neuron cell model of the possibility of flight. *Journal of Kharkov Air Force University, 1*(21), 68-74.

Shmelova, T., Miroshnitchenko, M., & Timoshenko, G. (2014). Analysis of a passenger mass influence on the model of identifying the aircraft takeoff weight and its takeoff balance data. *Journal of Kirovograd National Technical University, 27*, 260–265.

Sikirda, U. (2004). Modeling of the decision support system of air traffic controller in freelance flight situations. Kyiv: Academic Press.

Tarasov, V., Gerasimov, V., Levin, I., & Kornijchuk, V. (2007). *Intelligent Decision Support Systems: Theory, Synthesis, Efficiency*. Kyiv, Ukraine: MAKNS.

Tomashevskiy, V. (2005). *System modeling*. Kyiv, Ukraine: BHV.

Tymoshenko, G. (2012). Simulation of aircraft loading planning system. *Journal of Kirovograd National Technical University, 25*(1), 405–412.

Tymoshenko, G., & Osadcha-Fic, L. (2014). Simulation of the control of the take-off mass and centering of the plane. *Materials of the XII International scientific conference "Control and Management in complex systems - 2014"*, 57.

Tymoshenko, G., & Saganovska, L. (2016). Automation of the decision making a process for the flight after loading an airplane. *Materials of the XIII International scientific conference "Control and Management in complex systems - 2016"*, 331-332.

Voronin, A. (2009). *The information system of decision-making*. Kyiv, Ukraine: The National Aviation University.

Zaks, L. (1976). *Statistical estimation*. Moscow, Russia: Statistic.

ADDITIONAL READING

Osadchiy, S. I., & Tymoshenko, A. S. (2016). Generation of takeoff weight and balance test model. *Automatics, 2016*, 151–152.

Tymoshenko, A.S. (2013). Analysis of a passenger mass variation influence on the model of identifying the aircraft takeoff weight and takeoff balance data. *Engineering in agricultural production, branch engineering, automation, 26*, 188-192.

KEY TERMS AND DEFINITIONS

Center of Gravity (CG): Is the point about which an airplane would balance if it were possible to suspend it at that point. It is the mass center of the airplane or the theoretical point at which the entire weight of the airplane is assumed to be concentrated. It may be expressed in inches from the reference datum, or in percent of the mean aerodynamic chord (MAC).

Datum (Reference Datum): Is an imaginary vertical plane or line from which all measurements of arm are taken. The datum is established by the manufacturer. Once the datum has been selected, all moment arms and the location of CG range are measured from this point.

Dry Operating Weight: Operational empty weight. Basic weight plus operational items such as crew and pantry (equipment, food, beverages).

Fuel Load: Is the expendable part of the load of the airplane. It includes only usable fuel, not fuel required to fill the lines or that which remains trapped in the tank sumps.

GDEA: The subsystem of calculation of gravimetric descriptions of an empty airplane.

Landing Weight: Takeoff weight minus trip fuel.

Mean Aerodynamic Chord (MAC): Is the average distance from the leading edge to the trailing edge of the wing.

Takeoff Weight: Gross weight of aircraft at brake release for takeoff (i.e., actual zero fuel weight plus takeoff fuel).

Useful Load: Is the weight of the pilot, copilot, passengers, baggage, usable fuel, and drainable oil. It is the empty weight subtracted from the maximum allowable gross weight. This term applies to general aviation aircraft only.

Weight, Basic Empty: Consists of the airframe, engines, and all items of operating equipment that have fixed locations and are permanently installed in the airplane.

Zero Fuel Weight: Dry operating weight plus total traffic load.

Chapter 14
Modeling With Colored Petri Nets:
Specification, Verification, and Performance Evaluation of Systems

Dmitry A. Zaitsev
Vistula University, Poland

Tatiana R. Shmeleva
A. S. Popov Odessa National Academy of Telecommunications, Ukraine

ABSTRACT

Aviation and aerospace systems are complex and concurrent and require special tools for their specification, verification, and performance evaluation. The tool in demand should be easily integrated into the standard loop of model-driven development. Colored Petri nets represent a combination of a Petri net graph and a functional programming language ML that makes it powerful and convenient tool for specification of real-life system and solving both tasks: correctness proof i.e. verification and performance evaluation. This chapter studies basic and advanced features of CPN Tools – a powerful modeling system which uses graphical language of colored Petri nets. Starting with a concept of colored hierarhical timed Petri net, it goes through declaration of color sets and functions to peculiarities of hierarchical design of complex models and specification of timed characteristics. The authors accomplish the chapter with a real-life case study of performance evaluation for switched Ethernet network.

INTRODUCTION

Automated systems in the aviation and aerospace industries represent concurrent systems which behavior correctness should be verified in a formal way as well as its performance, on average and in the worst case, carefully estimated. Thus, a model-driven development is an appropriate approach to the design of such systems. Petri nets family represents an appropriate choice of the corresponding modeling para-

DOI: 10.4018/978-1-5225-7709-6.ch014

digm since Petri nets specify concurrent processes and allow their formal verification via state space and algebraic methods as well as their statistical analysis for the performance evaluation.

A colored Petri net represents a combination of a Petri net graph with a functional programming language ML (Meta Language) which constructs load nodes, arcs, and tokens of a Petri net. It offers facilities for the modular approach application via the creation of hierarchical nets via the transition substitution. Besides, it allows simulating timed characteristics of a system using a wide range of random value distribution functions. Special ergonomic interface that allows work even with two mouse manipulators makes the process of models (programs) development fast, convenient, and more reliable.

For the correctness proof, we use a model-checking approach via creation and analysis of the model state space, while for the performance evaluation, a classic simulation approach is applied. To avoid storing bulky raw statistical data, we offer our technique of measuring fragments when certain parts of a colored Petri net calculate statistical characteristics "on flight" directly in the process of simulation.

Thus the chapter offers a new powerful and convenient toolset for the development of automated systems in the aviation and aerospace industries which applicability is acknowledged by a few successful examples of application described in recent publications. The chapter is accomplished with a case study where ARINC-429 and switched Ethernet (a prototype for AFDX) applications are modeled.

BACKGROUND

Petri nets represent a unification of a graphical language introduced by Frank and Lillian Gilbreth for production processes specification. Their most significant advantage consists in the possibility of applying formal methods for their analysis. Though, general Petri nets are too abstract to be a convenient tool for a specialist in some application area. Developed by Jensen and Kristensen (2009) colored Petri nets are loaded by a functional programming language that makes their application less laborious and more concise. A colored Petri net represents a computationally universal system that allows their wide application not only for specification and modeling but for programming and control as well.

Dodd (2006) started a professional application of colored Petri nets for modeling in an avionics mission computer. Then a few works appeared which develop the application of colored Petri nets in the aviation and aerospace industries with modeling a landing detector (Kordon, 2016), auto flight control system (Bourdil et al, 2016) and others.

The authors of this chapter have gained a strong experience in colored Petri nets with teaching them for more than a decade, developing enterprise-level models, and applying them in research projects. The first paper (Zaitsev, 2004) that develops a switched Ethernet model represents a personal record of citation. Note that, the modern avionics protocol AFDX (ARINC, 2009) represents a specialization of a general switched Ethernet technology. Then a series of papers have been published which present models of MPLS networks (Zaitsev & Sakun, 2008), develop a new parametric modeling paradigm (Zaitsev & Shmeleva, 2009), and analyze threats in modern computing grids (Zaitsev et al, 2014) and (Zaitsev et al, 2016). Supplied with the measuring fragments, the technique was presented for the first time in (Zaitsev, 2004) and further developed with recent works, our library of models represents a condensed experience and is distributed via official site of CPN Tools http://cpntools.org.

MODELING WITH COLORED PETRI NETS

Colored Petri Nets and General Organization of CPN Tools

CPN Tools is a modeling system, that offers a very powerful class of Petri nets for models' description, namely hierarchical timed colored Petri nets. It was proved that they are equivalent to Turing machine and constitute a computationally universal system. Thus, an arbitrary object (algorithm) can be specified by this nets.

A simple concept of a *colored Petri net* uses different types of tokens. The type of a token is given by a natural number and represented visually as a color: 1-red, 2-blue, 3-green, etc. The concept of a colored Petri net of CPN Tools is more complicated with a token represented by an object of an abstract data type, like in programming languages.

Timed Petri nets use a concept of model's time to represent durations of actions in real-life objects. As distinct from classical Petri nets, where the firing of a transition occurs instantly, the firing of a transition in a timed net is concerned with a definite duration or a time delay. It allows us the analysis of time characteristics of real-life objects, for instance, a response time as a characteristic of the network's QoS.

Hierarchical nets provide a composition of compound models. In such nets, an element can be represented by another net. In CPN Tools, a transition can be substituted by a net. Thus we have a nested construction: a net inside a net.

Functions of CPN Tools

The basic functions of CPN Tools are following: creation (editing) of a model; analysis of the model behavior via its simulation; construction and analysis of the model state space.

For the creation of models, a special graphics editor of colored Petri nets is provided. The editor allows drawing Petri nets on the computer's screen and inputting attributes of nets' elements and additional declarations written in CPN ML language. A model can consist of a few pages. Pages are connected with each other to provide a hierarchical structure.

For simple models, generating its complete state space (reachability graph) is possible. It is the best way, for instance, for verification of telecommunication protocols. CPN Tools provides creation of state space and an automatic report on it where the conclusions about standard properties of Petri nets, such as boundedness and liveness, are drawn. Moreover, a special language on the base of CPN ML is provided for description of queries regarding nonstandard properties of the state space a user is interested in. Unfortunately, for complicated models, their state space may be huge and its creation is unfeasible.

The only way to analyze complicated models is a simulation of its behavior. CPN Tools provides step-by-step simulation for debugging of a model as well as automatic simulation of specified numbers of steps. Simulation on big time intervals is a way for statistical analysis of models' behavior. It is useful for estimation of networks' characteristics such as throughput (bandwidth) and QoS.

The Interface of CPN Tools

The main window of CPN Tools (Figure 1) contains two areas: a workspace – the right part and an index – the left part. The index consists of a Tools box, Help and Options; below them, descriptions of nets are situated. In the workspace area, pages of nets are visualized. There is a graphical cursor in the main window for the interaction with CPN Tools.

To open a tool's palette, we drag it from the index to the workspace. To take an instrument from a tool's palette, we click on it. Then the cursor takes shape of the corresponding instrument. To abandon a tool, we drag it back on the palette or push Esc bottom on the keyboard.

Models are called nets within CPN Tools. Their descriptions are situated in the index under standard items. Each net in CPN Tools possesses:

- **Name:** The name of the correspondent file with type .cpn;
- **Step:** The number of steps that have been executed in a simulation;
- **Time:** The current model time;
- **History:** A list of commands that have been performed on the net;
- **Declarations:** Declarations of color sets, functions, variables, and constant values;
- **Pages:** Names of net's pages.
- **CPN Tools Has Two Kinds of Help:** Speech bubbles and online help. Speech bubbles appear on the screen when we are keeping the cursor on the corresponding item for a few seconds. To start the help we drag it from the index to the workspace.

Figure 1. The main window of CPN Tools with opened palettes of tools

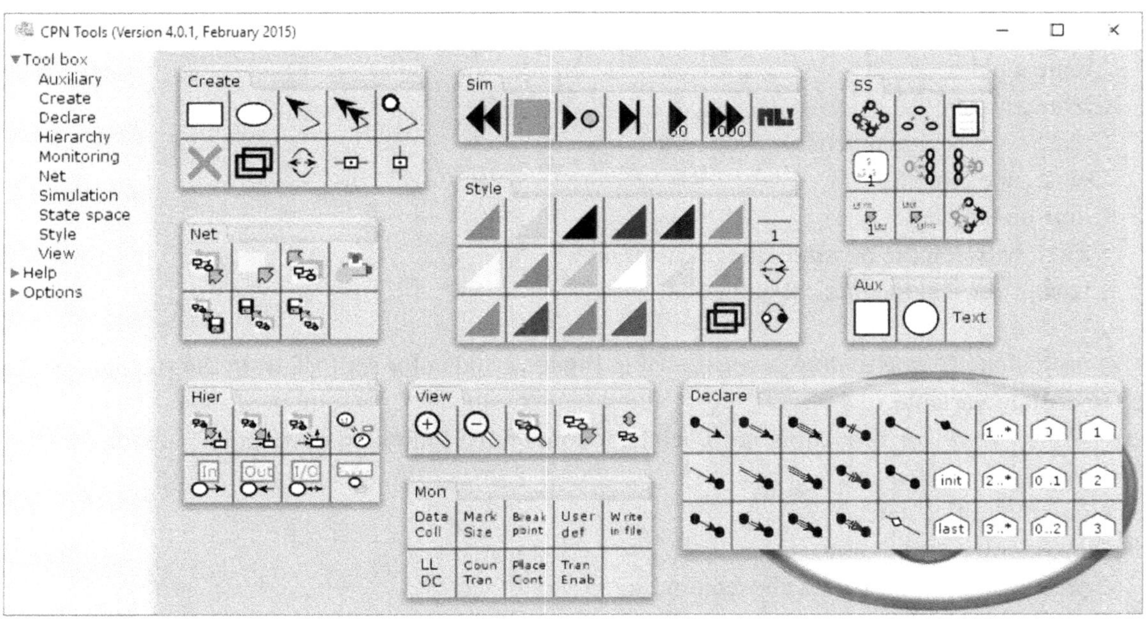

Tools of CPN Tools

The Toolbox provides ten palettes of tools which are opened together in Figure 1:

- **Net Tools:** For operations with entire nets;
- **Create Tool:** For drawing and editing Petri nets;
- **Simulate Tools:** For simulation of net's behavior;
- **State Space Tools:** For creation and analysis of state space;
- **Hierarchy Tools:** For the creation of multilevel nets;
- **Style Tools:** For peculiarities of nets' appearance;
- **View Tools:** For scale choice and highlighting groups of elements;
- **Auxiliary Tools:** For improving nets' readability.

Net Tools contain the following items in its palette (from left to right and from up to down):

- Create a new net;
- Create a new page;
- Close a net;
- Load a net;
- Save a net;
- Save a net with a new name;
- Print a net.

Create Tools contain the following items in its palette:

- Create a transition;
- Create a place;
- Create an arc;
- Create a vertical magnetical guideline;
- Delete an element;
- Clone an element;
- Switch between the possible directions of arc;
- Create a vertical magnetical guideline.

We use palette Create to draw a net shown in Figure 2 and color set INT from the standard declarations to declare variable x.

Simulation Tools for simulating net's behavior contain the following items in its palette:

- Go to the initial state;
- Stop ongoing simulation;
- Execute a transition with a chosen binding;
- Execute a transition;
- Execute the specified number of transitions showing intermediate markings;

Figure 2. An example of a simple model

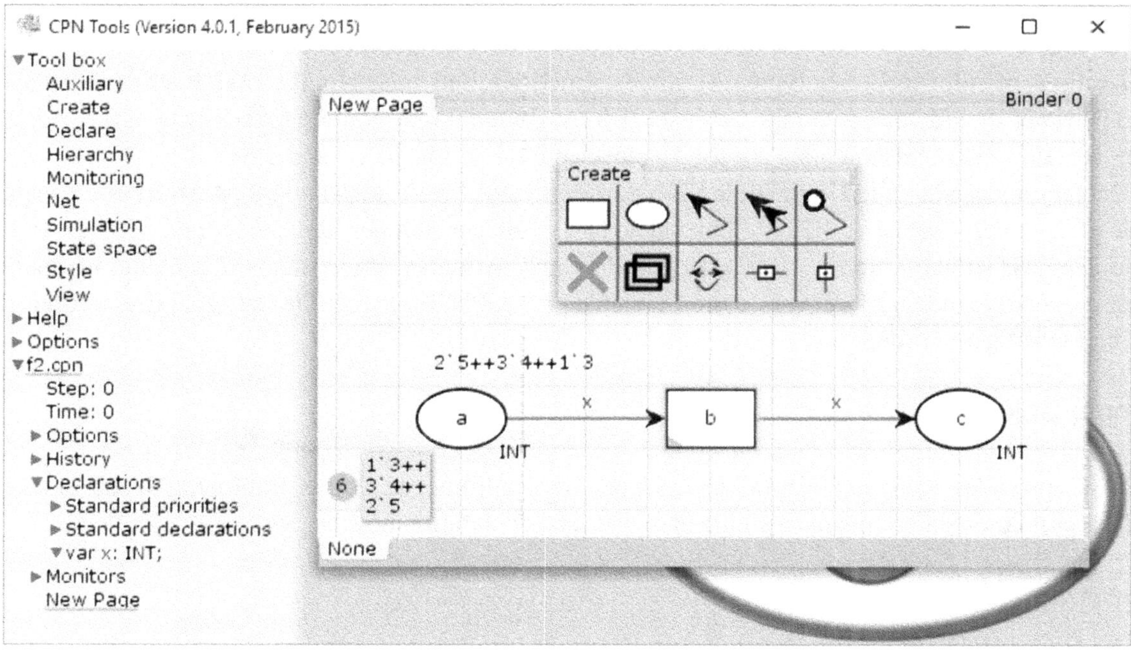

- Execute the specified number of transitions without showing intermediate markings;
- Evaluate a text as ML code.

Items that “execute a transition” are aimed at the debugging of nets with step-by-step simulation. We click on a firable transition to fire it. In Figure 3, token 4 has been chosen by variable x in a random way from the place *a* and moved by the transition *b* to the place *c*.

As for the “execution of a specified number of transitions”, they are chosen randomly by CPN Tools. The only difference between the two modes is that CPN Tools stops and shows intermediate markings in the first case and shows only the final marking in the second case. The result is the same but the mode without showing intermediate markings is much faster. It is used for simulation at large intervals of time for accumulation of statistical information.

Figure 3. An example of firing a transition

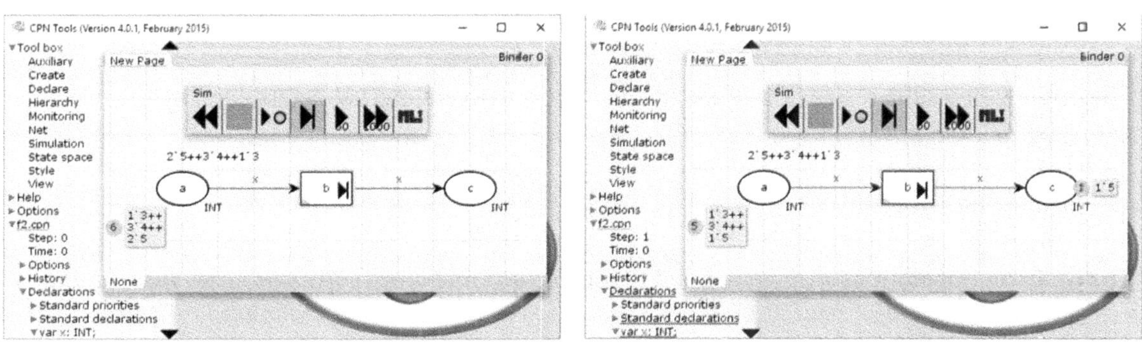

Basics of CPN ML

CPN Tools uses CPN ML for declarations and inscriptions of graphical elements. CPN ML provides declarations of color sets (data types), variables, functions, and values (constants). Each place of a colored Petri net should have a definite color set as its attribute; it may contain tokens only of the specified color set. Variables and functions are used as inscriptions of transitions and arcs.

Declarations are situated in the index as a part of the net. There are standard predetermined declarations of such color sets as UNIT – elementary, INT – integer number, REAL – a real number, BOOL – Boolean, STRING – string. User's declarations may be added after standard declarations using a context-sensitive menu. Moreover, for complicated nets, CPN Tools provides external declarations that can be loaded from a file.

Simple Color Sets

CPN ML provides such simple color sets as Unit, Boolean, Integer, String, Enumerated, and Index.

The *unit* color set comprises a single element. The declaration has the syntax:

colset name = unit [with new_unit];

Without an option, the token name is "()". An example of a specific name declaration follows:

colset p=unit with packet;

The *boolean* values are *true* and *false*. The declaration has the syntax:

colset name = bool [with (new_false, new_true)];

The option allows the new names for *true* and *false*, for instance, *yes* and *no*:

colset Answer = bool with (no, yes);

The following operations can be applied to Boolean variables: *not b, b1 andalso b2, b1 orelse b2,* which correspond to negation, conjunction, and disjunction, respectively.

Integers are numbers without a decimal point. The declaration has the syntax:

colset name = int [with int-exp1...int-exp2];

The option allows the restriction of the integer color set to an interval determined by the two expressions in *int-exp1* and *int-exp2*:

colset Dozen = int with 1..12;

The following operations may be applied to integer variables: +, -, *div, mod, abs, Int.min, Int.max.*

Strings are specified by sequences of printable ASCII characters surrounded with double quotes. The declaration has the syntax:

colset name = string [with string-exp1..string-exp2 [and int-exp1..int-exp2]];

The option specifies the ranges of valid characters:

colset LowerString = with "a".."z";

The following operations can be applied to string variables: ^ - concatenation, String.size, Substring.

Enumerated values are explicitly named as identifiers in the declaration. The declaration has syntax:

colset name = with id0 | id1 | ... | idn;

For example for ARINC-429 specification:

closet BNR_cs = with FailureWarning | NoComputedData | FunctionalTest | NormalOperation;

Indexed values are sequences of values comprised of an identifier and an index-specifier. The declaration has syntax:

colset name = index id with int-exp1..int-exp2;

Indexed values have the form: *id i* or *id(i)* where *i* is an integer and *int-exp1* <= *i* <= *int-exp2*. For example in the task about dining philosophers, we declare philosophers as:

colset PH = index ph with 1..5;

And philosopher *ph(2)* takes forks *fork(1)* and *fork(2)*.

Compound Color Sets

Compound color sets constitute a combination of simple color sets. CPN ML provides such compound color sets as *products, records, unions, lists, subsets, and aliases*.

Products and records represent corteges of data formed by Cartesian products of components' color sets. The only difference between them is that components of a product color set are unnamed while components of a record color set have their names.

Declaration of *product* color set has syntax:

colset name = product name1 * name2 * ... * namen;

Values of this color set have form:

(v1, v2, ..., vn) where *vi* has type *namei* for *1<=i<=n*.

To extract ith element of a product the following operation is used:

#i name

Declaration of *record* color set has syntax:

colset name = record id1:name1 * id2:name2 * ... * idn:namen;

Values of this color set have form:

{id1=v1, id2=v2, ..., idn=vn} where *vi* are values of type *namei* for *1<=i<=n.*

To extract *ith* element of a product, the following operation is used:

#idi name

Examples of products and records are considered in Section 7.

Declaration of Variables and Constants

A variable is an identifier whose value can be changed during the execution of the model. Variables are used in inscriptions of Petri net elements.

Declaration of the *variable* has the syntax:

var id1, id2, ..., idn: cs_name;

where *idi* is an identifier, *cs_name* is the name of a previously defined color set. For instance:

var f1, f2: frame;

A *value* declaration binds a value to an identifier (which then works as a constant). Declaration of value has the syntax:

val id = exp;

where *id* is an identifier and *exp* is a CPN ML expression. The expression represents the value to be associated with the identifier.

Functions

Functions of CPN ML implement standard control structures of a programming language such as "if" and "case" operators. Since ML constitutes a functional programming language, its benefits are revealed with recursive functions which are used instead of loops as well.

Declaration of a *function* has the following syntax:

```
fun id pat1 = exp1

| id pat2 = exp2

| ...

| id patn = expn;
```

where *pat1, pat2, ..., patn* are patterns and *exp1, exp2, ..., expn* all have the same type. The declaration means that in the case when actual arguments satisfy pattern *pati*, the value of the function is calculated as *expi*. For instance, the following function calculates a factorial of an integer number using recursion:

```
fun fact (0) = 1

| fact (i) = i * fact(i-1);
```

if-then-else control structure is available for functions' description:

```
if bool-exp then exp1 else exp2;
```

where *exp1* and *exp2* have the same type.

Their meaning is as usual like in other programming languages. For instance, a function that calculates the sign of a number may be written:

```
fun sign (x) = if x>0 then 1 else if x<0 then ~1 else 0;
```

Random Functions

Random functions provide facilities for modeling statistical characteristics. For instance, they allow us the description of traffic's intensity or faults' frequency in telecommunication systems. There are a few ways for random choice in CPN Tools:

- Free variables;
- Function *ran*;
- Special random distribution functions.

Free variables, inscribed on an output arc of a transition, are not bound on an input arc or in the guard. They are assigned random values when executing a CP-net. Function *ran* allows obtaining a uniform distribution in the range of the variable's color set.

CPN Tools provides also a series of a special random distribution function for such well-known distributions as Bernoulli, binomial, Erlangen, exponential, normal, Poisson, student, uniform (discrete and continuous). For instance, function

erlang(n:int, r:real): real

where $n>=1$ and $r>0.0$, returns a drawing from an n-Erlangen distribution with intensity r.

Graphical Language of Models' Description

In CPN Tools, each element of a Petri net has its attributes described in CPN ML language. Using Create palette, we put an element onto a page of the model. Then, the element's attributes should be added. For this purpose, we click by mouse on the corresponding element and use Tab key for switching among attributes. Pressing Esc key lets us leave the chosen element; the same result may be obtained by clicking the mouse at some other spot of the model.

Multi-Sets

Multi-sets are widely used in CPN Tools for place's marking representation and other purposes. As distinct from a usual set, it contains each element with a definite multiplicity, in other words, in a few copies. Multi-sets are also named bags.

The back-quote operator (denoted by `) is the multi-set constructor. For example, 3`5 specifies a multi-set with three appearances of the color 5. The description of a multi-set has the syntax:

i`c

The integer *i* should be non-negative. If this is not the case, then an empty multi-set is returned. The multi-set operator combined with multi-set addition (++) and subtraction (--) provide an explicit method for specifying multi-sets.

The following Constants, Operations, and Functions are available for multi-sets: *empty, ==, <><>, >>, >>==, <<, <<==, ++, --, **, size, random, cf, filter*, which correspond to emptying, equality, inequality, greater than, greater than or equal to, less than, less than or equal to, addition, subtraction, multiplication, size of, count of color set, filtering, respectively.

In CPN tools, initial and current marking of a place is represented with a multi-set of the place color set; at the choice of a token by a variable in the inscription of place's output arc, CPN Tools provides a random choice like with the function *ran*.

Timed Multi-Sets

Timed multi-sets represent timed delays in the model within CPN Tools. Declaration of the corresponding color set is accomplished with the modifier *timed*. The @, @+, and @@+ operators are used to add timestamps to tokens. Adding a *time delay* of *x* to a color *c* will attach a time stamp with a value that is equal to the *current model time* + *x* to the color *c*. The following operations are valid for timed multi-sets: @, @+, +++, which mean attach a timestamp, delay, and addition, respectively.

For instance the declaration

colset tint = int timed;

var t: tint;

t = 1`2@100 ++ 1`3@200 ++ 1`4@300;

means that token 2@100 might be taken by CPN Tools only after model time instance equaling to 100, token 3@200 – only after model time instance equaling to 200 etc. Application of a time delay is illustrated in Figure 4.

Note that, CPN Tools implements very simple and powerful class of Timed Petri nets due to the time stamp concept. Each token *k* is supplied with its time stamp *t* (*k@t*). For moments of the model time *Time<t*, the token is not processed by the simulation system; it is being in so named void state and does not take part in firing any transitions. After a moment of time *t*, token *k* wakes up and participates in the firing of transitions. This way of times representation allows the instantaneous firing of transitions. Output tokens supplied with time delays (*k@t+d*) wait their time of activation inside corresponding output places.

Place Inscriptions

There are three inscriptions that may be associated with a place. Two are optional and one is required:

- Color set inscription (required);
- Initial marking inscription (optional);
- Place name inscription (optional).

On the initial marking, CPN Tools automatically creates the current marking which is written in green color and shows the total number of tokens and markings' detail.

Fusion places make models more clear and define connections between pages of a net. Fusion places can be considered as the first step to hierarchical Petri nets, so item "Fusion" is situated in Hierarchy palette. Each place of a fusion set is supplied with the fusion tag of the same name; all places in a fusion set are considered by CPN Tools as the same place. Visually the changing of marking for one of the places causes in the changing of other fusion places marking. Fusion places should be of the same color set.

Figure 4. Timed tokens

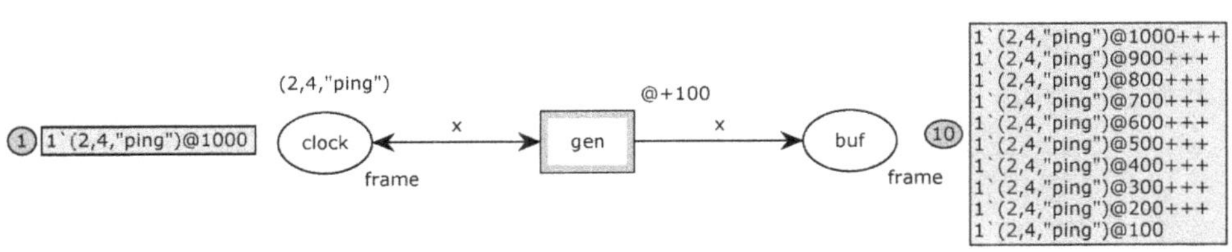

Arc Inscriptions

A color set of the arc inscription should match the color set of the place attached to the arc. There is an essential difference between inscriptions of input and output arcs of a transition.

The transition's input arc expression constitutes a pattern for the token choice. This pattern is described by a predicate which may be applied in function *filter* for the corresponding color set. In the simplest case, this predicate consists of a single variable of the corresponding color set as in the above examples.

The transition's output arc expression constitutes a constructor for new tokens creation. This constructor often uses variables of the input arcs' inscriptions and in a simple case coincide with one of them.

Transition Inscriptions

There are four inscriptions that may be associated with a transition. All of them are optional:

- Transition name inscription;
- Guard inscription;
- Time inscription;
- Code segment inscription.

A transition *delay* must be a positive integer expression. The expression is preceded by @+, and this means that the time inscription has the form @+ *delay-expr*. Before a time inscription has been added, the default text for the inscription is @+. A time delay is always added relative to the current time.

A *guard* is a CPN ML Boolean expression that evaluates to *true* or *false*. Before a guard has been added, the default text for the inscription is *[]*. Guard may be a single Boolean expression or a list of Boolean expressions *[b-exp1, b-exp2, ..., b-expn]*. The transition fires only in the case its guard is true and a guard restricts the choice of input tokens. Moreover, a guard allows the comparison of parameters of tokens taken from different places using their combinations in the expressions.

Each transition may have an attached *code segment* which contains ML code. Code segments are executed when their parent transition occurs. Code segments use CPN variables and bind CPN variables located on output arcs that are not bound elsewhere.

Peculiarities of Timed Nets in CPN Tools

CPN Tools provides both Petri net without times and timed Petri nets. If none of the color sets of a model has a modifier *timed*, the net is considered untimed. In this case, CPN Tools uses only internal Step variable that means the number of executed transitions. The algorithm of CPN Tools simulation is represented in Figure 5.

Figure 5. Simulation in the untimed mode

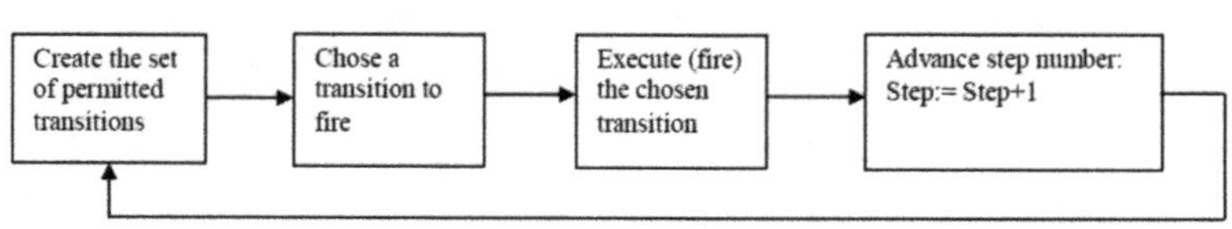

Note that, the transition's choice is implemented either manually for step-by-step simulation or automatically in a random way by CPN Tools for execution of a specified number of steps.

For timed Petri nets, the algorithm is more complicated because the way of time advancing (Figure 6). The next instant of the model time will not be the instant *Time+1*. The system advances time to the nearest event in the future *Time:=time-of-nearest-future-event* and then executes all the events that may occur in this instant of time advancing the variable *Step* in the described above way. So we should be careful combining timed and untimed nets because permitted transitions fire until the set of permitted transitions becomes empty. It may cause so named infinite looping in models' behavior.

For the description of time delays, either delays of transitions or delays of arcs are used. In the most cases, delays of transitions are more comprehensible because transitions model actions of an object. But using arcs' delays gives us more flexibility for models construction. When modeling traffic, it is convenient to use random distribution functions as time delays.

Construction of Hierarchical Models

The application of hierarchy is very common for engineering. A telecommunication device consists of blocks, blocks consist of boards, boards consist of chips, capacities, resistances etc. In programming, a program is assembled of modules (procedures, functions). A hierarchical model means a nested construction: a net inside a net. Generally, any element of a Petri net can be substituted by a nested net but CPN Tools uses substitution of transitions only.

Basics of Transition Substitution

When substituting a transition, we have a pair of nets at least. A transition of the upper-level net is substituted by a lower-level net. Let us consider a model of LAN described in subsection 7.2. At the top page *LAN*, transition *WS3* is substituted by net *WS*. The substitution is pointed by the tag *WS* drawn near transition *WS3*. Logically, the behavior of a whole net is the same as when subnet *WS* was put inside net *LAN*. Places of a lower-level net which are used in connection with the upper-level net are named contact places or *ports* and marked with a special tag (I/O). Corresponding places of the upper-level net are named *sockets*. In the above example, we have ports *LANin*, *LANout*, *Own* in page *WS* and sockets *p3out*, *p3in*, *aWS3* on page *LAN*.

To prepare the substitution of a transition, we should:

- Create both upper-level and lower-level nets situated in separate pages of the model (*LAN, WS*);
- Point out ports of the lower-level net (*LANin, LANout, Own*);
- Provide the number of sockets equal to the number of ports (3).

Figure 6. Simulation in the timed mode

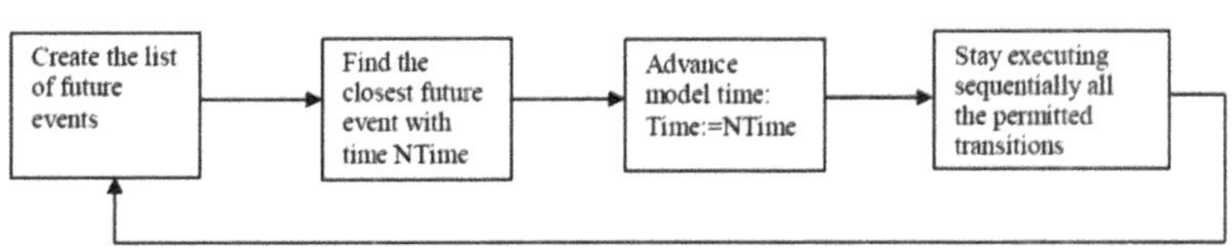

To implement the substitution of a transition, we should:

- Assign a net to the transition (*WS3->WS*);
- Assign ports to sockets (*LANin->p3out, LANout->p3in, Own->aWS3*).

Ports of the lower-level net are pointed out using the Hierarchy palette. Three types of tags are available: *In, Out, I/O*. Ports of type *In* are used when the corresponding place has only input arcs in the upper-level net and only output arcs in the lower-level net. Ports of type *Out* are used when the corresponding place has only output arcs in the upper-level net and only input arcs in the lower-level net. They can be considered as input/output variables in a programming language routine. When there are no restrictions on arcs, a port of type *I/O* is used. Such a place can have incidental arcs of both directions in the upper-level net as well as in the lower-level net.

To assign a page to a transition, the second tool of Hierarchy palette is used. We click with it on a transition (*WS3*) and then on the subpage (*WS*). The corresponding tag of subpage will be attached to the transition. The assignment of ports to sockets is more complicated because the operation should be executed for each port of subpage. The third tool of Hierarchy palette "assign a port to a socket" is used. We should have both pages on the screen. To make the assignment of a port, we click on it and then click on the corresponding socket.

Note that the hierarchical nets make the development of models considerably easy like modules in programming languages. At first, it allows hiding of details and management of models' complexity. At second, it provides reuse of submodels. For instance, in the example of LAN model, there are 5 workstations and 2 servers but we have single subpages for a workstation (*WS*) and a server (*S*).

Bottom-Up Development

When creating a hierarchical net "bottom-up", we start by creating separate pages. In contrast to top-down development, this approach involves creating the most detailed parts of the net first. Later, existing pages are assigned as subpages for substitution transitions as it was described in the previous subsection. It looks like the application of libraries in programming languages.

For instance, in LAN example, described in subsection 7.2., we create pages *WS, S, MWS, SWI* then create top page *LAN* and then assign substitution of transitions (*WS1-WS4->WS, S1-S4->S, WS5->MWS*) and the corresponding port/socket mapping.

Top-Down Development

When creating a hierarchical net "top-down", we start by creating the top-level page that represents a lookout of the subpages and how they are connected. For instance, in LAN example we should create top page *LAN* first. There is special operation "move a transition to a subpage" in Hierarchy palette (the first tool). As a result, the subpage tag is added to the transition, which is now a substitution transition. A new page is created with a copy of the places surrounding the target transition. The page is named after the target transition. This page is a pattern for the creation of a lower-level net.

We should edit the pattern to create the submodel of a workstation. For instance, rename page *WS3* in *WS*, rename its places *p3out* in *LANin*, *p3in* in *LANout*, *aWS3* in *Own*, delete transition *WS3* and draw the net *WS* using created ports. Described tool executes the assignment of the subpage tag, port type tags and assignment of ports to sockets automatically. So it allows us to avoid routine operations.

Note that, the development of models requires both bottom-up and top-down approaches as the tool "move a transition to a subpage" may be applied only once for each required subpage. For instance in such a way we can create *SWI, MWS* then *WS* for *WS1* and *S* for *S1*. But for *WS2-WS4* and *S2*, we use early created *WS* and *S* in a bottom-up way.

Analyzing Models

CPN Tools provides two basic ways of models' analysis: *simulation* of net behavior and generation of *state space*. Moreover, we should be confident that a model is adequate to the object and works in the proper way. That involves a preliminary stage very common for programming languages and usually called a debugging. At this stage, we acquire confidence that our model works properly correcting errors. We offer also a special way for the models' analysis named *measuring fragments*. For estimation of models' characteristics, we create special fragments of nets which calculate characteristics during the simulation process. Such a fragment is created in subsection 7.2 for estimation of Ethernet response time.

Debugging Models

Debugging of models involves syntax checking and step-by-step simulation. When we load a net, the syntax check takes a couple of minutes to complete. During this phase, the elements will change aura from orange to yellow to no aura (or red, if there is an error). If the orange aura stays, it is probably because either there is something missing or there is an error on a related net element. A yellow glow indicates that the place/transition/arc/page/net is currently being checked. If there is an error, the corresponding element (component) is underlined with red. To see an error message, we move the cursor over the element and observe a speech bubble that appears.

Step-by-step simulation is used to trace movements of tokens in a model. For instance, we can trace frames in the Ethernet model described in subsection 7.2 on their way from workstation to server and backward. We can also choose manually the bindings of firing transition parameters, for instance, make a choice of an input token among available ones. For further debugging, the execution of the specified number of transitions is useful to estimate the behavior of the model during larger intervals of time.

State Space Analysis

State space of a colored Petri net is more complicated than reachability set or reachability graph of a classical Petri net. In a classical Petri net, marking of places is represented by a vector of natural numbers but in a colored Petri net – by a vector of multi-sets (timed multi-sets).

Analysis of state space is possible for rather small or simple models because of the well-known effect of state space explosion. The number of states for l-bounded Petri net with *m* places is estimated 2^m. In telecommunications, analysis of state space is applicable mainly at verification of protocols when we need information about formal properties of nets such as boundedness, safeness, liveness etc.

State Space palette contains such tools as:

- Enter State Space;

- Calculate State Space;
- Calculate SCC graph;
- State Space to Sim;
- Sim to State Space;
- Save Report.

Before a state space is calculated and analyzed, it is necessary to generate the state space code. This code is generated when we apply *Enter state space* tool. Then, if the state space is expected to be small, we can simply apply *Calculate state space* tool to the net's page. If the state space is expected to be large, we may need to change the options for *Calculate state space* tool. The options for the Calculate state space tool allow us to determine when the calculation of a state space stops. In addition, strongly connected components (SCC) of the state space graph can be calculated using the corresponding tool. Calculated state space is stored in CPN Tools temporary files. There are two ways to analyze it:

- Save report into a file;
- Create State Space queries.

To save a report, we apply the *Save state space report* tool to the net's page. The contents of a report are determined by the options for *Save state space report* tool. Queries are used to investigate properties of a CP-net by writing special CPN ML functions. They are quite complicated and use special predefined functions. CPN Help contains a reference to State Space Queries Manual.

Let us consider the well-known example of dining philosophers shown in Figure 7.

The model in CPN Tools has compact representation due to the application of indexed color sets for description of philosophers (*PH*) and chopsticks (*CS*) and function *Chopsticks* that returns the numbers of sticks used by philosopher *ph(i)*.

In the obtained report, statistics section describes the state space size and SCC graph. Boundedness section gives the upper and lower bounds of markings in numerical and multi-set forms. Home properties section lists home markings. Liveness properties section describes deadlocks and live transitions. Fairness properties section describes the type of net's fairness.

Note that CPN Tools does not give a way to save the complete generated state space. It is kept internally in CPN Tools temporary files. To investigate it beyond the bounds of the standard report we should write special queries to state space in CPN ML language.

Simulation of Net Behavior

CPN Tools may be used as a typical simulation system. When the behavior of a net is rather complicated, we can simulate it at large intervals of time and draw conclusions about characteristics of the modeled system. Especially when random functions are widely used within a model, we are more interested in its statistical properties than in its state space. For instance, we can simulate the behavior of Ethernet model during one day of real-time and make conclusions about such its characteristics as average response time, the percentage of collisions etc.

From a standard simulation system point of view, CPN Tools gives us scarce facilities for such analysis. But it gives us a language to describe the processes of accumulation and calculation of characteristics. It is the same language of colored Petri nets and it can be applied for estimation of statistical

Figure 7. Dining philosophers model

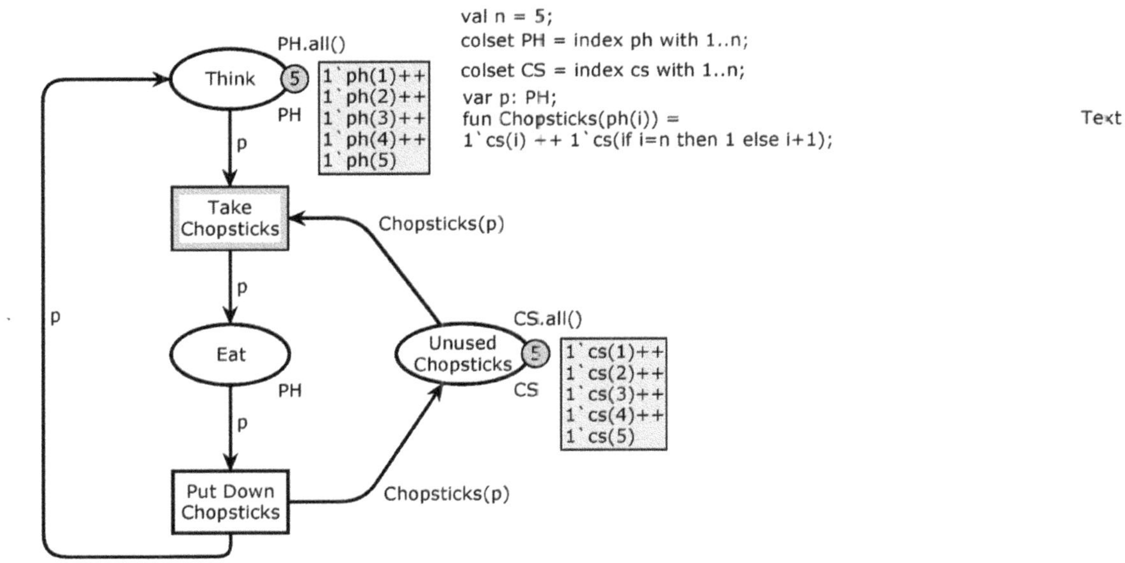

characteristics of models. Such additional nets or their parts are called *measuring fragments* and studied in the next subsection.

The simulation itself implies special consideration of experiments with the model. The first thing is *scaling of times*. Time in CPN Tools is measured in *Model Time Units* (MTU) that has no dimension and represented as a natural number. That's why we are interested in scaling of times to make a model realistic. The model times scaling is described and illustrated by examples (Zaitsev, 2013).

The second thing is the existence of the *state stable mode* of the model's behavior. If the state stable mode exists, the model is balanced. The increase of modeling time duration does not cause the significant change of its statistical characteristics. The simplest way to determine state stable mode is a sequential enlargement of modeling time duration. If characteristics do not change after a certain moment of model time then the state stable mode has been achieved.

The third thing is an *estimation of averages* (or higher statistical moments) of characteristics in the state stable mode. Suppose our model calculates characteristics though results taken in a single experiment with the model are not valuable. A few experiments should be provided with a model and according to mathematical statistics, its number should be about 20 to achieve a credible confidence interval.

Measuring Fragments

Since the colored Petri nets language constitutes a complete algorithmic system, it can be applied for accumulation and calculation of statistical characteristics. Extra parts of the net, added to the source model for accumulation and calculation of statistical characteristics, are called *measuring fragments*. Let us study a few simple measuring fragments. Suppose we have a buffer, and transition *put* increases it's marking, and transition *get* decreases its marking as shown in Figure 8.

We can easily calculate the current number of tokens in *buffer* with a measuring fragment added to the basic net in Figure 9.

Figure 8. Model of a buffer

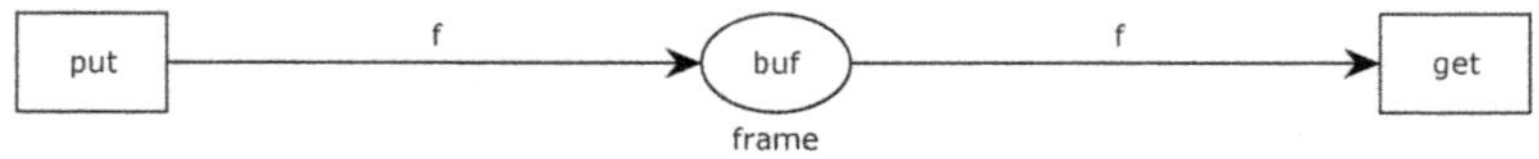

Figure 9. Buffer with its current length counter

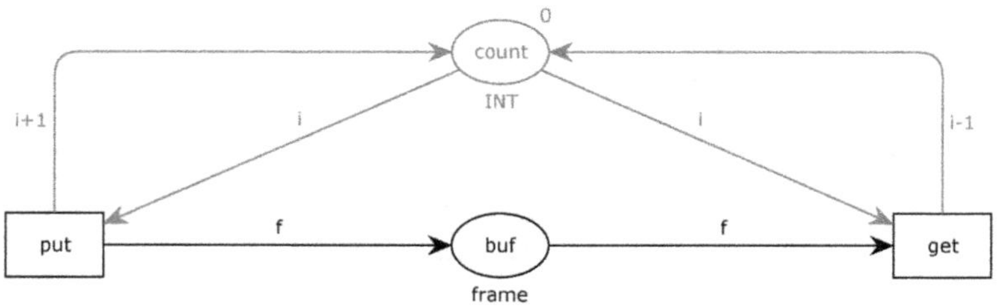

To calculate the maximal number of tokens in place *buffer*, the measuring fragment shown in Figure 10 is applied.

To calculate an average number of tokens, time intervals should be taken into consideration because average of distribution is calculated with formulae:

ac=(c1*dt1 + c2 *dt2 + … + ck*dtk) / dt

where *ac* is the sought average, *ci* is a value on time interval *dti* and *dt* is the total interval of time. The measuring fragment shown in Figure 11 calculates an average number of tokens in place *buffer*.

The recalculations are started both at the increase (*put*) and at the decrease (*get*) of the tokens number. Function *cT()* returns the current value of model time. Place *sum* keeps the current sum of products. Place *prevt* keeps the previous moment of time value when marking of place *buffer* had been changed. Place *average* keeps an average number of tokens in place *buffer*.

Figure 10. Buffer with its maximal length counter

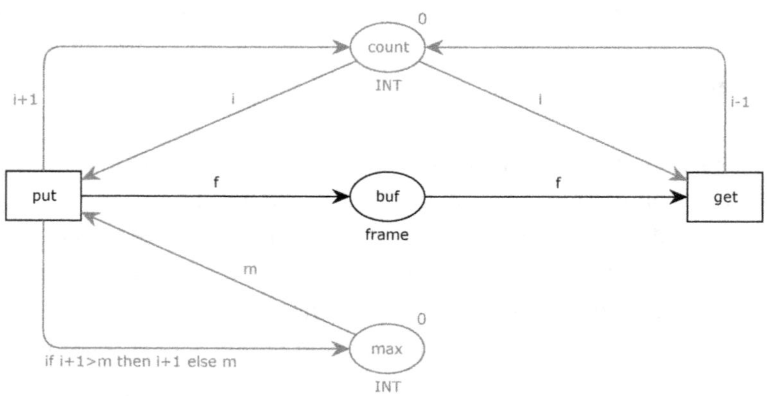

Figure 11. Buffer with its average length counter

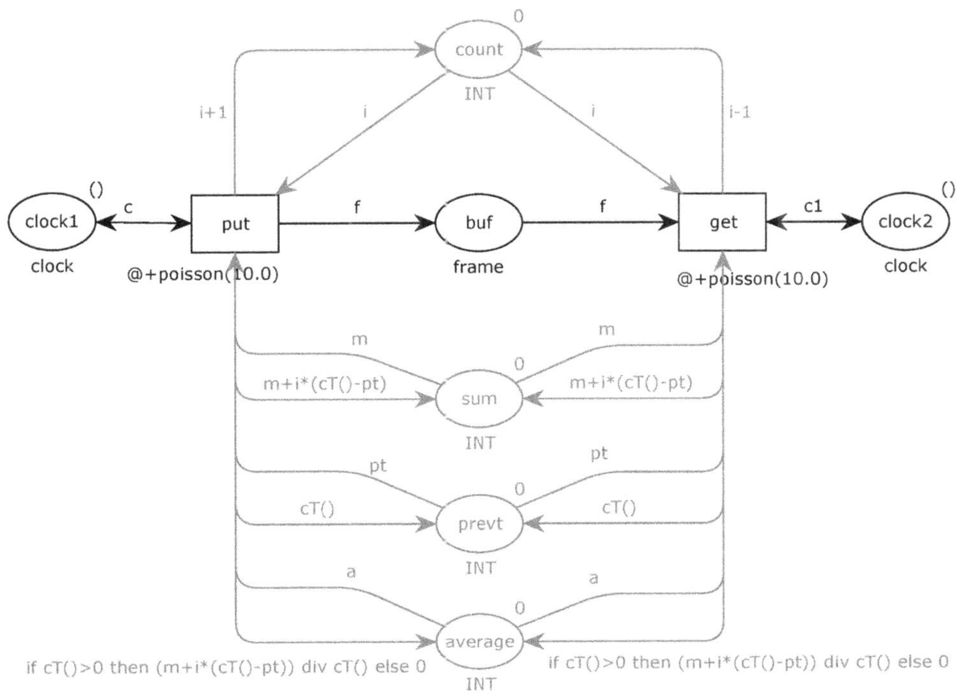

Case Study: Modeling Protocols and Networks

In aviation, colored Petri Nets can be applied for specification and verification of either organizational or telecommunication protocols and performance evaluation of flight control systems. Sometimes organizational procedures, for instance, specified in (ICAO, 2016), are rather sophisticated and deadlocks in them may induce delays which are especially dangerous for operational instructions on board.

Let us consider two examples of systems specification using CPN Tolls in the avionics area. The first example applies traditional ARINC-429 protocol (ARINC, 2004; Martinec, 2001) for the specification of a simple altitude switch (ASW); some fault-tolerance issues can be studied using the constructed model. Though a recent trend is the application of more powerful modern protocols based on well-known switched Ethernet technology. Full-Duplex Switched Ethernet (AFDX) specified in (ARINC, 2009) is a data network, patented by international aircraft manufacturer Airbus for safety-critical applications that utilizes dedicated bandwidth while providing deterministic quality of service (QoS). The second example is devoted to QoS parameters evaluation in a general switched Ethernet that can be adjusted to AFDX standard as well.

Model of Protocol ARINC-429

Protocol ARINC-429 (ARINC, 2004; Martinec, 2001) is the most widespread in avionics. It is based on one-way (simplex) transmission of signals via a shielded twisted pair of wires on a star or bus-drop topology, or multiple bus design; for two-way communication, two twisted pairs are applied to have a symmetric scheme for the opposite direction of transmission. ARINC-429 protocol uses a point-to-point format, transmitting data from a single source on the bus to up to 20 receivers. Most messages contain one 32 bits data word having 5 fields: Parity – 1 bit, Sign/Status Matrix (SSM) – 2 bits, Data – 19 bits, Source/Destination Identifier (SDI) – 2 bits, Label – 8 bits.

Let us construct a model of simplified Altitude Switch, considered in (Miller&Tribble, 2001). For simplicity, we do not check parity and do not encode fractional part and negative values of the altitude. A digital altimeter transmits data words with the label value of 116 (decimal), the altitude value sign occupies bit 29, bits from 28 to 16 contain the integer part of altitude value in binary, bits 15, 14, 13 represent values equal to ½, ¼, 1/8, respectively, for the fractional part. A simplified operation of ASW consists in waking up a device of interest (DOI) when the altitude value becomes less than a given threshold value.

A CPN Tools window containing the model of ASW and the simulation system feedback is shown in Figure 12, declarations of color sets, variables, values, and functions are situated in the index. Transition *Fly* simulates flight using a constant value of vertical speed given by place *Vertical Speed* and recalculating the actual altitude value in place *Altitude* according to the timer represented by place *Time*. The measured altitude value is put into place *Val*, encoded by transition *Encode* into ARINC429 data word, and is stored in place *Tx* for transmission. Transition *Transmit* transmits data words with a fault probability value given by place *Fault Probability*. The received data word is situated in place *Rx*, the altitude value decoded from it is compared to the threshold altitude given by place *Threshold Altitude*. In case the received value is less than the threshold, DOI wakes up with a token put into place *Wake Up*.

The ARINC-429 data word is specified by color set ARINC429, functions *f_alt_encode* and *f_alt_decode* provide the encoding and decoding of the altitude actual value into the *Value* field of the data word. As for actual data scaling, the altitude value is measured in feet, the model time unit is equal to 320 ms that corresponds to the bitrate of 100kbps. The model can be applied for studying fault-tolerance issues.

Figure 12. Model of ASW

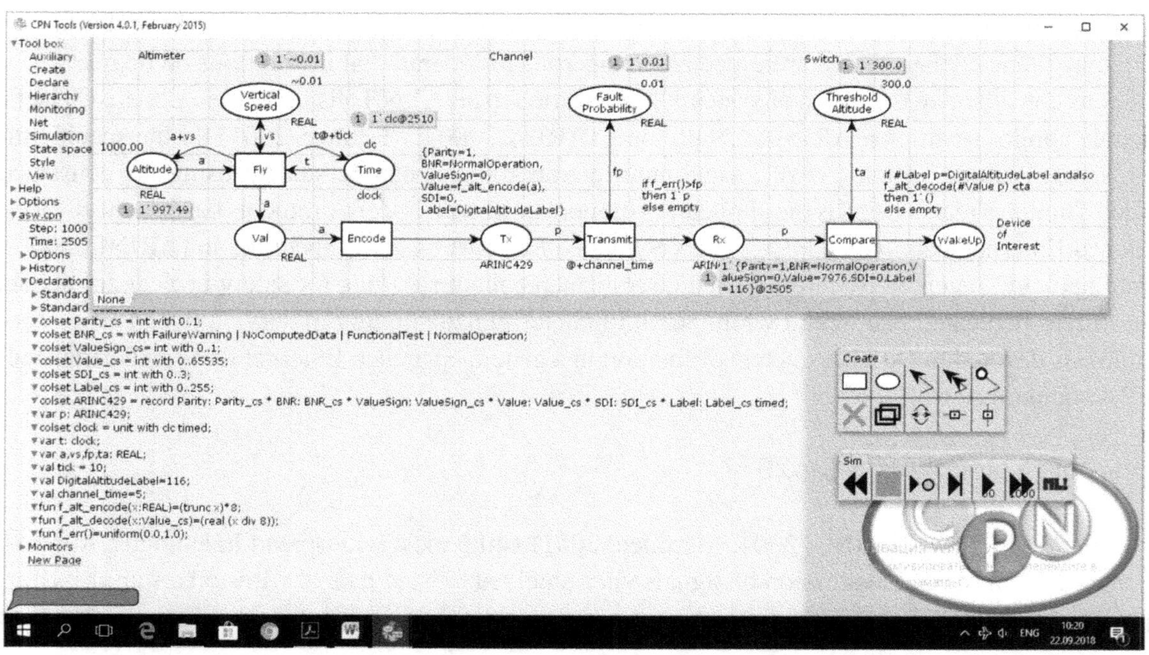

Figure 13. Model of LAN

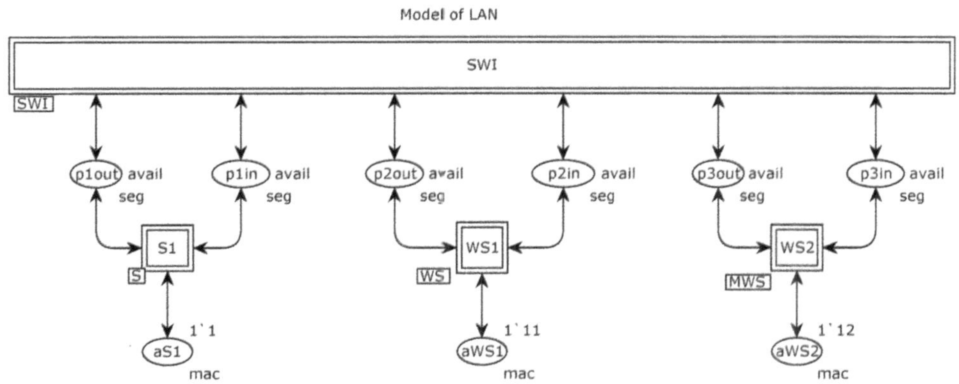

Model of a Switched Network

We consider a generalized switched network with the frame (packet) header having two addresses – sender (*src*) and destination (*dst*); the frame (packet) content is represented with the frame number (*nf*). Addresses are represented as positive integer numbers. It resembles the switched Ethernet features stipulated by such standards as IEEE 802.3 and IEEE 802.1D.

We model a network which contains a switch and three computers attached to it: one server and two workstations. We suppose that the workstations generate periodically requests to servers represented by a single frame. The server processes a request and answers with a few frames of response The main page of the model is shown in Figure 14. A switch is denoted as *SWI*, a workstation is denoted as *WS* or *MWS*, a server is denoted as *S*. Places with the prefix "a" contain addresses of computers. Places with prefix "p" represent ports; each port is specified with two places having suffixes "in" and "out" which represent independent input and output tracts, respectively, of a port working in a full-duplex mode.

The following listing contains declarations of color sets, variables, constants, and functions which load elements of the Petri net graph:

```
color mac = INT timed;
color portnum = INT;
color nfrm = INT;
color sfrm = product nfrm * INT timed;
color frm = product mac * mac * nfrm timed;
color seg = union f:frm + avail timed;
color swi = product mac * portnum;
color swf = product mac * mac * nfrm * portnum timed;
color remsv = product mac * nfrm timed;
var src, dst, target: mac;
var port: portnum;
var nf, rnf: nfrm;
var t1, t2, s, q, r: INT;
fun Delay() = poisson(1500.0);
```

Figure 14. Model of workstation

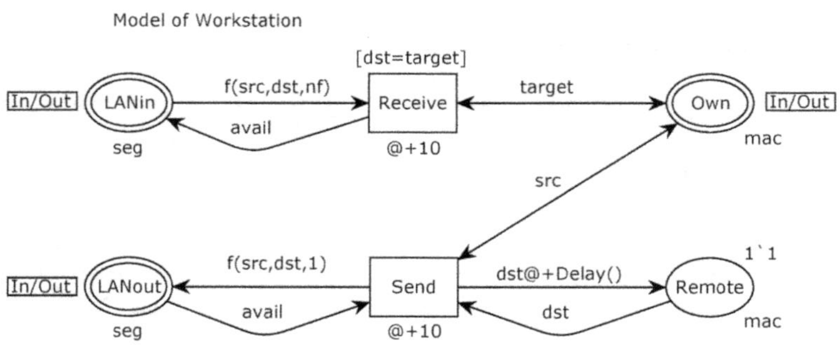

fun Dexec() = poisson(150.0);
fun Dsend() = poisson(15.0);
fun Nsend() = poisson(15.0);
fun cT()=IntInf.toInt(!CPN'Time.model_time);

The meaning of the basic color sets is the following: *mac* – the device address; *frm* – frame; *seg* – segment of network which is either available or busy with transmission of a frame; *swi* – a switching table record that maps MAC addresses into ports' numbers; *swf* – switched frames waiting for their forwarding. As for random values, a simple Poisson distribution is applied in the following functions: *Delay* – the interval between workstation requests to a server; *Dexec* – a time of a request execution by a server; *Dsend* – delay to send a separate frame by a server; *Nsend* – the number of frames in the server reply.

The workstation model shown in Figure 15 periodically sends requests to the servers which MAC addresses are listed in the place *Remote*. It also absorbs received frames.

The server model shown in Figure 16 receives a request, stores a pair consisting of a sender address and the frame number in place *Remote*, and generated a few frames of reply after the interval of time specified by the function *Dexec()*; then it sends frames one-by-one into its output tract.

The central element of the model is a switch of frames shown in Figure 17. For the convenience of drawing, it is organized as a collection of the required number of ports. A static switching table in place

Figure 15. Model of server

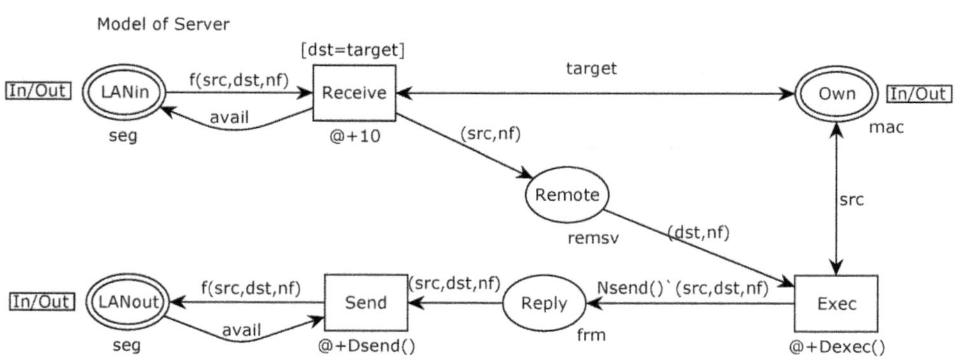

Figure 16. Model of switch

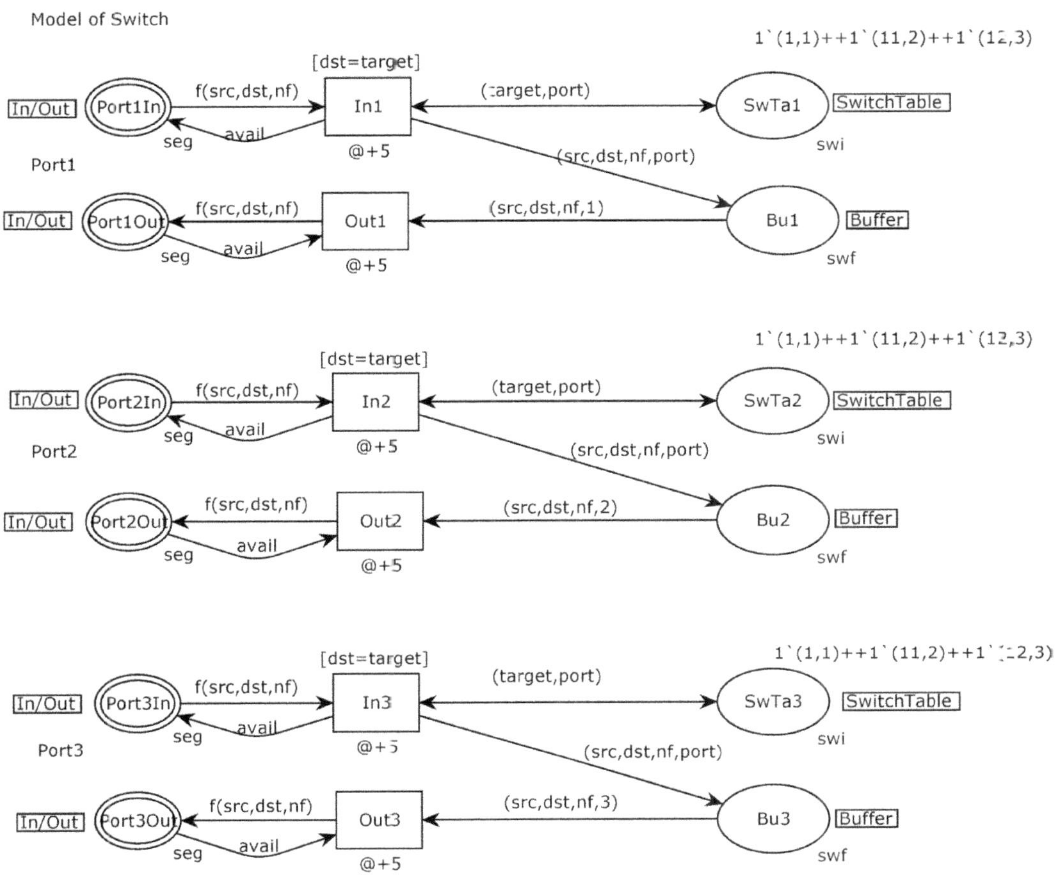

Figure 17. Model of a measuring workstation with the CPN Tools feedback

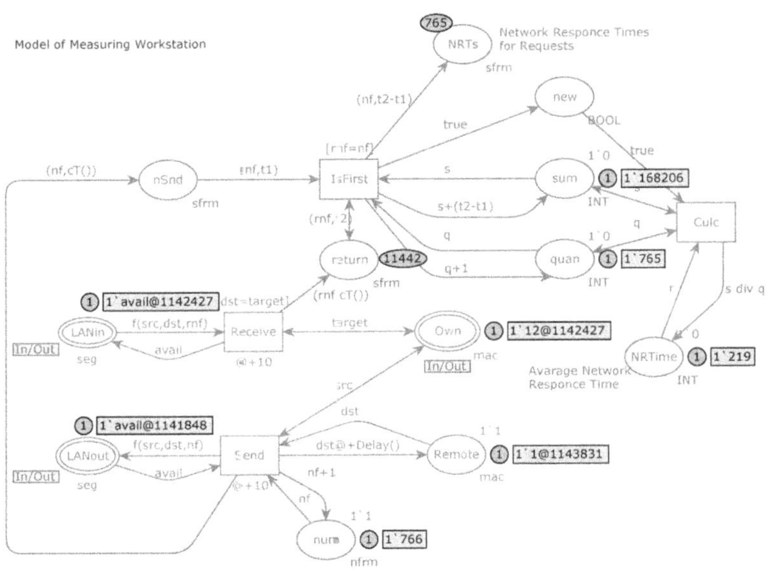

SwitchTable specifies attachment of devices to the switch ports. After the switching decision, a frame waits for the availability of the destination port in the internal buffer modeled by place *Buffer*.

From a simple simulation point of view, the model is complete though merely useless save some educational purposes to explain basic procedures of switching networks. Recently we can use CPN Tools as GPSS collecting statistical information with special monitors to process it when a simulation is finished. Our approach consists in the creation of special measuring fragments of colored Petri nets which accomplish the basic model calculating the required statistical characteristics "on-flight" – directly in the process of simulation.

A measuring workstation model, shown in Figure 18 together with the CPN Tools feedback, calculates an average network response time. At first, it saves the frame number and a timestamp when a request is sent to a server in place *nSnd*. At second, it stores the received frame timestamp together with the frame number in the place *return*. And finally, it recalculates the average value in the place *NRTime*; for the presented simulation it equals 219.

Note that inside CPN Tools times are discrete and represented by non-negative integer numbers. We call them MTU – model time units. To estimate time in real life units, we apply time scaling procedure described in (Zaitsev 2004) and (Zaitsev 2013).

SOLUTIONS AND RECOMMENDATIONS

We have developed a technique for performance evaluation of networks by colored Petri nets and have built a library of models which includes such modern networking technology as Ethernet, IP, MPLS, Bluetooth, PBB, VoIP etc. The models are recommended by the CPN Tools developers and distributed via official site of CPN Tools http://cpntools.org.

FUTURE RESEARCH DIRECTIONS

Recently a new direction of building re-enterable models in parametric form allows us to construct a model for a networking technology. Its structure is invariant and a given network topology is represented as a marking of dedicated places. Re-enterable models of Ethernet, IP, MPLS have been constructed and we proceed with PBB and VoIP.

Further, we plan to apply the developed technique to the specification of airplane control systems.

CONCLUSION

Thus colored Petri nets is a convenient, powerful, and a vivid tool for specification, verification, and performance evaluation of systems which is acknowledged by their numerous application in telecommunication and networking. Recent experimental papers discuss the advantages of their application for the airplane control systems design which opens wide prospects of using colored Petri nets in the aviation and aerospace industries

REFERENCES

ARINC. (2004). *Mark 33 Digital Information Transfer System (DITS), Part 1: Functional Description, Electrical Interface, Label Assignments and Word Formats, ARINC Specification 429 Part 1-17*. Annapolis: ARINC.

ARINC. (2009). *Aircraft Data Network, Part 7: Avionics Full-Duplex Switched Ethernet Network (AFDX)*. ARINC Industry Activities.

Bourdil, P.-A., Berthomieu, B., & Jenn, E. (2014). Model-Checking Real-Time Properties of an AutoFlight Control System Function. *Proceedings of the 2014 IEEE International Symposium on Software Reliability Engineering Workshops (ISSREW'14)*, 120–123. 10.1109/ISSREW.2014.40

Dodd, R. B. (2006). *Colored Petri Net Modeling of a Generic Avionics Mission Computer*. Technical Report DSTO-TN-0692, Air Operations Division, Defense Science and Technology Organization, Australian Government, Department of Defense.

ICAO. (2016). *2016-2030 Global Air Navigation Plan*. Montreal: International Civil Aviation Organization.

Jensen, K., & Kristensen, L. M. (2009). *Coloured Petri Nets — Modeling and Validation of Concurrent Systems*. Springer-Verlag Berlin. doi:10.1007/b95112

Kordon, F. (2016). Modeling and verification of critical components in airplane systems: Model of a landing detector. *Model Checking Contest*. Retrieved from https://mcc.lip6.fr/index.php

Martinec, D. A. (2001). ARINC 429. In C. R. Spitzer & B. Raton (Eds.), *The Avionics Handbook*. CRC Press LLC.

Miller, S. P., & Tribble, A. C. Extending the Four-Variable Model to Bridge the System-Software Gap. In *Proceedings of the 20th Digital Avionics Systems Conference (DASC01)*. 10.1109/DASC.2001.963383

Zaitsev, D. A. (2004). Switched LAN simulation by colored Petri nets. *Mathematics and Computers in Simulation*, *65*(3), 245–249. doi:10.1016/j.matcom.2003.12.004

Zaitsev, D. A., & Sakun, A. L. (2008). An Evaluation of MPLS Efficacy using Colored Petri Net Models. *Proceedings of International Middle Eastern Multiconference on Simulation and Modelling (MESM'2008)*, 31-36.

Zaitsev, D. A., & Shmeleva, T. R. (2009). Parametric Petri Net Model for Ethernet Performance and QoS Evaluation. In *Proc. of 16th Workshop on Algorithms and Tools for Petri Nets*. University of Karlsruhe.

Zaitsev, D. A., Shmeleva, T. R., Retschitzegger, W., & Pröll, B. (2014). Blocking Communication Grid via Ill-Intentioned Traffic, Proceedings of 14th Middle Eastern Simulation & Modelling Multiconference, 63-71.

Zaitsev, D. A., Shmeleva, T. R., Retschitzegger, W., & Pröll, B. (2016). Security of grid structures under disguised traffic attacks. *Cluster Computing*, *19*(3), 1183–1200. doi:10.100710586-016-0582-9

ADDITIONAL READING

Jebali, F., & Jenn, E. (2016). Auto Flight Control System, *Model Checking Contest* https://mcc.lip6.fr/index.php

Petrucci, L., Billington, J., Kristensen, L. M., & Qureshi, Z. H. (2003). "Developing a formal specification for the mission system of a maritime surveillance aircraft," In *Proceedings of Third International Conference on Application of Concurrency to System Design*, 2003, pp. 92-101. 10.1109/CSD.2003.1207703

Qureshi, Z. H. (2006). Formal modelling and analysis of mission-critical software in military avionics ystems, In *Proceedings of the eleventh Australian workshop on Safety critical systems and software* - Volume 69, Melbourne, Australia, p. 67-77.

Shmeleva, T. R., Zaitsev, D. A., & Zaitsev, I. D. (2009). Verification of square communication grid protocols via infinite Petri nets, In *Proceedings of MESM 2009 - 10th Middle Eastern Simulation Multiconference*, September 27-29, Beirut, Lebanon, p. 53-59.

van der Aalst, W. M. P., & Stahl, C. (2011). *Modeling Business Processes — A Petri Net-Oriented Approach*. The MIT Press.

Zaitsev, D. A. (2005). Enterprise Petri net based CAM software Opera-Topaz, In *Proceedings of 3rd International Industrial Simulation Conference (ISC'2005)*, Berlin, p. 124-128.

Zaitsev, D. A. (2013). *Clans of Petri Nets: Verification of Protocols and Performance Evaluation of Networls*. LAP Lambert Academic Publishing.

Zaitsev, D. A., & Shmeleva, T. R. (2006). *Simulating Telecommunication Systems with CPN Tools: Students' book*. Odessa: ONAT.

KEY TERMS AND DEFINITIONS

Airlines Electronic Engineering Committee (ARINC): ARINC-429 is a standard for avionics data transmission based on a point-to-point one-way connection via twisted pair.

Altitude Switch (ASW): A facility to wake up a device of interest in case the altitude value is less than a given threshold.

Avionics Full-Duplex Switched Ethernet Network (AFDX): A modern standard for avionics data transmission based on switched Ethernet technology.

Colored Petri Net (CPN): A combination of a petri net graph and a functional programming language ML.

Measuring Fragment: A part of a colored petri net which calculates statistical characteristics directly in the process of simulation.

Meta Language (ML): A functional programming language.

Multiset: A set of elements supplied with their multiplicities (for example, two books and three pencils).

Petri Net: A directed bipartite graph with a dynamic process defined on it.

State Space: A set of valid marking of a petri net and transitions between them.

Chapter 15
Analysis of Aviation Incidents Using the GERT-Network Taking Into Account the Psychological Characteristics of the Operator

Iryna Yakunina
Kirovograd Flight Academy of the National Aviation University, Ukraine

Abdel-Badeeh M. Salem
Ain Shams University, Egypt

Roman Yakunin
Kirovograd Flight Academy of the National Aviation University, Ukraine

ABSTRACT

In this chapter, the authors considered the construction of network models of aviation incidents taking into account the psychological characteristics of the operator. In particular, the GERT-network of aviation disaster was built. Applying of software to calculate the elements of the GERT-network was presented. The use of network models in the analysis of aviation incidents allows one to visualize the cause-and-effect relationship in aviation incidents and gives the opportunity to obtain quantitative characteristics of the incident for subsequent analysis.

BACKGROUND

Civil aviation is an area of increased danger and responsibility. Civil Aviation solves the problem of ensuring the growing demand for all types of air transportation, increasing the regularity and cost-effectiveness of flights, providing a high level of safety of flights. Due to the fact that conducting natural experiments in aviation is not permissible, and the safety of flights needs to be improved, each aviation event is subject

DOI: 10.4018/978-1-5225-7709-6.ch015

to a posteriori analysis. The a posteriori analysis allows us to develop recommendations for the future, as well as provides initial data for building different mathematical models of aviation events In particular, the use of GERT-networks for modeling aviation events can analyze and predict the development of a special case in flight, and also allows to build adequate models of flight situations development.

The report on the investigation of the aviation event is the basis for taking measures to ensure the safety of flights necessary to prevent the subsequent aerial events for the same reasons. Therefore, the final report on the aviation event must clearly identify what happened, how it happened and why it happened. The conclusions, causes and / or concomitant factors presented in the final report should facilitate the preparation of safety recommendations so that the necessary precautionary measures can be taken. The analytical part of the final report should include an assessment of the data obtained during the collection of actual information and the analysis of the circumstances and events that have occurred or could have occurred. The rationale should be logical and promote assumptions that will then need to be discussed and verified by matching existing evidence (ICAO, 2013, ICAO, 2014, ICAO, 2015). The use of GERT-networks for quantitative analysis of the above-mentioned factors and causes allows us to elucidate the logical and causal relationships that contributed to the occurrence of this aviation event. Formalizing the actions of the operator of the air navigation system as a human operator in special cases of flight with the help of a network planning device allows determining the optimal sequence and time of execution of operational procedures aimed at parrying special flight situations. The use of network charts in the a posteriori analysis allows qualitative and quantitative analysis of aviation events to improve flight safety.

In the event of a special case in flight, the operators of the aeronautical system - the air traffic controller and commander fall into the conditions of an acute shortage of time and significant psychophysiological load, which greatly complicates the decision-making process. In addition, information on the occurrence of a special case in flight is characterized by a high level of incompleteness and uncertainty of information. Therefore, it is expedient to construct mathematical models that would allow quantifying the possible options for completing the flight and, being components of the decision support system, would help to adopt the optimal solution for the given time conditions. In the general case, the guideline documents define the list of actions that must be performed in accordance with the pilot's and the controller, if a special case has occurred with the aircraft. But the activity of aeronautical system operators in the a posterior analysis is worse for algorithmization due to the presence of a component of the psycho-emotional state of the operator.

Problems of professional selection, training and professional training of aeronautical system operators are described in many papers, in particular in the works of A. Vitryak and G. Kryzhanivsky, S. Melnychenko, V. Kharchenko, and V. Chynchenko (Adrianova, 1974, Vitryak, 2008, Kryzhanovsky, 1981, Zyryanov, 2012, Kharchenko, 2006, Kharchenko, 2012). Mathematical models based on graphs are presented in works by V. Karnaukhov, V. Grygi, D. Vinogradov, R. Dunets, J. Fax, A. Jadbabaie, V. N. Kasyanova, A. Kochkarova, S. Shahrampour, R. Olfati-Saber, T. Panyukova, M. Himsolt, I. Parasyuk, W. Ren, H. Tanner, Z. A. Felk (Himsolt, 1996, Olfati-Saber, 2010, Tanner, 2007, Grieg, 2008, Karnaukhov, 2009). The description of network models is presented in the works of D. Philips and A. García-Diaz, K. Neumann, A. Pritsker, A. Ventsel, A. Shibanova, D. Pismana, S. Semenova, A. Smirnov, M. Dorrera (Neumann, 1990, Pritsker, 1966, Wentzel, 2012, Zyryanov, 2012, Shibanov, 2003). Data mining technology in e-learning, ontological engineering in e-learning and intellectual e-learning systems are presented in works by Abdel-Badeeh M. Salem (Abdel-Badeeh, 2008, Abdel-Badeeh, 2010, Abdel-Badeeh, 2011). The idea of using the GERT network was based on many researches. In particular, this is the work of Shmeliova, Pawlak, Romero, Ventura, Tyunin, Yakimov, Witten and Frank (Shmeliova, 2013, Pawlak,

1991, Romero, 2006, Tyunin, 2004, Witten, 2005). However, the question of quantifying the time of decision-making by operators of the air navigation system in special cases in the flight was not studied.

In view of the fact that in case of a special flight situation, operators of aeronautical systems need to make decisions in the event of a shortage of time, it is important to develop graph-analytical methods for analyzing the operators of operators for aeronautical systems for quantitative estimation of time and construction of deterministic and stochastic models on their basis.

Under a special case in flight, they understand the situation that arises in flight as a result of the influence of an adverse factor or a combination of adverse factors, and leads to a decrease in flight safety (Kryzhanovskiy, 1984).

An unexpected message of a special occurrence in the flight, an acute shortage of time, sudden awareness of an extremely dangerous emergency can lead members of the crew of the aircraft to a series of erroneous actions, usually associated with the emergence of the effect of antagonism mechanisms of memory and oppression of intellectual and motor activity; the inability to reproduce the reasons for refusal in memory and to carry out in due course the necessary actions, which are indicated in normative documents; loss of ideas about the ratio of course, flight altitude and terrain; disorientation in the form of illusions about the flow of time and location of the aircraft (Strelkov, 2004). One of the chief tasks of the dispatcher in such cases is the timely provision of assistance to the crew of an aircraft that is in distress.

The GERT stochastic network is defined as a network that can only be executed when performing a certain subset of arcs (operating procedures). In this case, the execution time of each arc (operating procedure) is selected in accordance with the probabilistic distribution. In stochastic networks, it is not necessary to execute all the arcs included in it for the execution of the node; therefore, such models allow the existence of cycles and loops (Phillips, 1984, Golenko-Ginzburg, 2015).

Let you perform operating procedures (*i, j*) is the operator of air navigation system is to accidental largest Y_{ij}. By definition operating procedure (*i, j*) can be performed only if the node is executed and r_{ij} is the likelihood that the operating procedure (*i, j*) be executed provided that the knot and executed. For random size Y_{ij} *W*-function is defined by the formula:

$$W_{ij}\left(s\right) = p_{ij} M_{ij}\left(s\right) \tag{1}$$

Network G consists of two arcs (*i, j*) and (*k, j*), which are characterized by random values X_1 and X_2, which the specified by generating function points $M_1(s)$ for X_1 and $M_2(s)$ for X_2. For random variables X_1 and X_2 you can calculate numeric characteristics – expectation $M(X_i)$ and variance $D(X_i)$.

The GERT network can be replaced by one equivalent arc, that is, a random variable *Y* that characterizes the flow of the GERT network can be calculated.

The generated torque functions for three basic equivalent arc transformations are as follows:

$$M_E\left(s\right) = \frac{W_E\left(s\right)}{p_E} = \frac{p_1 M_1\left(s\right) \cdot p_2 M_2\left(s\right)}{p_1 p_2} = M_1\left(s\right) \cdot M_2\left(s\right), \tag{2}$$

for successive arcs;

$$M_E(s) = \frac{W_E(s)}{p_E} = \frac{p_1 M_1(s) + p_2 M_2(s)}{p_1 + p_2}, \tag{3}$$

for parallel arcs;

$$M_E(s) = \frac{W_E(s)}{p_E} = \frac{p_1 M_1(s)}{1 - p_2 M_2(s)} \cdot \frac{1 - p_2}{p_1} = \frac{(1 - p_2) \cdot M_1(s)}{1 - p_2 M_2(s)}, \tag{4}$$

for arc and loop.

Numerical characteristics $M(Y)$ and $D(Y)$ of the random value Y of GERT-network, which is equal to the combination of random variables X_1 and X_2 for the equivalent transformation of successive arcs

$$M(Y) = M(X_1) + M(X_2),\ D(Y) = D(X_1) + D(X_2). \tag{5}$$

Numerical characteristics $M(Y)$ and $D(Y)$ of the random value Y of GERT-network, which is equal to the combination of random variables X_1 and X_2 for the equivalent transformation of parallel arcs

$$M(Y) = \frac{p_1 M(X_1) + p_2 M(X_2)}{p_1 + p_2}; \tag{6}$$

$$D(Y) = \frac{p_1 D(X_1) + p_2 D(X_2)}{p_1 + p_2} + \frac{p_1 p_2 \left(M(X_1) - M(X_2)\right)^2}{(p_1 + p_2)^2}. \tag{7}$$

Numerical characteristics $M(Y)$ and $D(Y)$ of the random value Y of GERT-network, which is equal to the combination of random variables X_1 and X_2 for the equivalent transformation of arc and loop

$$M(Y) = M(X_1) + \frac{p_2}{1 - p_2} M(X_2) \tag{8}$$

and

$$D(Y) = D(X_1) + \frac{p_2}{1 - p_2} D(X_2) + \frac{p_2}{(p_2 - 1)^2} \left(M(X_2)\right)^2, \tag{9}$$

where X_1 and X_2 are the random values of the arcs of the GERT network, Y the random value of the flow of GERT-network, p_1 and p_2 are the probability of executing the arcs X_1 and X_2 respectively. Since in GERT-network all random variables are independent, they do not correlate with each other.

The transformation of a homogeneous GERT-network to an equivalent arc is reduced to the recognition of successive, parallel arcs, arcs and loops, and their replacement into equivalent arcs with corresponding mathematical expectations and dispersions.

For two or more successive arcs: the *W*-function for an equivalent arc equals the product of the *W*-functions for these arcs.

For two or more parallel arcs: the *W*-function for an equivalent arc equals the sum of *W*-functions for these arcs.

Thus, if the GERT network consists of parallel and sequential arcs or loops, then it can be transformed into an equivalent network that consists of only one arc (Phillips, 1984).

The purpose of the GERT network is to calculate the mathematical expectation and dispersion of the time of execution of the whole network – the complex of elementary actions of the operator of the air navigation system.

Let's consider the stochastic network model GERT of the development of the flight situation $G = (N; A)$ with the set of nodes N and the set of arcs A. The time t_{ij} for the transition from the i-th to the j flight situation is a random variable. The transition (i, j) can be executed only if the second node is executed. To determine the time t_{ij} for the transition from the i-th to the j flight situation, it is necessary to know the conditional probability (in the discrete case) or the distribution density (in the continuous case) of the random variable Y_{ij}. This allows conducting a $G= (N; A)$ survey to determine the moments of time distribution for t_{ij} network G that can be used to calculate the mathematical expectation of $M(E)$ and the dispersion δ^2 of the network runtime G in the event of complicated, complicated, emergency or catastrophic situation.

In order to determine the probability of occurrence of a particular network event – Q, the mathematical expectation - $M(E)$ and the dispersion of the time before the occurrence of events – D[T], it is necessary to simplify the initial model by combining successive and parallel circuits into a single branch with equivalent output parameters P_{ij}, $M_{ij}\left(S\right)$ and their transformation $W_{ij}^{*}\left(S\right)$ – the coefficient of transmission of the dynamic systems.

The general expression for the transfer function of the network will be as follows:

$$Q = \sum_{i=1}^{a} \prod_{j=1}^{D_i} P_{ij} \leq 1 \tag{10}$$

where a - the number of minimum throughputs inside the graph;

D_i – the number of arcs connecting the nodes of the graph in i-th connection;

P_{ij} – probability of signal transfer (transition of the indicator) between j -th arcs of the graph in their i-th combinations (Phillips, 1984).

Case-Based Reasoning (CBR) Methodology for Developing IEDSSs. CBR is an analogical reasoning method provides both a methodology for problem solving and a cognitive model of people (Kolonder,

1993). CBR means reasoning from experiences or "old cases" in an effort to solve problems, critique solutions, and explain anomalous situations. It is consistent with much that psychologist have observed in the natural problem solving that people do. People tend to be comfortable using CBR methodology for decision making, in dynamically changing situations and other situations were much is unknown and when solutions are not clear.

From knowledge engineering point of view, the "case" is a list of features that lead to a particular outcome. (e.g. The information on a patient history and the associated diagnosis). Determining the appropriate case features is the main knowledge engineering task in case-based IEDSS software. This task involves defining the terminology of the domain and gathering representative cases of problem solving by the expert. Determining the appropriate case features is the main knowledge engineering task in developing case-based IEDSS. This task involves defining the terminology of the domain and gathering representative cases of problem solving by the experts. Representation of cases can be in any of several forms (predicate, frames, and scribes).From the computational point of view, CBR refers to a number of algorithms and techniques useful in solving new cases when they are presented (as shown in Fig. 5).. In addition, there are techniques that can be used to modify earlier cases to better match new cases and other techniques to synthesize new cases when they are needed. The methodology of CBR is becoming popular in developing case-based IEDSSs because it automates applications that are based on precedent or that contain incomplete causal models (Kolonder, 1993). In a Case-Based IEDSS an incomplete mode or an environment which does not take into account all variables could result in either an answer built on incomplete data or simply no answer at all. Case-based methodologies attempt to get around this shortcoming by inputting and analysing problem data, then retrieving a similar case from the case memory library, and finally displaying a solution based on examination of these previous cases .CBR has already been applied in a number of different applications in many domains, e.g., medicine, industry, law, banking.

The system's methodology of CBR can be summarized in the following steps:

1. The system will search its case-memory for an existing case that matches the input problem specification.
2. If we are lucky (our luck increases as we add new cases to the system), we will find a case that exactly matches the input problem and goes directly to a solution.
3. If we are not luck, we will retrieve a case that is similar to our input situation but not entirely appropriate to provide as a completed solution.
4. The system must find and modify small portions of the retrieved case that do not meet the input specification. This process is called "case-adaptation".
5. The result of case adaptation process is (a) completed solution, and (b) generates a new case that can be automatically added to the system's case-memory for future use.
6. Benefits of CBR Approach to IEDSSs (Salem, 2008).

Research reveals that users learn best when they are presented with examples of problem-solving knowledge and are then required to apply the knowledge to real situations. The case-memory of examples and exercises capture realistic problem-solving situations and presents them to the learners as virtual simulations.

On the other hand, there are several benefits where user/learners should be able to perform better using CBR methodology, e.g.

1. With more cases available, users will be able to recognize more situations and the solutions that go with these cases include failure cases, students will be able to benefit from the failures of others.
2. Retrieval cases will allow users to better recognize what is important in a new situation. Cases indexed by experts would recall and will show the user ways of looking at a problem that he might not have the expertise for without the system.
3. Users will have access to obscure cases that they otherwise would not able to make use of. These obscure cases can help with any of the tasks previously listed.
4. During a training period CBR system provides the user with a model of the way decision making ought to be done, for example, what things ought to be considered and provides them with concrete examples on which to hang their more abstract knowledge (Salem, 2011).

Illustrative Example

This example, like other similar ones, takes as its basis the statistical information obtained from experienced aviation specialists. The characteristics of the distribution laws are obtained on the basis of the calculation formulas of mathematical statistics that are included in the standard course of higher mathematics. On the example of the aircraft of the model under consideration, a method is studied that can be applied to aircraft of more modern models.

For example, the GERT-network for the aviation event, which occurred on December 5, 1995 at the airport of Nakhichevan, was constructed and calculated. Causal consequences of events that contributed to the emergence of an accident, are reflected in Table. 1.

Determine separately each loop of the Li-GERT-network of the Tu-134B disaster and the corresponding equivalent of the transmission (Fig. 1).

Transmittance equivalents for each network loop:

$$T(L_1) = W_1(T) \cdot W_3(T) \cdot W_6(T) \tag{11}$$

$$T(L_2) = W_1(T) \cdot W_2(T) \cdot W_5(T) \cdot W_7(T) \tag{12}$$

$$T(L_3) = W_1(T) \cdot W_2(T) \cdot W_4(T) \cdot W_8(T) \cdot W_{10}(T) \cdot W_{12}(T) \tag{13}$$

$$T(L_4) = W_1(T) \cdot W_2(T) \cdot W_4(T) \cdot W_8(T) \cdot W_9(T) \cdot W_{11}(T) \cdot W_{13}(T) \times$$

$$\times W_{14}(T) \cdot W_{16}(T) \cdot W_{17}(T) \cdot W_{18}(T) \cdot W_{19}(T) \cdot W_{20}(T) \cdot (1/W_E(T)). \tag{14}$$

Masonic's topological equation

$$H = 1-L_1-L_2-L_3-L_4 = 0 \Rightarrow L_4 = 1- L_1-L_2-L_3. \tag{15}$$

Coefficient of passing a fictitious arc

Figure 1. CBR Methodology. Boxes represent processes and ovals represent

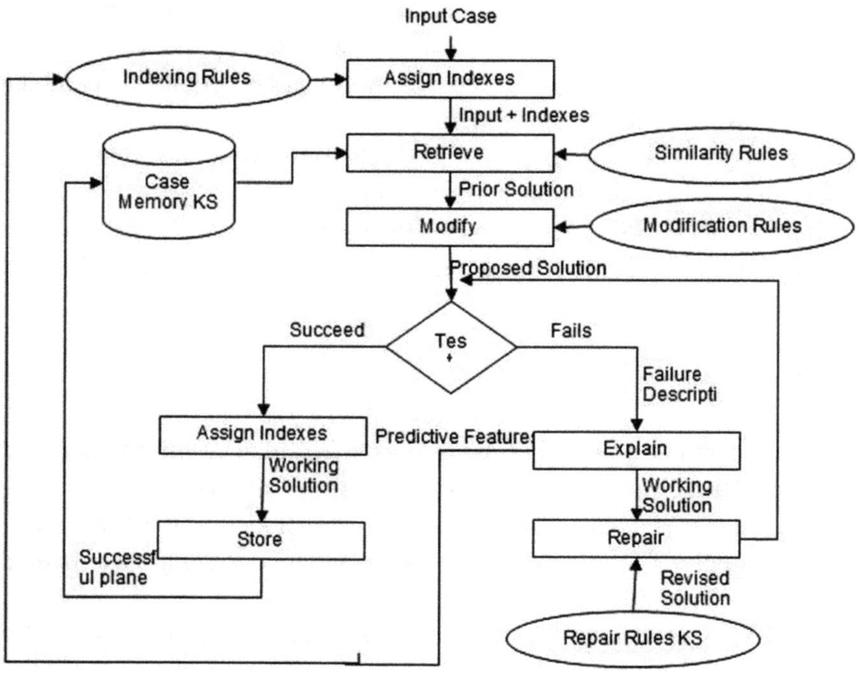

$$W_E = \frac{W_1(T)\cdot W_2(T)\cdot W_4(T)\cdot W_8(T)\cdot W_9(T)\cdot W_{11}(T)\cdot W_{13}(T)}{1-T(L_1)-T(L_2)-T(L_3)}\times$$

$$\times\frac{W_{14}(T)\cdot W_{16}(T)\cdot W_{17}(T)\cdot W_{18}(T)\cdot W_{19}(T)\cdot W_{20}(T)}{1-T(L_1)-T(L_2)-T(L_3)}. \tag{16}$$

Parameters for each arc GERT-network of the Tu-134B disaster (Dec. 5, 1995, Nakhichevan airport):

$1 \to 2$: $M_1(T) = 8$ c, $\sigma_1 = 3$ c, $p_1 = 0{,}02$, $W_1(T) = 0{,}02\cdot e^{8*t+\frac{3^2 t^2}{2}}$;

$2 \to 4$ $M_2(T) = 4$ c, $\sigma_2 = 1{,}1$ c, $p_2 = 0{,}3$, $W_2(T) = 0{,}3\cdot e^{4*t+\frac{(1{,}1)^2 t^2}{2}}$;

$2 \to 5$: $M_3(T) = 6$ c, $\sigma_3 = 1{,}1$ c, $p_3 = 0{,}7$, $W_3(T) = 0{,}7\cdot e^{6*t+\frac{(1{,}1)^2 t^2}{2}}$;

$4 \to 6$: $M_4(T) = 8$ c, $\sigma_4 = 2$ c, $p_4 = 0{,}3$, $W_4(T) = 0{,}3\cdot e^{8*t+\frac{2^2 t^2}{2}}$;

Table 1 Causal and consequential connections of events that contributed to the emergence of an aviation accident

Arc	Event
1	The flight takes place under normal conditions
1-2	Denied the left engine
1-3	There were no engine failures
2-4	Flight mechanic has determined the failure of the right engine (by mistake)
2-5	The crew determined the failure of the left engine
4-6	The aircraft was piloted by a second pilot
4-7	The aircraft was piloted by the commander of the aircraft
6-8	The second pilot does not inform the aircraft commander of the mistakenness of the definition of the engine refused
6-9	The second pilot informs the commander of the aircraft about the mistakenness of the definition of the engine refused
8-10	The commander of the aircraft gives the command about the regular shutdown of the right engine, and after 10 seconds about the emergency shutdown of the right engine
8-11	The commander of the aircraft gives the command to transfer the engine to a small gas mode, let it work so two minutes, and then turn off
10-12	The flight mechanic turns off the right engine
10-13	The flight mechanic notices the error before turning off the right engine
12-14	Flight mechanic reports "Refusal is complete, the second refused"
14-15	The commander of the aircraft decides to make a forced landing outside the airfield
15-16	There are no obstacles ahead of course
15-17	There are obstacles in front of the course (residential quarters)
17-18	The commander of the aircraft turns the aircraft to the right with a roll of 36 °
18-19	The commander of the aircraft turns the ship to the right with a roll of maximum 15 °
18-20	The aircraft lost its height and crashed to the ground with its right wing
20-21	Catastrophe
3-22 9-22 13-22 16-22	The end of the flight without sacrifices

$$4 \rightarrow 7:\ M_5(T) = 8\ c,\ \sigma_5 = 2{,}6\ c,\ p_5 = 0{,}7,\ W_5(T) = 0{,}7 \cdot e^{8*t+\frac{(2{,}6)^2 t^2}{2}};$$

$$5 \rightarrow 1:\ M_6(T) = 7\ c,\ \sigma_6 = 1{,}3\ c,\ p_6 = 0{,}9,\ W_6(T) = 0{,}9 \cdot e^{7*t+\frac{(1{,}3)^2 t^2}{2}};$$

$$7 \rightarrow 1:\ M_7(T) = 7\ c,\ \sigma_7 = 2{,}3\ c,\ p_7 = 0{,}9,\ W_7(T) = 0{,}9 \cdot e^{7*t+\frac{(2{,}3)^2 t^2}{2}};$$

$$6 \rightarrow 8:\ M_8(T) = 3\ c,\ \sigma_8 = 1{,}4\ c,\ p_8 = 0{,}3,\ W_8(T) = 0{,}3 \cdot e^{3*t+\frac{(1{,}4)^2 t^2}{2}};$$

Figure 2 GERT-network of Tu-134B disaster (December 5, 1995, Nakhichevan airport)(a), and loop L_i (i=1,2,3,4) GERT-network of the Tu-134B disaster (b)

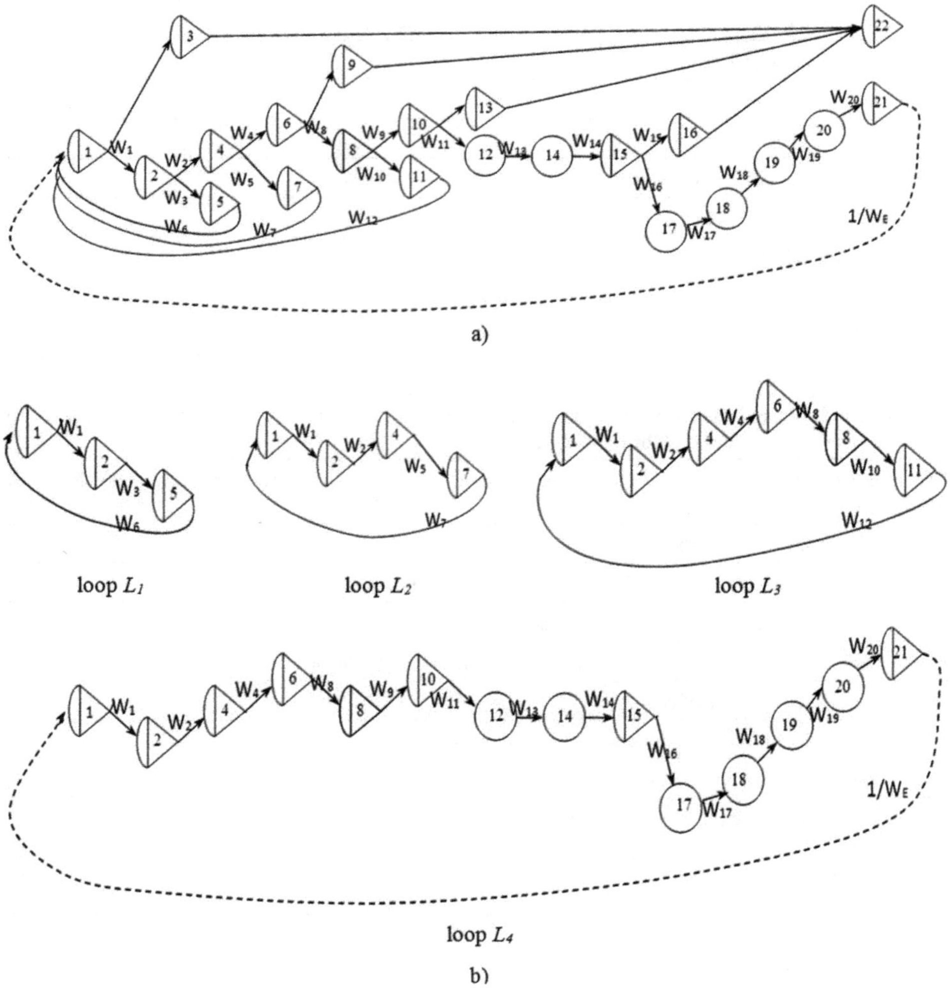

$$8 \rightarrow 10:\ M_9(T) = 13\ c,\ \sigma_9 = 2{,}6\ c,\ p_9 = 0{,}3,\ W_9(T) = 0{,}3 \cdot e^{13^*t + \frac{(2{,}6)^2 t^2}{2}};$$

$$8 \rightarrow 11:\ M_{10}(T) = 3\ c,\ \sigma_{10} = 1\ c,\ p_{10} = 0{,}7,\ W_{10}(T) = 0{,}7 \cdot e^{3^*t + \frac{1^2 t^2}{2}};$$

$$10 \rightarrow 12:\ M_{11}(T) = 3\ c,\ \sigma_{11} = 0{,}9\ c,\ p_{11} = 0{,}3,\ W_{11}(T) = 0{,}3 \cdot e^{3^*t + \frac{(0{,}9)^2 t^2}{2}};$$

$$11 \rightarrow 1:\ M_{12}(T) = 120\ c,\ \sigma_{12} = 8{,}6\ c,\ p_{12} = 0{,}9,\ W_{12}(T) = 0{,}9 \cdot e^{120^*t + \frac{(8{,}6)^2 t^2}{2}};$$

$12 \rightarrow 14$: $M_{13}(T) = 2$ c, $\sigma_{13} = 1{,}6$ c, $p_{13} = 1$, $W_{13}(T) = 1 \cdot e^{2*t+\frac{(1{,}6)^2 t^2}{2}}$;

$14 \rightarrow 15$: $M_{14}(T) = 2$ c, $\sigma_{14} = 2{,}3$ c, $p_{14} = 1$, $W_{14}(T) = 1 \cdot e^{2*t+\frac{(2{,}3)^2 t^2}{2}}$;

$15 \rightarrow 16$ $M_{15}(T) = 2$ c, $\sigma_{15} = 6$ c, $p_{15} = 0{,}5$, $W_{15}(T) = 0{,}5 \cdot e^{2*t+\frac{6^2 t^2}{2}}$;

$15 \rightarrow 17$: $M_{16}(T) = 2$ c, $\sigma_{16} = 1{,}7$ c, $p_{16} = 0{,}5$, $W_{16}(T) = 0{,}5 \cdot e^{2*t+\frac{(1{,}7)^2 t^2}{2}}$;

$17 \rightarrow 18$: $M_{17}(T) = 4$ c, $\sigma_{17} = 1{,}3$ c, $p_{17} = 1$, $W_{17}(T) = 1 \cdot e^{4*t+\frac{(1{,}3)^2 t^2}{2}}$;

$18 \rightarrow 19$: $M_{18}(T) = 5$ c, $\sigma_{18} = 2$ c, $p_{18} = 1$, $W_{18}(T) = 1 \cdot e^{5*t+\frac{2^2 t^2}{2}}$;

$19 \rightarrow 20$: $M_{19}(T) = 4$ c, $\sigma_{19} = 1{,}7$ c, $p_{19} = 1$, $W_{19}(T) = 1 \cdot e^{4*t+\frac{(1{,}7)^2 t^2}{2}}$;

$20 \rightarrow 21$: $M_{20}(T) = 1$ c, $\sigma_{20} = 0{,}9$ c, $p_{20} = 1$, $W_{20}(T) = 1 \cdot e^{1*t+\frac{(0{,}9)^2 t^2}{2}}$.

The law of distribution is defined as normal, in accordance with the central limit theorem (with the addition of a sufficiently large number of independent random variables distributed according to any distribution laws with approximately the same order of dispersion, the law of the distribution of the sum is quite close to normal) (Wentzel, 1999). The mathematical expectation, variance and probability for each arc is determined by the expert survey method.

We got W(0)=2,4713 · 10-5 – the probability that aviation engine failure on take-off would lead to a disaster;

$M(T_0) = \mu_1 = \frac{d}{dt} M(0) = 66{,}4106$ – mathematical expectation of the time of the onset of the disaster;

σ2 = 7,721 – dispersion.

In order for the constructed mathematical model to reflect the most realistic situation in flight, one must take into account the presence of factors that have the greatest impact on the functioning of the aeronautical system. There are two main factors that can be distinguished: this is a disaster as a control and crew of an aircraft, which has the most direct impact on the operation of aircraft mechanisms.

In constructing the previous mathematical model, we did not take into account the psycho-emotional state of the operator of the aeronautical system. By default, it was considered that in the event of a special case in flight, the operator will act in accordance with the requirements of the guidance documents. In fact, a special case in flight is an event characterized by a high degree of extreme, uncertainty and incompleteness of information. When an aircraft begins to suffer disasters, the operator of the aeronautical system falls into a stressful situation that leads to the excitation of the nervous system. The optimum degree of excitation of the nervous system of the operator of the aeronautical system contributes to the successful parrying of a special case in flight. But if there is an over excitation of the nervous system or vice versa apathy, then the probability of successful parrying of a special case in the flight is rapidly falling. Thus, the problem of quantitative assessment of the mental state of the operator of the aeronautical system in case of occurrence of a special case in flight and taking into account this assessment when constructing a mathematical model of activity is relevant.

In case of a special case in flight there is a change in the type of mental activity and there is a transition from formalized to unsynchronized regimes (Zadorozhny, 2011). Psychologists are basically accustomed to the idea that predicting human-operator activity in a stressful situation can be a source of anxiety of personality (Martinov, 2012, Strelkov, 2004, Strelkov, 2008, Bodrov & Lukyanova, 2007). There are many methods for determining the level of anxiety of a person. But the most accurate tests of the study of human mental qualities, in particular the level of anxiety, are professional testing techniques, such as Minnesota Multiphasic Personality Inventory or Keethel's Personality Test.

In psychology, there are three levels of anxiety: low, moderate, high (Astapov, 2008). Anxiety, as a permanent feature of the personality that can disrupt activities under extreme conditions, causes uncertainty in oneself and thereby assess the effectiveness of performing professional actions (Bodrov & Lukyanova, 2007).

Due to the volume and complexity of calculating parameters of GERT - network, the computer program "GERT - analysis 2018" is offered for use. It is proposed to apply this program for the study of the laws of the human-operator of the air navigation system in the complex of assessment and forecasting of the development of the flight situation in the conditions of risk and uncertainty in the event of a freelance situation. The computer program "GERT - analysis 2018" allows you to calculate the probability of execution of the node, the mathematical expectation and dispersion of the time of execution of the node. Full software would occupy a large amount of our article, so only the main fragments of the incident analysis algorithm are offered.

The New button opens a new document for constructing the GERT network. The preparation of a new document includes the left-hand side of the table of cause and effect of the events from which the aviation event took place, the field for network construction.

The Appoint button allows you to enter mathematical expectation, dispersion, probability and distribution law for each arc of the network. To enter arc parameters, it is necessary to activate the corresponding arc of the network by clicking on it 1 time with the left button of the manipulator. Active arc is indicated by green color. In the Select field, you have the option to select a distribution law, by default, the program considers the distribution law normal. After entering all the parameters, you need to click the "add" button.

Figure 3 "GERT - analysis 2018". New button

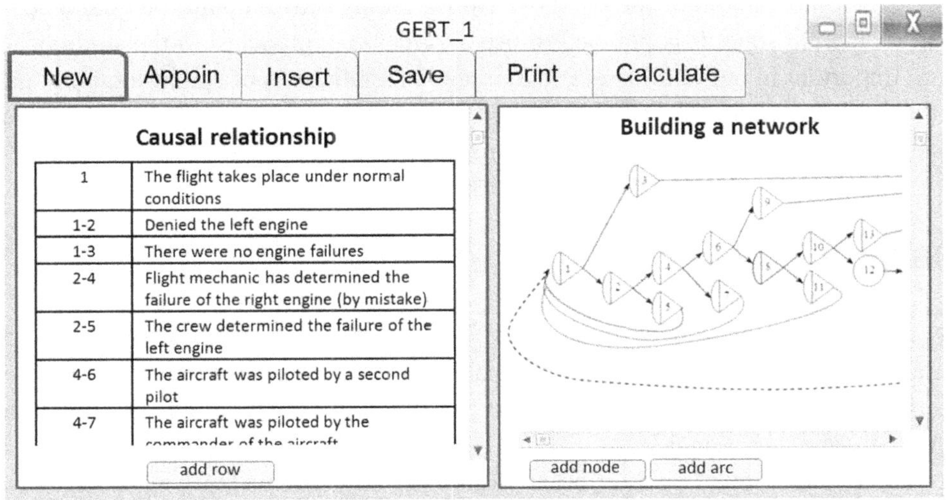

Figure 4 "GERT - analysis 2018". Appoint button

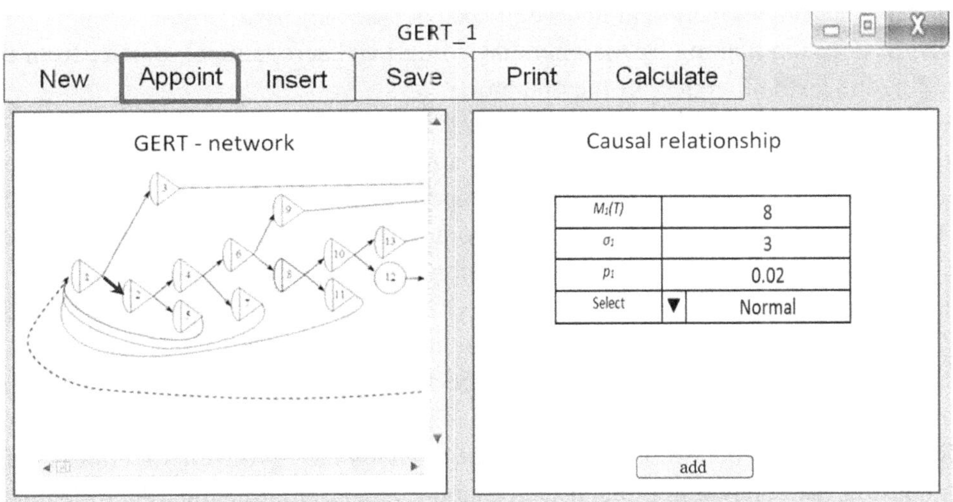

After entering the data for all arcs, you can select the button "Calculate", where the program will automatically calculate the probability that a special case in flight will develop before the disaster, the mathematical expectation of the time of disaster and dispersion.

FUTURE RESEARCH DIRECTIONS

In the future, it is planned to investigate the correlation between the level of anxiety and the false actions of the operator of the aeronautical system. It is also planned to build an algorithm for interpretation qualitative indicators of a operators' mental state in a special case occurs in flight to quantitative indicators, to modify the construction of the GERT-network, taking into account the quantitative indicators of the operator's mental state. It is planned to use alternative approaches to the study of the problem. Among them, important in our opinion are such issues as application of spatio-temporal graphs for the synthesis of specialized multipliers (Greer, 1995; Griga, 2008). It is planned to further study the value of the results obtained in practice in order to improve the safety of flights.

CONCLUSION

In this way, using the characteristic functions of the time for performing elementary operations by the aeronautical system operator, we obtain an analytical expression for the distribution density of the execution time of the entire set of actions aimed at parrying a special case in flight and numerical characteristics for a given aviation incident.

Improved author's method of a posteriori analysis of aviation events for using stochastic networks of type GERT, which allows to visualize logical and causal links that contributed to the occurrence of the aviation event, and allows to determine its quantitative time indices.

The improvement of the author-made graph-analytical methods for making decisions by operators was continued an air navigation system in case of special cases in flight.

The model of decision making by the operator is improved aeronautical system, from the point of view of studying the level of anxiety of the operator.

Knowledge engineering, computational intelligence, artificial intelligence and data mining paradigms give decision support systems added intelligence capabilities, allowing them to exhibit more intelligent behaviour. Knowledge engineering offers robust computational techniques for accumulating, updating, managing and representing knowledge. These techniques enable users with learning mechanisms that help to induce knowledge from raw data. Data mining techniques enable users with learning mechanisms that help to induce knowledge from raw data. Ontological engineering offers a promising way to develop efficient e-learning systems capable to facilitate knowledge sharing, refine, search, and reuse. Moreover, the convergence of artificial intelligence technologies, data engineering, computational intelligence, machine learning, and web technologies is enabling the creation of a new generation of web-based intelligent expert decision support systems. Among the envisaged directions for future work we include: (i)using the behavioral biometrics technology for user authentication / identification (this technology is based on the user data of electrocardiogram and electroencephalogram bio-signals); (ii) the investigation of using ontological engineering approach in developing IEDSSs that provide users with intelligent browsing and searching support in their request for relevant material on the web and (iii) using the distributed artificial intelligence methodology in developing a new versions of IEDSSs as an open intelligent information systems.

REFERENCES

Adrianova, V. E. (1974). Human activity in management systems. Leningrad: Academic Press.

Arkhangelskii, V. I., Boganenko, I. N., Grabovsky, G. G., & Ryumshin, N. A. (2000). *Human Automation Systems: Quality, Safety and Reliability Management*. Kiev: KIA.

Astapov, V. M. (2008). Anxiety and anxiety. Moscow: Litagent "Kogito-Center."

Bodrov, V. A., & Lukyanova, N. F. (2007). Personal characteristics of pilots and professional efficiency. *Psychological foundations of professional activity*. Moscow. *Logos*, 423–430.

Gallay, M. L. (1970). *Flight of an aircraft with incomplete and asymmetrical traction*. Moscow: Mechanical Engineering.

Golenko-Ginzburg, D.I. (2010). *Stochastic network models of planning and development management*. Voronezh: Scientific Book.

Greer, J. (1995). *Proceedings of AI-ED 95, World Conference on Artificial Intelligence in Education*. Association for Advancement of Computing in Education (AACE).

Griga, V. M., & Dunets, R. B. (2008). Application of spatio-temporal graphs for the synthesis of specialized multipliers. *Radioelectronic Computer Systems*, *7*, 113-117.

Himsolt, M. (1996). The Graphlet system (system demonstration). *Lecture Notes in Computer Science*, *1190*, 233–240. doi:10.1007/3-540-62495-3_51

ICAO. (2013). Safety Management Manual (SMM) (3rd ed.). Doc. ICAO 9859-AN 474. Author.

ICAO. (2014). *Safety Report*. Montreal, Canada: Author.

ICAO. (2015). *Doc 9756, Manual of Aircraft Accident and Incident Investigation*. Montreal: ICAO.

Karnaukhov, V. A. (2009). *The theory of graphs and networks in the modeling of ATC processes*. Ulyanovsk: UAVU CA.

Kharchenko, V.P., & Chinchenko, Y.V. (2006). Automation of the process of professional selection of aviation civil aviation personnel. *Visnyk NAU, 2*, 34-36.

Kharchenko, V. P., Shmelova, T. F., & Sikirda, Yu. V. (2012). *Making decisions by the operator of aeronautical systems*. Kirovograd: CLA NAU.

Kolonder, J. (1993). *Case-Based Reasoning*. Morgan Kaufmann.

Kryzhanovsky, G. A., Darymov, Yu. P., & Solodukhin, V. A. (1981). Automation of air traffic control processes. Moscow: Academic Press.

Kryzhanovsky, G. A., Kupin, V. V., & Plyasovskikh, A. P. (2008). Theory of transport systems. St. Petersburg: Academic Press.

Makarov, R. N., Nedelko, S. N., Bamburkin, A. P., & Grigoretskii, V. A. (2005). *Aviation pedagogy*. Moscow: Academic Press.

Martinov, A. I. (2012). *Mental training pilots of ultralight aircraft*. Moscow: Himalaia.

Neumann, K. (1990). *Stochastic project networks: temporal analysis, scheduling and cost minimization*. Berlin: Springer-Verlag. doi:10.1007/978-3-642-61515-3

Olfati-Saber, Fax, & Murray. (2010). Reply to comments on consensus and cooperation in networked multi-agent systems. *Proceedings of the IEEE, 7*, 1354–1355.

Pawlak, Z. (1991). Theoretical Aspects of Reasoning about Data. Kluwer Academic Publishers. doi:10.1007/978-94-011-3534-4

Phillips, D., & Garcia-Diaz, A. (1984). *Methods of network analysis*. Moscow: The World.

Pritsker, A. A. (1966). *GERT: graphical evaluation and review technique*. RAND Corporation.

Romero, C., & Ventura, S. (2006). *Data mining in e-learning*. Southampton, UK: Wit Press. doi:10.2495/1-84564-152-3

Salem. (2008). Data Mining Technology in e-Learning. *Proceedings of 6th International Conference on Emerging e-learning Technologies and Applications, Information and Communication Technologies in Learning, (ICETA2008)*.

Salem. (2010). Ontological Engineering in e-Learning. *Proceedings of 8th International Conference on Emerging e-learning Technologies and Applications, Information and Communication Technologies in Learning*.

Salem. (2011). Intellectual E-Learning Systems. In *Proceedings of the Practical Conference "VR/AR Applications in Training"*. Vidzeme University of Applied Sciences.

Shibanov, A.P. (2003). Finding the distribution density of the execution time of the GERT network on the basis of equivalent simplifying transformations. *Automation and Telemechanics, 2*, 117-126.

Shibanov, A. P. (2003). *Generalized GERT-networks for modeling protocols, algorithms and programs of telecommunication systems*. Ryazan.

Shmelova, T. (2013). Scientific and methodological basis of modeling decision support Air Navigation System.. Kiev: Academic Press.

Strelkov, Y. K. (2004). *Psychology of work, engineering and professional psychology*. Moscow: Academic Press.

Tanner, Jadbabaie, & Pappas. (2007). Flocking in fixed and switching networks. *Transactions on Automatic Control, 5*, 863–868.

Tyunin, V., & Yakimov, V. (2004). Theory and practice of flight training. Kiev: Academic Press.

Vitryak, A. M. (2008). *Reliability of professional activity of flight and dispatching staff on international airways*. Moscow: Kirovograd.

Wentzel, E.S. (1999). *Theory of Probability*. Moscow: Vysshaya shkola.

Wentzel, E. S., & Ovcharov, L. A. (2003). *Theory of probability and its engineering applications*. Moscow: Academy.

Witten, I., & Frank, E. (2005). Data Mining. In *Practical Machine Learning Tools and Techniques* (2nd ed.). Elsevier.

Zadorozhny, V. D. (2011). *The methodology of flight training*. Ulyanovsk.

Zyryanov, A. A., & Dorrer, M. G. (2012). Algorithm for translating the model of business processes into the GERT-network model. Bulletin of KSAU, 12, 13-18.

Zyryanov, A.A. & Dorrer. M.G. (2012). Forecast of the dynamics of event-driven models of business processes based on GERT-networks. *Informatization and Communication, 7*, 124-127.

KEY TERMS AND DEFINITIONS

Aeronautical System: A complex human-machine system that, through the use of special technical means, ensures the organization of air traffic with safe, regular, and efficient aeronautical services.

Aviation Accidents: Occurrence associated with the operation of an aircraft, which takes place between the time any person boards the aircraft with the intention of flight until all such persons have disembarked, where a person is fatally or seriously injured, the aircraft sustains damage or structural failure, or the aircraft is missing or is completely inaccessible. If the aircraft is destroyed or severely damaged so that it must be written off, it is further defined as a hull loss accident.

GERT-Network: A stochastic network that allows you to simulate an accident both in the direction of deterioration, and in the direction of improving the flight situation. GERT is an alternative probabilistic network planning method used in case of organization of activities, when subsequent actions can begin after the completion of only a certain number of previous actions, so it allows cycles and loops to be present.

Operating Procedure: A set of elementary actions that an aeronautical system operator must perform in accordance with regulatory documents in the event of a special case in flight.

Personal Anxiety: The propensity to perceive, in practice, all situations as threatening, and to react to these situations with a state of intense anxiety. High personal anxiety can be the reason for the disorientation of the human operator in space and time.

Posteriori Analysis: Performed after an accident has already occurred. The purpose of this analysis is to develop recommendations for the future.

Chapter 16
Applications of the Elementary Theory of Catastrophes in Aviation

Olha Bondar
Kirovograd Flight Academy of the National Aviation University, Ukraine

ABSTRACT

In this chapter, the author considers the connection between the mathematical theory of catastrophes and certain characteristics of the aviation system. The basis for this connection is the similarity of the concept of "catastrophe" in aviation and mathematics. With small perturbations of some systems, their state changes little. At small perturbations of other systems, a sharp transition occurs to another state. This transition is called a jump or a catastrophe. A catastrophe can also be called a loss of stability. Comparison of the notion of "stability" in aviation and mathematics is part of this study. The mathematical representation of the simplest aviation systems and their disasters is shown in elementary examples. It does not use a deep mathematical apparatus as well as special studies of aviation systems. The aim of the work is to direct the aviation specialist's eye to the mathematical theory of catastrophes, and to direct the mathematician's attention to the problems of aviation safety.

BACKGROUND

The mathematical theory of catastrophes has been accepted into analytics' standard toolbox in ship stability, fluid geometry, optics, thermodynamics, and other applications. The problem is that among these applications there is practically no research of automated systems in aviation. It is known that any important aviation system must have a serious mathematical apparatus. Therefore, the study of its functioning should contain mathematical methods. But aviation specialists, as a rule, do not have deep mathematical knowledge. As a rule, mathematicians are not familiar with the basics of the functioning of aviation systems. Therefore, it makes sense to consider the proposed theory. One of its scientific fields can be the analysis of air disasters or in other words aviation catastrophes.

The air disaster or catastrophe is an event that occurred during the flight operation of an aircraft. This event leads to severe consequences for persons associated with exploitation, and destruction or damage

DOI: 10.4018/978-1-5225-7709-6.ch016

to the aircraft itself. In other words, the air disaster is a significant change in the state of people and aircraft in the process of their standard functioning (Yakunina, 2017). In this sense, the definitions of the catastrophe are similar in aviation and in the mathematical theory of catastrophes. Catastrophes in mathematics are called spasmodic changes in the state of a mathematical system that arise as a sudden response of the system to a smooth change in external conditions.

First of all, we made an attempt to show the relationship between the concept of "stability" in aviation (Pashkovsky, 2018). and mathematics. We also showed the simplest examples of the application in aviation of the elementary concepts of catastrophe theory. At the same time, we do not penetrate deeply into any of these interesting areas of human activity. The aim of the work is to acquaint specialists in the field of aviation systems with some concepts of the mathematical theory of catastrophes. At the same time, we want to acquaint researchers of the theory of catastrophes with the problems of aviation safety.

The theory of catastrophes, as a separate section of research, appeared in the second half of the 20th century. Its sources were (Arnold, 1990, Golubitsky & Guillemin, 1973, Milnor, 1963) the theory of singularities of smooth mappings and the theory of bifurcations of dynamical systems. The theory of singularities of smooth mappings is a generalization of the investigation of functions to a maximum and a minimum. The theory of singularities of smooth mappings has found applications in the theory of stability of motion of dynamical systems, the theory of equilibrium position bifurcations, geometric and wave optics. Then this theory, along with the applications, was called the theory of catastrophes.

Among the works on catastrophe theory (Poston & Stewart, 1978, Brocker & Lander, 1975, Zeeman, 1976) there are studies of ship stability, fluid geometry, optics, thermodynamics, modeling of brain activity and so on.

Terminology and Scheme of Application of the Theory of Catastrophes

As in any new theory, the mathematical theory of catastrophes has not yet established a single terminology. It is important that as many aviation experts as possible get to know this theory. Therefore, it makes sense to bring here some competing terminology systems (Poston & Stewart, 1978). Their terms are used by researchers in applications, depending on various needs.

In catastrophes theory, the mapping is considered:

$$f : R^n \times R^r \rightarrow R$$

Elements of this mapping can be called (in various combinations) according to the terminology table 1. Other terms are sometimes used.

The scheme of most applications of the theory of catastrophes is as follows (Arnold, 1990):

- The process being studied is described with the help of a number of control and internal parameters.
- The states of equilibrium of the process form the surface of a particular number of dimensions in this space.
- The projection of the surface of equilibrium on the plane of the control parameters can have singularities that are assumed to be singularities of general position.
- Then the theory prompts the geometry of "catastrophes", that is, the theory describes jumps from one state of equilibrium to another when the control parameters change.

Table 1. Terminology table

<table>
<tr><th>f</th><th colspan="2">$\mathbf{R}^n$</th><th>n</th><th>$\mathbf{R}^r$</th><th>r</th></tr>
<tr><td rowspan="2">Function that controls the process in R^n</td><td rowspan="2">State space</td><td>The space of internal variables</td><td>Internal dimension</td><td>The space of external variables</td><td>External dimension</td></tr>
<tr><td>The behavioral space</td><td>The behavioral space</td><td>The control space</td><td>The number of control variables (parameters) or the dimension of control</td></tr>
<tr><td>r - parametric family of functions</td><td colspan="2">Scope of function definition</td><td>Dimension of the domain of definition of functions</td><td>Deformation space (parameters)</td><td>Deformation dimension</td></tr>
<tr><td>Deformation of the germ $F(x)=f(x,0)$</td><td colspan="2">The domain of definition of the germ</td><td>Dimension of the domain of definition of the germ</td><td>The space (parameters) of the deformation</td><td>The number of strain parameters</td></tr>
</table>

Examples of Elementary Catastrophes

The founder of the catastrophe theory is the American mathematician H. Whitney. In 1955 he published a paper "Mappings of the Plane into the Plane" (Whitney, 1955). He proved in it that there are only two kinds of "typical" (occurring in the vast majority of cases) features when mapping the plane into the plane. These singularities are preserved for small deformations of the mapping.

Thanks to the subsequent research (Poston & Stewart, 1978) of many scientists, so-called elementary catastrophes are now widely known. Elementary catastrophes are deformations of the system that describe possible bifurcations (sets of jumps of qualitative properties) that occur when the system is small deformation. These are "typical" r-parameter ($r \leq 5$) families of functions that include functions with degenerate critical points. Detailed consideration of them requires a serious mathematical apparatus. Therefore, here we confine ourselves to only some fragments of research. These fragments contain models of the state and functioning of some aviation systems. Here models are equations that describe the dependence of the specified system parameters. There is no need to provide drawings or diagrams, because the study of functions on extremum is included in the standard course of higher mathematics and, therefore, it has classic illustrations.

So, you can understand the meaning of the term catastrophe from the simplest examples. Recall that a catastrophe we understand as a jump-like change in the state of the system, which arises as a sudden response of the system to a smooth change in external conditions.

Example 1: Let the flight trajectory of an aircraft be specified in a rectangular Cartesian coordinate system by parametric equations:

$$x = \ln t \,/\, t + a\,,\ y = -t \ln t + b\,,$$

where x is the horizontal component (distance) of the flight path, y is the vertical component (height) of the flight path, $t > 0$ is the time. It is obvious that with increasing time, when the horizontal component $x \leq a$, the vertical component does not change significantly:

$$b \leq y \leq b + 1 \,/\, e\,.$$

It means, that y varies from b to an approximate value of $b + 0.37$. But for $t > 1$, when the range of the trajectory $x > a$ differs from a by a negligible amount, the height y of the trajectory will decrease sharply. For example, with the horizontal component $x \approx a + 0.23$, the vertical component it is $y \approx b - 23$. In this case, the speed dy / dx will increase indefinitely in absolute value, making a jump for $x = 1 / e + a$. Such a sharp increase in speed and a decrease in altitude can lead to an air crash.

Example 2: A visual geometric illustration of a mathematical catastrophe (in applications - aircraft crashes (Kitty, 2017).) can be observed using the example of the motion of an unmanned aerial vehicle along the so-called Cassini oval. This is a flat algebraic curve of the fourth order. Its equation in Cartesian rectangular coordinates has the form

$$(x^2 + y^2) - 2c^2(x^2 - y^2) = a^4 - c^4 .$$

If the parameter a changes slowly according to the law with time t

$$a = c(2\sqrt{2} - 1 - t(\sqrt{2} - 1)),$$

then for $0 \le t \le 1$ the curve is an oval. For $1 < t < 2$, the oval will contract along the axis Oy. At $t = 2$, it will turn into Bernoulli's lemniscate. And for $t > 2$ it falls into two ovals. We see that for a smooth small $\varepsilon > 0$, the time variation in the interval $[2 - \varepsilon, 2 + \varepsilon]$ trajectory makes a jump from the oval to the two ovals. If in this interval the parameter t (not necessarily the time) changes according to the sinusoidal law, then the curve will repeat the jumps from one oval to two and back.

These examples show that abrupt changes in the state of the system can depend both on the internal characteristics of the system itself and on the properties of the external conditions acting on the system.

Sustainability in Aviation: Stable and Unstable Balance

In order for the mathematical theory of catastrophes to serve as a tool for studying the functioning of aviation systems, it is first necessary to compare the meaning of concepts common to these areas of research. And the first of them is the concept of stability, which is found in aviation and in mathematics.

The state of the aircraft in flight, in which the forces acting on it and their moments do not cause its rotation and do not violate uniform rectilinear motion, is called the airplane equilibrium (Nikolayev, 1990).

In many elastic structures with the same external loads, several equilibrium positions are possible. For example, a gabarit freight is immersed in a cargo plane, separated by an air cushion from its elastic package, which has the form of a rectangular parallelepiped. If the upper base of the package is free of other loads, then it is in a state of equilibrium, resembling the top of the inflated bubble. If another load is placed on this base, then a “catastrophe” or “cotton” will occur at a certain moment with a smooth increase in its weight: the upper part of the package will jump from one equilibrium position to another - the convex part becomes concave. The theory of catastrophes is applicable to the study of such claps, and a similar mathematical catastrophe (geometric interpretation) can mean an aviation catastrophe - damage or loss of cargo or even an aircraft.

The number of equilibrium positions can be not only an integer. For example, a set of equilibrium states can be represented by segments of lines, regions on a plane, and manifolds.

Example 3**:** Let the take-off area of an aircraft in a mountainous terrain represents a surface schematically represented by a graph of the function given in the neighborhood (0, 0)

$$f = x^2 y\ .$$

The surface resembles a concave trough for positive y and an inverted trough for negative y, glued along the Ox axis at $y = 0$. If only the force of gravity acts on the aircraft, then the set of equilibrium states is the points of the axis Oy in the neighborhood (0, 0). Of these, only at the points of the Oy axis for $y > 0$ will the apparatus be stable if the other forces are sufficiently influenced by it. These external forces or conditions can be a small side wind, a slight displacement of the center of gravity due to the movement of cargo or passengers inside the aircraft and other factors.

After the impact of these forces, the aircraft will again return to the positive semi-axis Oy. The points of this semi-axis will thus be points of stable equilibrium. The remaining points of the set of equilibrium states will be points of unstable equilibrium, because the aircraft in them, under the influence of gravity and external disturbances, will have to fall down along the inverted trench.

These examples illustrate the definition of (Nikolayev, 1990) of stable equilibrium: the position of the aircraft, in which, despite the influence of external disturbing forces, the aircraft will be near the equilibrium position, is called stable equilibrium. If under the action of even insignificant external disturbances, the aircraft leaves far from the equilibrium position, then this equilibrium position is called unstable.

By the stability of the aircraft, its ability to independently, without the intervention of the pilot to restore the balance, broken as a result of some disturbing effect, that is, the property to establish a stable equilibrium. At the same time external to the functioning of the aircraft disturbances can be gravity and other forces acting on the aircraft, runway irregularities, gusts of wind, asymmetrical braking of the wheels, pilot control and other effects.

Thus, the stability of the movement of the aircraft means its ability to establish a stable equilibrium. And it depends (Nikolayev, 1990) on the actual flight conditions, the features of the aerodynamic configuration, the position of the center of gravity, the configuration, the operating mode of the engines, the degree of rigidity of the aircraft structure, the operation of the onboard automatic control systems, and the degree of the pilot's influence on the control mechanisms.

Stability in Mathematics: Structural Stability

The concept of stability in mathematics refers primarily to critical points of functions. The critical point of the function is stable (Arnold, Varchenko & Gusein-zade, 2014).), if the point does not disappear with a small perturbation of the function, but only slightly shifts.

Example **4:** The function $f = x^2$ has one critical point $x = 0$, because at this point the derivative is zero. This is the minimum point. We add to the function a small value εx. The new function $f = x^2 + \varepsilon x$ also has one critical point $x = -\varepsilon/2$. This is also the minimum point, the coordinate of which differs slightly from zero. Therefore, the critical point $x = 0$ of $f = x^2$ is stable. Similarly, the point $x = 0$ of the maximum of the function $f = -x^2$ is stable.

Note that the critical point $x = 0$ is non-degenerate for both functions. This means that the second derivative of each function is not zero at this point.

A function is often considered as a special case of a mapping. In the theory of singularities of differentiable mappings the stability of the mapping (function) *f* means:

- Its stability at each point, that is, all critical points of *f* are non-degenerate;
- All the critical values of the function *f* are different.

The concept of stability in catastrophe theory refers, generally speaking, to critical points of mappings of manifolds. For further discussion, we recall (e. g. Arnold, Varchenko & Gusein-zade, 2014, Poston & Stewart, 1978) the corresponding definitions.

A manifold is a multidimensional analog of a smooth curve or surface. A smooth submanifold in R^n is any subset *M* in R^m having the following properties:

1. it looks locally like a piece of R^m;
2. it is imbedded in R^n smoothly, that is, under the appropriate assumptions, has at each point a unique tangent plane.

Condition (1.) means the existence of a map (or local coordinate system) at the point $x \epsilon M$. Note that the concept of the map appeared as a designation for a fragment of the geographical map of the area. And matching cards meant binding one fragment to another. So, a map is a local diffeomorphism (a one-to-one map differentiable together with its inverse maps) $f : U \to M$, where *U* is some open subset of R^m and *f*(*U*) is a neighborhood of *x* with respect to *M*.

Standard coordinates in R^n induce local coordinates in *M*. In applications, they are often called curvilinear or generalized coordinates, and the word "parameter" is used instead of the word "coordinate".

Critical points of smooth mappings. Let there be given a differentiable (smooth) map of an *m*-dimensional manifold *M* into an *n*-dimensional manifold *N*:

$$f : M \to N .$$

Let $x_1, \ldots, x_m$ be local coordinates in a neighborhood of a point *x* on *M* and $y_1, \ldots, y_n$ in a neighborhood of *f*(*x*) on *N*. The map *f* is given in these coordinates by *n* smooth functions of *m* variables:

$$y_1 = f_1(x_1, \ldots, x_m), \ldots, y_n = f_n(x_1, \ldots, x_m) .$$

Then the point *x* is called the critical point of the map *f* if the rank of the Jacobi matrix $(\partial f_i / \partial x_j)$ at this point is not maximal, that is, it is less than the smaller of the dimensions of the manifolds *M* and *N*:

$rank(\partial f_i / \partial x_j) < \min(\dim M, \dim N)$.

In this paper we confine ourselves to the case when the manifold N is a line. Then the point x will be critical if the rank of the Jacobi matrix is zero, that is, $\partial f / \partial x_j(x) = 0$ for all $j = 1, ..., m$. A critical point is non-degenerate if the determinant of the Hessian matrix at this point is not equal to zero.

Thus, the stability of the mapping $f : M^m \to R$ at the point x_0 is equivalent to the existence in the neighborhoods of the points $x_0 \in M^m$ and $y_0 = f(x_0) \in R$ of coordinates $x_1, ..., x_n, y$ such that the map can be represented in one of the forms:

1. $y = x_1$ if the point x_0 is regular;
2. $y = x_1^2 + ... + x_k^2 - x_{k+1}^2 - ... - x_n^2$, ($k$=0,1,…,$n$), if the critical point x_0 is non-degenerate.

The concepts of "stability" and "structural stability" are found in various applications. In some applications, their values are the same, in other applications they have their own values.

For example, from the point of view of flight dynamics, a stable equilibrium point corresponds to the stable critical point $x = 0$ of the trajectory function $f = x^2$. At the same time, a stable critical point $x = 0$ of the function $f = -x^2$ corresponds to an unstable equilibrium under certain conditions. The inflection point also corresponds to an unstable equilibrium.

We recall that a regular or critical point of a function is stable if, for a small perturbation of a function, the point remains a regular or critical point of the same type, respectively. In the application to aviation systems this may mean that sufficiently small external disturbances of the system do not significantly affect its functioning. Hence it follows that the mathematical description of the dynamics of the system must have a corresponding insensitivity to perturbations, that is, stability.

Consequently, the mathematical formulation of this requirement depends on two things:

1. What kind of disturbance we allow;
2. What we are ready to be insensitive.

It is said (Poston & Stewart, 1978) that a stable map is *structurally stable* if sufficiently small changes in the conditions of the system's operation do not significantly affect its state. Therefore, in the theory of catastrophes in determining structural stability:

1. Small smooth perturbations of the corresponding family of functions are allowed;
2. Requires a differentiable equivalence of their domain of definition.

Recall (Arnold, Varchenko & Gusein-zade, 2014) that differentiable mappings $f_1 : M_1 \to N_1$ and $f_2 : M_2 \to N_2$ are said to be *differentially equivalent* if $f_2 = h \circ f_1 \circ H^{-1}$, where $h : N_1 \to N_2$ and $H : M_1 \to M_2$ are diffeomorphisms (mutually one-valued mappings differentiable together with their inverses). In the language of local coordinates, this means that one mapping can be transformed into another by means of smooth changes of variables.

In particular, the function f is structurally stable if for all sufficiently small smooth functions p the critical points of the functions f and $f + p$ are of the same type. In other words, the functions f and $f + p$ are equivalent, after a suitable transfer of the beginning.

Manifold of Catastrophe and Isotopic Functions

The state space of the aviation system can be represented as the domain of the functions that describe these states. Then the set of equilibrium positions of the system can be interpreted as the set of critical points of these functions. In other words, the vector applied at the point of the system state space indicates the rate of change of state. The point at which the vector vanishes will be the equilibrium position. The state of the system at this point does not change with time. The equilibrium surface consisting of equilibrium points is called (Poston & Stewart, 1978) the *manifold of the catastrophe*.

In examples 1, 2 and 4, manifolds of catastrophe have points: $x = 1/e + a$ in the first example, $t = 2$ in the second example, $x = 0$ or $x = -\varepsilon/2$ in the fourth example.

In Example 3, the function $f = x^2 y$ has the derivatives $\partial f / \partial x = 2xy$, $\partial f / \partial y = x^2$, which simultaneously vanish on the segment of the Oy axis that belongs to the neighborhood of the point (0, 0). This means that the manifold of catastrophe is this segment.

We note that many authors do not call a nondegenerate critical point a point of catastrophe. They explain this by the fact that the standard representation of such a point does not contain a dependence on the parameter. But we propose to characterize some critical points as points of catastrophes. These are the points for which a smooth transition through critical values, the level surface "jump" changes the topological type. A dependence on the parameter is visible from their representation by isotopic functions (see below). Thus, it makes sense to call nondegenerate critical points the points of catastrophes.

We recall the statement of H. Whitney, who showed (Whitney, 1955) that under the mapping of a plane onto a plane, only two kinds of singularities are stable ("typical"), the others are destroyed under a small perturbation. By singularities are understood the manifolds of catastrophe - a set of critical points: points of minima and (or) maxima and (or) excesses of the function.

In a neighborhood of a point that is not a position of equilibrium, the family of level surfaces (the state of the system is characterized by a constant) can be transformed by a smooth substitution of coordinates into a family of parallel lines or planes (Arnold, 1990).

The change of coordinates is often viewed as the result of a transition from one coordinate system to another. It is said that one function is converted into an equivalent function. But in many real physical systems, a certain continuous transformation must correspond to the replacement of coordinates. It must represent not only the result, but also the process of transition of one state of the system to another. In the process of this transformation, it is important that a stable map at a point has a similar characteristic or feature for close mappings. This means that in a neighborhood of a point the initial and final equivalent mappings must be written by similar formulas.

This process can be analytically viewed using isotopic functions introduced in (Bondar, 2015).

The functions f_0 and f_1 on a manifold M are called isotopic (C^r-isotopic or continuously isotopic or isotopic Morse functions) if there exist an isotopy H fixed on X:

$$H : (M, X) \times I \to (M, X) \times I,\ H_t \in Iso_0(M, X),$$

and an isotopy

$$h : [0,1] \times I \to [0,1] \times I \, , \; h_t \in Iso_0^+([0,1]) \, ,$$

such that for all t∈[0,1]

$$f_t = h_t \circ f_0 \circ H_t^{-1}$$

functions of the same form (C^r -smooth or continuous or Morse functions). In this paper, *X* is assumed to be an empty set. More precisely, let $f_0 : R^n \to R$ be a differentiable function and $x_0 = (x_0^1, \ldots, x_0^n)$ be a nondegenerate critical point of this function.

Then it is possible to indicate coordinate isotopy representations

$$H : U_0 \times [0,k] \to U_k \times [0,k], \; k \in N \, ,$$

neighborhood U_0 of the point x_0 on some neighborhood U_k of the origin 0 of the space R^n and isotopy

$$h : V_0 \times [0,k] \to V_k \times [0,k]$$

neighborhood V_0 of the point $f_0(x_0)$ on the neighborhood V_k of the origin 0 of the space *R* such that the differentiable mappings

$$H_k \subset Iso_0(U_k) \, , \; H_0 = id_{U_k} \, ,$$

$$h_k \subset Iso_0^+(V_k) \, , \; h_0 = id_{V_k} \, ,$$

and for all points $y = (y^1, \ldots, y^n) \in U_k$, for which $y^i(x_0) = 0$, *i*=1,…,*n*, the function

$$f_k = -(y^1)^2 - \ldots - (y^\lambda)^2 + (y^{\lambda+1})^2 + \ldots + (y^n)^2$$

will be a locally isotopic function f_0 :

$$f_k = h_k \circ f_0 \circ H_k^{-1} \, ,$$

that is, it is possible to specify local isotopic transformations of coordinate systems such that the function f_0 is locally differentiable isotopic to the function f_k .

Examples of the Simplest Isotopic Transformations

The local graph of a function of two variables can most clearly be represented in the form of a terrain relief. Let us consider the simplest aircraft, to which only gravity and elementary control act *f*. For example, the control may be the flight altitude, which is specified by the function from the horizontal component x of the flight path.

Example 5: Let $f = x^2 + a$, (a >0) in a neighborhood of zero. As mentioned above, the manifold of catastrophe is the critical point $x = 0$. For simplicity, we take $a = 0$ and move the control $f = x^2$:

$$f_0 = x^2 + \varepsilon x .$$

The following isotope transformations

$$f_0(x) = \{z : z = x^2 + \varepsilon x\},$$

$$H_t(x) = \{u_t : u_t = x + \varepsilon t / 2\},$$

$$h_t(z) = \{s_t : s_t = z - \varepsilon^2 t(t-2) / 4\},$$

$$f_t(u_t) = \{s_t : s_t = u_t^2 + \varepsilon u_t(1-t)\}$$

bring the function f_0 to the equivalent function $f_1 = u_1^2$. The flight dynamics is schematically displayed during the transformation of coordinates. The point $x = 0$ of stable equilibrium passes to the point $u_1 = \varepsilon / 2$, which is not an equilibrium point. The aircraft flying at this point under the action of gravity will move to the point $u_1 = 0$ of stable equilibrium:

$$(x = 0, z = 0) \Leftrightarrow (u_1 = \varepsilon / 2,\ s_1 = \varepsilon^2 / 4) \Rightarrow (u_1 = 0, s_1 = 0).$$

Example 6: Let $f = -x^2$. The manifold of catastrophe is also a critical point $x = 0$. Let us move the control *f*: $f_0 = -x^2 + \varepsilon x$. The following isotope transformations

$$f = -x^2 + \varepsilon x ,$$

$$f_0(x) = \{z : z = -x^2 + \varepsilon x\},$$

$$H_t(x) = \{u_t : u_t = x - \varepsilon t / 2\},$$

$$h_t(z) = \{s_t : s_t = z - \varepsilon^2 t(2-t)/4\},$$

$$f_t(u_t) = \{s_t : s_t = -u_t^2 + \varepsilon u_t(1-t)\}$$

lead the function to the equivalent function $f_1 = -u_1^2$. The point $x = 0$ of the unstable equilibrium passes to the point $u_1 = -\varepsilon/2$, which is not an equilibrium point. The aircraft flying at this point under the action of gravity will fall, moving farther from the point of equilibrium:

$$(x = 0, z = 0) \Leftrightarrow (u_1 = -\varepsilon/2, s_1 = -\varepsilon^2/4) \Rightarrow (u_1 \to -\infty, s_1 \to -\infty).$$

In this example, the mathematical concept of a "point of catastrophe" can also denote an aviation catastrophe.

Example 7: Let the control *f* be given by a function $f = x^3$, which has a degenerate critical point x=0. This point is not a point of catastrophe, because when passing through it the level surface does not change the topological type. Small stirring function

$$f = x^3 + \varepsilon x, \varepsilon > 0,$$

turns a critical point into a regular point, and the function turns into a function $f_1 = u_1$. The corresponding isotopic transformations show this:

$$f_0(x) = \{z : z = x^3 + \varepsilon x\},$$

$$H_t(x) = \{u_t : u_t = (1-t)x + t(x^3 + \varepsilon x)\},$$

$$h_t(z) = \{s_t : s_t = z\}.$$

Thus, in a sufficiently small neighborhood of zero the isotopic transformation allows us to get rid of the critical point.

Equation of the Disturbed Motion of the Aircraft

Let us consider (Kitty, 2017, Nikolayev, 1990) a rectilinear steady-state horizontal flight of an aircraft, as its supporting motion. Let the aircraft have longitudinal static stability overload.

The aircraft is called *stable overload*, if it independently, without pilot intervention, seeks to maintain the overload of the initial flight regime and return to it after the disturbance ceases.

Longitudinal static overload resistance does not mean that the aircraft is stable, that is, it does not guarantee the return of the aircraft to the initial flight regime. We start from the simplified theory of short-period motion. Then we can assume that at the beginning of the disturbed motion the speed does not have time to change noticeably

$$\Delta V = 0 .$$

Consequently, the solution of the system of equations of the perturbed motion of an airplane can be reduced to the form

$$\Delta\alpha = ce^{-nt}\sin(\nu t + \phi) ,$$

where $\Delta\alpha$ is the increment of the angle of attack, the parameter c is the amplitude of the oscillations with respect to the angle of attack at the initial time, φ is the angle determined by the initial conditions, n is the damping degree of the longitudinal oscillations (damping coefficient), and ν is the circular frequency of the natural oscillations. These parameters depend on the perturbing action on the aircraft, the design features of the aircraft, the degree of static stability, the altitude and speed of flight, the density of the air environment. The larger the value of n is, the more vibrations attenuate, the greater the stability of the supporting motion the aircraft possesses.

From the point of view of catastrophe theory, the map $\Delta\alpha$ depends on five parameters: c, n, ν, t and φ. The choice of control parameters depends on the features of the functioning of the aviation system and the purpose of the study.

Let, for example, the state of a system be described by one variable, the time t. Then the remaining 4 parameters must belong to the control space R^4. Mapping

$$\Delta\alpha : R \times R^4 \to R$$

is a function of t, depending on 4 parameters:

$$\Delta\alpha_{c,n,\nu,\phi}(t) = \{f : f = ce^{-nt}\sin(\nu t + \phi)\} .$$

The critical points of this function are $t = \nu \cdot arctg(\nu t + \phi) \; / \; n + m\pi$, where $m = 0, 1, 2, \ldots$ In them, $\Delta\alpha$ has local minima and maxima. These points constitute the manifold of catastrophe. When passing through them, the direction of the rate of change of the function $\Delta\alpha$ is reversed.

Together with the time t, an important characteristic of the system is the damping coefficient n. The larger it is, the smaller the amplitude of the oscillations of the function $\Delta\alpha_{c,\, n,\, \nu,\, \varphi}(t)$.. Therefore, it makes sense to consider the map $\Delta\alpha$ as a function of two variables:

$$\Delta\alpha : R^2 \times R^3 \to R ,$$

$$\Delta\alpha_{c,\nu,\phi}(n,t) = \{f : f = ce^{-nt}\sin(\nu t + \phi)\} .$$

The partial derivatives of this function with respect to n and t do not simultaneously vanish. This means that there are no critical points for the function. Note that the function will not have critical points when any other set of parameters is included in the state space along with the time. Thus, one can get rid of the manifold of catastrophes by expanding the state space and carrying out isotopic transformations, if necessary.

FUTURE RESEARCH DIRECTIONS

Numerous features, bifurcations and catastrophes (jumps) arise in all problems of finding extremums (maxima, minima), optimization problems, control and decision making. Therefore, the prospect of further research is, on the one hand, the expansion of the field of applications of catastrophe theory to aviation systems. This will allow aviation specialists to apply the mathematical theory of catastrophes to the study of aviation systems. On the other hand, this will allow mathematicians to see applications in aviation and, therefore, will expand the mathematical apparatus of these studies. In particular, we plan to improve the chapter flow by adding connection of formulae with computational methods, algorithms and estimations of time complexity to check their applicability in real-time restrictions.

CONCLUSION

The phrase "catastrophe theory" is an important concept in mathematics and in aviation systems. But in each of these areas of human activity the theory of catastrophes has its own definitions, its components, its connections and component dependencies, its methods of study. In this paper, the author tried to give an idea of the elementary mathematical theory of catastrophes and to show some examples of its application to aviation systems. He made an attempt to find a connection between the mathematical theory of catastrophes and certain characteristics of the aviation system.

Thus, the definition of a catastrophe in aviation and mathematics was stated;

- Concepts of stability, stable and unstable equilibrium are considered;
- Examples of structural stability and varieties of catastrophe in aviation are indicated;
- A description of the mathematical apparatus (isotopic functions, isotopic transformations), connected with the varieties of the catastrophe, has been stated;
- Variants of the equation of the disturbed motion of the aircraft from the point of view of catastrophe theory are considered.

Joint research of mathematicians and aviation specialists should be directed, first of all, to the problems of aviation transport safety.

REFERENCES

Arnold, V. (1990). *The theory of catastrophes*. Moscow: Nauka.

Arnold, V., Varchenko, A., & Gusein-zade, S. (2014). *Singularities of differentiable mappings*. Moscow: Nauka.

Bondar, O. (2015). On the determination of isotopic functions. *International conference "Geometry in Odesa-2015"*, 67.

Brocker, T., & Lander, L. (1975). *Differentiable Germs and Catastrophes*. Cambridge, UK: Cambridge University Press. doi:10.1017/CBO9781107325418

Golubitsky, M., & Guillemin, V. (1973). *Stable Mappings and their Singularities*. Berlin: Springer. doi:10.1007/978-1-4615-7904-5

Kitty, M. (2017). Dynamics of take-off and landing. Moscow: Allert.

Milnor, J. (1963). Morse Theory. *Annals of Mathematics Studies*, 51.

Nikolayev, L. (1990). Aerodynamics and dynamics of the flight of transport aircraft. Moscow. *Transport*, 392.

Pashkovsky, I. (2018). Stability and controllability of the aircraft. Moscow: Anton.

Poston, T., & Stewart, I. (1978). *Catastrophe theory and its applications*. London: Pitman.

Whitney, H. (1955). On singularities of Mappings of Euclidean Spaces I. Mappings of the Plane into the Plane. *Annals of Mathematics*, *62*(3), 374–410. doi:10.2307/1970070

Yakunina I. (2017). Analysis methods of operator's activity of the aeronautical system in special cases of flight for reducing the decision-making time. *Kropyvnytskyi, 120.*

Zeeman, E. (1976). Catastrophe theory. *Scientific American*, *234*(4), 65–83. doi:10.1038cientificamerican0476-65

Chapter 17
Use of Big Data in Aviation:
New Opportunities, Use Cases, and Solutions

Roman Odarchenko
National Aviation University, Ukraine

Zohaib Hassan
IQRA National University, Pakistan

Abnash Zaman
IQRA National University, Pakistan

ABSTRACT

The expansion of data and its efficient handling is becoming a more popular tendency in recent times bringing new difficulties to learn new avenues. Data analytics can be done more proficiently with the availability of distributed architecture of not only SQL (NoSQL) databases. Technological advancements around us are changing very rapidly, and major shift is being carried out, a switch from relational to non-relational world. When moving from relational to non-relational models, database administrators face common problems due to the fact that NoSQL is a no-schema database. The purpose of conducting this research is to propose a mechanism by which the schema of a relational database management system and its data can be transformed into big data by following some standardize guidelines. This model can be quite useful for relational database administrators by enabling them to give attention to logical modeling rather than procedural writing for each and every SQL to NoSQL transition.

INTRODUCTION

Establishing communication with the aircraft is a priority task, which depends on the work of the administrative center (the so-called back-end airline) and the operation of departments directly related to customers (front-end). This includes improving the level of security, improving operational efficiency and, as a result, increasing revenue for the carrier. Already today, the analysis of the data is predicted to

DOI: 10.4018/978-1-5225-7709-6.ch017

be the main factor that can change the organization of airline operations, and there are several reasons for this. Firstly, with the help of Big Data, ground preparations are faster due to a reduction in the processing cycle time (turnaround time). Secondly, data analysis offers the best solutions in the field of airspace management, which is the driving force for increasing efficiency. Thirdly, data analysis allows airlines to find an individual approach to each passenger. This is an opportunity for an organized re-registration of passengers during the cancellation of the flight and a direct way to increase customer loyalty to the brand.

In the air transport industry, huge amounts of data are circulated daily - from flights, tariffs and transactions to information about existing and potential customers. They change so quickly that it is impossible to process them in the traditional way. For these purposes, airlines around the world resort to the so-called Big Data technologies. They allow you to do something that a person can not do. Computer algorithms are quickly sorted out with a huge amount of chaotic data. Their analysis allows you to optimize the company's business processes. In particular, Big Data technologies help to integrate the internal systems of the air carrier with the airport system, get real-time weather information, predict future aircraft failures. In addition, "large data" provides information on each client, which allows the airline to find an individual approach to the passenger, as well as conduct targeted marketing campaigns that increase customer loyalty to the brand.

Air carriers use Big Data technologies for different purposes. For example, the low-cost Ryanair applies them for targeted advertising. Airlines such as KLM or SWISS use "large data" to improve the quality of customer service. British Airways collects information about passengers using its own application, which uses to personalize the service. This works as follows: if the client reported his allergy, flight attendants will take this fact into account on all the following flights. If the passenger has raised his level in the loyalty program, it will also be known about this, for example, a glass of champagne can be offered to him. And if the client flies for the first time in business class, the aircraft personnel will receive the relevant information and will do everything to ensure that the passenger wants to repeat this experience. American airline Delta has created for its customers an application that allows you to track luggage. And Turkish Airlines with the help of Big Data will help passengers navigate the new Istanbul airport, which will open in 2018. In addition, if the passenger parked his car in the parking lot before departure, the special application of the airline remembers the location.

CIVIL AIRCRAFT BIG DATA PLATFORM

Large plane platforms are shown in Figure 1. During the flight, the flight data will move according to civil air craft in real time transferring large data platforms to the Civil Aircraft platform or through an open-ended data link. Statistics include aircraft position data and aircraft health monitoring data. After logging in, data from all parameter data will be sent offline. In real-time flight data is analyzed by Data Center and the plane's health decision's resultssent back. The database sends data recovery data and recovery recommendations to the airport maintenance center, which results in the maintenance of data center. To provide the aircraft's real-time tracing the aircraft flight data and the airport's operational data is used by data centerand provide advice for flight settings.

Real-time aircraft monitoring using the Big Data platform - SAP Predictive Maintenance. Maintenance is the repair, inspection or modification of an aircraft or its component. Aircraft maintenance is regulated to ensure safe and correct flight operation of the aircraft. National regulations are coordinated

Figure 1. Scenario of civil aircraft big data platform

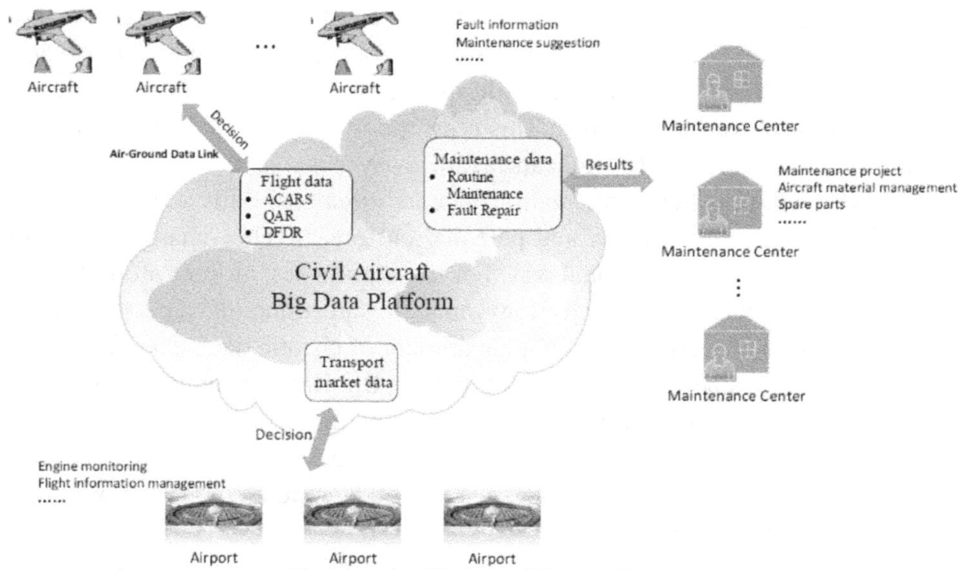

in line with international standards, maintained by bodies such as the International Civil Aviation Organization (ICAO) (ICAO, 1999).

Predictive Maintenance Technique is designed to help determine the loss of an aircraft equipment and to predict when maintenance is to be done. The key to predictive maintenance is "correct information at the right time" In the above Figure 2, with the help of the SAP Predictive Maintenance program, the Big Data from an aircraft in flight is analyzed in real time. Figure 2 gives us the following information:

Figure 2. SAP predictive maintenance

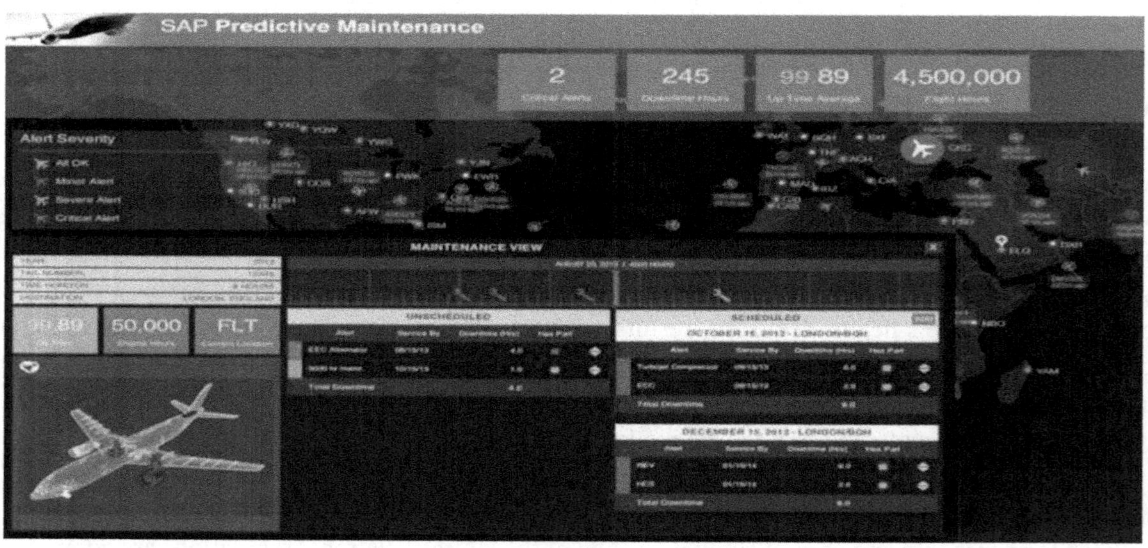

the aircraft have to make in October the scheduled maintenance for the engine compressor, also we can see that the repair time is 6 hours, and the replacement part is in stock. As unscheduled maintenance, the airplane has to replace an engine electrical component - EEC, the repair time is 4 hours, but the piece is not yet in stock. As well as unscheduled maintenance, the aircraft has to perform the necessary maintenance after 5,000 hours of flight; this operation can be seen it will take an hour.

REAL-TIME MONITORING OF AIRCRAFT PROPULSION SYSTEMS (ENGINES)

Air transport is necessary for modern life. Maintenance of aircraft propulsion systems is costly, requiring a very careful approach. Propulsion systems are equipped with sensors that monitor the engine's operating conditions during the flight. In the Figure 3 below are a few sensors that fit an aircraft engine (Jin Chen and others, 2016).

By combining data from these Big Data Sensors with advanced analyzes, it is possible to monitor the behavior of propulsion systems in real time, so that we can predict the remaining useful life of a component belonging to the engine so we can schedule maintenance in a timely manner to avoid unnecessary mechanical malfunctions, but also a very high cost generated by defective maintenance (Maja Harfman Todorovic and others, 2016). This monitoring system includes data entry, data storage, data processing and analysis, all of this is essential for creating a predictive maintenance solution. Although this example is customized for aircraft engine monitoring, the solution can be easily generalized for other predictive maintenance scenarios such as landing gear, structural integrity of the aircraft, etc. By reducing downtime and ensuring effective engine performance, this solution allows the aircraft fleet to remain operational in the most cost effective way. If an aircraft is in flight, poor maintenance of propulsion systems can have very serious consequences. The best case may be to land the aircraft after a failure in the propulsion system, in a safe condition, until the malfunction is remedied. This involves problems that can be particularly costly, given that the aircraft is in a location where there are no suitable technicians or spare parts.

Figure 3. Simplified view of some of the sensors

AVIATION PHM SYSTEM FRAMEWORK

The aviation PHM system research framework based on PHM big data center is shown in Figure 1, which describes its basic implementation framework and shows the functional requirements of aviation PHM system. PHM center mainly gathers all kinds of information from aviation participants such as aircraft, airports, spare parts warehouse, repair factory, maintenance training, overhaul base, accident rescue, etc., which builds a big aviation database. Based on the cloud computing server, it carries out data mining to obtain various experience and knowledge about failure which are needed in PHM. Then, it provides decision supports for aviation participants, government or industries. And ultimately, the PHM functions of independent monitoring, health management and emergency response are realized with the efficient linkage of aviation participants (Guigang Zhang and others, 2015; Bofei Zhou, 2014). In details, PHM center gets flight data from the aircrafts, obtains aviation operation data from the airports, collects spare parts supply information from the spare parts warehouses, acquires maintenance information and experience from repair factories, maintenance training points and overhaul bases, and receives accident disposal tracking information from the accident rescue sites. Then, the PHM Center carries on the comprehensive analysis of the information, digs into the relevance of data to get the statistical rules of failure rate and the experience related to fault. Based on the analysis results obtained, PHM center can provide the technology supports of state monitoring and fault prediction for aircraft, the aircraft health status and operation suggestions for the airport, allocation suggestions for the spare parts warehouse, fault diagnosis and maintenance recommendations for repair factory and overhaul base, personnel training suggestions for maintenance training, and aviation accident emergency response programmer for accident rescue. Moreover, it can provide suggestions on the development of aviation for both the government and industry. PHM center provides integrated services for the aviation participants based on big data mining, so as to realize the efficient use of information and improve the operation efficiency of the aviation industry chain. For example (Figure 4 and Figure 5), it can make the airport directly obtain the aircrafts' health status and assign flight missions reasonably to the aircrafts to improve the aircraft utilization rate. It can make the reasonable choice for repair or overhaul according to the health state of aircraft. And it can also make the spare parts warehouse prepare materials ahead of the aircraft arrival, so as to shorten the maintenance period and send the security available aircraft in time to return to the airport (Anandavel Murugan Chandramohan and others, 2014).

BACKGROUND

For RDBMS, the architecture that is typically followed is based on a client-server architecture through which the server manages database application tasks. This architecture proved to be the best and it is highly optimized. However, as data increases, RDBMS does not provide the best results, especially for-reading process. Log-based and fast growing databases require other types of databases that are sufficient for high-speed reading operations. NoSQL is the answer to this situation. The NoSQL database proves to be the most suitable for those situations and provides a lot of control over the scheduling and high availability through distributed architecture by distributing data in different nodes (Nitnaware, 2014).

We have studied various NoSQL databases and, based on this, for readingoprationsHBaseis best suited (Abramova and others, 2014).

Figure 4. Aviation PHM system research framework based on PHM Big Data center

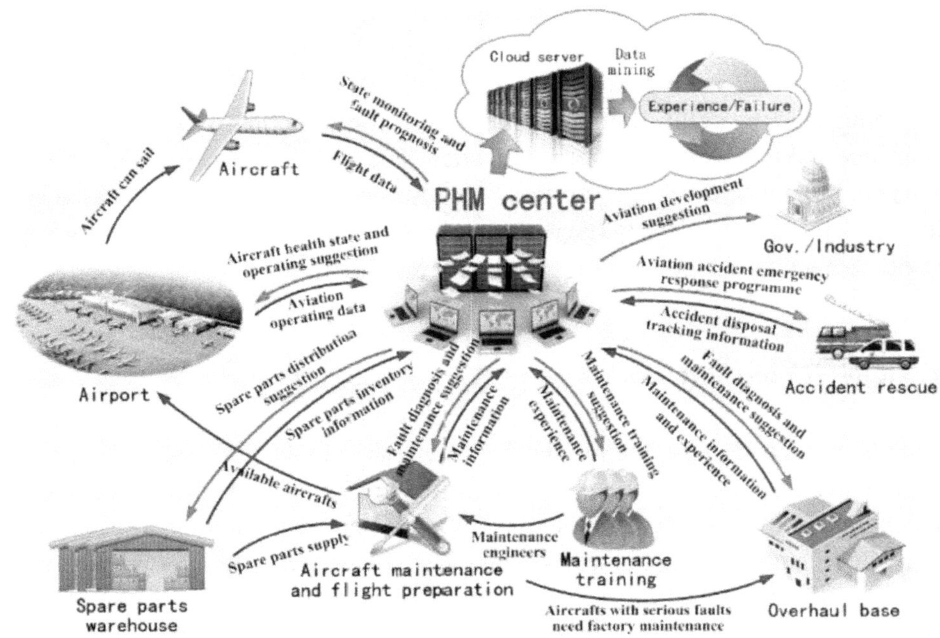

Figure 5. The key technologies, scientific problems and application systems of aviation PHM system

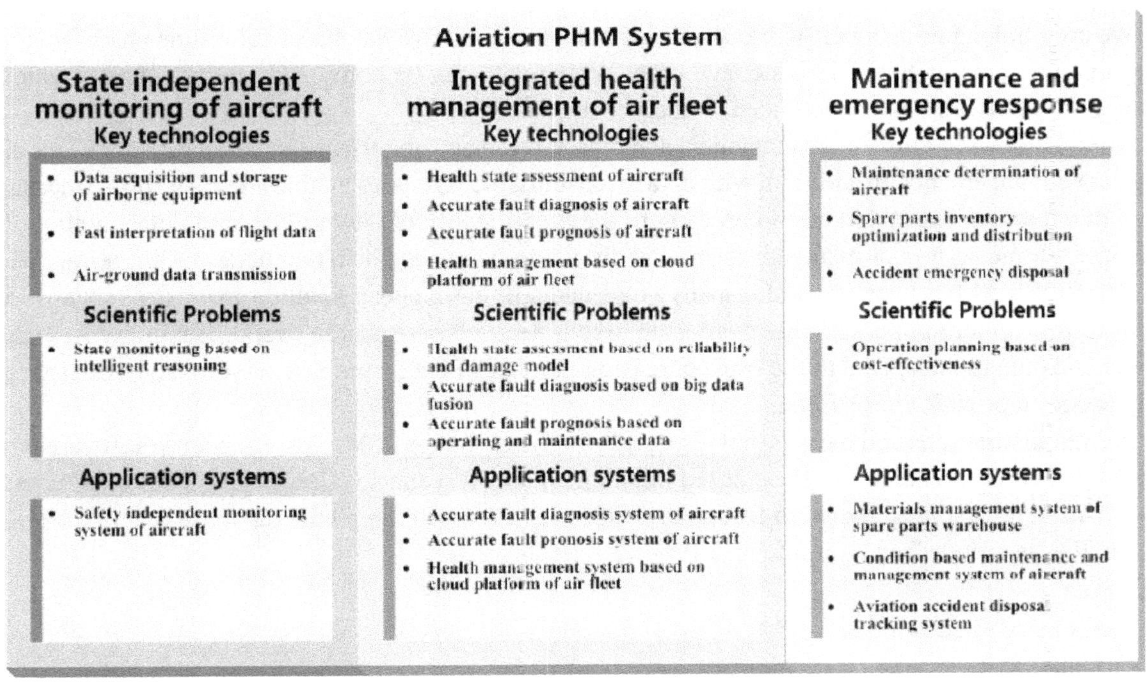

Considering the HBase table design, there are basically two ways. The first is a "high-narrow" design that can be used for informal collections with a large number of tables but fewer columns. The second method is "FlatWide" design, which can take the issue that the number of table columns is large, but the number of rows is small (Lee, 2014).

While developing schema architecture for relative databases, another method will be based on log files to maintain operation, configuration, modification, and question method. For tracking database wide proceduresandto monitor the databaseusually the log files are used. In addition, the answer to this question is that the database can identify the recording process. So, checking the log files, we can track the tables that often reach by the processing of quires (Hsu, 2014).

INTRODUCTION TO NoSQL

There is four points of interest in the NoSQL database compared to the traditional compatible database: It is easy to expand, high efficiency, availability is high and present flexible data model (Sharma, 2012). Previous research on this topic is very common and does not create a combination of actual data from any particular related database management system (RDBMS) product or any real-time system (Liu, 2015). Therefore, the uniqueness of this research is that structured query language or SQL is not only one of the finest attributes of RDBMS. When we talk about mass of information, we are talking about the greater impact of SQL on NoSQL (Chang, 2008). NoSQL has been mentioned as the alternative of the traditional RDBMS model using SQL, which is used by industry leaders such as Oracle, DB2, and MS-SQL (Jung, 2015). On the other side, we now have SSL models and products., such asApache Spark, Hadoop, Taren tool, Cassandra, MongoDB, Cassandra, and. Hbase (Sharma, 2015).

We are implementing a real-world application that uses relative databases for online processing and use large amounts of RDBMS generated data. Which data can be converted into Big data to gain its benefits? Specifically, when you need to analyze the data?

NoSQL allows application management and reduce the neeed ofsoftware for database schema modification or enhancements. In addition, with data extensions, NoSQL is a scalable and easy-to-use measurement in database. These databases can capture the functionality of new groups and nodes without the database administrators or manual information distribution between different nodes (Abramova, 2014).

The RDBMM has been in use for many years and still has a cheap solution for many usage issues. However, for modern applications, there is an urgent need to increase the overall flexibility in scaling model and data models. Due to the ever-increasing number of data, complex server resources are regularly added to provide more users..

Present instructions and models, but also goals for the exchange of Microsoft-specific software products.SQL Server to Hadoop H-Base based upon a real-world scenario (Lee, 2015). The NoSQL terms and ideas were first used by Carlo Strozziin 1998, which refer to the media database that do not relay on the SQL user interface (Abramova, 2014).

Figure 6. Cap theorem and NoSQL databases

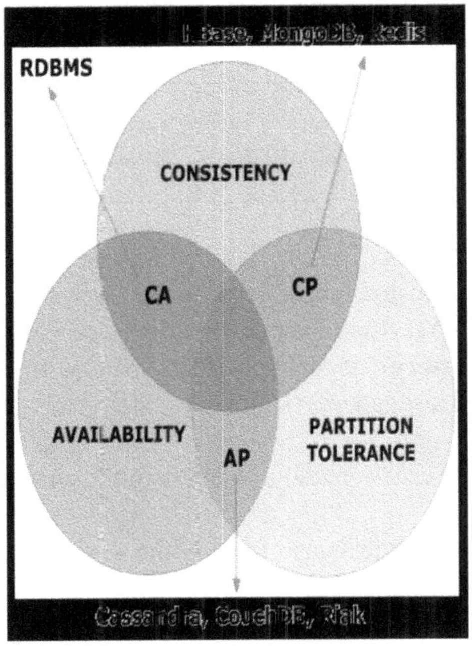

CAP THEORAM

When it comes to related and incomplete relative databases, the two terms are basically talking about them there.

- ACID Properties (Relational Databases)
- BASE Properties (Non-Relational Databases)

ACID SOLUTION stands for Atomicity, Persistence, Isolation and Durability (Figure 6). This is the basic principle behind the RDMBM database, mainly for transactions. Transactions can be explained to work properly on data. The ACID features the key to ensuring the integrity of the information (Brewer, 2000).

BASE stands for Basically Obtainable, Stable state, eventually consistent. It utilizes split tolerance and emphasizes onScalability and availability, is a good option to use simple and fast to use, where availability and rate of data are very concerned.

THE HBASE HUSH DATABASE

We studied one of the HUSH databases, both SQL and NoSQL versions are provided schemes (George, 2011). HUSH can be used to configure a specific set of data tables that have a balanced data set.This feature helps to simply understand what is provided for the transformation and how to perform a specific

Table 1. Dataset database tables

S.No.	Table Name	Row Count	Size (MB)
1	pp.production	1.728,460	81
2	pp. packing	1, 276537	176
3	pp. pieces	1,037,953	160
4	pp. audit log	780,041	50

process. Therefore, evaluating the Hush database helps our research clearly understand the schema model of NoSQL and SQL databases. RAD (Rapid application development): It

General Guidelines is the necessary data for applications focused on the market and large data models, so conversions can be performed fast and efficiently (Lee, 2015).

- **Scalability:** Is definitely the user-demand to meet with the constantly growing throughput of the data and also to access it.
- **Consistent Performance:** Low response time is the key to success when dealing with large amounts of data, and is critical to its growth.
- **Operational Reliability:** Built-in high availability.

DATABASE GROWTH

Table 1 Specify the development of database tables per day in the same transaction, and just enter large tables. When we calculate the growth during this time, a very large amount of tera byte data is generated. This data needs analysis for reading operation to get data deeper knowledge and data analysisThe actual use case shows that once the transaction is prepared from the RDBMS system, the data only needs to be read. Therefore, based on the analysis, we choose HBase as the non-relational model, and use our proposed program to plan the large table SQL schema to NoSQL mode.

HBASE TABLE SCHEMA

From a good perspective, the HBase model consists of four main components (Vora, 2011).

- A Table is a collection of rows and columns family collection.
- Colom Family is a logical physical group of columns related to basic storage products.
- Compared to the traditional RDBMS model, the columns are very different in HBase. Columns exist only when data is inserted, and can have multiple variants based on timestamps that are automatically generated during data insertion in this way, for a row key, there may be multiple versions based on timestamps, and only the last available version cannot be questioned unless any other version is clearly answered. Each row on the table may have a different set of different columns set by its row keys.

- The row key is a clear starting key that clearly identifies the record. This is an important order and therefore provides effective question processing.

```
Table {
        "Row-Key1": {//Row Key
                "ColumnFamily1": {//Column Family
                        "Column-Key1": {Value1} //Column Qualifier
                                "Column-Key2": {Value2}
                                        ...
                                }
                        }
                }
```

Note: HBase Column & Column Family Schema [1]

RELATIONSHIP TYPE

One-to-One

For the conversion of a one-to-one relationship, there are two possible ways to convert it to an HBase table.

- **Single HTable:** The first way is to collect two tables in an HTable. Later HT can have two generations.
  1. Single Column Family.
  2. Multiple or Two Columns families.
- **Two or Multiple HTables:** The second possible sequence is to create two HTables, each of which is a column family, and includes all SQL columns in each column's family. Finally, enter every HTable row key in two HTables.

One-to-Many

In one-to-Many relationship conversion, there are two possible ways to convert it into HBase.

- **Single HTable:** The first way is to collect two tables in an HTable. Later HT can have two generations
  1. Multiple or Two Columns families
  2. Single Column Family
- **Two or Multiple HTables:** Each SQL table is included in two separate HTables with a separate column family. The main HTable original relationship is included and the second column series, in which column keys are referred to in the second HTable.

EXPERIMENT AND DESIGN CONSIDERATIONS

Provided, the sample table [pp. Production] of RDBMS, experiment is done in which Table 1 is used on the database listed in Table 2. According to Table 3, the four sampling sample shows the data generated during the process. When converting to HBase model, this RDBMS table model can be used by using our

Step 1:SQL Server table columns are marked by method.Add metadata information 'description'. In a Table 4 single table only those columns which requiredto be DATASET: The database table is imported and the map is mounted on the HBase table described in "Description" meta data.

Step 2: Our script checks for the metadata information based on a SQL query run on the information metadata of a SQL Server database. The {Pu, 2016 #12}t script picks only those columns and generates a list.

Step 3: As explained in Table 3, based on that, we take the Row#1 and Row#5 for the same machine that metadata information, we generate the CREATE generates different outputs during the production scheduling TABLE HBASE scripts for those SQL server table based on the shifts. This helps developers and administrators to have pre-generated scripts available for them to run on the HadoopHBase environment

WORKLOADS OVERVIEW

Table 4 shows the workloads on base of which we recommend performing performance tests and our future research work.

Table 3 is a mapping of the structure of Table 4 from SQL Server, which can be compared with the RDBMS query time and the HBase table. Here we explain how to use single-column and multi-column series and timestamp columns.

UNIQUESNESS ON THIS STUDY

There is very small number of database schema conversions between non-relational and relational databases. And a few examples of smart conversion models to help you get started with such programs (Lee, 2015). This situation is common to every developer and database administrator during the transition

Table 2. Metadata of a SQL server database/mapping of RDBMS table to hbase table

Row#	Machine ID	ShiftProduction	SpeedRPM	Timestamp
1	10	1,650	850	20160324
2	24	1,354	850	20160324
3	105	985	650	20160324
4	240	835	650	20160324
1	24	1,430	850	20160325

Table 3. Workloads Overview

Workload	Description
Workload A	Loads data from RDBMS to NoSQL
Workload B	100% read operation
Workload C	Short Ranges – Intervals of data

Table 4. SQL Server can compare it with RDBMS

Row Key	Value (Column Family, Column, Version and Cell)
1	info: {'machine make': 'Macpro'}{Chen, 2016 #8} prod: {'machineid':'10','Macpro':'1650'@ts='20160324','Macpro': '1430'@ts=' 20160325'}
2	info: {'machine make': 'Jet'} prod: {'machine id': '24', 'Macpro ': '1354'}{Sukhum, 2016 #6}

from a relational model to a non-relational model (Gajendran, 2012). It prompted us to study a more systematic approach to mobility. Migration can be challenging and trivial in two things; architecture translation and data transformation. The simplest export and import issues cannot solve the problem between the same data store because you need to know what and why it was imported.

PROPOSED APPROACH

Rules will follow by experts that are why migration can be trivial result. The purpose of this study is to make the change process easier and standardized. In the following sections, we demonstrate the use of rules and perform experiments that can be verified by the NoSQL community.

Studies and experiences on the database are based on 04 out of 400 computerized machines, which have been developed about 0.1 million models per day. Each unit can be ranked in sub-models of average 100 key units. It concludes that approximately 0.1 * 100 records per day occur from the same production environment. You will have to increase such big data by taking advantage of parallel and scalable data. There is an important success in creating an alert based on the decision.

- There are two basic goals for a standard schema change:
- Reduce database administrators' work, and
- Set conversion requirements to achieve a smooth transition from SQL-Server tables to HBase tables.

SQL SERVER TABLE SCHEMA AND METADATA

We first standardize the SQL Server table metadata to implement the conversion process (Figure 7). The metadataof table contains several fields (computed value, datatype, column name, description of column, default value)

- Each of our models is based on metadata information stored in the Description field.
- In any SQL server table, to keep the name we set the ‘Description’ column which is required to be generalized to the HBase column.
- Each of our pattern maps is shown in a field that contains meta-data column information and only those fields are selected.

PROBLEM STATEMENT AND MOTIVATION

There are currently some examples of the database schema conversion between the relational and non-relational databases present in the literature, and there are few examples of computerized conversion

Figure 7. SQL server column metadata settings

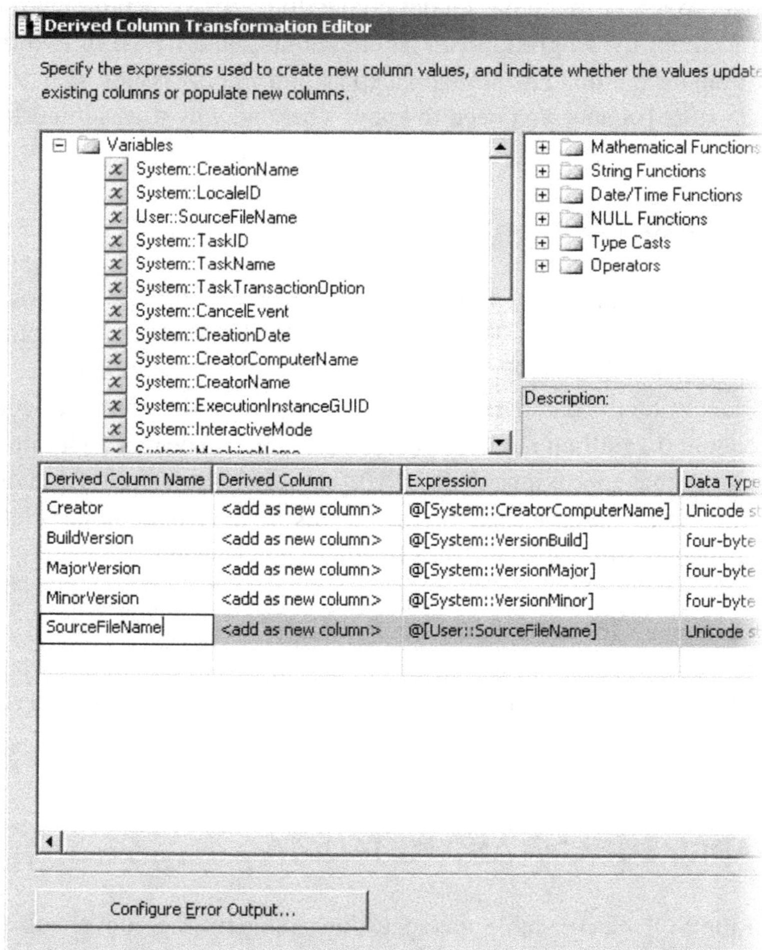

devices, even if anything that can help you get started (Lee, 2015). The subject of a big data is very broad and many investigations have been made, but there are some ways to find it with ongoing research.

This research consists two different ways of novelty:

- Specific RDBMS platforms transformation is targeted in it, i. e. Microsoft SQL server.
- A specific business domain's a real-time data that operates automated machines is targeted. These types of machines generate large amounts of data during the production cycle.

After organizing literature survey, we found that such research and experiments were not made for specific industries like textile industry. Collect data later; perform business statistics selection by performing statistical analysis. The size of the data running 24 hours the actual machine unit can be checked (unless any strike machine is due to deterioration or inadequate power).

CONCLUSION AND FUTURE WORK

In this study, we looked at the problem of converting SQL to NoSQL, providing a common solution that can be used further in real-world scenarios. Our preliminary results show that this approach has great potential to convert large amounts of data from SQL Server to big data based on schema conversion, which is easier when using our methods (Figures 8 and 9).

We want to increase the research so that the general data can be converted into large data structure and can create a high-scale analysis solution, especially to create a large amount of data for traffic manufacturing industry,

Figure 8. Clouderahadoop environment – configured

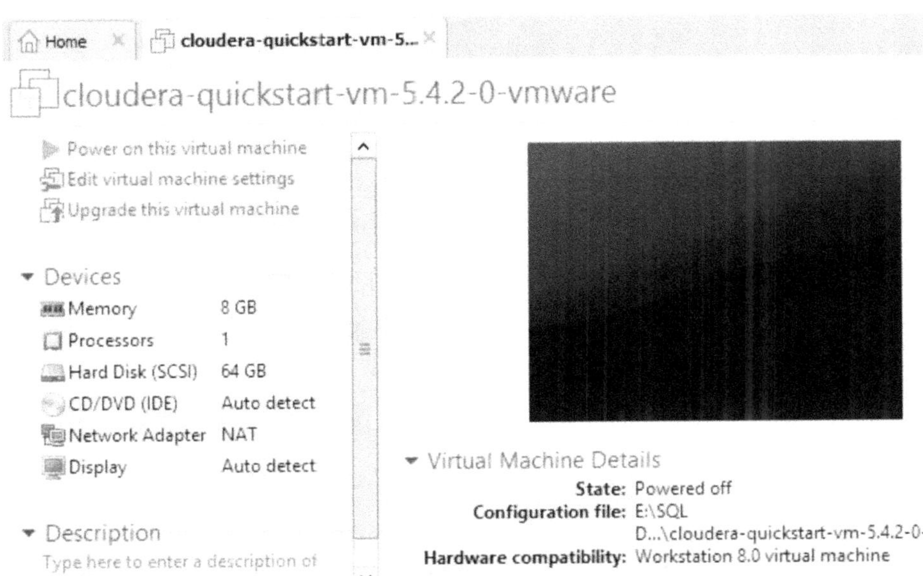

Figure 9. Transforming structures

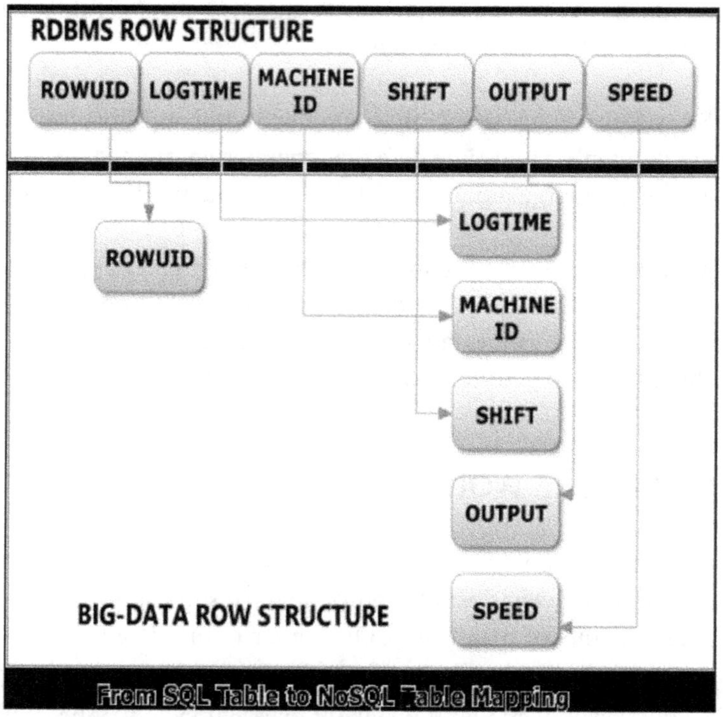

FUTURE RESEARCH DIRECTIONS

As a future work, this research can be extended to create a business solution that may be associated with a relational database engine so that data can be converted into big data very effectively. Industry, which can manufacture products on automatic machines around the world, can benefit from this solution.

REFERENCES

Abramova, V., Bernardino, J., & Furtado, P. (2014). Experimental evaluation of NoSQL databases. *International Journal of Database Management Systems*, *6*(3), 2014. doi:10.5121/ijdms.2014.6301

Apache HBaseProject. (n.d.). Retrieved from http://hbase.apache.org

Binani, S., Gutti, A., & Upadhyay, S. (2016). SQL vs. NoSQL vs. NewSQL-A Comparative Study. *Database*, 36–39.

Borthakur, D. (2007). *The hadoop distributed file system: Architecture and design*. Hadoop Project Website.

Brewer, E. A. (2000). *Towards robust distributed systems* (Vol. 7). PODC.

Chandramohan, Mylaraswamy, Xu, & Dietrich. (2014). Big Data Infrastructure for Aviation Data Analytics. *2014 IEEE International Conference on Cloud Computing in Emerging Markets (CCEM)*, 1-6.

Chang, F., Dean, J., Ghemawat, S., Hsieh, W. C., Wallach, D. A., Burrows, M., & Gruber, R. E. (2008). Bigtable: A distributed storage system for structured data. *ACM Transactions on Computer Systems, 26*(2), 2008. doi:10.1145/1365815.1365816

Chen, J., Zhi, Lv., Liu, Y., Huang, J., Zhang, G., Wang, J., & Chen, X. (2016). A Big Data Analysis and Application Platform for Civil Aircraft Health Management. *2016 IEEE Second International Conference on Multimedia Big Data (BigMM)*, 404-409. 10.1109/BigMM.2016.54

Gajendran, S. K. (2012). A survey on NoSQL databases. University of Illinois.

George, L. (2011). HBase: the definitive guide. O'Reilly Media, Inc.

Hsu, J. C., Hsu, C. H., Chen, S. C., & Chung, Y. C. (2014). Correlation Aware Technique for SQL to NoSQL Transformation. *Ubi-MediaComputing and Workshops (UMEDIA), 2014 7th International Conference*, 43-46. 10.1109/U-MEDIA.2014.27

International Civil Aviation Organization (ICAO). (1999). *Manual of technical provisions for the aeronautical telecommunication network (ATN)*. Retrieved from https://www.icao.int/safety/acp/repository/_%20Doc9705_ed2_1999.pdf

Jung, M. G., You, S. A., Bae, J., & Choi, Y. L. (2015). A Study on Data Input and Output Performance Comparison of MongoDB and PostgreSQL in the Big Data Environment. *2015 8th International Conference on Database Theory and Application (DTA)*, 14-17. 10.1109/DTA.2015.14

Lee, C. H., & Zheng, Y. L. (2015). SQL-to-NoSQL Schema Denormalization and Migration: A Study on Content Management Systems. *Systems, Man, and Cybernetics (SMC), 2015 IEEE International Conference*, 2022-2026.

Lee, C. H., & Zheng, Y. L. (2015). Automatic SQL-to-NoSQL schema transformation over the MySQL and HBase databases. *Consumer Electronics-Taiwan (ICCE-TW), 2015 IEEE International Conference*, 426-427.

Liao, Y. T., Zhou, J., Lu, C. H., Chen, S. C., Hsu, C. H., Chen, W., & Chung, Y. C. (2016). Data adapter for querying and transformation between SQL and NoSQLdatabase. *Future Generation Computer Systems*, 17–23.

Liu, C., Fu, Z., Yang, Z., & Xiu, J. (2015). General Research on Database Migration from RD{Pu, 2016 #12}BMS to Hbase. In *2015 International Symposium on Computers & Informatics*. Atlantis Press.

Nitnaware, C., & Khan, A. (2015). A multidimensional data storage model for location based application on Hbase. *Innovations in Information, Embedded and Communication Systems (ICIIECS). 2015 International Conference*, 1-5. 10.1109/ICIIECS.2015.7193237

Sharma, S., Tim, U. S., Gadia, S., Wong, J., Shandilya, R., & Peddoju, S. K. (2015). Classification and comparison of NoSQL big data models. *International Journal of Big Data Intelligence, 2*(3), 201–221. doi:10.1504/IJBDI.2015.070602

Sharma, V., & Dave, M. (2012). SQL and NoSQL Databases. *International Journal of Advanced Research in Computer Science and Software Engineering, 2*(8), 2012.

Todorovic, M. H., Datta, R., Stevanovic, L., She, X., Cioffi, P., Mandrusiak, G., . . . Frangieh, T. (2016). Design and Testing of a Modular SiC based Power Block. *PCIM Europe 2016 International Exhibition and Conference for Power Electronics Intelligent Motion Renewable Energy and Energy Management*, 1-4.

Vora, M. N. (2011). *Hadoop-HBase for large-scale data. In Computer science and network technology (ICCSNT), 2011 international conference* (Vol. 1, pp. 601–605). IEEE. doi:10.1109/ICCSNT.2011.6182030

Wang, Y., Xu, Y., Liu, Y., Chen, J., & Hu, S. (2015). QMapper for Smart Grid: Migrating SQL-based Application to Hive. *2015 ACM SIGMOD International Conference on Management of Data*, 647-658. 10.1145/2723372.2742792

Xu, B., & Kumar, S. A. (2015). Big Data Analytics Framework for System Health Monitoring. *2015 IEEE International Congress on Big Data*, 401-408. 10.1109/BigDataCongress.2015.66

Zhang, G., Wang, J., Zhi, Lv., Yang, Y., Su, H., Yao, Q., ... Huang, J. (2015). A integrated vehicle health management framework for aircraft - A preliminary report. *2015 IEEE Conference on Prognostics and Health Management (PHM)*, 1-8. 10.1109/ICPHM.2015.7245034

Zhou, B. (2014). *China Eastern Airlines Big Data platform.* Available: https://www.behance.net/gallery/34149479/china-eastern-airlines-big-data-platform

KEY TERMS AND DEFINITIONS

ACID: Atomicity, consistency, isolation, durability.

BASE: Basically available, soft-state, eventual consistent.

Big Data: A term used to refer to the study and applications of data sets that are too complex for traditional data-processing application software to adequately deal with.

CAP: Consistency, availability, partition tolerance.

Database: An organized collection of data stored and accessed electronically.

HBase: An open-source, non-relational, distributed database modeled after Google's Bigtable and written in Java.

HDFS: Hadoop distributed file system.

HUSH: HBase URL shorten.

Metadata: Is data that provides information about other data.

NoSQL: Not only structured query language.

PHM: Prognostics and health monitoring.

RDBMS: Relational database management systems.

SQL: Structured query language.

Structured Query Language: A domain-specific language used in programming and designed for managing data held in a relational database management system (RDBMS), or for stream processing in a relational data stream management system (RDSMS).

Compilation of References

Abramova, V., Bernardino, J., & Furtado, P. (2014). Experimental evaluation of NoSQL databases. *International Journal of Database Management Systems*, *6*(3), 2014. doi:10.5121/ijdms.2014.6301

Adrianova, V. E. (1974). Human activity in management systems. Leningrad: Academic Press.

Air Pilot Manual. (2011). *Human Factors and pilot performance*. Pooleys, England Alderson.

Akdeniz, A. (2001). The impact of mandated aging airplane programs on jet transport airplane scheduled structural inspection programs. *Aircraft Engineering and Aerospace Technology*, *73*(1), 4–16. doi:10.1108/00022660110366827

Aleven, V. (2010). Rule-Based Cognitive Modeling for Intelligent Systems. Advances in Intelligent Tutoring Systems, 33-62. doi:10.1007/978-3-642-14363-2_3

Aleven, V., Nkambou, R., Bourdeau, J. & Mizoguchi, R. (2010). Rule-Based Cognitive Modeling for Intelligent Systems. *Advances in Intelligent Tutoring Systems*, 33-62.

Aleven, V., McLaren, B., Roll, I., & Koedinger, K. (2016). Toward Meta-cognitive Tutoring: A model of Help-Seeking with a Cognitive Tutor. *International Journal of Artificial Intelligence in Education*, *16*, 101–130.

Alexander, D., & Merkert, R. (2017). Challenges to domestic air freight in Australia: Evaluating air traffic markets with gravity modeling. *Journal of Air Transport Management*, *61*, 41–52. doi:10.1016/j.jairtraman.2016.11.008

Alfredson, J., & Ohlander, U. (2016). System Characteristics and Contextual Constraints for Future Fighter Decision Support. *International Journal of Information System Modeling and Design*, *7*(1), 1–17. doi:10.4018/IJISMD.2016010101

Amalberti, R. (2000, September). The paradox of the ultra-safe systems. *Flight Safety Australia*.

Aman, S., & Szpakowicz, S. (2008). Using Roget's Thesaurus for Fine-Grained Emotion Recognition. *Proceedings of the 3rd International Joint Conference on Natural Language Processing*, 296–302.

Amarosicz, M., Psiuk, K., Rogala, T., & Rzydzik, S. (2016). Diagnostic Shell Expert Systems. *Diagnostica*, *17*(1), 33–40.

Ambrosia, V., & Zajkowski, T. (2015). Selection of Appropriate Class UAS/Sensors to Support Fire Monitoring: Experiences in the United States. In Handbook of Unmanned Aerial Vehicles (pp. 2723–2754). Springer Netherlands.

Amelin, K., Amelina, N., Granichin, O., Granichina, O., & Andrievsky, B. (2013). Randomized Algorithm for UAVs Group Flight Optimization. In *11th IFAC International Workshop on Adaptation and Learning in Control and Signal Processing*. University of Caen Basse-Normandie. 10 3182/20130703-3-FR-4038.00130

Amin, M., & Shebl, D. (2014). Reasoning dynamic fuzzy systems based on adaptive fuzzy higher order Petri nets. *Information Sciences*, *286*, 161–172. doi:10.1016/j.ins.2014.07.011

Anderson, H. (1982). *Predicting Wind Driven Wildland Fire Size and Shape.* Research Paper INT-305. USDA Forest Service, Intermountain Forest and Range Experiment Station.

Anderson, E. W. (1994). Cross-category variation in customer satisfaction and retention. *Marketing Letters, 5*(1), 19–30. doi:10.1007/BF00993955

Anderson, E. W., & Sullivan, M. (1993). The antecedents and consequences of customer satisfaction for firms. *Science, 12*(2), 125–143.

Anderson, J. R., Corbett, A. T., Koedinger, K. R., & Pelletier, R. (1995). Cognitive Tutors: Lessons Learned. *Journal of the Learning Sciences, 4*(2), 167–207. doi:10.120715327809jls0402_2

Antonio, C. (2010). *Virtual Room: a case study in the training of pilots.* Cape Canaveral: HCI Aeroconference.

Antonio, C. (2011). Human-centred design in aviation. *Proceedings of the Fourth Workshop on Human Centered Processes.*

Antonio, C. (2011). Resilience and Ergonomics in Aviation. In *Proceedings of the fourth Resilience Engineering Symposium.* Mines ParisTech.

Apache HBaseProject. (n.d.). Retrieved from http://hbase.apache.org

Approach and Landing Accident Reduction (ALAR). (n.d.). Retrieved from http://flightsafety.org/current-safety-initiatives/approach-and-landing-accident-reduction-alar

ArcGIS 3D analiz, eğitim dökümanı. (2008). İşlem şirketler qrubu, Ankara.

ARINC. (2004). *Mark 33 Digital Information Transfer System (DITS), Part 1: Functional Description, Electrical Interface, Label Assignments and Word Formats, ARINC Specification 429 Part 1-17.* Annapolis: ARINC.

ARINC. (2009). *Aircraft Data Network, Part 7: Avionics Full-Duplex Switched Ethernet Network (AFDX).* ARINC Industry Activities.

Arkhangelskii, V. I., Boganenko, I. N., Grabovsky, G. G., & Ryumshin, N. A. (2000). *Human Automation Systems: Quality, Safety and Reliability Management.* Kiev: KIA.

Arkhangelskiy, V., Bohaenko, I., Grabowskiy, G., & Ryumshyn, N. (1999). *Neural Networks in Systems of Automation.* Kiev: Techniques.

Arnold, V. (1990). *The theory of catastrophes.* Moscow: Nauka.

Arnold, V., Varchenko, A., & Gusein-zade, S. (2014). *Singularities of differentiable mappings.* Moscow: Nauka.

Arroyo, I., Woolf, B. P., Burelson, W., Muldner, K., Rai, D., & Tai, M. (2014). A Multimedia Adaptive Tutoring System for Mathematics that Addresses Cognition, Metacognition and Affect. *International Journal of Artificial Intelligence in Education, 24*(4), 387–426. doi:10.100740593-014-0023-y

Artemenko, O. (2010). *Modeling of the automated pre-flight information preparation system* (Doctoral dissertation). Kiev Institute of Automatic, Ukraine.

Artemenko, O., & Shmeleva, T. (2005). Modeling the decision-making process for departure in the automated pre-flight information preparation system. *Artificial Intelligence, 4*, 441–446.

Astapov, V. M. (2008). Anxiety and anxiety. Moscow: Litagent "Kogito-Center."

ATCsimulator. (n.d.). Retrieved from http://www.atcsimulator.com/

Atkin, J., Burke, E., & Greenwood, J. (2011). A comparison of two methods for reducing take-off delay at London Heathrow airport. *Journal of Scheduling*, *14*(5), 409–421. doi:10.100710951-011-0228-y

ATTower® ATC Tower Simulation. (n.d.). Retrieved from http://www.ufainc.com/products/attowertowsim-tower-simulation/

Austin, R. (2010). *Unmanned Aircraft Systems*. West Sussex, UK: Wiley. doi:10.1002/9780470664797

Autodesk 3ds-max visualization, simulation program. (2016). Retrieved from https://www.autodesk.com/products/3ds-max/overview

Aviation Accident Statistics. (2018). *National Transportation Safety Board*. Retrieved from www.ntsb.gov/aviation/aviation.htm

Aviation Accident Statistics. (2018). *National Transportation Safety Board*. Retrieved from www.ntsb.gov/aviation/aviation.htm

Babaev, N. (1967). Calculation and design of elements of gyroscopic devices. *Engineering*, 480.

Bagnara, S., & Pozzi, S. (2008). Fondamenti, Storia e Tendenze dell'HCI. In A. Soro (Ed.), *Human Computer Interaction. Fondamenti e Prospettive*. Monza, Italy: Polimetrica International Scientific Publisher.

Bainbridge. (1987). The Ironies of Automation. New Technology and Human Error.

Balakrishnan, K., & Honavar, V. (1998). Intelligent Diagnosis Systems. *Journal of Intelligent Systems*, *8*(3), 237–290.

Balasubramanian, A. (2017). *Digital elevation model (DEM) in GIS. Technical Report*. University of Mysore.

Ballard, D. H. (1981). Generalizing the Hough transform to detect arbitrary shapes. *Pattern Recognition*, *13*(2), 111–122. doi:10.1016/0031-3203(81)90009-1

BAO Tower – New Version. (n.d.). Retrieved from http://www.airliners.net/aviation-forums/aviation_hobby/read.main/73050/

Baolin, F., Lu, L., Lu, X., & Zhu, K. (2017). An Optimized Cooperative Control Algorithm of Multi-Agent Array Based on the Skeletal Mechanism of Chrysopelea. *Proceedings of the 2017 IEEE International Conference on Robotics and Biomimetics*, 2003 – 2008.

Barnes, C., Elliott, L.R., Coovert, M.D., & Harville, D. (2004). Effects of Fatigue on Simulation-based Team Decision Making Performance. Ergometrika, 4.

Basic principles on human factors in the user manual Audit Aviation Safety. (2010). Doc. ICAO 9806.

Bayramov, A. A., & Hashimov, E. G. (2018). Assessment of invisible areas and military objects in mountainous terrain. *Defence Science Journal*, *68*(4), 343–346. doi:10.14429/dsj.68.11623

Bayramov, A. A., Hashimov, E. G., & Khalilov, B. M. (2017). GIS technology and terrain orthophotomap making for military application. *Journal of Defense Resources Management.*, *8*(2), 81–90.

Bernardini, P. (2017). First data from the DAMPE space mission. *Nuclear And Particle Physics Proceedings*, *291-293*, 59–65. doi:10.1016/j.nuclphysbps.2017.06.014

Bervery, P. W. (2009). *Building Intelligent Interactive Tutors*. Burlington, MA: Morgan Kaufmann.

Bessing, S. B., Gilbert, S. B., Ourada, S., & Ritter, S. (2009). Authoring Model-Tracing Cognitive Tutors. *International Journal of Artificial Intelligence in Education, IOS Press*, *19*, 189–210.

Bhagwat, A. (2015). Knowledge Based Service Diagnosis System. *International Journal of Computer Science and Technology*, *3*(5), 182–184.

Bijuraj, L. V. (2013). Clustering and its Applications. *Proceedings of National Conference on New Horizons in IT – NCNHIT 2013*, 169-172.

Binani, S., Gutti, A., & Upadhyay, S. (2016). SQL vs. NoSQL vs. NewSQL-A Comparative Study. *Database*, 36–39.

Biswas, G., Segedy, J. R., & Bunchongchit, K. (2016). From Design to Implementation to Practice a Learning by Teaching System: Betty's Brain. *Journal of Artificial Intelligence in Education*, *26*(1), 350–364. doi:10.100740593-015-0057-9

Bodrov, V. A., & Lukyanova, N. F. (2007). Personal characteristics of pilots and professional efficiency. *Psychological foundations of professional activity*. Moscow. *Logos*, 423–430.

Bondar, O. (2015). On the determination of isotopic functions. *International conference "Geometry in Odesa-2015"*, 67.

Borovikov, V. P. (2008). *Neural Networks. STATISTICA Neural Networks: Methodology and Technology of Modern Data Processing* (2nd ed.). Moscow: Hotline-Telecom.

Borthakur, D. (2007). *The hadoop distributed file system: Architecture and design*. Hadoop Project Website.

Bouabdallah, S., & Siegwart, R. (2005). Backstepping and Sliding-mode Techniques Applied to an Indoor Micro Quadrotor. *IEEE International Conference on Robotics and Automation*, 2259 – 2264. 10.1109/ROBOT.2005.1570447

Bourdil, P.-A., Berthomieu, B., & Jenn, E. (2014). Model-Checking Real-Time Properties of an AutoFlight Control System Function. *Proceedings of the 2014 IEEE International Symposium on Software Reliability Engineering Workshops (ISSREW'14)*, 120–123. 10.1109/ISSREW.2014.40

Boy, G. (2011). The Handbook of human Machine Interface – A Humancentered design approach. Ashgate.

Bradski, G., & Kaehler, A. (2008). *Learning OpenCV: Computer vision with the OpenCV library*. O'Reilly Media, Inc.

Breen, J. (2011). *Twitter Sentiment Analysis Tutorial*. Retrieved from https://github.com/jeffreybreen/twitter-sentiment-analysis-tutorial-201107/tree/master/data/opinion-lexicon-English

Brewer, E. A. (2000). *Towards robust distributed systems* (Vol. 7). PODC.

Brocker, T., & Lander, L. (1975). *Differentiable Germs and Catastrophes*. Cambridge, UK: Cambridge University Press. doi:10.1017/CBO9781107325418

Buromensky, N. G. (2016). Survivability of Military Communication Systems. *Military Thought*, *25*(3), 117–121. doi:10.21557/MTH.47912151

Butin, V. V. (2013). Remote manned flying objects as Remote Sensing data source. *Geomamics*, *3*, 24–27.

Chandramohan, Mylaraswamy, Xu, & Dietrich. (2014). Big Data Infrastructure for Aviation Data Analytics. *2014 IEEE International Conference on Cloud Computing in Emerging Markets (CCEM)*, 1-6.

Chang, F., Dean, J., Ghemawat, S., Hsieh, W. C., Wallach, D. A., Burrows, M., & Gruber, R. E. (2008). Bigtable: A distributed storage system for structured data. *ACM Transactions on Computer Systems*, *26*(2), 2008. doi:10.1145/1365815.1365816

Chaput, J. A. (2004). *Lecture Notes, Conceptual Design of UAV Systems*. University of Texas at Austin.

Chaput, J. A. (2010). Rapid Air System Concept Exploration - A Parametric Physics Based System Engineering Design Model. In *AIAA Aviation Technology*. Integration, and Operations Conference. doi:10.2514/6.2010-9301

Chen, J., Zhi, Lv., Liu, Y., Huang, J., Zhang, G., Wang, J., & Chen, X. (2016). A Big Data Analysis and Application Platform for Civil Aircraft Health Management. *2016 IEEE Second International Conference on Multimedia Big Data (BigMM)*, 404-409. 10.1109/BigMM.2016.54

Chi, M., VanLehn, K., & Litman, D. (2010). Do micro-level tutorial decisions matter: Applying reinforcement learning to induce pedagogical tutorial tactics. In V. Aleven, J. Kay & J. Mostow (Ed.), *Intelligent Tutoring Systems: 10th International Conference, ITS 2010* (pp. 224-233). Heidelberg, Germany: Springer. 10.1007/978-3-642-13388-6_27

Chialastri, A. (2015). *Human factor – Il rapporto uomo-macchina*. Roma: IBN.

Chukhray, A. G. (2013). *Metodology for learning algorithms*. Kharkiv, Ukraine: National Aerospace University, KhAI.

Cinar, G., Emeneth, M., & Mavris, D. N. (2016). A Methodology for Sizing and Analysis of Electric Propulsion Subsystems for Unmanned Aerial Vehicles. *54th AIAA Aerospace Sciences Meeting*. 10.2514/6.2016-0216

Clarke, K., Brass, J., & Riggan, P. (1994). A Cellular Automaton Model of Wildfire Propagation and Extinction. *Photogrammetric Engineering and Remote Sensing*, *60*, 1355–1367.

Commander and Staff Officer Guide. Army Tactics, Techniques, and Procedures ATTP 5-0.1. (2011). Headquarters Department of the Army.

Conati, C., Gertner, A., & VanLehn, K. (2002). Using Bayesian Networks to Manage Uncertainty in Student Modeling. *User Modeling and User-Adapted Interaction, Kluwer Academic Publishers*, *12*(4), 371–417. doi:10.1023/A:1021258506583

Cooper, G. E., White, M. D., & Lauber, J. K. (Eds.). (1980). Resource management on the flightdeck. *Proceedings of a NASA/Industry Workshop (NASA CP-2120).*

Dalkilic, S. (2017). Improving aircraft safety and reliability by aircraft maintenance technician training. *Engineering Failure Analysis*, *82*, 687–694. doi:10.1016/j.engfailanal.2017.06.008

Damangir, H., & Abedini, M. (2014). System identification and subsequent discharge estimation based on level data alone gradually varied flow condition. *Flow Measurement and Instrumentation*, *36*, 24–31. doi:10.1016/j.flowmeasinst.2014.01.002

Debar, H., Thomas, Y., Cuppens, F., & Cuppens-Boulahia, N. (2007). Enabling automated threat response through the use of a dynamic security policy. *Journal In Computer Virology*, *3*(3), 195–210. doi:10.100711416-007-0039-z

Decision Systems Laboratory GeNIe 2.0. (n.d.). Decision Systems Laboratory, University of Pittsburg. Retrieved from http://genie.sis.pitt.edu

Dekker, S. (2000). Sharing the Burden of Flight Deck Automation Training. *The International Journal of Aviation Psychology*, *10*(4), 317–326.

Dekker, S. (2003). *Human Factor in aviation: a natural history*. Technical Report 02. Lund University School of Aviation AF 447.

Dekker, S. (2005). *Why We need new accident models*. Lund University School of Aviation, Technical Report 2005-02.

Dergachov, K. Y., Krasnov, L.O., & Radomskyi, O.M (2017). Methods of local navigation of mobile robots based on of technical vision. *Electrotechnic and Computer Systems*, (25), 295-300.

Dinc, A. (2010). *Optimization with elitist genetic algorithm method in preliminary design of unmanned air vehicle & its propulsion system simultaneously* (PhD thesis). Graduate School of Sciences, Anadolu University. (in Turkish)

Dinc, A. (2015a). Sizing of a Turboprop Unmanned Air Vehicle and its Propulsion System. *Journal of Thermal Science and Technology, 35*(2), 53–62.

Dnipropetrovsk regional branch, Zaporizhzhya air traffic service system. (2013). Operating instructions of the aerodrome control tower. *Zaporizhzhia, 91.*

Dodd, R. B. (2006). *Colored Petri Net Modeling of a Generic Avionics Mission Computer.* Technical Report DSTO-TN-0692, Air Operations Division, Defense Science and Technology Organization, Australian Government, Department of Defense.

Donner-Banzhoff, N., & Hertwig, R. (2014). Inductive foraging: Improving the diagnostic yield of primary care consultations. *The European Journal of General Practice, 20*(1), 69–73. doi:10.3109/13814788.2013.805197 PMID:23815336

Doyle, J.C. (2010). Contrasting view of Complexity and Their Implications For Network-Centric Infrastructures. *IEEE Transactions on Systems, Man, and Cybernetics. Part A, Systems and Humans, 40*(4).

Dzhuma, L. N., Pilipyonok, O. N., & Timoshenko, A. V. (2012). Analysis of the structure of Tower controller workplace and peculiarities of its modeling. *Materials of the first International scientific and practical conference Management of high-speed moving objects and professional training of operators of complex systems*, 32-33.

Dzhuma, L.N., & Pascal, O.N. (2010). Automation of labor processes in air traffic control system. *Journal of Eastern Ukrainian National University of Vladimir Dal. Scientific Journal, 9*(151), 155-160.

Dzhuma, L.N., & Pilipyonok, O.N. (2014). Dataware model improvement of the intellectual training system «Tower controller». *Scientific Proceeding of Ukrainian Research Institute of Communication. Scientific and Production Digest, 5*(33), 48-57.

Dzhuma, L.N., & Pilipyonok, O.N. (2015). Modeling of the air traffic flow in the area of Tower controller's responsibility. *Scientific Proceeding of Ukrainian Research Institute of Communication. Scientific and Production Digest, 5*(39), 93-97.

Dzhuma, L.N., Pascal, O.N., & Lishavskaya, V.L. (2011). Application of information technologies for design of Tower controller's workplace. *Journal of Easton Ukrainian National University of Vladimir Dal. Scientific Journal, 7*(161), 241-246.

Dzikovska, M., Steinhauser, N., Farrow, E., Moore, J., & Campbell, G. (2014). BEETLE II: Deep Natural Language Understanding and Automatic Feedback Generation for Intelligent Tutoring in Basic Electricity and Electronics. *International Journal of Artificial Intelligence in Education, 24*(3), 284–332. doi:10.100740593-014-0017-9

Erik, H. (2009). *The ETTO Principle – Efficiency-Thoroughness Trade-Off.* Ashgate.

European Organization for the Safety of Air Navigation (Eurocontrol). (2000). *Simulations Facilities for Air Traffic Control Training. HUM.ET1.ST07.3000-REP-02.* Brussels, Belgium: Author.

European Organization for the Safety of Air Navigation (Eurocontrol). (2004a). ATM Services' Personnel (2nd ed.). ESARR 5. Belgium, Brussels: Author.

European Organization for the Safety of Air Navigation (Eurocontrol). (2004b). *EATM Training Progression and Concepts.* Brussels, Belgium: Author.

European Organization for the Safety of Air Navigation (Eurocontrol). (2015). *Specifications for the ATCO Common Core Content Initial Training* (2nd ed.). Brussels, Belgium: Author.

Evlanov, L. (1979). Control of dynamic systems: monograph. Nauka, 132.

Fahlstrom, P. G., & Gleason, T. J. (2012). *Introduction to UAV Systems* (4th ed.). Wiley.

Federation of American Scientists. (1996). *Air Combat Command Concept of Operations for Endurance Unmanned Aerial Vehicles*. Retrieved from http://www.fas.org/irp/doddir/usaf/conops_uav/part03.htm

Ferber, J., Gutknecht, O., & Michel, F. (2004). From Agents to Organizations: An Organizational View of Multi-Agent Systems. *LNCS*, *2935*, 214–230.

Ferlazzo, F. (2005). *Metodi in ergonomia cognitiva*. Roma: Carocci.

Filaretov, V., Yukhimets, D., & Bidyadhar, S. (2017). Information-Control System for Cooperative Control of a Mobile Robots Group. *International Conference on Industrial Engineering, Applications and Manufacturing (ICIEAM)*, 1 – 6. 10.1109/ICIEAM.2017.8076169

Finney, M. (1999). Mechanistic modeling of landscape fire patterns. In Spatial Modelling of Forest Landscape Change: Approaches and Applications (pp. 186–209). Cambridge University Press.

Fioriti, M., Vercella, V., & Viola, N. (2018). Cost-Estimating Model for Aircraft Maintenance. *Journal of Aircraft*, 1–12. doi:10.2514/1.c034664

Flemisch, F., Canpolat, Y., Altendorf, E., Weßel, G., Flemisch, F., … Schutte, P. (2017). Shared and cooperative control of ground and air vehicles: Introduction and General overview. *IEEE International Conference on Systems, Man, and Cybernetics (SMC)*, 858 – 863. 10.1109/SMC.2017.8122717

Flight Manual. (n.d.). Retrieved from http://www.avialogs.com/index.php/en/aircraft/europe-and-consortiums/airbus/a319-320-321/

Flight Safety Foundation. (2003). The Human Factors Implications for Flight Safety of Recent Development In the Airline Industry. In *Flight Safety Digest, March April Hawkins Frank (1987), Human factor in Flight*. Aldershot, IL: Ashgate.

FMI 3-04.155. (2006). *Army unmanned aircraft system operations headquarters*. FMI.

Franchi, A., Giordano, P. R., Secchi, C., Son, H. I., & Bulthoff, H. H. (2011). A Passivity-Based Decentralized Approach for the Bilateral Teleoperation of a Group of UAVs with Switching Topology. *IEEE International Conference on Robotics and Automation*, 898 – 905. 10.1109/ICRA.2011.5980368

Frank, J., & Thürmer, J., Lukas, & Gollwitzer Peter, M. (2015). Promoting the translation of intentions into action by implementation intentions: Behavioral effects and physiological correlates. *Frontiers in Human Neuroscience*, *9*, 395. doi:10.3389/fnhum.2015.00395 PMID:26236214

Gajendran, S. K. (2012). A survey on NoSQL databases. University of Illinois.

Gallay, M. L. (1970). *Flight of an aircraft with incomplete and asymmetrical traction*. Moscow: Mechanical Engineering.

Gardner, M. W., & Dorling, S. R. (1998). Survey and Critique of Techniques for Extracting Rules from Trained Artificial Neural Networks. *Atmospheric Environment*, *32*(14–15), 2627–2636. doi:10.1016/S1352-2310(97)00447-0

General Atomics Aeronautical Systems. (2014). *Product Brochures*. Retrieved from http://www.ga-asi.com/products/aircraft/pdf/Predator_B.pdf

George, L. (2011). HBase: the definitive guide. O'Reilly Media, Inc.

Gerasimov, B. (2007). Intelligent decision support systems. European University.

Gluhih, I. (2000). Integrated automated systems of intelligent decision support in air traffic control (theory, models, algorithms, decision-making). Samara, 34.

Goedeking, P. (2010). *Networks in Aviation: Strategies and Structures*. Springer. doi:10.1007/978-3-642-13764-8

Goldberg, Y. (2016). A Primer on Neural Network Models for Natural Language Processing. *Journal of Artificial Intelligence Research, 57*, 345–420. doi:10.1613/jair.4992

Golenko-Ginzburg, D.I. (2010). *Stochastic network models of planning and development management.* Voronezh: Scientific Book.

Golubitsky, M., & Guillemin, V. (1973). *Stable Mappings and their Singularities.* Berlin: Springer. doi:10.1007/978-1-4615-7904-5

Gong, Y., Li, J., Zhou, Y., Li, Y., Chung, H. S., Shi, Y., & Zhang, J. (2016). Genetic Learning Particle Swarm Optimization. *IEEE Transactions on Cybernetics, 46*(10), 2277–2290. doi:10.1109/TCYB.2015.2475174 PMID:26394440

Gorbunov, A.L., Eliseev, B.P., & Nechaev, E.E. (2010). Innovative approach to the creation of air traffic controller simulations on the basis of mixed reality: the need and the possibility. *Scientific Proceeding of Moscow State Technical University of Civil Aviation, 161*, 5-14.

Greer, J. (1995). *Proceedings of AI-ED 95, World Conference on Artificial Intelligence in Education.* Association for Advancement of Computing in Education (AACE).

Griga, V. M., & Dunets, R. B. (2008). Application of spatio-temporal graphs for the synthesis of specialized multipliers. *Radioelectronic Computer Systems, 7*, 113-117.

Guevara, J., Jiménez, A. R., Prieto, J. C., & Seco, F. (2012). Auto-localization algorithm for local positioning systems. *Ad Hoc Networks, 10*(6), 1090–1100. doi:10.1016/j.adhoc.2012.02.003

Gulevich, S. P., Melnikov, D. N., & Suvorov, A. P. (2016). *Mathematical modeling of UAV flight in the regime of the terrain following. In Perspectives of UAVs development and application. Pros. of Scientific Practical conference* (pp. 70–76). Kolomna: Ministry of Defense of the Russian Federation.

Gulevitch, S. P., Veselov, Y. G., Pryadkin, S. P., & Tyrnov, S. D. (2012). Analysis of factors affecting safety of flight of unmanned aerial vehicles (UAV). Causes of accidents of UAVs and methods of preventing them. Science and education Russia, Bauman Moscow State Technical University.

Gundlach, J. (2014). *Designing Unmanned Aircraft Systems: A Comprehensive Approach* (2nd ed.). AIAA. doi:10.2514/4.102615

Han, L., Cheng, X., Xu, W., & Tan, G. (2017). Reconfigurable Wireless Control System for a Dual-arm Cooperative Robotic System. *Proceedings of the 2017 IEEE International Conference on Robotics and Biomimetics*, 202 – 207. Retrieved October 09, 2018, from https://ieeexplore.ieee.org/document/8324418/

Hao, Y., Pan, S., Qiao, Y., & Cheng, D. (2018). Cooperative Control via Congestion Game Approach. *IEEE Transactions on Automatic Control, 9*. Retrieved from https://ieeexplore.ieee.org/document/8334284/

Harchenko, V., Shmelova, T. & Sikirda, U. (2012). *Decision making by the operator of the air navigation system: the monograph.* Kirovograd Flight Academy of National Aviation University.

Harchenko, V., Shmelova, T., & Sikirda, U. (2011). Graph-analytical models of decision-making by a person of an air navigation system operator. *Journal of the National Aviation University, 1*, 5–17.

Hashimov, E.G., Bayramov, A.A., Abdullayev, F. & Mammadli A. (2017). Development of the multirotor Unmanned Aerial Vehicle. *National Security and Military Sciences, 3*(4), 21-31.

Hashimov, E. G., & Bayramov, A. A. (2017). Investigation of the observation conditions on the terrain of war operation using GIS technology. *Advanced Information Systems, 1*(1), 65–69. doi:10.20998/2522-9052.2017.1.12

Hashimov, E. G., Nasibov, Ya. A., Bayramov, A. A., & Amanov, R. R. (2015). Application of the relief digital model for battle operations plannig. *Herbi Bilik.*, *4*, 63–69.

Himsolt, M. (1996). The Graphlet system (system demonstration). *Lecture Notes in Computer Science*, *1190*, 233–240. doi:10.1007/3-540-62495-3_51

Hoehndorf, R., & Queralt-Rosinach, N. (2017). Data Science and symbolic AI: Synergies, challenges and opportunities. *Data Science*, *1*(1-2), 27–38.

Hoek, W., & Wooldridge, M. (2008). Multi-Agent Systems. In *Handbook of Knowledge Representation* (pp. 1–41). Amsterdam: Elsevier. doi:10.1016/S1574-6526(07)03024-6

Hollnagel, E. (2008). Critical Information Infrastructures: should models represent structures or functions. In *Computer Safety, Reliability and Security*. Heidelberg, Germany: Springer.

Hollnagel, E., Woods, D., & Leveson, N. (2006). *Resilience Engineering – Concepts and Precepts*. Ashgate.

Honeywell. (2014). *Product Brochures*. Retrieved from http://www51.honeywell.com/aero/common/documents/myaerospacecatalog-documents/BA_brochures-documents/TPE331-10_PredatorB_0292-000.pdf

Hsu, J. C., Hsu, C. H., Chen, S. C., & Chung, Y. C. (2014). Correlation Aware Technique for SQL to NoSQL Transformation. *Ubi-MediaComputing and Workshops (UMEDIA), 2014 7th International Conference*, 43-46. 10.1109/U-MEDIA.2014.27

Huang, A. S., Bachrach, A., Henry, P., Krainin, M., Maturana, D., Fox, D., & Roy, N. (2017). Visual odometry and mapping for autonomous flight using an RGB-D camera. In *Robotics Research* (pp. 235–252). Cham: Springer. doi:10.1007/978-3-319-29363-9_14

Hu, J., & Hong, Y. (2007). Leader-following Coordination of Multi-Agent Systems with Coupling Time Delays. *Physica A*, *374*(2), 853–863. doi:10.1016/j.physa.2006.08.015

Hutchins, E. (1995). How a cockpit remembers its speeds. *Cognitive Science*, (19): 265–288.

IATA. (1994). *Aircraft Automation Report, Safety Advisory Sub-Committee and Maintenance Advisory Sub-committee.* ICAO - Human Factors Digest No. 5, Operational Implications of Automation in Advanced Technology Flight Decks (Circular 234).

ICAO. (1998). Doc. 9683-AN/950. Montreal, Canada: ICAO.

ICAO. (2013). Safety Management Manual (SMM) (3rd ed.). Doc. ICAO 9859-AN 474. Author.

ICAO. (2014). *Safety Report*. Montreal, Canada: Author.

ICAO. (2015). *Doc 9756, Manual of Aircraft Accident and Incident Investigation*. Montreal: ICAO.

ICAO. (2016). *2016-2030 Global Air Navigation Plan*. Montreal: International Civil Aviation Organization.

International Air Transport Association (IATA). (2018). *Artificial intelligence (AI) in aviation exploring: The fundamentals, threats and opportunities of AI in the aviation industry*. Montreal: Author.

International Civil Aviation Organization (ICAO). (1999). *Manual of technical provisions for the aeronautical telecommunication network (ATN)*. Retrieved from https://www.icao.int/safety/acp/repository/_%20Doc9705_ed2_1999.pdf

International Civil Aviation Organization (ICAO). (2002). Human Factors Guidelines for Safety Audits Manual (1st ed.). Doc. ICAO 9806-AN/763. Montreal: Author.

International Civil Aviation Organization (ICAO). (2004). *Cross-Cultural Factors in Aviation Safety: Human Factors Digest Nº 16. Circ. ICAO 302-AN/175.* Montreal: Author.

International Civil Aviation Organization (ICAO). (2012). *Manual on Flight and Flow Information for a Collaborative Environment. Doc. ICAO 9965-AN/483.* Montreal: Author.

International Civil Aviation Organization (ICAO). (2013a). Safety Management Manual (SMM) (3rd ed.). Doc. ICAO 9859-AN/474. Montreal: Author.

International Civil Aviation Organization (ICAO). (2013b). *State of Global Aviation Safety.* Montreal: Author.

International Civil Aviation Organization (ICAO). (2016). *Business Plan (2017-2019).* Montreal: Author.

International Civil Aviation Organization. (1998). Human Factors Training Manual (1st ed.). Doc. ICAO 9683-AN/950. Canada, Montreal: Author.

International Civil Aviation Organization. (2002). Human Factors Guidelines for Safety Audits Manual (1st ed.). Doc. ICAO 9806-AN/763. Canada, Montreal: Author.

International Civil Aviation Organization. (2005). *Global Air Traffic Management Operational Concept. Doc. ICAO 9854-AN/458.* Montreal, Canada: Author.

International Civil Aviation Organization. (2007). Air Traffic Management (15th ed.). Doc. ICAO 4444-ATM/501. Canada, Montreal: Author.

International Civil Aviation Organization. (2007). ICAO. *Environment Reporter, 2007.* Retrieved from http://www.icao.int/environmental-protection/Documents/Env_Report_07.pdf

International Civil Aviation Organization. (2008). *Threat and Error Management (TEM) in Air Traffic Control. Cir. ICAO 314-AN/178.* Montreal, Canada: Author.

International Civil Aviation Organization. (2009). *Manual on Global Performance of the Air Navigation System. Doc. ICAO 9883.* Montreal, Canada: Author.

International Civil Aviation Organization. (2012). *Manual on the Approval of Training Organizations. Doc. ICAO 9841-AN/456.* Montreal, Canada: Author.

International Civil Aviation Organization. (2016a). Manual on Air Traffic Controller Competency-based Training and Assessment (1st ed.). Doc. ICAO 10056-AN/519. Canada, Montreal: Author.

International Civil Aviation Organization. (2016b). *Procedures for Air Navigation Services – Training (PANS-TRG). Doc. ICAO 9868.* Montreal, Canada: Author.

Isermann, R. (2006). *Fault-diagnosis systems: an introduction from fault detection to fault tolerance.* Springer Science & Business Media. doi:10.1007/3-540-30368-5

Jain, M. C. (2009). *Textbook of Engineering Physics (Part I).* Delhi, India: PHI Learning Private Ltd.

James, R. (1990). *Human error.* Cambridge, UK: Cambridge University Press.

James, R. (2008). *The human contribution.* Farnham, UK: Ashgate.

Jao, C. (2011). *Efficient decision support systems.* Rijeka, Croatia: InTech.

Jensen, K., & Kristensen, L. M. (2009). *Coloured Petri Nets — Modeling and Validation of Concurrent Systems.* Springer-Verlag Berlin. doi:10.1007/b95112

Ji, W., Li, X., & Chen, J. (2013). Remote Update on Flash-Based Hydrologic Remote Terminal through GPRS Technology. *Advanced Materials Research, 846-847*, 1711-1715. Retrieved from www.scientific.net/amr.846-847.1711

Jin, X., Chen, S., Jia, Z., & Zeng, Z. (2016). The Research of Compose Method for Removal/Install Part in Aircraft Maintenance Manual. *Applied Mechanics and Materials, 835*, 828-834. Retrieved from www.scientific.net/amm.835.828

Jin-Yu, Z., Yan, C., & Xian-Xiang, H. (2009, April). Edge detection of images based on improved Sobel operator and genetic algorithms. In *Image Analysis and Signal Processing, 2009. IASP 2009. International Conference on* (pp. 31-35). IEEE.

Jung, M. G., You, S. A., Bae, J., & Choi, Y. L. (2015). A Study on Data Input and Output Performance Comparison of MongoDB and PostgreSQL in the Big Data Environment. *2015 8th International Conference on Database Theory and Application (DTA)*, 14-17. 10.1109/DTA.2015.14

Kala, R., & Warwick, K. (2012). Planning Autonomous Vehicles in the Absence of speed Lanes using Lateral Potentials. *Proceedings of the 2012 IEEE Intelligent Vehicles Symposium*, 597-602. 10.1109/IVS.2012.6232148

Karafyllidis, I., & Thanailakis, A. (1997). A model for predicting forest fire spreading using cellular automata. *Ecological Modelling*, *99*(1), 87–97. doi:10.1016/S0304-3800(96)01942-4

Karmanov, A. G. (2012). *GIS tutorial. Sankt-Peterburg Naional Reseach Universityof Information Technologies, mechanics and optics.*

Karnaukhov, V. A. (2009). *The theory of graphs and networks in the modeling of ATC processes*. Ulyanovsk: UAVU CA.

Kasak, V. (2010). *The system of methods for the integration of living entities in special situations during the flight: the monograph.* The National Aviation University.

Kharchenko, V., Chynchenko, Yu., & Raychev, S. (2007). Threat and Error Management in Air Traffic Control. *Proceedings of the National Aviation University, 3–4*, 24-29. 10.18372/2306-1472.33.1491

Kharchenko, V.P., & Chinchenko, Y.V. (2006). Automation of the process of professional selection of aviation civil aviation personnel. *Visnyk NAU, 2*, 34-36.

Kharchenko, V. P., Shmelova, T. F., & Sikirda, Yu. V. (2012). *Making decisions by the operator of aeronautical systems.* Kirovograd: CLA NAU.

Kharchenko, V., Shmelova, T., Artemenko, O., & Otryazhiy, V. (2011). *A. r. Computer Program "Choosing the Pre-Flight Information and Decision Making about Departure for Automated System of Pre-Flight Information Preparation (AS PIP)"*. Kyiv: State Department of Intellectual Property.

Kharchenko, V., Shmelova, T., Sikirda, Yu., & Gerasimenko, O. (2011). *A. r. Computer Program "Optimizing the Choice of Alternative Variant of Aircraft Flight Completion in Flight Emergencies "Prompt"*. Kyiv: State Department of Intellectual Property.

Kitty, M. (2017). Dynamics of take-off and landing. Moscow: Allert.

Kolodziej, K. W., & Hjelm, J. (2006). *Local positioning systems: LBS applications and services.* CRC Press. doi:10.1201/9781420005004

Kolonder, J. (1993). *Case-Based Reasoning*. Morgan Kaufmann.

Komashynskiy, V., & Smirnov, D. (2002). *Neural Networks and Its Use in Systems of Management and Communication.* Moscow: Hotline-Telecom.

Koo, T. J., & Shahruz, S. M. (2001). Formation of a Group of Unmanned Aerial Vehicles (UAVs). *Proceedings of the American Control Conference Arlington*, 69-74. Retrieved October 09, 2018, from https://ieeexplore.ieee.org/document/945515/

Koonce, J. M. (2002). *Human Factors in the Training of Pilots*. London: Taylor & Francis.

Коростелев, А. (2016). Five-step preparation of the pilot of civil aviation in a system of continuous training. *Proceedings Of The National Aviation University*, *6*(1-2). doi:10.18372/2306-1472.6.9721

Kordon, F. (2016). Modeling and verification of critical components in airplane systems: Model of a landing detector. *Model Checking Contest*. Retrieved from https://mcc.lip6.fr/index.php

Korn, G. A., & Korn, T. M. (1968). *Mathematical Handbook for scientist and engineers*. McGraw-Hill Company.

Kotov, A. A. (2016). *Methodical estimation of task success of the combat support when robot system application. In Perspectives of UAVs development and application. Pros. of Scientific Practical conference* (pp. 135–140). Kolomna: Ministry of Defense of the Russian Federation.

Kotov, A. A., & Dmitriev, S. Yu. (2014). Main statements of the method of successful forming of data base taking into account aging and update. *Quality. Innovation. Education*, *5*, 51–55.

Kozlovskyy, V., Parkhomey, I., Odarchenko, R., Gnatyuk, S., & Zhmurko, T. (2016). Method for UAV trajectory parameters estimation using additional radar data. *Proceedings IEEE 4th International Conference Methods and Systems of Navigation and Motion Control (MSNMC)*, 39 – 42. 10.1109/MSNMC.2016.7783101

Krutova, V. & Popova, V. (1989). Fundamentals of Scientific Research. *High school,* 400.

Kryzhanovsky, G. A., Darymov, Yu. P., & Solodukhin, V. A. (1981). Automation of air traffic control processes. Moscow: Academic Press.

Kryzhanovsky, G. A., Kupin, V. V., & Plyasovskikh, A. P. (2008). Theory of transport systems. St. Petersburg: Academic Press.

Kucherov, D. P. (2014). Agent-based approach to the problem movements by the route, *Visnik of the Volodymyr Dahl East Urk. National University*, *4*(211), 168–171.

Kucherov, D. P., & Kozub, A. N. (2015). Control System Objects with Multiple Stream of Information. *Proceedings IEEE 3rd International Conference "Actual Problems of Unmanned Aerial Vehicles Developments (APUAVD)"*, 290 – 293. 10.1109/APUAVD.2015.7346623

Kucherov, D. P., Kozub, A. N., & Kostyna, O. N. (2016). Group Behavior of UAVs in Obstacles Presence. *Proceedings IEEE 4th International Conference "Methods and Systems of Navigation and Motion Control" (MSNMC)*, 51 – 54. 10.1109/MSNMC.2016.7783104

Kulik, A. (2016). *Intelligent Transport System in Aerospace Engineering, in Intelligent transportation systems-problems and perspectives.* Springer International Publishing. Retrieved October 09, 2018, from http://www.springer.com/us/book/9783319191492

Kulik, A. S. (1991, September). Fault diagnosis in dynamic Systems via signal-parametric approach. In *IFAC/IMACS Symp. Baden-Baden* (*Vol. 1*, pp. 157-162). Academic Press.

Kulik, A. S. (2014). Rational control of aerospace objects operability under the influence of the destabilizing actions. *Aerospace Engineering and Technology*, (1), 31-38.

Kulik, A.S., & Radomskyi, O.M. (2017). Usage of a technical vision unit in a mobile walking robot control system. *Electrotechnic and Computer Systems*, (25), 347-356.

Kulik, A. S. (1991). Fault diagnosis in dynamic Systems via signal-parametric approach. *IFAC/IMACS Symposium of fault detection, supervision and a technical process, SAFE PROCESS 91*, *1*, 157-162.

Kurien, J. & R-Moreno, M. (2008). Costs and Benefits of Model-based Diagnosis. *Proc. of the 2008 IEEE Aerospace Conference*, 1–14.

Kurzke, J. (2007). Design and Off-Design Performance of Gas Turbines. GasTurb 11 User Manual, Munich, Germany.

Le Grange, L., & Buys, A. (2012). The management of maintenance technology transfer in the South African aviation industry. *South African Journal of Industrial Engineering*, *13*(2). doi:10.7166/13-2-312

Lee, C. H., & Zheng, Y. L. (2015). Automatic SQL-to-NoSQL schema transformation over the MySQL and HBase databases. *Consumer Electronics-Taiwan (ICCE-TW), 2015 IEEE International Conference*, 426-427.

Lee, C. H., & Zheng, Y. L. (2015). SQL-to-NoSQL Schema Denormalization and Migration: A Study on Content Management Systems. *Systems, Man, and Cybernetics (SMC), 2015 IEEE International Conference*, 2022-2026.

Leychenko, S., Malishevskiy, A., & Mikhalic, N. (2006). *Human Factors in Aviation: monograph in two books. Book 1st*. Kirovograd: YMEKS.

Li, T., Liu, P., & Ke, Y. (2013). Battlefield Awareness Network Research Based on Intelligence Role Division and Wireless Sensor Network. *Applied Mechanics and Materials, 300-301*, 580-584. Retrieved from www.scientific.net/amm.300-301.580

Liao, Y. T., Zhou, J., Lu, C. H., Chen, S. C., Hsu, C. H., Chen, W., & Chung, Y. C. (2016). Data adapter for querying and transformation between SQL and NoSQLdatabase *Future Generation Computer Systems*, 17–23.

Liu, C., Fu, Z., Yang, Z., & Xiu, J. (2015). General Research on Database Migration from RD{Pu, 2016 #12}BMS to Hbase. In *2015 International Symposium on Computers & Informatics*. Atlantis Press.

Liu, L., Liang, X., Zhu, C., & He, L. (2017). Distributed Cooperative Control for UAV Swarm Formation Reconfiguration Based on Consensus Theory. *2nd International Conference on Robotics and Automation Engineering*, 264 – 268. Retrieved October 09, 2018, from https://ieeexplore.ieee.org/document/8291392/

Liu, B. (2012). Sentiment Analysis and Opinion Mining. *Synthesis Lectures on Human Language Technologies*, *5*(1), 64–72. doi:10.2200/S00416ED1V01Y201204HLT016

Li, W. (2018, March). The Constrained Rayleigh Quotient with a General Orthogonality Constraint and an Eigen-Balanced Laplacian Matrix: The Greatest Lower Bound and Applications in The Greatest Lower Bound and Applications in Cooperative Control Problems. *IEEE Transactions on Automatic Control*, *12*. Retrieved from https://ieeexplore.ieee.org/abstract/document/8314760/

Long, Y., & Aleven, V. (2015). Active Learners: Redesigning and Intelligent Tutoring Systems to Support Self-Regulated Learning. *Proceedings of the 8th European Conference on Technology Enhanced Learning*, 490-495.

Loukopoulos. (2008). The limits of expertise. Ashgate.

Lucas, P. (2001). Bayesian model-based diagnosis. *International Journal of Approximate Reasoning*, *27*(2), 99–119. doi:10.1016/S0888-613X(01)00036-6

Luhn, H. P. (1957). A Statistical Approach to Mechanized Encoding and Searching of Literary Information. *IBM Journal of Research and Development*, *1*(4), 309–317. doi:10.1147/rd.14.0309

Lvovskiy, E. (1989). Statistical methods for constructing empirical formulas. *High school*, 239.

Ma, C.-Q., & Zhang, J.-F. (2010). Necessary and Sufficient Conditions for Consensusability of Linear Multi-Agent Systems. *IEEE Transactions on Automatic Control*, *55*(5), 1263–1268. doi:10.1109/TAC.2010.2042764

Mahdavi, I., & Shirazi, B. (2010). A Review of Simulation-based Intelligent Decision Support System Architecture for the Adaptive Control of Flexible Manufacturing Systems. *Journal Of Artificial Intelligence*, *3*(4), 201–219. doi:10.3923/jai.2010.201.219

Makarov, R. N., Nedelko, S. N., Bamburkin, A. P., & Grigoretskii, V. A. (2005). *Aviation pedagogy*. Moscow: Academic Press.

Manning, Ch. (2009). *Introduction to Information Retrieval.* Cambridge University Press.

Martinec, D. A. (2001). ARINC 429. In C. R. Spitzer & B. Raton (Eds.), *The Avionics Handbook*. CRC Press LLC.

Martinez Bastida, J. P., & Chukhray, A. G. (2016). An active diagnostic algorithm for a gyroscopic sensors unit. *2016 II International Young Scientists Forum on Applied Physics and Engineering (YSF)*, 29-32. doi: 10.1109/YSF.2016.7753793

Martínez de Dios, J., Arrue, B., Merino, L., Ollero, A., & Gómez-Rodríguez, F. (2007). Computer vision techniques for forest fire perception. *Image and Vision Computing*, *26*(4), 550–562. doi:10.1016/j.imavis.2007.07.002

Martinez, M. (2017). Knowledge Engineering for Intelligent Decision Support. *Proc. of the Twenty-Sixth International Joint Conference on Artificial Intelligence*, 5131–5135. 10.24963/ijcai.2017/736

Martinov, A. I. (2012). *Mental training pilots of ultralight aircraft*. Moscow: Himalaia.

Maurino, D., & Salas, E. (2010). *Human Factor in aviation*. Academic Press.

Merino, L., Caballero, F., Martinez-de-Dios, J., Maza, I., & Ollero, A. (2012). An Unmanned Aircraft System for Automatic Forest Fire Monitoring and Measurement. *Journal of Intelligent & Robotic Systems*, *65*(1-4), 533–548. doi:10.100710846-011-9560-x

Merino, L., Martínez de Dios, J., & Ollero, A. (2015). Cooperative Unmanned Aerial Systems for Fire Detection, Monitoring, and Extinguishing. In *Handbook of Unmanned Aerial Vehicles* (pp. 2693–2722). Springer Netherlands. doi:10.1007/978-90-481-9707-1_74

Miller, S. P., & Tribble, A. C. Extending the Four-Variable Model to Bridge the System-Software Gap. In *Proceedings of the 20th Digital Avionics Systems Conference (DASC01)*. 10.1109/DASC.2001.963383

Milnor, J. (1963). Morse Theory. *Annals of Mathematics Studies*, 51.

Ministry of Transport and Communications of Ukraine. (2007). *The Rules for Issuing Certificates of Aviation Personnel in Ukraine*. Kyiv: Author.

Ministry of Transport of Ukraine. (2003). *About Approving the Flight Rules of Aircrafts and Air Traffic Services in the Classified Airspace of Ukraine*. Kyiv: Author.

Morin, E. (1989). Il Metodo – La conoscenza della conoscenza. Feltrinelli editore.

Morin, E. (2001). Il Metodo – La natura della natura. Raffaello Cortina editore.

Morin, E. (2004). Il Metodo – La vita della vita. Raffaello Cortina editore.

Morin, E. (2008). *Il Metodo – Le idee: habitat, vita organizzazione, usi e costumi*. Raffaello Cortina editore.

Moskvichov, V. (1998). Information support for controller decisions for controlling the forced landing of aircraft outside the aerodrome. Kyiv: Academic Press.

Multipurpose controller simulator "Expert". (n.d.). Retrieved from http://www.nita.ru/catalog/kompleksnyy-dispetcherskiy-trenazher-ekspert/

Murray, R.M. (2007). Recent Research in Cooperative Control of Multi-Vehicle Systems. *ASME Journal of Dynamic Systems, Measurement, and Control*, 1 – 27.

National Bureau of Air Accidents Investigation of Ukraine. (2014). *Analysis of flight safety data from the results of the investigation of events with civil aircraft of Ukraine that took place in 2013.* Kiev: Author.

National Bureau of Air Accidents Investigation of Ukraine. (2015). *Analysis of flight safety data from the results of the investigation of events with civil aircraft of Ukraine that took place in 2014.* Kiev: Author.

National Bureau of Air Accidents Investigation of Ukraine. (2016). *Analysis of flight safety data from the results of the investigation of events with civil aircraft of Ukraine that took place in 2015.* Kiev: Author.

Navon, D. (1977). Forest before trees: The precedence of global features in visual perception. *Cognitive Psychology*, *9*(9), 353–383. doi:10.1016/0010-0285(77)90012-3

Nazarov, B. (1964). Gyro on a rocket. *Voenizdat*, 104.

Nazarov, A. S. (2010). *Photogrammetry*. Minsk: TetraSystem.

Nedelko, V. (2002). Ensuring the effectiveness of information support of decision-making in automated systems of air traffic services with elements of artificial intelligence. Kyiv: Academic Press.

Neumann, K. (1990). *Stochastic project networks: temporal analysis, scheduling and cost minimization*. Berlin: Springer-Verlag. doi:10.1007/978-3-642-61515-3

NexSim Tower Simulator. (n.d.). Retrieved from https://www.umes.edu/Engineering/Default.aspx?id=26070

Nikolayev, L. (1990). Aerodynamics and dynamics of the flight of transport aircraft. Moscow. *Transport*, 392.

Nitnaware, C., & Khan, A. (2015). A multidimensional data storage model for location based application on Hbase. *Innovations in Information, Embedded and Communication Systems (ICIIECS). 2015 International Conference*, 1-5. 10.1109/ICIIECS.2015.7193237

Ni, W., & Cheng, D. (2010). Leader-following consensus of multi-agent systems under fixed and switching topologies. *Systems & Control Letters*, *59*(3-4), 209–217. doi:10.1016/j.sysconle.2010.01.006

Noh, S., & Choi, C. (2012). Predicting the Operational Effectiveness of Aircraft Survivability Equipment Suite. *IACSIT International Journal of Engineering and Technology*, *4*(4), 372–375. doi:10.7763/IJET.2012.V4.387

Norman, D. (1988). *The psychology of everyday things*. New York, NY: Basic Books.

Northrop Grumman. (2010). *Global Hawk Product Brochure*. Retrieved from: http://www.northropgrumman.com/Capabilities/RQ4Block10GlobalHawk/Documents/GHMD-New-Brochure.pdf

Novák, M., Kameníková, I., & Podhadský, J. (2016). Modern Communication Devices in the General Aviation Aircraft Maintenance. *MAD - Magazine of Aviation Development*, *4*(17), 5. doi:10.14311/mad.2016.17.01

Odarchenko, R. S., Gnatyuk, S. O., Zhmurko, T. O., & Tkalich, O. P. (2015, October). Improved method of routing in UAV network. In *Actual Problems of Unmanned Aerial Vehicles Developments (APUAVD), 2015 IEEE International Conference* (pp. 294-297). IEEE. 10.1109/APUAVD.2015.7346624

Ögren, P., Fiorelli, E., & Leonard, N. E. (2004). Cooperative Control of Mobile Sensor Networks: Adaptive Gradient Climbing in a Distributed Environment. *IEEE Transactions on Automatic Control*, *49*(8), 1292–1302. doi:10.1109/TAC.2004.832203

Olaf, C. (2010). *SAGA: System for Automated Geoscientific Analyses. Physical Geography, University Hamburg*. GEOSTAT.

Olfati-Saber, Fax, & Murray. (2010). Reply to comments on consensus and cooperation in networked multi-agent systems. *Proceedings of the IEEE, 7*, 1354–1355.

Olfati-Saber, R. (2001). *Nonlinear control of underactuated mechanical systems with application to robotics and aerospace vehicles* (Ph.D. thesis). MIT.

Olfati-Saber, R., Dunbar, W. B., & Murray, R. M. (2003). Cooperative Control of Multi-Vehicle Systems using Cost Graphs and Optimization. *Proceedings of the 2003 American Control Conference*, 1 – 15. Retrieved October 09, 2018, from https://ieeexplore.ieee.org/document/1243403/

Olfati-Saber, R., Fax, J. A., & Murray, R. M. (2007). Consensus and Cooperation in Networked Multi-Agent Systems. *Proceedings of the IEEE*, *95*(1), 215–233. doi:10.1109/JPROC.2006.887293

Olsson, H., Egberth, M., Engberg, J., Fransson, J., & Pahlén, T. (2005). Current and Emerging Operational Uses of Remote Sensing in Swedish Forestry. In *Proc. of the 5th Annual Forest Inventory and Analysis Symposium* (pp. 39–46). US Forest Service.

Oravec, M., & Jadlovská, A. (2017). Sensors Fault Diagnosis Algorithm Design of a Hydraulic System. *Acta Electrotechnica Et Informatica*, *17*(2), 30–37. doi:10.15546/aeei-2017-0013

Osadchiy, S., & Tymoshenko, G. (2017). Methods for Determine the weight and the Center of gravity of UAV. *Materials of the fourth International conference "Actual problems of Unmanned Aerial Vehicles Developments"*, 139-142.

Paquette, L., Lebeau, J. F., Beaulieu, G., & Mayers, A. (2015). Designing a Knowledge Representation Approach for the Generation of Pedagogical Interventions by MTTs. *International Journal of Artificial Intelligence in Education*, *25*(1), 118–156. doi:10.100740593-014-0030-z

Parasumaran, R., & Wickens, C. (2008). Humans: Still Vital After All These Years of Automation. *Human Factors*, *50*(3), 511–520. doi:10.1518/001872008X312198 PMID:18689061

Parasumaran, S. (2000). A Model for Types and Levels of Human Interaction with Automation. *IEEE Transactions on Systems, Man, and Cybernetics. Part A, Systems and Humans*, *30*(3).

Pashkovsky, I. (2018). Stability and controllability of the aircraft. Moscow: Anton.

Pawlak, Z. (1991). Theoretical Aspects of Reasoning about Data. Kluwer Academic Publishers. doi:10.1007/978-94-011-3534-4

Pawlak, Z., Jerzy, W., Slowinski, R., & Ziarko, W. (1995). Rough Sets. *Communications of the ACM*, *38*(11), 88–95. doi:10.1145/219717.219791

Pelánek, R., & Jarusek, P. (2015). Student Modeling Based on Problem Solving Times. *Journal of Artificial Intelligence in Education*, *25*(4), 493–519. doi:10.100740593-015-0048-x

Peng, H., Dan, M., & Qu, H. (2011). Research of Aero-Engine Life Prediction Based on Take-Off EGTM. *Applied Mechanics and Materials, 99-100*, 286-292. Retrieved from www.scientific.net/amm.99-100.286

Peter, W., Steven, P., & Bob, F. (2002). *The Prescription and Practice of Work on the Flight Deck*. Department of Computer Science University of York.

Petrov, M. V. (2013). UAV practical application experience Swinglet Cam. *Geoprofi, 2*, 42–44.

Phillips, D., & Garcia-Diaz, A. (1984). *Methods of network analysis*. Moscow: The World.

Pilipyonok, O. M. (2017). *Method of improving the aircraft management quality of operators of the navigation service systems and traffic control* (Doctoral dissertation). Kircvograd Flight Academy of the National Aviation University.

Pilipyonok, O. N. (2012a). Development of the module of information support air traffic controller activity. *Abstracts of the 12 International scientific and practical conference for young scientists and students Flight. Modern problems of science*, 35.

Pilipyonok, O.N. (2012b). Peculiarities of modeling of the operating conditions of Tower controller. *Materials of the 32 All-Ukrainian scientific and practical conference for young scientists and students Aviation and space: status, achievements and prospects*, 151-152.

Pilipyonok, O.N. (2013). Development of the training system prototype Tower controller. *Materials of the All-Ukrainian scientific and practical conference for young scientists and students Problems of navigation and traffic management*, 32.

Pilipyonok, O.N. (2014). The decision making model as the main component of the intelligent learning system. *Scientific Proceeding of Ukrainian Research Institute of Communication, 4*(32), 54-61.

Pipattanasomporn, M., Feroze, H., & Rahman, S. (2009). Multi-Agent Systems in a Distributed Smart Grid: Design and Implementation. *Proceeding IEEE PES 2009 Power Systems Conference and Exposition (PSCE'09)*, 1 – 8. 10.1109/PSCE.2009.4840087

Podinovski, V. V. (2007). *Introduction in the theory of importance measure*. Moscow: Fizmatgiz.

Poston, T., & Stewart, I. (1978). *Catastrophe theory and its applications*. London: Pitman.

Pritsker, A. A. (1966). *GERT: graphical evaluation and review technique*. RAND Corporation.

Qin, C., Wu, X., Jiang, J., & Zhu, A. (2016). Case-based knowledge formalization and reasoning method for digital terrain analysis – application to extracting drainage networks. *Hydrology and Earth System Sciences, 20*(8), 3379–3392. doi:10.5194/hess-20-3379-2016

Ralli, M. (1993). Fattore umano ed operazioni di volo. Libreria dell'orologio.

Rasmussen, J. (1983). Skills, Rules, Knowledge: Signals, Sign and Symbol and Other Distinctions in Human Performance Models, AF 447: a paradigmatic accident? The role of automation. *IEEE Transactions on Systems, Man, and Cybernetics*, SMC-13.

Raymer, D. P. (1999). *Aircraft Design: A conceptual approach*. Washington, DC: AIAA.

Ren, W., Beard, R. W., & Atkins, E. M. (2005). A Survey of Consensus Problems in Multi-Agent Coordination. *American Control Conference*, 1859 – 1864.

Retscher, G. (2016). Indoor navigation. Encyclopedia of Geodesy, 1-7. doi:10.1007/978-3-319-02370-0_9-1

Rezaee, H., & Abdollah, F. (2016). Consensus Problem in General Linear Multi-Agent Systems Under Stochastic Topologies. *IFAC-PapersOnLine, 49*(13), 13–18. doi:10.1016/j.ifacol.2016.07.919

Rieck, K., Trinius, P., Willems, C., & Holz, T. (2011). Automatic analysis of malware behavior using machine learning. *Journal of Computer Security, 19*(4), 639–668. doi:10.3233/JCS-2010-0410

Rizun, N., & Kucharska, W. (2018). Text Mining Algorithms for Extracting Brand Knowledge. The Fashion Industry Case. In K.S. Soliman (Ed.), Innovation Management and Education Excellence through Vision 2020 (pp. 1972-1983). Milan, Włochy: International Business Information Management Association (IBIMA).

Rizun, N. (2018). Semantic Analysis and Text Summarization in Socio-Technical Systems. In T. Shmelova, Yu. Sikirda, N. Rizun, A.-B. M. Salem, & Yu. Kovalyov (Eds.), *Socio-Technical Decision Support in Air Navigation Systems: Emerging Research and Opportunities* (pp. 243–281). Hershey, PA: IGI Global. doi:10.4018/978-1-5225-3108-1.ch008

Rizun, N., Taranenko, Y., & Waloszek, W. (2017). The Algorithm of Building the Hierarchical Contextual Framework of Textual Corpora. *Eighth IEEE International Conference on Intelligence Computing and Information System*, 366-372. 10.1109/INTELCIS.2017.8260064

Rizun, N., & Taranenko, Yu. (2018). The Methodology of Constructing and Analyzing the Hierarchical Contextually-Oriented Corpora. *Proceeding of Federated Conference on Computer Science and Information Systems*. 10.15439/2018F69

Rizun, N., & Waloszek, W. (2018). Methodology for Text Classification using Manually Created Corpora-based Sentiment Dictionary. *Proceeding of the 10th International Conference on Knowledge, Discovery and Information Retrieval*. 10.5220/0006932602120220

Robert, B. A. (2008). Basic probability theory. Dover Publications, Inc.

Robert, W., & Kendall, W. L. (2011). Fundamental on Situational Surprise: a case study with Implications for Resilience. In *Proceedings of the fourth Symposium on Resilience Engineering*. Mines-Tech.

Romero, C., & Ventura, S. (2006). *Data mining in e-learning*. Southampton, UK: Wit Press. doi:10.2495/1-84564-152-3

Rothermel, R. (1972). *A Mathematical Model for Predicting Fire Spread in Wild Land Fuels*. Research Paper INT-115. USDA Forest Service, Intermountain Forest, and Range Experiment Station.

Russel, S., & Norving, P. (2010). *Artificial Intelligence: A Modern Approach*. Prentice Hall.

Salas, E., & Maurino, D. (2010). *Human factors in aviation*. Burlington, MA: Elsevier. doi:10.1016/B978-0-12-374518-7.00001-8

Salem. (2008). Data Mining Technology in e-Learning. *Proceedings of 6th International Conference on Emerging e-learning Technologies and Applications, Information and Communication Technologies in Learning, (ICETA2008)*.

Salem. (2010). Ontological Engineering in e-Learning. *Proceedings of 8th International Conference on Emerging e-learning Technologies and Applications, Information and Communication Technologies in Learning*.

Salem. (2011). Intellectual E-Learning Systems. In *Proceedings of the Practical Conference "VR/AR Applications in Training"*. Vidzeme University of Applied Sciences.

Salton, G. (1989). *Automatic Text Processing: Transformation, Analysis, and Retrieval of Information by Computer*. Addison-Wesley Longman Publishing.

Sanchez, G., Salazar, R., Hassanalian, M., & Abdelkefi, A. (2018). Sizing and performance analysis of albatross-inspired tilt-wing unmanned air vehicle. AIAA/ASCE/AHS/ASC Structures, Structural Dynamics, and Materials Conference, Kissimmee, FL.

Sayed-Mouchaweh, M., & Lughofer, E. (2015). Decentralized fault diagnosis approach without a global model for fault diagnosis of discrete event systems. *International Journal of Control*, *88*(11), 2228–2241. doi:10.1080/00207179.2015.1039594

Sebastiani, F. (2002). Machine Learning in Automated Text Categorization. *ACM Computing Surveys*, *34*(1), 1–47. doi:10.1145/505282.505283

Shahbazi, M., Théau, J., & Ménard, P. (2014). Recent Applications of Unmanned Aerial Imagery in Natural Resource Management. *GIScience & Remote Sensing*, *51*(4), 339–365. doi:10.1080/15481603.2014.926650

Shanabrook, D. H., Arroyo, I., Woolf, B. P., & Burleson, W. (2012). Visualization of Student Activity Patterns within Intelligent Tutoring Systems. *Proceeding of the 11th International Conference on Intelligent Tutoring Systems (ITS'2012)*, 46-51. 10.1007/978-3-642-30950-2_6

Sharifi, F., Mirzaei, M., Zhang, Y., & Gordon, B. (2015). Cooperative Multi-vehicle Search and Coverage Problem in an Uncertain Environment. *Unmanned Systems*, *3*(1), 35–47. doi:10.1142/S230138501550003X

Sharma, S., Tim, U. S., Gadia, S., Wong, J., Shandilya, R., & Peddoju, S. K. (2015). Classification and comparison of NoSQL big data models. *International Journal of Big Data Intelligence*, *2*(3), 201–221. doi:10.1504/IJBDI.2015.070602

Sharma, V., & Dave, M. (2012). SQL and NoSQL Databases. *International Journal of Advanced Research in Computer Science and Software Engineering*, *2*(8), 2012.

Sherstjuk, V., Zharikova, M., & Sokol, I. (2018). Forest Fire-Fighting Monitoring System Based on UAV team and Remote Sensing. *Proc. of the 38th Int. Conf. "Electronics and Nanotechnology"*, 99-106. 10.1109/ELNANO.2018.8477527

Shibanov, A.P. (2003). Finding the distribution density of the execution time of the GERT network on the basis of equivalent simplifying transformations. *Automation and Telemechanics*, *2*, 117-126.

Shibanov, A. P. (2003). *Generalized GERT-networks for modeling protocols, algorithms and programs of telecommunication systems*. Ryazan.

Shibl, R., Lawley, M., & Debuse, J. (2013). Factors influencing decision support system acceptance. *Decision Support Systems*, *54*(2), 953–961. doi:10.1016/j.dss.2012.09.018

Shmeleva T., Artemenko O., & Tymoshenko G. (2012). Construction of a neuron cell model of the possibility of flight. *Journal of Kharkov Air Force University*, *1*(21), 68-74.

Shmelova, T. (2013). Scientific and methodological basis of modeling decision support Air Navigation System.. Kiev: Academic Press.

Shmelova, T. F., & Bondarev, D. I. (2015, October). Graph theory applying for quantitative estimation of UAV's group flight. In *Actual Problems of Unmanned Aerial Vehicles Developments (APUAVD), 2015 IEEE International Conference* (pp. 328-331). IEEE.

Shmelova, T., Sikirda, Yu., Zemlyanskiy, A., & Danilenko, O. (2015). Fuzzy Assessment of the Situation's Complexity during Simulation Training of Specialists in Air Traffic Services. In *Materials of International Scientific-Practical Conference "Problems of Energy Efficiency and automation in industry and agriculture"*. Kirovohrad: Kirovohrad National Technical University.

Shmelova, T., Lazorenko, V., & Bilko, A. (2014). Estimation of Pre-Simulating Training Tasks Complexity. *Proceedings of the National Aviation University, No.*, *1*(62), 17–22.

Shmelova, T., Miroshnitchenko, M., & Timoshenko, G. (2014). Analysis of a passenger mass influence on the model of identifying the aircraft takeoff weight and its takeoff balance data. *Journal of Kirovograd National Technical University*, *27*, 260–265.

Shmelova, T., Sikirda, Y., Assaul, A., & Stasiuk, O. (2015). Structural Analysis of Management Environment of Aviation Enterprise from the Point of Systematic Approach. *Proceedings of the National Aviation University, 2*, 3–12.

Shmelova, T., Sikirda, Y., Zemlyanskiy, A., Lazoprenko, V., & Danilenko, O. (2016). Artificial neural network for air traffic controller's pre-simulator training. *Proceedings of the National Aviation University*, (3), 13 – 23. 10.18372/2306-1472.68.10905

Shmelova, T., Sikirda, Yu., Zemlyanskiy, A., Danilenko, O., & Lazorenko, V. (2016). Artificial Neural Network for Air Traffic Controller's Pre-Simulator Training. *Proceedings of the National Aviation University, No., 3*(68), 13–23.

Shmelova, T., Yakunina, I., Moyseenko, V., & Grinchuk, M. (2014). *A. r. Computer Program "Network Analysis of Flight Emergency"*. Kyiv: State Intellectual Property Service of Ukraine.

Shmelova, T., Yakunina, I., & Yakunin, R. (2013). *A. r. Computer Program "Test Extraordinary Incident"*. Kyiv: State Intellectual Property Service of Ukraine.

Sikirda, U. (2004). Modeling of the decision support system of air traffic controller in freelance flight situations. Kyiv: Academic Press.

Skalický, P., & Palasiewicz, T. (2017). Intelligence Preparation of the Battlefield as a Part of Knowledge Development. *International Conference Knowledge-Based Organization, 23*(1), 276-280. 10.1515/kbo-2017-0045

Sładkowski, A. (Ed.). (2013). *Some Actual Issues of Traffic and Vehicle Safety. Faculty of Transport*. Silesian University of Technology.

Social Media Data. (2016). *Twitter US Airline Sentiment. U.S. Airlines Sentiment Analysis*. Retrieved from https://data.world/socialmediadata/twitter-us-airline-sentiment.

Software and Solutions for Air Traffic Control and Management. Product Focus – TATIC TWR. (n.d.). Retrieved from http://www.airport-suppliers.com/supplier/Saipher_ATC/press_release/Software_and_Solutions_for_Air_Traffic_Control_and_Management_1/

Solovtsova, N. E. (Ed.). (2009). *Encyclopaedia of missile strategic army*. Moscow: RVSN.

Sokolov, A. V., & Tixonov, M. L. (2008). GIS military application. *OBSERVEF, 5*, 37–45.

State Aviation Administration of Ukraine. (2013). *Analysis of flight safety level and identification of potential accident factors with civil aircraft of Ukraine in 2012*. Kiev: Author.

Statistical Summary of Commercial Jet Airplane Accidents. Worldwide Operations 1959-2011. (n.d.). Retrieved from http://www.boeing.com/news/techissues/pdf/statsum.pdf

Strelkov, Y. K. (2004). *Psychology of work, engineering and professional psychology*. Moscow: Academic Press.

Suzuki, K. (2013). *Artificial Neural Networks – Architectures and Applications*. InTech. doi:10.5772/3409

Svenmarckt, P., & Dekker, S. (2003). Decision support in fighter aircraft: From expert systems to cognitive modeling. *Behaviour & Information Technology, 22*(3), 175–184. doi:10.1080/0144929031000109755

Symeonidisa, A., Chatzidimitriouc, K., Athanasiadisd, I., & Mitkas, P. (2007). Data mining for agent reasoning: A synergy for training intelligent agents. *Engineering Applications of Artificial Intelligence, 20*(8), 1097–1111. doi:10.1016/j.engappai.2007.02.009

Tan, C. F., Wahidin, L. S., Khalil, S. N., Tamaldin, N., Hu, J., & Rauterberg, G. W. M. (2016). The Application of Expert System: A Review of Research and Applications. *Journal of Engineering and Applied Sciences (Asian Research Publishing Network), 11*(4), 2448–2453.

Tanner, Jadbabaie, & Pappas. (2007). Flocking in fixed and switching networks. *Transactions on Automatic Control, 5*, 863–868.

Tarasov, V., Gerasimov, V., Levin, I., & Kornijchuk, V. (2007). *Intelligent Decision Support Systems: Theory, Synthesis, Efficiency*. Kyiv, Ukraine: MAKNS.

Тамаргазін, О. (2016). The technique of periodicity choice of maintenance control of aviation engineering products. *Proceedings Of The National Aviation University, 6*(1-2). doi:10.18372/2306-1472.6.9716

The FIRSTplusTM Tower Simulator. (n.d.). Retrieved from http://www.network54.com/Forum/248068/thread/1269639662/5/Turkish+HAVELSAN+Simulator

Todorovic, M. H., Datta, R., Stevanovic, L., She, X., Cioffi, P., Mandrusiak, G., . . . Frangieh, T. (2016). Design and Testing of a Modular SiC based Power Block. *PCIM Europe 2016 International Exhibition and Conference for Power Electronics Intelligent Motion Renewable Energy and Energy Management*, 1-4.

Tomashevskiy, V. (2005). *System modeling*. Kyiv, Ukraine: BHV.

Tower Simulator. (n.d.). Retrieved from http://www.towersimulator.com/

Trimble UX5 HP Unmanned Aircraft System. (2015). Retrieved September 6, 2015, from http://www.kmcgeo.com/Datasheets/UX5HP.pdf

Turney, P. (2002). Thumbs Up or Thumbs Down? Semantic Orientation Applied to Unsupervised Classification of Reviews. In *Proc. ACL-02, 40th Annual Meeting of the Association for Computational Linguistics*. Association for Computational Linguistics.

Tymoshenko, G., & Osadcha-Fic, L. (2014). Simulation of the control of the take-off mass and centering of the plane. *Materials of the XII International scientific conference "Control and Management in complex systems - 2014"*, 57.

Tymoshenko, G., & Saganovska, L. (2016). Automation of the decision making a process for the flight after loading an airplane. *Materials of the XIII International scientific conference "Control and Management in complex systems - 2016"*, 331-332.

Tymoshenko, G. (2012). Simulation of aircraft loading planning system. *Journal of Kirovograd National Technical University, 25*(1), 405–412.

Tyunin, V., & Yakimov, V. (2004). Theory and practice of flight training. Kiev: Academic Press.

Ursavas, E. (2014). A decision support system for quayside operations in a container terminal. *Decision Support Systems, 59*, 312–324. doi:10.1016/j.dss.2014.01.003

VanLehn, K. (2008). Intelligent Tutoring Systems for Continuous, Embedded Assessment. In The future of assessment: Shaping teaching and learning (pp. 113–138). Lawrence Erlbaum Associates.

VanLehn, K., Burleson, W., Chavez Echeagaray, M. E., Christopherson, R., Gonzalez Sanchez, J., Hastings, J., ... Zhang, L. (2011). The Affective Meta-Tutoring Project: How to motivate students to use effective meta-cognitive strategies. In *Proceedings of the 19th International Conference on Computers in Education, ICCE 2011* (pp. 128-130). Chian Mai, Thailand: ICCE.

VanLehn, K., Burleson, W., Girard, S., Chavez-Echeagaray, M. E., Hidalgo-Pontet, Y., & Zhang, L. (2014). The Affective Meta-Tutoring project: Lessons learned. Intelligent Tutoring Systems. *12th International Conference, ITS 2014*, 94-103.

Varvarina, E. A. (2012). Remote manned flying objects using for construction of orthophotoplanes of the linear objetcs. *Earth Science*, *2*, 66–70.

Verbert, K., De Schutter, B., & Babuska, R. (2015). Reasoning under uncertainty for knowledge-based fault diagnosis: A comparative study. *Proc. of the 9th IFAC Symposium on Fault Detection, Supervision and Safety of Technical Processes (SafeProcess 2015)*, 422–427. 10.1016/j.ifacol.2015.09.563

Vitryak, A. M. (2008). *Reliability of professional activity of flight and dispatching staff on international airways*. Moscow: Kirovograd.

Vora, M. N. (2011). *Hadoop-HBase for large-scale data. In Computer science and network technology (ICCSNT), 2011 international conference* (Vol. 1, pp. 601–605). IEEE. doi:10.1109/ICCSNT.2011.6182030

Voronin, A. (2009). *The information system of decision-making*. Kyiv, Ukraine: The National Aviation University.

Walker, E., Rummel, N., & Koedinger, K. R. (2014). Adaptive Intelligent Support to Improve Peer Tutoring in Algebra. *International Journal of Artificial Intelligence in Education*, *24*(1), 33–61. doi:10.100740593-013-0001-9

Walsh, P. P., & Fletcher, P. (2004). *Gas Turbine Performance* (2nd ed.). Oxford, UK: Blackwell. doi:10.1002/9780470774533

Wang, Y., & Heffernan, N. (2012). Leveraging first response time into the knowledge tracing model. In *Proceedings of the International Conference on Educational Data Mining* (pp. 176-179). Academic Press.

Wang, Y., Xu, Y., Liu, Y., Chen, J., & Hu, S. (2015). QMapper for Smart Grid: Migrating SQL-based Application to Hive. *2015 ACM SIGMOD International Conference on Management of Data*, 647-658. 10.1145/2723372.2742792

Wardihani, E., Ramdhani, M., Suharjono, A., Setyawan, T., Hidayat, S., Widodo, S., ... Saifullah, F. (2018). Real-time forest fire monitoring system using unmanned aerial vehicle. *Journal of Engineering Science and Technology*, *13*(6), 1587–1594.

Weckman, G., Marvel, J., & Shell, R. (2006). Decision Support Approach to Fleet Maintenance Requirements in the Aviation Industry. *Journal of Aircraft*, *43*(5), 1352–1360. doi:10.2514/1.19749

Wenger, N., Antoniev, A., & Gorod, A. (2013). The International Space Station: Applying system of systems methodology. *Proceedings of 2013 8th International Conference on System of Systems Engineering*, 235 - 242. 10.1109/SYSoSE.2013.6575273

Wentzel, E.S. (1999). *Theory of Probability*. Moscow: Vysshaya shkola.

Wentzel, E. S., & Ovcharov, L. A. (2003). *Theory of probability and its engineering applications*. Moscow: Academy.

Whitney, H. (1955). On singularities of Mappings of Euclidean Spaces I. Mappings of the Plane into the Plane. *Annals of Mathematics*, *62*(3), 374–410. doi:10.2307/1970070

Wieber, F., Lukas, J., & Gollwitzer, P. M. (2015). Promoting the translation of intentions into action by implementation of intentions: Behavioral effects and physiological correlates. *Frontiers in Human Neuroscience*, 9. PMID:26236214

Witayangkurn, A., Nagai, M., Honda, K., Dailey, M., & Shibasaki, R. (2011). Real-time monitoring system using unmanned aerial vehicle integrated with sensor observation service. *International Archives of the Photogrammetry, Remote Sensing and Spatial Information Sciences, XXXVIII-1/C22, ISPRS 2011 Workshop*.

Witten, I., & Frank, E. (2005). Data Mining. In *Practical Machine Learning Tools and Techniques* (2nd ed.). Elsevier.

Wolfgang, K. (1967). Gestalt psychology. *Psychological Research, 31*(1), 1830.

Woods, D. D. (1990). Modeling and predicting human error. In Human performance models for computer aided engineering (pp. 248-274). Academic Press.

Woods, D., Dekker, S., Cook, R., Johannesen, L., & Sarter, N. (2010). *Behind human error.* Aldershot, UK: Ashgate publishing.

Xiao, F., Wang, L., Chen, J., & Gao, Y. (2009). Finite-time formation control for multi-agent systems. *Automatica, 45*(11), 2605–2611. doi:10.1016/j.automatica.2009.07.012

Xie, Y., Shi, Z., Sun, C., Deng, Y., & Wu, X. (2017). The Integration of Adaptive Cooperative Control for Position and Attitude of Distributed Spacecraft. *Proceedings of the 36th Chinese Control Conference*, 3409 – 3413. 10.23919/ChiCC.2017.8027886

Xu, B., & Kumar, S. A. (2015). Big Data Analytics Framework for System Health Monitoring. *2015 IEEE International Congress on Big Data*, 401-408. 10.1109/BigDataCongress.2015.66

Yakunina I. (2017). Analysis methods of operator's activity of the aeronautical system in special cases of flight for reducing the decision-making time. *Kropyvnytskyi, 120.*

Yu, K., & Wang, F. (2013). Research and Design on Intelligent Decision Support System Frame for Vehicle Maintenance. *Advanced Materials Research, 765-767*, 3-7. Retrieved from www.scientific.net/amr.765-767.3

Yu, X., Guo, Ge., & Gong, G. (2017). Cooperative control for a platoon of vehicles with leader-following communication strategy. *Chinese Automation Congress (CAC)*, 6721 – 6726. Retrieved October 09, 2018, from https://ieeexplore.ieee.org/document/8243988/

Yuan, C., Liu, Z., & Zhang, Y. (2017). Aerial Images-Based Forest Fire Detection for Firefighting Using Optical Remote Sensing Techniques and Unmanned Aerial Vehicles. *Journal of Intelligent & Robotic Systems, 88*(2-4), 635–654. doi:10.100710846-016-0464-7

Yuan, C., Zhang, Y., & Liu, Z. (2015). A Survey on Technologies for Automatic Forest Fire Monitoring, Detection and Fighting Using UAVs and Remote Sensing Techniques. *Canadian Journal of Forest Research, 45*(7), 783–792. doi:10.1139/cjfr-2014-0347

Yun, L., & Jian, Z. (2018). Design and Implementation of Cooperative Turning Control for the Towing System of Unpowered Facilities. *IEEE Access: Practical Innovations, Open Solutions, 6*, 18713–18722. doi:10.1109/ACCESS.2018.2819692

Zadeh, L. (1965). Fuzzy sets. *Information and Control, 8*(3), 338–353. doi:10.1016/S0019-9958(65)90241-X

Zadorozhny, V. D. (2011). *The methodology of flight training.* Ulyanovsk.

Zagorski, M.Y., Bogdanov, V.L., Garmanov, V.V., Koroleva, V.P., & Ryabov, Y.V. (2015). Mathematical model of stereomate films and the renewal algorithm of terrestial relief on the basis on their. *Modern Problems of Space Earth Remote Sensing, 3,* 56.

Zaitsev, D. A., Shmeleva, T. R., Retschitzegger, W., & Pröll, B. (2014). Blocking Communication Grid via Ill-Intentioned Traffic, Proceedings of 14th Middle Eastern Simulation & Modelling Multiconference, 63-71.

Zaitsev, D. A. (2004). Switched LAN simulation by colored Petri nets. *Mathematics and Computers in Simulation, 65*(3), 245–249. doi:10.1016/j.matcom.2003.12.004

Zaitsev, D. A., & Sakun, A. L. (2008). An Evaluation of MPLS Efficacy using Colored Petri Net Models. *Proceedings of International Middle Eastern Multiconference on Simulation and Modelling (MESM'2008)*, 31-36.

Zaitsev, D. A., & Shmeleva, T. R. (2009). Parametric Petri Net Model for Ethernet Performance and QoS Evaluation. In *Proc. of 16th Workshop on Algorithms and Tools for Petri Nets*. University of Karlsruhe.

Zaitsev, D. A., Shmeleva, T. R., Retschitzegger, W., & Pröll, B. (2016). Security of grid structures under disguised traffic attacks. *Cluster Computing*, *19*(3), 1183–1200. doi:10.100710586-016-0582-9

Zaks, L. (1976). *Statistical estimation*. Moscow, Russia: Statistic.

Zaritski, V. N., & Charkevich, L. A. (2007). *General Tactics*. Tambov: TGTU.

Zbrutsky, O. V., Prokhorchuk, O. V., Kolesnichenko, V. B., & Marynoshenko, O. P. (2013). Unmanned Aerial Vehicles and Technologies by NTUU "KPI". *IEEE 2nd International Conference "Actual Problems of Unmanned Air Vehicles Developments" Proceedings*.

Zeeman, E. (1976). Catastrophe theory. *Scientific American*, *234*(4), 65–83. doi:10.1038cientificamerican0476-65

Zhang, H., & Chen, J. (2014). Bipartite Consensus of Linear Multi-Agent Systems over Signed Digraphs: An Output Feedback Control Approach. *Proceedings of the 19th World Congress the International Federation of Automatic Control*, 4681-4686. 10.3182/20140824-6-ZA-1003.00608

Zhang, G., Wang, J., Zhi, Lv., Yang, Y., Su, H., Yao, Q., ... Huang, J. (2015). A integrated vehicle health management framework for aircraft - A preliminary report. *2015 IEEE Conference on Prognostics and Health Management (PHM)*, 1-8. 10.1109/ICPHM.2015.7245034

Zharikova, M., Sherstjuk, V., & Baranovskij, N. (2015). The Plausible Wildfire Model in Geoinformation Decision Support System for Wildfire Response. Proc. of the 15th Int. Multidisciplinary Sc. Geoconference SGEM-2015, 2(3), 575–583.

Zharikova, M., & Sherstjuk, V. (2015). Development of the Model of Natural Emergencies in Decision Support System. *Eastern European Journal of Enterprise Technologies*, *4*(1), 62–69. doi:10.15587/1729-4061.2015.37801

Zharikova, M., & Sherstjuk, V. (2017). Threat assessment method for intelligent disaster decision support system. *Advances in Int. Systems and Computing*, *512*, 81–99.

Zhou, B. (2014). *China Eastern Airlines Big Data platform*. Available: https://www.behance.net/gallery/34149479/china-eastern-airlines-big-data-platform

Zuo, Z., Han, Q.-L., Ning, B., Ge, X., & Zhang, X.-M. (2018). An Overview of Recent Advances in Fixed-Time Cooperative Control of Multi-Agent Systems. *IEEE Transactions on Industrial Informatics*, *14*(6), 2322–2334. doi:10.1109/TII.2018.2817248

Zyryanov, A. A., & Dorrer, M. G. (2012). Algorithm for translating the model of business processes into the GERT-network model. Bulletin of KSAU, 12, 13-18.

Zyryanov, A.A. & Dorrer. M.G. (2012). Forecast of the dynamics of event-driven models of business processes based on GERT-networks. *Informatization and Communication, 7*, 124-127.

About the Contributors

Tetiana Shmelova, Ph.D., DSc., Professor of Department of Air Navigation Systems in National Aviation University (Ukraine). Thesis in the field of Scientific and Methodological Basis of Decision-Making in the Air Navigation System (2012). Areas of Scientific Interests: Mathematical Models of Decision-Making by the Human operator, especially in Emergency Situation; Decision Support System; Research of Air Navigation system as a Socio-technical system; Aviation Sociometry and Socionics; Management and Marketing in aviation; Information technology and Informatics of Decision-Making; Problems of Human factors in aviation. Author of more than 250 scientific articles and guides (about 200 articles is in Ukrainian and Russian, about 35 articles in English, 14 methodical manuals, 12 copyright certificates for computer programs), 6 monographs in fields of aviation, economics, mathematics, the theory of system. Teaching courses: Theory of decision-making, Mathematical Programming, Effective-ness of Air Traffic Management.

Yuliya Sikirda, Associate Professor of Management, Economy, Law and Tourism Department at the Flight Academy of National Aviation University. Has graduated from the Flight Academy of National Aviation University (former State Flight Academy of Ukraine) in 2001, received a Master degree on specialty "Air Traffic Services" specialization "Air Traffic Control". Has got a diploma of PhD in Technical Sciences on specialty "Automated Control Systems and Advanced Information Technologies" (2004). PhD thesis "Modelling of Decision Support System for an Air Traffic Controller in Flight Emergency Situations", the thesis has defended in Kyiv Automation Institute. In 2006 has received an attestat of Associate Professor of Management and Economy Department. Since 2014 is the member of the specialized scientific council for the defense of dissertations for PhD in Technical Sciences on the specialty "Navigation and Traffic Control". The author and co-author of about 180 scientific works, including three manuscripts and more than 40 articles in specialized scientific publications, three copy-right certificates on computer programs. Together with the scientific work has actively engaged in the methodological provision of the educational process, during the years of labor activity above 80 teach-ing and methodological works have been issued. Areas of scientific interests: increasing of the decision making efficiency by the human-operator of the Air Navigation System in flight emergencies; assess-ment of the influence of aviation enterprises' internal and external management environment factors on the activity of the Socio-technical System's operator. Teaching courses: "Introduction to the Specialty", "Transport and Transport Infrastructure", "Information Systems in Management", "Methodology of Scientific Researches", "Mathematical Modelling of Professional Tasks".

* * *

Azad Bayramov, ScD, professor Azad Bayramov has completed his MSc in Azerbaijan State University, Physical faculty, Baku, Azerbaijan in 1975. He has defended his D.'s dissertation in 2000. He is currently serving as a professor in War College of Armed Forces of the Azerbaijan Republic. His research areas include development of technical systems, development of new detection methods of military objects and armoured fighting vehicles, mathematical modelling of combat activities. His contribution to the current study include quantitative assessment of invisible areas and military objects, modelling of UAV flights, comparison data.

Antonio Chialastri is an airline captain who have been flying for 30 years. He flew on several airplanes, and was appointed Type Rated Instructor, writing several training program for his airline. He got a degree in Philosophy, one degree in Epistemology and a Master in Bioethics. From 2006 he has been participating to several international conferences, where He was invited as keynote speaker and as a chairman. He wrote five books on Human factors, plus other chapters in collective books. He is member of the STASA think-tank, based in Rome. At the moment he teaches human factor at the Master in civil aviation at Rome University La Sapienza.

Andriy Chukhray is a head of Computer Science department, and Professor of Aircraft Control Systems department. In 2000 he graduated from National Aerospace University "Kharkov aviation institute (KhAI)" with honors diploma in Software Engineering. He received his Ph.D. in Automatized Control Systems and Progressive Information Technologies in National Aerospace University "KhAI" (2004). He received his Doctor of Science in Information Technologies in 2014 National Aerospace University "KhAI". The field of scientific research of PhD and doctoral thesis is fundamentals of creating Intelligent Tutor Systems that teach algorithmic tasks. He is author of more than 120 scientific publications and patents, among those 3 are in SCOPUS, scientific advisor of 3 PhD completed dissertations. In 2006 he won in competition for getting scientific grant of President of Ukraine, in the field of Intelligent Tutor Systems. In 2011 he won competition for getting grant of Ukrainian Ministry of Education and Science for scientific traineeship in Carnegie Mellon University (Pittsburgh, USA: Certificate of successful traineeship in Carnegie Mellon University in the field of Intelligent Tutor Systems, mentor – Dr. Bruce M. McLaren http://www.cs.cmu.edu/~bmclaren). He has experience in coordination and participation EU Tempus Project («Communication and Information technology for Improvement Safety and Efficiency of Traffic flows: EU-RU-UA Master and PhD Programs in Intelligent Transport Systems, CITISET», 2011-2014). Developed under his guidance Intelligent Tutor Systems has been introduced in Ukraine, England and Austria.

Ali Dinc was born in 1970, Turkey. He graduated from the department of Aeronautical Engineering, Middle East Technical University (METU) in 1992. He earned his MSc degree in 1995 within the scope of a cooperation project between university and aerospace industry. Then, he worked in different positions of design engineering and management levels in aircraft engine design area. In this context, he has worked in Turkey, USA and Spain for the development of engine parts and components in aerospace industry. He received his PhD degree in 2010. His areas of expertise are aircraft engine design, engine cycle analysis, numerical modeling of systems, optimization, unmanned air vehicles sizing and performance. He currently works as Assistant Professor in College of Engineering and Technology, American University of the Middle East, Kuwait.

Minglei Fu was born in March 1981. He got the bachelor degree of communication engineering from Zhejiang University of Technology in 2004. He got the master degree of communication and information system from Zhejiang University of Technology in 2007. And he got the Ph.D of control theory and control engineering from Zhejiang University of Technology in 2010. Now, he is an associate professor and master tutor in the department of applied physics. He was selected as the candidate for young and middle-aged academic leaders in universities and colleges in Zhejiang Province in 2017. He was selected as the candidate of the "133 Innovative Talents Project" of Yiwu City in 2017. He has hosted or participated in 10 national and provincial research projects, published 17 journal articles indexed by SCI and 38 journal articles indexed by EI, published 2 monographs, and authorized 15 invention patents. He won the second prize of Zhejiang Provincial Technological Invention Award in 2015, the second prize of Scientific Research Outstanding Achievement Award (Science and Technology) in 2015, and the second prize of Zhejiang Provincial Science and Technology Award in 2012. His research interests are deep learning, network information dissemination model. His technical interests are optcelectronic signal processing, intelligent hardware.

Olena Havrylenko in 2001 was graduated from National Aerospace University KhAI and had got magister degree from "Software of Automated Systems and Computer Engineering". In 2005 defended thesis on the specialty "Information Systems and Progressive Information Technologies" and had got PhD degree. Now docent of Aircraft Control Systems department of KhAI.

Ramgopal Kashyap's areas of interest are image processing, pattern recognition, and machine learning. He has published many research papers, and book chapters in international journals and conferences like Springer, Inderscience, Elsevier, ACM, and IGI-Global indexed by Science Citation Index (SCI) and Scopus (Elsevier). He has Reviewed Research Papers in the Science Citation Index Expanded, Springer Journals and Editorial Board Member and conferences programme committee member of the IEEE, Springer international conferences and journals held in countries: Czech Republic, Switzerland, UAE, Australia, Hungary, Poland, Taiwan, Denmark, India, USA, UK, Austria, and Turkey. He has written many book chapters published by IGI Global, USA.

Andrei M. Kozub was born in Kiev, Ukraine. He graduated from the Kharkov Nuclear Missile College, Ukraine, in 1988, received specialty on software of automated control systems and a qualification engineer and mathematician. Completed postgraduate studies at the Kiev institute of the Air Force in 2001, Ph.D. in 2002; thesis is devoted to technical issues of building guidance system of aircraft guided missiles. Since 2003 he received the academic rank of senior researcher and graduated from the National Defence Academy of Ukraine in 2004.Since 2012 Associate Professor in the area of uses of space systems and GIS software the National University of Defence of Ukraine. He is a co-author of two textbooks on remote sensing of the Earth and spacecraft control of dual-use. His current teaching and research interests are in area of space systems, devices of opto-electronic surveillance, aerospace monitoring, optimizing, complexation, efficiency, geoinformatics, pattern recognition, remote sensing.

Jarosław Kozuba, born on February 13, 1963 in Lublin. In the years 1986-2009 he was a professional military service as a professional soldier - a pilot. He ended his military service as a colonel in the Department of Transformation of the Ministry of National Defense. In 2007 - 2010 he was a lecturer at the State Higher Vocational School in Chełm, at the same time acting as the Deputy Director of the Institute of Technical Sciences and Director of the Aviation Training Center. From 2010 he was a lecturer at the Military Academy of Air Forces, where he served, among others, the functions of the Director of the Center for Training and Improvement of Aviation Personnel (2012-2015) and the Dean of the Department of National Security and Logistics (2015-2016). From September 1, 2016, he was the Vice-Rector for Scientific Affairs of the Air Force Officers School in Dęblin. In 2014 he obtained the title of doctor habilitated in technical sciences, in the discipline of transport at the University of Defense in Brno (Czech Republic). Currently, he is a professor at the Silesian University of Technology (Faculty of Transport).

Roman Odarchenko, Associate professor, Vice-director of Educational and Research Institute of Air Navigation, Electronics and Telecommunications in National Aviation University, PhD in Telecommunications with broad experience in teaching, scientific work, and managing. Extensive knowledge of electronics and telecommunications, especially 2G/3G/4G/5G cellular networks. Strengths in problem-solving and project management. Current research interests are the creation of new technological solutions for LTE networks, moving to 5G cellular networks, raising the quality of service level. Dr. Odarchenko received PhD in National Aviation University in 2013

Sergei Osadchiy, Professor, Doctor of Technical Sciences; in 1979 graduated from The Kiev Institute of Civil Aviation Engineers; from 1982 to 2001 worked in The State Flight Academy of Ukraine; from 2001 worked in Central Ukrainian National Technical University; from 2004 worked as The Head of the Department of Automation of Production Processes; in 2013 defended his doctoral dissertation on the topic "Methodology of maximizing the quality of stabilization of the cruising motion of an autonomous high-speed object in a cavernous environment".

Larisa Saganovskaia graduated from the Kirovograd Institute of Agricultural Engineering. In 2003 she graduated from the post-graduate course. Since 2003 has worked as a senior lecturer of the Department of Information Technologies in the Flight Academy in the city of Kropiwnickiy (moa).

Vladimir Sherstjuk is currently Full Professor of Information technologies department at Kherson National Technical University, Ukraine. He earned his Master of Software Engineering at Kherson Industrial Institute in 1987, then Ph.D. in Computer Science at Kherson State Technical University in 1986 and Doctor in Computer Science at Kherson National Technical University in 2014. His thesis dwelled on scenario-case control of dynamic objects. After completing his Ph.D., Vladimir Sherstjuk accepted an Associate Professor position of Information Technologies department, and then in 2015, he accepted a Full Professor position in the same department. Vladimir Sherstjuk has over 250 refereed publications.

T. R. Shmeleva received the Eng. degree in applied mathematics from Moscow Rail- road Engineer Institute, Moscow, Russia, in 1990; the Ph.D. degree in telecommunications from Odessa National Academy of Telecommunications, Odessa, Ukraine, in 2008. T. R. Shmeleva is author of more than 30 papers and chapters published in journals, books, and conference proceedings. Her research interest is Petri net theory and its application in networking, especially in verifications of protocols, and performance evaluation. She developed the analysis of infinite Petri nets with regular structure and parametric colored Petri net models of networks. Since 2005, she has been with the Odessa National Academy of Telecommunications, where she is currently an Associate Professor of Switched systems department. Tatiana is a member of SIAM.

Aleksander Sładkowski was born on the 23rd of November 1956 in Dnepropetrovsk (Ukraine) in the Polish-Russian family. In 1973 he entered the Dnepropetrovsk State University (DSU) in Mechanics and Mathematics Faculty. In 1987 A. Sładkowski defended his dissertation receiving a PhD degree in physics and mathematics in specialty 01.02.04 "Mechanics of deformable solids". Defense of work was held on the specialized council of DSU. In 1997 A. Sładkowski defended his dissertation for the habilitate degree on technical sciences in specialty 05.15.16 "Mining machinery" on the specialized council of the National Mining Academy of Ukraine. Since 2000 the activity of A. Sładkowski associated with the Silesian University of Technology, where he has the position of a full professor and head of Department of Logistics and Aviation Technologies at the Faculty of Transport. He is professor honoris causa of USURT, academician of Russian Academy of Transport, doctor honoris causa of EUNU, member of the Union of Scientists in Bulgaria, honorary member of Sci. Council of Inst. of Machine Mechanics of GNAS.

Roman Yakunin was born in 1981. In 2004 was graduated from Kirodohrad Teachers Training University, got Master Degree in Pedagogical Sciences. Now work as software developer, robo-designer on Robotic lessons.

Dmitry A. Zaitsev (IEEE M'10–SM'11) was born in May 13, 1964, in Konstantinovka, Ukraine. He received the Eng. degree in applied mathematics from Donetsk Polytechnic Institute, Donetsk, Ukraine, in 1986, the Ph.D. degree in automated control from Kiev Institute of Cybernetics, Kiev, Ukraine, in 1991, and the Dr.Sc. degree in telecommunications from Odessa National Academy of Telecommunications, Odessa, Ukraine, in 2006. He was an Associate Professor and then a Professor with Donetsk Polytechnic Institute from 1992 to 1996, with Surgut State University, Surgut, Russia, from 1997 to 1998, and with Odessa National Academy of Telecommunications from 2003 to 2009. From 1999 to 2003, he was a leading Engineer with the Computer Center of Odessa Railway. In March 2005, he was a Visiting Professor with Paris-Dauphine University, Paris, France. From 2007 to 2008, he was a Codirector in a project on verification of networking protocols supported by North Atlantic Treaty Organization (NATO). Since 2009, he has been with the International Humanitarian University, Odessa, where he is currently a Professor of computer engineering. Since 2011, he has been a Codirector in the China-Ukraine research project on production control supported by governments of both countries. He published more than 80 papers. His major field of study is Petri net theory and its application in networking, production control, and computing. He developed the analysis of infinite Petri nets with regular structure, the decomposition of Petri nets in clans, and the method of synthesis of fuzzy logic function given by tables. Prof. Zaitsev is a member of the Association for Computing Machinery and the Society for Industrial and Applied Mathematics and an associate member of Gesellschaft fur Informatik.

Maryna Zharikova is currently Associate Professor of Information technologies department at Kherson National Technical University, Ukraine. She earned her Bachelor of Software at Kherson National Technical University in 1999, and her PhD in Mathematical Modeling and Computational Methods from

Kharkiv National University of Radioelectronics in 2004. Her thesis dwelled on forest fire modeling. After completing her PhD, Maryna Zharikova accepted a position as Senior Lecturer of the same department. In 2008, she accepted an Associate Professor position at Information Technologies department. Maryna Zharikova has over 60 refereed publications.

Index

D

E

F

G

H

I

K

L

M

N

O

P

Q

R

www.ingramcontent.com/pod-product-compliance
Lightning Source LLC
LaVergne TN
LVHW062105300926
849161LV00010B/2377